Tokenizing the Future

Wolfgang Prinz · Daniel Trauth
Editors

Tokenizing the Future

A Guide to Web3 and the Metaverse

Springer

Editors
Wolfgang Prinz
Fraunhofer Institute for Applied
Information Technology FIT
Sankt Augustin, Nordrhein-Westfalen,
Germany

Daniel Trauth
Fraunhofer Institute for Applied
Information Technology FIT
Sankt Augustin, Nordrhein-Westfalen,
Germany

ISBN 978-3-031-91404-1 ISBN 978-3-031-91405-8 (eBook)
https://doi.org/10.1007/978-3-031-91405-8

© The Editor(s) (if applicable) and The Author(s), under exclusive license to Springer Nature Switzerland AG 2025

This work is subject to copyright. All rights are solely and exclusively licensed by the Publisher, whether the whole or part of the material is concerned, specifically the rights of translation, reprinting, reuse of illustrations, recitation, broadcasting, reproduction on microfilms or in any other physical way, and transmission or information storage and retrieval, electronic adaptation, computer software, or by similar or dissimilar methodology now known or hereafter developed.
The use of general descriptive names, registered names, trademarks, service marks, etc. in this publication does not imply, even in the absence of a specific statement, that such names are exempt from the relevant protective laws and regulations and therefore free for general use.
The publisher, the authors and the editors are safe to assume that the advice and information in this book are believed to be true and accurate at the date of publication. Neither the publisher nor the authors or the editors give a warranty, expressed or implied, with respect to the material contained herein or for any errors or omissions that may have been made. The publisher remains neutral with regard to jurisdictional claims in published maps and institutional affiliations.

This Springer imprint is published by the registered company Springer Nature Switzerland AG
The registered company address is: Gewerbestrasse 11, 6330 Cham, Switzerland

If disposing of this product, please recycle the paper.

Contents

Introduction

Introduction 3
Wolfgang Prinz

Decentralization

The Web's Evolution: The Socio-Technical Transition Towards a Decentralized Web3 11
Lisa Klug, Max Halbwachs, and Leif Oppermann

Trustworthy Tokenization in the Metaverse with Secure Digital Identities Through eIDAS 2.0? 35
Ignacio Alamillo and Steffen Schwalm

Modern Blockchain Wallets 51
Mikolaj Pawel Radlinski-Konas, Wolfgang Prinz, and Daniel Trauth

Sustainable Blockchains 65
Mikolaj Pawel Radlinski-Konas, Wolfgang Prinz, and Daniel Trauth

The Purpose and Difficulties of Decentralization 85
Arpad Djuraki

Decentralization Levels of L2 Scaling Solutions on Ethereum 103
Lorenz Raphael Lehmann

Smoothing Blockchain Adoption with Abstracted Smart Accounts 111
Felix Hildebrandt

Decentralized Account Management: A Web3 Solution 133
Felix Hildebrandt

Web3—arsNFT Badges as an Application Option 147
Alexander Robert Skurka

Web3 Security 163
José Carlos Ramírez and Isaac Agudo

Decentralised Finance (DeFi)

DeFi and Its Implications on Enterprise Software and Business Processes 175
Simon Engel

Programmable Money: Aligning Your Money with Your Values 189
Selin Sezer

A Short History of Decentralized Finance (DeFi) 201
Marcelo Emmerich

Real-World Assets (RWAS) Tokenization in Web3: Transforming Finance and Ownership 215
Phulchand Saraswati

Digital Economy

From Sensors to Solutions: The Role of Helium Blockchain and LoRaWAN in AI Innovations in Environmental Monitoring, Smart Parking, Crowd Management, and Urban Gardening 229
Daniel Trauth, Wolfgang Prinz, and André Heryschek

Token-Based Economies in Decentralized Societies 245
Felix Hildebrandt

Increasing Economic Performance Through Digital Application 267
Arpad Djuraki

The Potential of Web3 in the Data Economy and AI Opportunities 289
Kai Schmitz-Hofbauer

Web3 Gaming and NFT-Based In-Game Items 309
Marcus Rump and Oliver Nolden

Decentralized Science (DeSci): How Web3 is Revolutionizing Science 321
Lukas Weidener

Quantifying MEV NFT Arbitrage 337
Matthias Franz Krekeler

Blockchain-Based Data Security and Enhanced Transparency in the Digital Signage Industry 357
I. Dimitrov, D. Trauth, and W. Prinz

Contents

**The Impact of Blockchain on Transparency and Trust
in Sustainable Agri-Food Supply Chains** 369
Thuy Tien Nguyen Thi, Mandana Gharehdaghi,
Maximilian Austerjost, and Axel T. Schulte

Metaverse

Shared Manufacturing 385
Patrick Stuckmann-Blumenstein, Larissa Krämer, Dominik Bons,
Patrick Keitzl, and Eugen Burov

Virtual Workspaces and Collaboration 399
Erik Jarne Prinz and Cedric Muschick

**The Role of Blockchain and Distributed Ledger Technologies
in the Industrial Metaverse** 407
Orhan Küpeli, Alexander Grünewald, Tan Gürpinar, Max Schwarzer,
and Austin King

**Potentials and Applications of the Industrial Metaverse Using
the Example of Synthetic Data Generation** 423
Oliver Petrovic, Josefine Monnet, Petar Tesic, Yannick Dassen,
and Werner Herfs

Tokenization

Introduction to Tokenisation 439
Lukas Wagner

ERC Token Standards Powering NFTs: An Overview 451
Lorenz Raphael Lehmann

Utilizing Tokenized Real-World Assets in DeFi 461
Lorenz Raphael Lehmann

Tokenisation of Tangible Assets 471
Markus Fehn

**Engagement Reimagined: Translating Psychological Ownership
into Token-Based Engagement Models** 485
Lea Horn

NFT—Non-Fungible Tokens 497
Diana Dabboussi-Gürman

**On the Role of Tokenization for Pursuing Environmental
Sustainability** 509
Vincent Schaaf, Jonathan Lautenschlager, Tobias Guggenberger,
Marc-Fabian Körner, Jens Strüker, and Nils Urbach

Legal

Crypto Art and Intellectual Property 527
Su-Zeong Fröhlich and Kerstin Gold

Introduction to Decentralization and Ownership in Web3 547
Gustav Hemmelmayr

Enforcement of Rights in the Metaverse 571
Simon J. Heetkamp and Ida Holschbach

Procurement of Industrial Machinery in the Metaverse 585
Natalia Broza-Abut, Tobias Jornitz, and Axel T. Schulte

SuppliedTrust: A Blockchain-Based Governance Framework to Establish More Trust in Consumer Products 599
Timucin Korkmaz

Introduction

Introduction

Wolfgang Prinz

In recent decades, the internet has undergone rapid development and profoundly changed our lives. From simple information exchange (known as Web 1.0) to complex social networks and global marketplaces (known as Web 2.0), the web has revolutionized our understanding of communication, economy, and society. Now, we stand at the beginning of a new era: Web3. This book aims to provide you with an in-depth insight into the concepts, technologies, and potential of Web3 and to show how it will shape the internet of the future.

When Tim Berners-Lee articulated his vision of the World-Wide Web (WWW), he envisioned it as a new infrastructure that would facilitate innovative interactions among users and promote new forms of collaboration, especially within the research community [1]. However, in its early days, the WWW and Web 1.0 respectively primarily served as a publishing platform where organisations, newspapers, or individuals created their initial home pages or digital publication outlets. Reflecting on these early stages, we refer to this first phase of the WWW as the information economy, characterised by the emergence of homepage toolkits, early content management systems, link-collection pages like Yahoo, and search engines such as Altavista or Google. Concurrently, printed web address books were still in circulation to assist users in locating information. During this period, contributing content to the web was feasible only for experts familiar with the necessary protocols and applications, leaving most users as passive consumers of the web. Accessing the Internet and the Web used to be far from spontaneous. Most users relied on dial-up modems with restricted bandwidth, necessitating deliberate planning for online activities like reading emails or visiting websites. This contrasts sharply with the impromptu internet use we see today.

W. Prinz (✉)
Fraunhofer FIT, Schloss Birlinghoven, Sankt Augustin, Germany
e-mail: wolfgang.prinz@fit.fraunhofer.de

With technical advancements, the web's evolution entered its second phase, the Web 2.0. Protocols enabled dynamic web pages with user-generated content, transforming the web from a one-way to a two-way medium. This led to new collaboration tools like BSCW [2] (the world's first "dropbox") and social media platforms such as Facebook, Flickr, and Germany's StudiVZ. User interaction shifted from passive reading to active participation. This period is now known as the ReadWriteWeb, Social Web [3], or Web 2.0. The web has shifted from a decentralized network of HTML pages linked via URLs and accessed by HTTP, to a centralized platform economy dominated by major sites like Facebook, eBay, Amazon, Twitter, Wikipedia, and YouTube, that enabled users to provide content to interact and to perform business.

Web3 aims to address some of the limitations and challenges posed by its predecessors by leveraging decentralized technologies, with blockchain being at the forefront. Blockchain technology introduces a paradigm shift in how data is managed, stored, and verified on the internet. At its core, blockchain is a distributed ledger technology that ensures transparency, security, and immutability of data. Unlike the centralized databases of Web2.0, blockchain operates on a decentralized network of nodes, each maintaining a copy of the ledger. This decentralization removes the need for a central authority, thereby reducing the risk of data breaches and monopolistic control.

One of the most profound impacts of blockchain is the establishment of trust without intermediaries. In Web 2.0, trust is typically centralized in the hands of powerful entities such as tech giants, which control and monetize user data. Blockchain disrupts this model by allowing peer-to-peer transactions where trust is built through consensus mechanisms like Proof of Work or Proof of Stake. Each transaction is verified and recorded on the blockchain, making it tamper-proof and transparent to all participants.

Blockchain empowers users by giving them control over their data. Through cryptographic keys, individuals can manage access to their personal information and assets without relying on third-party platforms. This aligns with the ethos of Web3, which prioritizes user sovereignty and privacy. Smart contracts, self-executing contracts with the terms directly written into code, further enhance this control by automating and securing agreements without intermediaries.

Centralized systems in Web 2.0 are prone to single points of failure, making them vulnerable to hacks and outages. Blockchain's decentralized architecture distributes data across multiple nodes, ensuring that even if one node fails, the system remains operational. This reduces the risk of large-scale data breaches and increases the overall resilience of the network.

The advent of blockchain has also paved the way for innovative applications such as cryptocurrencies, decentralized finance (DeFi), and non-fungible tokens (NFTs). These applications are transforming industries by providing new ways to transfer value, raise capital, and authenticate digital assets. For instance, DeFi platforms enable users to lend, borrow, and trade assets without traditional financial intermediaries, democratizing access to financial services.

In summary, blockchain is the cornerstone of Web3, addressing the shortcomings of Web 2.0 by decentralizing trust, enhancing data sovereignty, and improving security. As we move into this new era, blockchain's role as a game changer cannot be overstated [4].

While Web3 is often associated with decentralized technologies like blockchain, it is crucial to distinguish it from the concept of Web 3.0 as envisioned by Tim Berners-Lee. Berners-Lee's Web 3.0, also referred to as the Semantic Web, aims to create a more intelligent and interconnected internet by structuring data in ways that can be understood and processed by machines. This vision emphasizes data interoperability, semantic metadata, and artificial intelligence to enable a more intuitive and efficient web experience.

In contrast, Web3 focuses on decentralization and user empowerment through the use of blockchain technology. It aims to dismantle the centralized control held by tech giants, offering instead a network where individuals have sovereignty over their data and digital identities. This decentralization facilitates trustless peer-to-peer interactions and enhances security by distributing data across multiple nodes.

While both Web 3.0 and Web3 share a common goal of improving the internet, their approaches diverge significantly. Web 3.0 seeks to enhance data comprehension and connectivity, making the web smarter and more responsive to user needs. Web3, on the other hand, prioritizes user control, privacy, and the democratization of digital interactions through decentralized infrastructures.

Together, these paradigms illustrate the multifaceted potential of the internet's evolution, each addressing different aspects of the limitations and opportunities presented by earlier web technologies.

This book provides a comprehensive overview of the various aspects and applications of Web3 technologies, the digital economy, and the Metaverse. Topics such as decentralization and its significance for the digital age, the tokenization of assets and its impact on finance, as well as the legal frameworks for crypto art and trading in the Metaverse, are discussed.

The section on decentralization delves into different levels of decentralization and their importance for blockchain and Web3 technologies. The area of Decentralized Finance (DeFi) highlights the transformation of traditional financial systems through DeFi and the integration of smart contracts. Furthermore, the role of blockchain in the digital economy and its potential in various industries, such as urban gardening and environmental sensing, is described.

The part about the Metaverse [5] explores the potentials and applications of Virtual Reality (VR) and Augmented Reality (AR) in education, training, and collaborative work environments. Additionally, the legal implications and challenges related to enforcing rights in the Metaverse are discussed.

Tokenization, a central element of Web3, is explained in detail, including the standardization of ERC tokens and their use in various economic contexts.

Although this book focuses on the Web3, the digital transformation of organisations and society will depend on the integration of various technologies. To this respect, the combination of Web 3, XR/Metaverse, and generative AI is of primary

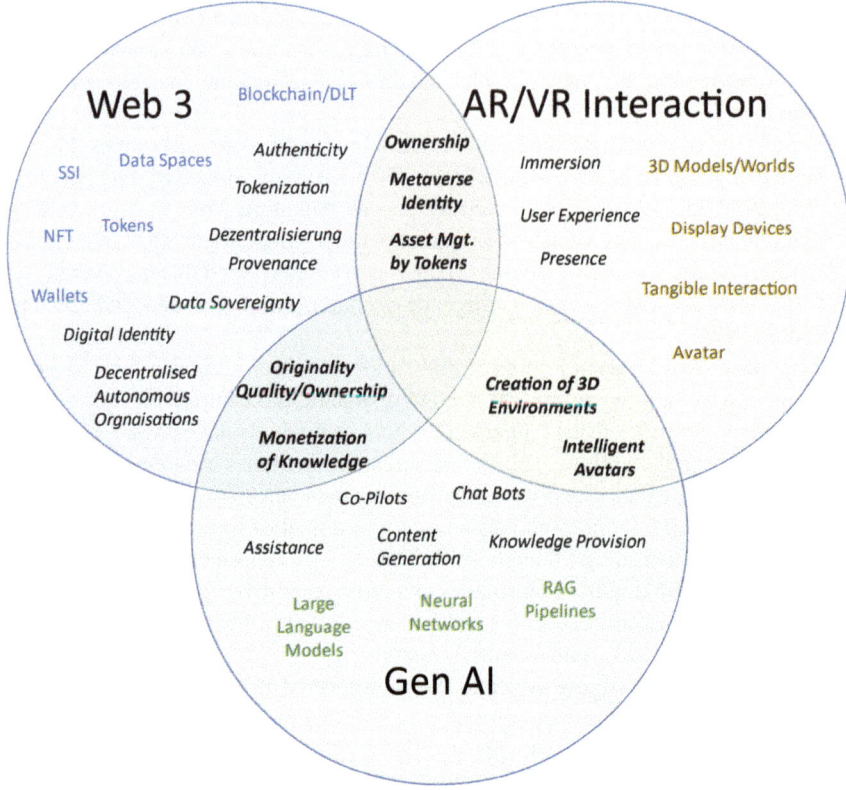

Fig. 1 The interplay of Web3, XR and generative AI

interest. Figure 1 depicts these technologies by highlighting their individual technological contributions as well as some of the opportunities and applications that exist in their combination.

Combining Web3 and XR will create new ways to signify ownership in 3D spaces, introduce new digital identities, or manage assets via tokens. Some of these concepts are already being implemented in "play to earn" games, where players receive tokens based on their performance. Other examples include the use of NFTs to authenticate real world products as well as their counterparts in a 3D environment. Generative AI can use web3 for data source authentication and monetizing valuable training data. XR applications gain from advanced avatars using speech models, and AI can assist in creating 3D environments.

These are merely initial implementations, and many more are expected to emerge in the coming years. The question remains what to call the field at the center. Will it be Web 5, the Metaverse, or something tied to a new economy? We believe the "Economy of Things," combining the Internet of Things, AI, and Web3, could be fitting. We hope this book and its articles offer guidance in this direction, providing valuable insights to both practitioners and experts interested in this field.

References

1. A. Berners-Lee, R. Cailliau, A. Luotonen, H.F. Nielsen, A. Secret, The World-Wide Web. Commun. ACM **37**, 8, 76–82 (1994). https://doi.org/10.1145/179606.179671
2. R. Bentley, T. Horstmann, K. Sikkel, J. Trevor, Supporting collaborative information sharing with the world wide web: the BSCW shared workspace system, in *Proceedings of the Fourth International Conference on World Wide Web (WWW4)*. (Association for Computing Machinery, New York, NY, USA, 1995), pp. 63–73. https://doi.org/10.1145/3592626.3592631
3. P. Hoschka, CSCW research at GMD-FIT: from basic groupware to the social Web. SIGGROUP Bull. **19**, 2, 5–9 (1998). https://doi.org/10.1145/290575.290576
4. C. Dixon, *Read Write Own Building the Next Era of the Internet, Random House* (2024), ISBN 978-0-593-73138-3
5. F. Buchholz, L. Oppermann, W. Prinz, There's more than one metaverse. i-com. **21**, 3, 313–324 (2022). https://doi.org/10.1515/icom-2022-0034

Prof. Wolfgang Prinz, PhD studied informatics at the University of Bonn and earned his Ph.D. in computer science from the University of Nottingham. He is vice chair of Fraunhofer FIT in Bonn, head of the Collaboration Systems research department, and a Professor at RWTH Aachen University. His research focuses on digitization, new cooperation platforms, mixed reality, and flexible communication infrastructure. In the Fraunhofer Blockchain Lab, he explores technical foundations and Blockchain-based applications. He has led various national and international research projects, including a significant European project on collaborative work environments, and serves as an editor for several journals and conferences.

Decentralization

The Web's Evolution: The Socio-Technical Transition Towards a Decentralized Web3

Lisa Klug, Max Halbwachs, and Leif Oppermann

1 Introduction to Transition Studies and Its Relevance in the Digital Evolution

This chapter describes the genesis of the platform-based Web 2.0 as it is present today as well as its possible transition towards a blockchain-based, more decentral Web3[1] in the future. To do so, it employs the "Multi-Level Perspective" (MLP)—a theory on so-called socio-technical transitions, which is based on evolutionary economics, the sociology of technology and neo-institutional theory [24]. While the MLP is mostly employed in the context of "sustainability transitions" of, for instance, the energy, transportation, or housing sectors [29, 34], it can explain the change of technologies and how they are used in society even in other contexts [19]. In this chapter, the MLP explains the emergence of the internet as the predominant socio-technical system of communication and the possible paths that Web3 can take to further transform this and other societal systems. Employing this theory provides a broad overview of the complex societal dynamics behind these technological changes rather than solely focusing on technological details.

---

[1] Please note the subtle difference between Web3 and Web 3.0. While Web3 aims at creating a decentralized, user-centered and sovereign Internet, the term Web 3.0 was popularized in the 2000s and aims at improving the efficiency and intelligence of the Internet through semantic technologies, where machines can process data intelligently. Web3 and Web3.0 are concepts that come from different times and backgrounds. In 1.3.1 we explain the concepts in more detail.

L. Klug (✉) · L. Oppermann
Fraunhofer Institute for Applied Information Technology FIT, Hürth/Sankt Augustin, Germany
e-mail: lisa.klug@fit.fraunhofer.de

L. Klug · M. Halbwachs
School of Business, Innovation and Sustainability, Halmstad University, Halmstad, Sweden

M. Halbwachs
Centre for Innovation Research (CIRCLE), Lund University, Lund, Sweden

© The Author(s), under exclusive license to Springer Nature Switzerland AG 2025
W. Prinz and D. Trauth (eds.), *Tokenizing the Future*,
https://doi.org/10.1007/978-3-031-91405-8_2

Key concepts of the MLP are the "socio-technical system" and the "regime". The socio-technical system denotes the configuration of technological artifacts and their societal contexts. This includes for example infrastructures, laws, or cultural manifestations constructed around a technology [20]. Such systems are constantly being reproduced by the users of the technology at focus because there are strong institutional arrangements such as informal rules, social norms, routines, or visions in place that make breaking out of the system difficult or even unfeasible. This institutional frame of socio-technical systems is what constitutes the "regime" [18, 20]. Intact regimes usually lead to stable socio-technical systems that only experience incremental innovation and are otherwise locked into their state [21]. They form a "selection environment" in which it is difficult for other socio-technical configurations to survive [22].

Systems change, i.e., a "socio-technical transition" can occur when external "landscape" pressures destabilize the regime. This can for example include demographic or environmental changes but even more short term crises, or political events [20, 23]. If such pressures are to the detriment of existing socio-technical configurations, this will weaken the regime, and in turn the system's reproduction. This opens up a "window of opportunity for new structures to innovate the system [21]. These new structures are developing in "socio-technical niches". Niches are "protective" societal spaces, in which innovations are being protected from the regime's selection environment and thus offer the opportunity for novel technologies or social practices to emerge [52]. One example for such developments is the emergence of steam engines in seafaring postal services which needed a way to transport mail within a set timeframe [19]. A very different and more recent example is the emergence of organically grown foods, which are protected from the regime's selection environment by individuals who are willing to pay more as a means to oppose food industrialization [26].

Niches undergo several stages of development before they can grow into a dominant socio-technical regime. Geels [21] describes this process in four phases, as shown in Fig. 1. In a first phase of "experimentation", new configurations slowly emerge and are experimented with on the fringes of society. Uncertainties are high in this phase and many of the niche-experiments will not stand the test of time. However, some might enter the second phase of a transition—"stabilization". Here, the single experiments become aligned and a "dominant design" of the new socio-technical configuration as well as a "proto-regime" emerge. This comes with significant improvements of the innovation's performance, price and uptake. However, only when the socio-technical system is being disturbed by landscape pressures, this mature niche can gain enough momentum to enter the third phase: "diffusion". Here the institutions of the old regime are in direct conflict with the rising niche. There may be political or cultural struggles surrounding the uptake of new technologies or practices. One such example is legal bans to call plant-based drinks "milk" [39]. If, and only if, the niche can win those struggles, it becomes "institutionalized", meaning that its proto-regime grows to alter or even replace the existing rules and institutions and in turn changes the socio-technical configurations of a society. The entire process of a socio-technical transition can take several decades [21]. Notably, this four-phase model of a transition is an ideal type and different transition pathways exist [23].

This chapter's analysis is based on the four phases described above. In the following sections, we will utilize the ideas of the MLP to describe and explain the emergence of the internet and the development of a radically different Web3 in a socio-technical niche. This will provide an overview of the broader societal context of the technologies discussed in this book.

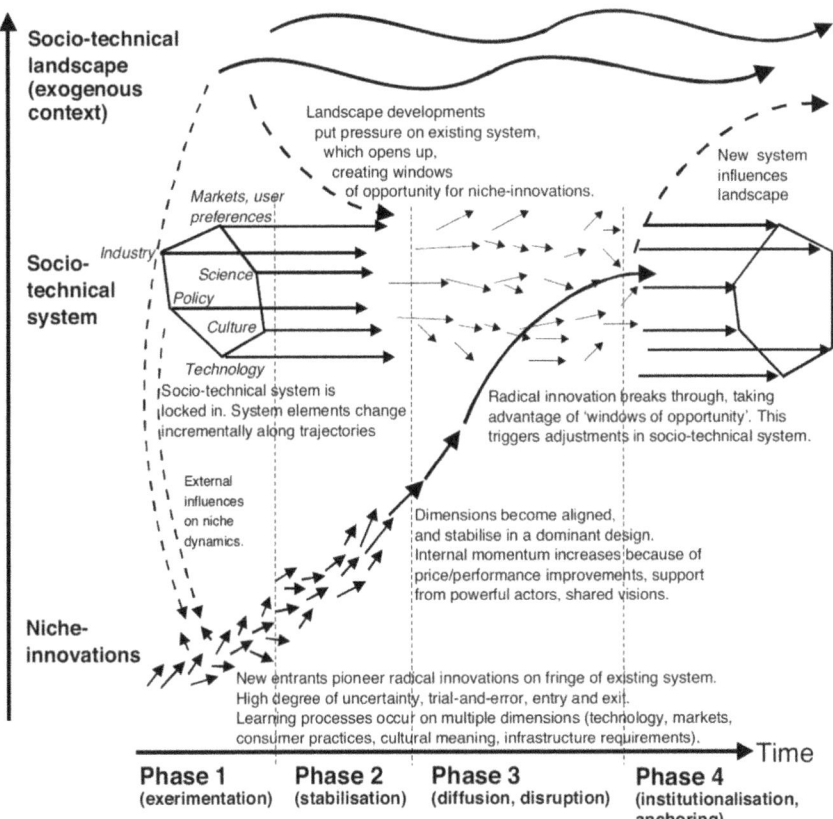

Fig. 1 Adapted from "Socio-technical transitions to sustainability: a review of criticisms and elaborations of the Multi-Level Perspective" by Geels, F., *Current Opinion in Environmental Sustainability, 2019, 39,* pp. 187–201, (https://doi.org/10.1016/j.cosust.2019.06.009). Copyright 2024 by Elsevier

2 Ex-post Analysis from ARPANET to the Web 2.0

2.1 Socio-technical System and Regime

Before the emergence of the internet, the socio-technical system that fulfilled the function of communication in large parts of the western, industrialized world, was characterized by linear and non-digital technological artifacts such as landline phones, paper and physical mail, printed media, linear TV and radio stations, transportation networks, centralized computing networks, and analogue banking. In this system knowledge was highly centralized to the providing companies of said technologies [8], which, especially in the case of mail and phone providers, often were in the hand of the state [1]. Thus, the related markets were hard to access for new entrants. Related to these artifacts, there were large infrastructures constructed to support their use. This includes, for instance, elaborate distribution systems for printed media or intercontinental connections with telegraph and telecommunication cables. User practices around the technologies included, as an example, the consumption of linear media at specific points in time (e.g., the morning newspaper, or evening news on TV), limited availability of information, sending telegrams and faxes, waiting times for official documents to be sent via mail, payments in cash or via analogue bank-transfers.

This socio-technical configuration was kept stable and was constantly reproduced due to the presence of a strong socio-technical regime. This regime was, for example, manifested through routines in households (e.g., reading said morning newspaper), but also in the media-and communication industries, or governments. Established procedures made using the established communication channels necessary to conduct regular business: government agencies were to be communicated with and legal documents to be sent via physical mail or fax in many countries, doctors' appointments were made via landline or personal contact, and those who wanted to stay updated on news had to resort to the existing channels of printed media, radio, and TV.

This strong socio-technical regime locked in the socio-technical system and only allowed for incremental innovation (such as color TV) to happen. Digital, decentralized communication technologies only existed in socio-technical niches, particularly in universities. How this regime was slowly destabilized by large-scale landscape developments is the subject of the next section.

2.2 Landscape Pressures and Socio-technical Niche-Innovation

In the following, technological developments relating to the internet as a communication technology are divided into the phases of experimentation, stabilization, diffusion and institutionalization. The phases are ideal–typical analytical concepts,

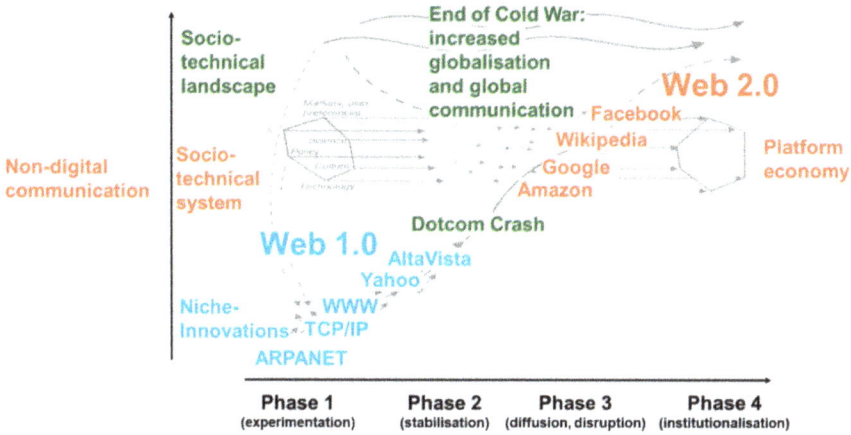

Fig. 2 The socio-technical transition from Web 1.0 to Web 2.0 using the model of [21], p. 191

and the distinction is therefore not always clear-cut. Figure 2 briefly summarizes the following description of the socio-technical transition.

2.2.1 Experimentation and Stabilization

The internet as we know it today was developed in the early 1990s, but most of the technical standards for the internet were developed in the 1960s [46]. At the time, the Cold War was prevalent. When the USSR successfully launched Sputnik I into space in October 1957, the Soviets' technological lead became apparent. This triggered the so-called Sputnik shock in the USA and the desire for military-scientific countermeasures [54]. The ARPA (Advanced Research Projects Agency) was founded, a research agency integrated into the Ministry of Defense. The aim of ARPA was to promote scientific projects and technologies and to utilize the results for military purposes.

Within ARPA, the ARPANET (Advanced Research Projects Agency Network) was then created at the end of the 1960s. Its development was influenced by the desire to build a resilient communications network that would remain operational even in the event of a nuclear attack [33]. Among other things, it was recognized that information should be transmitted in the form of small blocks of bits [2], and in 1969, the first computer network connected four research institutions (University of California Los Angeles, University of California Santa Barbara, University of Utah, Stanford Research Institute in Palo Alto) before the network grew and more nodes were gradually added [33, 54].

The U.S. Defense Department provided an active protective space [52] for the development of the ARPANET, as they provided significant funding and resources. Because of the military's focus on technological superiority, the researchers had freedom to experiment with innovative technologies and were less constrained by profit-driven motives—a key feature of a niche-experiment in the first phase of a socio-technical transition.

In the 1980s, the TCP/IP protocol (Transmission Control Protocol/internet Protocol) was introduced as the standard for all military computer networking [54]. This introduced IP addresses by which computers on the network are identified to this day. TCP/IP emerged early on as a dominant design of the emerging niche for networking protocols due to several factors, such its open architecture (allowed for interoperability), the decentralized design (which increased the fault tolerance as information can be routed around damaged nodes), and international collaborative efforts and standardization of TCP/IP, which helped driving the widespread adoption. With the introduction of TCP/IP and other protocols, the ARPANET became a more stable network. This work was continued and grew under the successor NSFNET (National Science Foundation Network). Throughout this period of research and education which lasted from 1985 to 1995, more universities, research institutions and some companies began to use the internet, mainly for scientific collaboration and data exchange [41].

A major key component in the development of the internet was the invention of the World Wide Web (WWW) by Tim Berners-Lee in the early 1990s. He based his idea on Ted Nelson's idea of Hypertext, simplified it, implemented it, and decided to make it available for free use. As with the ARPANET, this development also took place in a protected space, as Berners-Lee developed his idea at CERN. Berners-Lee also led the development of HTML (Hypertext Markup Language for creating websites and web applications), HTTP (Hypertext Transport Protocol), URL (Uniform Resource Locator), and the first browser to put it all together and display webpages. Thereby he initiated a shift from the internet as a messenger for information exchanges to the internet as a place for searching and retrieving documents on the web, counting around 50 million users in 1995 [5, 54] and billions today.

The WWW allowed non-experts to access the web. Although its design was originally "read" and "write", most people used it in a static fashion as consumers. Outside of emails and chats, it required technical skills to create websites and therefore, many users were only reading [38]. There were no mechanisms to incentivize content creation [44]. This first generation of the web up to the dotcom crash (a financial crisis in the year 2000 we will come back to shortly) is nowadays considered as Web 1.0 [38, 44].

The development of crucial technologies, including the TCP/IP protocol, HTTP, the World Wide Web, and user-friendly browsers supported the transition of the internet as a niche innovation from the experimentation phase to the stabilization phase. These technological innovations facilitated broader civilian use (as opposed to military) by improving data transmission and simplifying access to information. It transformed the internet from an experimental niche into a more stable platform for global communication and information exchange.

For the stabilization of the niche innovation internet and its further anchoring as a communication medium in society, sooner or later, the development of search engines played a decisive role. They have increased internal momentum through performance improvements in the application and have created ever more shared

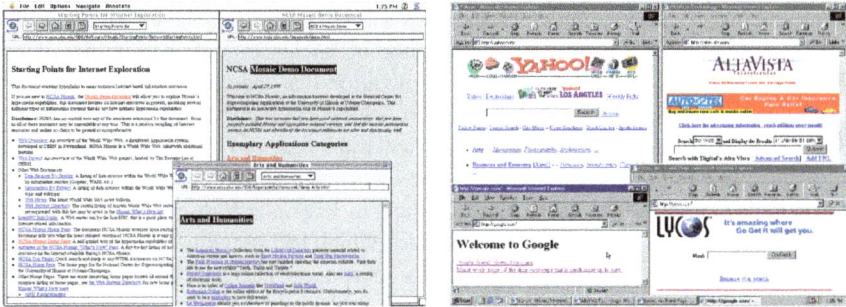

Fig. 3 Early curated index in NCSA Mosaic, the unofficial Netscape predecessor (left) and early web directories, search engines, and advertisement in Internet Explorer, which was based on NCSA Mosaic (right). (Images created using https://oldweb.today/)

visions in the evolving digital landscape by enabling people to find the content that others generated. In the first phase of the WWW, browsers came with curated web directories of links (e.g. Mosaic's "Starting Points for Internet Exploration", Fig. 3), curated indexes started to appear and even printed books serving as guides to interesting content. Yahoo was founded in 1994 and initially offered a list of links that could be thought of as a digital telephone directory, and also a search in that index. A short time later, AltaVista came onto the market and was one of the best-known search engines on the internet (alongside Lycos and others), until the emergence of Google[2] at the end of the 1990s.

Both the invention of the WWW and the emergence of search engines contributed to the rapid spread of the internet and paved the way for its diffusion. We have started with the analysis of the transition of non-digital communication, but at this point it already becomes apparent that it is no longer just a matter of a change in the communication system, but that other social functions, such as commerce, have also been transformed with the introduction of the internet.

2.2.2 Diffusion and Disruption

A core idea of the multi-level perspective is that socio-technical change is not based on convincing innovations alone, but also because existing regimes that maintain established routines come under pressure. The old socio-technical regime is weakened by events in the exogenous context that lead to changes in the landscape, institutions are weakened, and space is created for something new. A "window of opportunity" (visualized in Fig. 1 by the small arrows in phase 3) is created in which niche innovations can establish themselves [21]. In the case of the internet as a niche innovation, the end of the Cold War in the early 1990s contributed to creating this window of

---

[2] In the following, the name Google will be used regardless of whether the product or company is meant, as Google is more familiar in common usage than the name of its parent company, Alphabet Inc. which has served as the umbrella organization since 2015.

opportunity by triggering greater global networking. In its struggle against the structures of the incumbent socio-technical regime of linear communication, the niche benefited from landscape pressures of globalization that required an unprecedented need for international communication [25, 36] which was hard to accomplish with the incumbent structures of landline calls, physical mail, fax, and linear communication only, thus impeding the reproduction of the incumbent system. The rapid adoption of internet technology was further fueled when, in 1991, the National Science Foundation allowed commercial use of the internet, which had previously been restricted to government and academic communication purposes. This enabled the growth of commercial Internet Service Providers (ISPs), although the first ISP had already emerged in 1989. Subsequently, companies such as Netscape and Microsoft started a race for market share in the new economy and made web browsers as the underlying access technology available free of charge. This period from 1995 to around 2001 is also known as the First Browser War. This race was won by Microsoft with its Internet Explorer dominating the market at the turn of the millennium.

Another event on the landscape level contributing to the window of opportunity was the dotcom crash, a financial crisis in March 2000 that marked the bursting of a bubble. During the late 1990s, enormous hype had developed around the internet and new economy companies, leading to excessive investment and the dotcom bubble. This speculative bubble around the internet and new economy companies eventually burst, causing a crisis. Many internet companies went bankrupt and there was a shakeout in the market, but the technologies and infrastructure remained. The dotcom crash and the collapsed companies point to the failed experiments that are part of the trial-and-error process in the early stages of socio-technical transition. Although some of the start-ups were not successful with their ideas, by fighting to get their ideas accepted they paved the way for successors who successfully brought the same ideas to market after the regime changed (cmp. f. ex. German Tele-Info's "City Server" paved the way for Google's "Street View" feature [13] that has become a commodity for billions of users as an integral part of Google Earth and Google Maps). Ultimately, they could not prevail against others that refined their pioneering ideas to become a dominant design.

Following the development of the World Wide Web by Tim Berners-Lee in the late 1980s and early 1990s, the diffusion phase already began with the rapid increase in the number of internet users, the emergence of numerous commercial internet services and the expansion of internet use beyond traditional academic and scientific circles. However, the dotcom crash may have accelerated this diffusion, as it caused excessive speculation to burst and at the same time brought the technically and economically strongest internet companies to the fore, further institutionalizing the internet in society.

Two of these companies are Amazon and Google, which developed successful business models making them dominant actors on the internet. Google emerged as the leading search engine and created a solid foundation for online advertising, while Amazon expanded its business into the e-commerce sector.

Following the shakeout after the dotcom crash, technological developments like AJAX—Asynchronous JavaScript (a script language designed by Netscape for dynamic websites) and XML (an extensible markup language used to store and transport structured data in a human- and machine-readable format), and new business ideas centered around user contributions to platforms supported the emergence of Web 2.0 [43]. Web 2.0 represented a shift from just reading/ consuming information (as seen in the static Web 1.0) to both reading and writing, allowing users to create and share content on interactive web pages [44]. Platforms such as Blogger, Wikipedia, and later social networks such as Facebook and Twitter enabled users to create and share content in an ego-centric fashion. The design changed from connected computers to connected people through connected computers. This constitutes the emergence of the regime constituting the use of the internet as it is in place to this day.

2.2.3 Institutionalization and Anchoring

The transition of the socio-technical communication system found its end when the internet established itself as a new dominant communication system in the late 2000s. Companies such as Amazon, Google and Meta (formerly Facebook, now the parent company of Facebook, Instagram, and WhatsApp) have transformed the internet into a global economic platform on which companies and individuals can operate. This has led to a broad economic and social impact, whereby the platform economy has institutionalized the internet in almost all areas of life and ushered in a new era of the digital economy with platforms playing a central role. Previous socio-technical configurations have either largely disappeared (e.g., landline phones), or others, such as linear television, cash payments, or physical mail, have been demoted to secondary channels of communication in most countries, even in official settings such as government administration or business. Previously linear news media are present on their own websites and social media and are predominantly consumed in this way, with more people now accessing news via social media than via direct access to news websites/apps [48]. While the data ownership of such media channels still is highly centralized in firms such as Microsoft, Meta, ByteDance, Amazon, Google, or Alibaba and data are made accessible through e.g. membership fees or customized advertisements [30, 44], they allow for a more decentralized spreading of information throughout society. While this allows for the quick and widespread dissemination of information and the representation of diverse perspectives through user-generated content, it also poses potential threats through the loss of editorial standards, the spread of fake news, and the potential manipulation of user behavior.

The new socio-technical regime keeps this emergent configuration of technology and its use in place. Participation in society to a large degree necessitates usage of the internet for communication. Digital meetings with international business partners, the organization of social events through social media, online banking, digital IDs, and the need for organizations to access server infrastructures (that are largely owned by a few companies) are just a few examples of routines that have locked in this system and make it difficult for other practices and technologies to operate in its stead.

However, this regime is being challenged by new landscape pressures, such as increased awareness of data protection. How such new circumstances result in the creation of protective space for new niches and the potential for another socio-technical transition of the global communication system, is the subject of the next section.

3 Exploring the Road to Web3 with Ex-ante Analysis

Transitions theory, and the MLP in particular, can help understand what a future transition could look like. To do so, it needs to answer the question of what exactly would constitute a desirable socio-technical system, what niche-innovations are crucial to achieving such a system and how such innovations can be empowered to stabilize, grow, and align, to be mature enough to diffuse when a "window of opportunity" opens up. Particularly, the idea of Web3 constitutes a growing niche innovation that could be essential to such a future transition. We argue that Web3 is a new niche because it differs significantly from the current socio-technical system around Web 2.0 and the platform economy. Instead of incremental changes, Web3 concepts introduce radical innovations, such as decentralized data storage and user data sovereignty, which will affect both consumer usage patterns and corporate business models. For a transition to happen, its proponents should show awareness of the many processes that contribute to transitions in their different phases. The following subsections therefore seek to address these questions in an MLP-based ex-ante analysis of a possible socio-technical transition of the current communication system.

3.1 Web3 as a Niche-Innovation

The terms Web 1.0 and Web 2.0 were introduced above: Web 1.0 as introduced by Tim Berners-Lee was mainly used as a consumption platform for (static) documents [41]. The term "Information Economy" was coined at this time due to increased emphasis on and valuation of information as a capital good [10]. The term Web 2.0 was popularized by Tim O'Reilly in 2004 and characterizes a more dynamic online experience with user-generated content and increased interactivity, as opposed to Web 1.0 [44]. Web 2.0 is the internet most people are using today in a collaborative way, consuming and providing content. It is largely based on platform ecosystems that collect and aggregate user data centrally and often monetize it, for example in the form of personalized advertising [30].

The term Web3 was coined in 2014 by Gavin Wood, a co-founder of Ethereum, envisioning a decentralized internet powered by blockchain technology. Web3 builds upon the cryptographic foundations of Bitcoin, which was proposed in 2008 and introduced in 2009 [40] as a technical alternative to "money", but extends beyond cryptocurrency into decentralized applications and smart contracts. The 2007–2008 global financial crisis acted as a significant landscape pressure, revealing significant flaws in the global financial system, including excessive risk-taking by banks, lack of transparency, and the failure of regulatory bodies to prevent the collapse. This led to a widespread loss of trust in traditional financial institutions and central authorities. The crisis highlighted the dangers of centralized financial systems and raised interest in alternatives that could operate without the need for trusted intermediaries like banks or governments. These were favorable conditions to introduce an alternative asset that is not dependent on the traditional financial system. The interest in decentralized technologies, including Bitcoin, and, later, Ethereum, which laid the groundwork for Web3's development, was fueled. The first block of the Bitcoin blockchain (the "genesis block") contains a message referencing the financial crisis: "The Times 03/Jan/2009 Chancellor on brink of second bailout for banks." This was a clear critique of the existing financial system and a call for an alternative.

Bitcoin and the underlying blockchain technology were designed as a response to the centralized control of money and the lack of trust in traditional financial institutions. Bitcoin can be seen as one of the first applications of decentralized technologies in the context of the Web3, as it demonstrates the ability to transfer value peer-to-peer securely and transparently without a central authority. Many Web3 applications aim for such decentralization, and some involve Bitcoin or other cryptocurrencies. However, it is important to note that Web3 is a concept that goes beyond cryptocurrencies and blockchain and can encompass various decentralized technologies. While a standardized definition for Web3 has not yet been established, [38] consider applications such as cryptocurrencies, non-fungible tokens (NFTs), decentralized autonomous organizations (DAOs) and metaverse(s) as the backbones of Web3.

Web3 promotes decentralized storage systems, where data is not stored on central servers but distributed decentral in a network of nodes. Users regain data sovereignty, they are more independent from platforms and it is easier for them to monetize content they create on the internet [14, 38]. After users were primarily able to read in Web 1.0 and can also write in Web 2.0, a new feature of the emerging Web3 is that users can also own digital assets and provide cryptographic proof of that ownership (Fig. 4. The concepts of write and own differ in terms of control over digital property: while writing enables participation, it takes place according to the rules of the platform used, which has power over the content. Ownership on the Web3, on the other hand, gives users sovereignty and control over their own assets; users can take control and proof of their digital possessions into their own hands in a decentralized manner. This concept is made possible by the underlying blockchain technology, where ownership is recorded on an immutable ledger. While Web 2.0 represented a frontend revolution in that users could participate more easily in the network, Web3 is more of a backend revolution in which central data storages could be replaced with blockchain-based, decentralized ways of storing and organizing data [44].

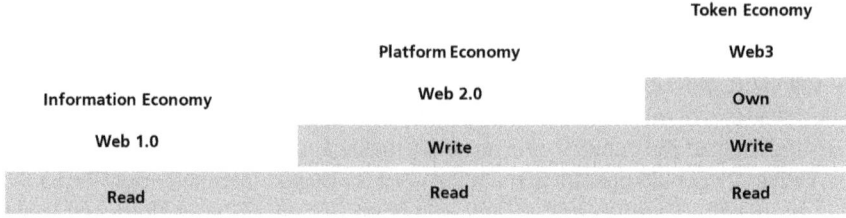

Fig. 4 The evolution of the web

Web3 can potentially lead to a token economy, "in which the community's revenue can be allocated to the actual content producers and service users who create value" [32], p. 773). Tokens can, for instance, represent currencies, ownership of assets,[3] voting rights or simply point systems for incentives with no conventional value [9, 32]. Besides the democratization of ownership and potentially more control over personal data, Web3 promotes the possibility that machines and devices become economically independent and create an "Economy of Things". In Web 2.0 platforms, machines serve a company, but in Web3, they could monetize their created value themselves and become more and more independent [45], shifting value creation from centralized to decentralized entities and unleashing interesting business innovations. One potential application of the Economy of Things can be smart city repair services. A connected streetlight detects through an embedded sensor that a light is out of order and sends a report to a decentralized network. An autonomous maintenance robot is dispatched to replace the defective part. After successful repair, payment is automatically processed between the connected devices. There are different models for organizing payment, such as a pay-per-use model in which a certain budget is provided by the municipality in a wallet, which is debited automatically, securely and transparently when a repair is actually carried out.

Before we continue our analysis, we would like to briefly address the important difference between Web 3.0 and Web3. Web 3.0 was designed by Tim Berners-Lee in 1999 and contains the idea of a "Semantic Web" that links the flood of data on the web with semantics and makes it machine-readable [44]. The inventor's vision was to let machines handle individual daily tasks (such as bureaucracy or trade) and at the same time assist the evolution of human knowledge as a whole [6, 44]. While Web3 aimed to create a decentralized, user-centric internet, Web 3.0 had the ideology of improving the efficiency and intelligence of the internet. The development of Web 3.0 was driven by academics and organizations such as the World Wide Web Consortium (W3C), while Web3 is closely linked to the blockchain community and is mostly developed by companies and crypto projects. Web 3.0 remains largely unrealized in its original vision, having been overtaken by the rapid development and commercialization of the internet, the lagging jurisdiction in the global network, and the resulting blossoming of platform capitalism [41]. However, Big Tech companies like Google or Meta are

---

[3] The technical possession of a token in the blockchain is separate from the question of whether this possession is legal in the sense of ownership [3].

excessively linking semantic data about their users, their behaviors and preferences as they use their services "for free", which is criticized as "surveillance capitalism" [30, 55].

Returning now to the framework and the four phases proposed by Geels [21], Fig. 1, Web3 is still mostly in the first phase, the experimentation phase, with many of its technologies and practices being actively developed and experimented with by a dedicated community of innovators. An exception here may be cryptocurrencies that already have a broad base of users and for which a dominant design has established itself and which are therefore more likely to be assigned to the stabilization phase. The bank, which was the subject of controversy during the emergence of Bitcoin, is now also trading cryptocurrency at the stock market. This can be seen as a settlement of technology and the transformation of a revolution into an evolution.

Unlike the development of the ARPANET, there is no singular protected space in which Web3 is currently being invented, but many independent initiatives in- and outside of academia that intend to advance the development of Web3. There are for instance government-funded projects that are creating smaller experimental spaces, or interest groups embedded in universities.

In addition, there is also a protected space for the technology among people and organizations who opt for Web3 for ideological reasons, such as commitment to decentralization and privacy, as well as people who have particularly high security standards that they feel are better met with Web3.

Alongside this, a large process of trial and error can be observed, with many start-ups emerging that have Web3 in their value proposition. At the same time, the stakeholders involved in Web3 are learning about multiple dimensions of Web3—not only the technology, but also markets, regulatory considerations, governance models, infrastructure requirements and consumer practices. For example, users must consider how they handle the keys to their wallet, and thus to their digital assets—unlike in Web 2.0, users have data sovereignty and must store the key pair securely; it is not possible to recover it, as is the case with a password in Web 2.0. This is a potential pitfall, as it contradicts good practice in established human–computer interaction, i.e. "How do I recover from mistakes?" [4], and users must learn how to manage their data sovereignty.

3.2 Socio-technical System and Regime

As our ex-post analysis in Sect. 1.2 has shown, a socio-technical system dominated by the internet as we know it shapes the way societies communicate today. This system is locked in and kept stable by a strong regime, with the influence of incumbent tech giants likely playing a role in this. This is, to name just one example, constituted in the entrenched routines of the average internet-user, who is very used to their data being owned and handled externally. Incremental innovation such as the use of VR for video conferencing and gaming, or digital IDs still happen in that system, but do not threaten its general structure and continuous reproduction.

However, there is criticism to the incumbent system regarding its performance in terms of upholding values of social justice, or democracy. For example, users themselves are often not rewarded for the content they produce, but the platform operators benefit financially from the data that users leave on the internet and the content that they actively produce [38].

Such new pressures could pose a serious challenge to the incumbent system in the coming years by questioning its very configuration, i.e., the underlying user habits, the market structure concentrated in a few large incumbent companies, or the location and ownership of the server infrastructures needed for all the acts of global communication in business, administration, culture, and personal life. This implies that a transition away from the incumbent socio-technical system towards one with more data privacy is not an unrealistic utopia but a real-world phenomenon in the making.

One opportunity that has arisen with the development of the Web3 and the associated technologies, particularly in relation to the communication system of the current socio-technical regime, is automation and the reduction of bureaucracy. Web3 technologies enable smart contracts, programmed self-executing contracts that are automatically carried out when a predefined condition is met. This could accelerate and decentralize processes that were previously regulated by central institutions and bureaucratic processes. The idea is further expanded in the Economy of Things, which was explained in the Smart City Repair example above, by making devices autonomous economic actors that could communicate with each other and process transactions. Both smart contracts and the Economy of Things represent an opportunity for a transition of the current socio-technical regime, as they can increase efficiency and create new opportunities for decentralized, direct communication.

3.3 Landscape Pressures on the Current Regime

In the exogenous context, there are several factors that have an influence on the development of Web3. First and foremost, a growing awareness of the limitations of centralized data and data privacy breaches, censorship, and the concentration of power in the hands of a few tech giants are leading to concerns about centralization.

Big Tech companies such as Google, Amazon, Apple, Meta, and Microsoft have transformed the internet into a global economic platform on which companies and individuals can operate. This has led to a broad economic and social impact, whereby the platform economy has institutionalized the internet in almost all areas of life and ushered in a new era of the digital economy with platforms playing a central role. However, existing legal requirements could change this regime. In July 2024, Apple had to open up the contactless payment technology in Apple Pay to rival pay systems, because the European Commission considered Apple abused its dominant position for mobile wallets on iOS (European Commission [15]). The underlying argument by the European Commission is that abusing the dominant position restricts competition and impedes innovation. Similar decisions around alternative app stores on Android and iOS that do not charge 30% fee of any transaction have been fought in court, most prominently by Epic Games.

The platform providers manage the data that companies or users place on the platforms [44]. The internet, which after Berners-Lee's invention was once open, decentralized and basically free of charge, is becoming increasingly gated, "with creators paying companies to publish their work and users paying companies for access to it" [38]. Payment is not always made in monetary form, but users can often use internet services free of charge by providing personal data that companies can use commercially. Landwehr (2023) argues that these highly profitable business models, which are largely free for end users, have moved the computer industry in a socially problematic direction. This regime contributes to what Lanier [31] calls "Siren Servers". Resistance to this has sparked new landscape developments, such as activism for data sovereignty, that questions the legitimacy of the incumbent regime, and in turn, triggered some legislative efforts to protect individuals' data [30]. This has been the case especially in the European Union, where policies such as the GDPR or the Digital Services Act have recently been implemented.

The introduction of the GDPR (General Data Protection Regulation) in 2018 has impacted the landscape by introducing stronger data protection standards. The GDPR places great emphasis on consent and transparent communication regarding the use of data, it has enshrined the right to be forgotten, and overall, it accompanies a trend towards greater data sovereignty. Although this is independent of the technological development of Web 2.0 or Web3, the trend toward data sovereignty supports the concepts of Web3.

The transatlantic data flow between Europe, the United States, and the United Kingdom (after the Brexit), which is used to support internet businesses, has been enabled by the implementation of data transfer agreements such as the "Safe Harbor" and "Privacy Shield" frameworks. However, divergent interpretations of data protection and privacy in the wake of 9/11 and the revelations by Edward Snowden in 2013 [53] have highlighted a legal discrepancy between Europe and the US. Subsequently, the Court of Justice of the European Union annulled those treaties in response to the historic rulings in the Schrems I [11] and Schrems II [50] cases. These rulings led to a realignment of transatlantic data protection policies [50], which have implications for businesses in the EU [42].

"Markets in Crypto-Assets" (MiCA), is a EU regulation that specifically promotes distributed ledger technologies such as blockchain [17]. It provides a comprehensive legal framework for crypto assets in the EU. Consumers will be better protected by establishing clear rules for crypto service providers, including the requirement to provide a detailed white paper and the specification of user rights. Crypto service providers of specific services which are under the scope of MICA must meet certain requirements regarding transparency, capital reserves and the protection of customer funds. Overall, the EU-wide harmonized introduction of this legal framework contributes to legal stability by avoiding national regulations and promotes innovation in the blockchain sector.

Perhaps the most profound and challenging rulings affecting the landscape concern the use of the vast amounts of personal data collected by dominant big tech companies such as Meta. According to the ruling of the European Court of Justice's in the Meta case (Case C-252/21, Court of Justice of the European Union [12]), such data must no longer be used indefinitely for "personalized user-experiences" i.e. micro-targeting. Instead, the digital sovereignty of users will be strengthened by offering ad-free subscription services [35] and potentially a new era will be ushered in through the self-determined, fair handling of personal data as the currency of the digital age [7].

3.4 What Does Web3 Need to Get Out of Its Niche?

We argue that Web3 is still largely in an experimental phase and that many applications have not yet developed a dominant design. Therefore, the question of what needs to happen for Web3 to emerge from its niche is particularly exciting in the current situation.

The literature on socio-technical transitions provides a list of principles that can help niches develop and mature so that they can transition when a window of opportunity opens [52].

Smith and Raven [52] explain that effective niche protection has three characteristics, which are explained below in relation to Web3:

1. Shielding path-breaking innovations against mainstream selection pressures.
2. Nurturing path-breaking innovations in protective spaces.
3. Empowering path-breaking innovations.

We will only cover a few interesting aspects here so as not to go beyond the scope of this book chapter. For a better insight, however, we recommend reading Smith and Raven [52] to all interested readers.

Shielding regards the protection of niche innovations from selection pressures of the regime (e.g., which technologies are able to amortize without subsidies) [47]. One example of such selection pressures are market processes, i.e., which technologies are able to amortize in mainstream markets. Here, subsidies, such as the European Commission's funding of several EU projects related to blockchain, are an example of actively shielding the Web3 niche (European Commission [16]).

Another common selection pressure are public policies. To strategically shield the niche, legal and political frameworks can be created to protect innovative developments on the Web3 and thus create a favorable environment for experimentation and growth. In Germany, this is made possible by the use of experimentation clauses, which allow deviations from a generally applicable regulation in order to test alternatives in practice.

Nurturing is conducted to help the innovation develop itself and consists of facilitation of learning, creation of networks and creation of a common vision [52]. In Web3, we observe that there are many fragmented developments that function independently of each other, but the challenge remains to connect them effectively to create a coherent and interoperable Web3 ecosystem. This challenge arises due to the diversity of protocols and missing standards, such as the lack of interoperability standards for wallets, with incompatibilities leading to difficulties in transferring data between different blockchains. This suggests that a shared vision for Web3 that aligns goals and helps to work collaboratively towards a common ecosystem has not yet been widely established.

Regarding the dimension of learning facilitation, blockchain and Web3 have potential in improving the usability to enhance broad learning. The process of setting up and managing a Web3 wallet is necessary to participate in many Web3 applications, but especially for users who are not technologically savvy, dealing with elements such as cryptographic keys, seed phrases and lack of account recovery options can be complex and intimidating. This uncertainty of users is a security issue as it facilitates frauds and scams. Possible strategies for learning facilitation include improving usability to make the creation process more intuitive as well as providing user-friendly tutorials.

Empowerment is about making sure that the protective shields of the niche are no longer needed, as the innovation is institutionalized. A distinction is made between fit-and-conform empowerment, in which the niche innovation is made competitive and conforms within unchanged selection environments, and stretch-and-transform empowerment, in which the niche is empowered to restructure mainstream selection environments [52].

Web3 is a technology that challenges the existing selection environment, which is characterized by centralized structures. It is already becoming apparent that Web3 can restructure the existing internet business models of the platform economy, as it stands for a high degree of data sovereignty and wants to prevent a few central players from having unrestricted access to and control over users' personal data. This argues for a stretch-and-transform logic dominating the discourse around Web3. Here, advocates of Web3 need to offer realistic solutions to instabilities and tensions experienced in the existing socio-technical regime in order to institutionalize Web3 practices [52]. Capabilities and resources are needed to enable participation in political debates about the design of future institutions [52]. In consequence, shielding measures such as the ones described above become part of the dominant regime and are thus seen as new norms, forming a selection environment of their own. Successful stretch-and-transform empowerment requires significant political support [47].

4 Discussion

Applying the Multi-Level Perspective to explain the socio-technical transition of the internet has enabled us to better understand the complex interactions between the three levels of technology itself, the socio-technical regimes (configuration of technologies, user-practices, markets, infrastructures, cultural manifestations) and the exogenous environment (with slow changing developments as well as shocks). The analysis delivers practical contributions for actors in Web3, as well as theoretical reflections on the framework.

The first practical contribution is based on the observation that Web3 is not a continuation of the transition process from analog communication and the ARPANET to today's platform economy with Web 2.0, but a new niche innovation, as it differs radically from Web 2.0 and the current socio-technical regime. Instead of an incremental innovation that fits into the existing regime, Web3 starts as a radical innovation with a background in finance and crypto that offers new ways of using the internet and thus attempts to challenge locked-in markets, user-practices, and infrastructures. It is only natural that niche innovations go through a process of trial and error and that it takes time for a dominant design to establish itself among the various possible applications. Going back to the development of Web 2.0, we saw that the development of features like search engines, which are crucial for the use of Web 2.0, took a long time. Many startups failed in the long run, but their ideas and products still contributed to establishing a dominant design. A comparison can be made here with wallets, which are needed to participate in Web3, but whose usability and lacking standardization is often criticized. Some wallets, such as MetaMask, use a pair of public and private keys to confirm transactions, while newer concepts, such as account abstraction, replace private keys with verification methods defined individually in a smart contract. The arrival of a dominant design for wallets still seems to be pending. Web3 comes with a number of inconveniences and novelties for the users of established technologies. Proponents of Web3 must therefore not only argue about technological superiority but must also offer solutions to tensions in the established socio-technical regime and facilitate broad scale second-order learning [47, 51].

The second practical insight relates to the interplay between niche-technology and regime or landscape: The presence of promising innovations such as blockchain technology alone is not sufficient to successfully establish the Web3 as a new, dominant regime. Exogenous landscape pressures are needed to provide windows of opportunity within the existing institutional and socio-technical structures and to overcome the dominance of the existing regime of a platform-based internet. As developments such as the covid-19 pandemic have shown, such exogenous landscape pressures are impossible to predict, especially when change is sudden and spreads across socio-technical systems [23]. However, a niche can be prepared to seize such opportunities, particularly through the measures of strategic niche management. Further, in this chapter we argued that shortcomings in data protection are a landscape pressure that could lead to tensions in the current socio-technical system, but without a respective data-based study this remains an assumption. We cannot say with certainty that society perceives centralized data storage in Web 2.0 as a problem that will put pressure on the socio-technical system.

This leads to the third insight from applying transition studies to Web3, that there are strategical possibilities for action to shield, nurture and empower the niche, thereby facilitating its evolution into a mainstream force. This knowledge can be used to actively influence the development of the Web3 niche. This chapter has only briefly touched on the possibilities, but the examples given above showcase the need for more proactive management of the Web3-niche. Some measures to shield and nurture the niche exist, for instance governmentally funded projects that offer learning opportunities around the usage of Web3. However, there is significant room for improvement. To successfully work for transformative change, powerful actors, in particular governments need to promote the institutional change needed to empower the existing innovations while providing shared infrastructures and protective spaces for the niche as long as needed. We thus see considerable potential for further research that can more systematically detail measures of strategic niche management of the Web3 niche.

From a theoretical perspective it can be noted that the transition from analogue communication to Web 2.0 changed not only the communication system, but also other social functions like finance and commerce have been transformed by emerging niche innovations. This showcases how a single niche-innovation can influence several socio-technical systems at once—a phenomenon that is recognized and discussed in the growing literature on multi-system interaction in transitions literature (see e.g. [28, 49]).

Also, this transition seems to have happened rapidly compared to other socio-technical transitions in history. For instance, the transition towards automobility is said to have taken until the 1950s-70 s, while having started before 1900 [27]. Regarding the internet, the development of the ARPANET began in the late 1960s, and if we take the establishment of the term Web 2.0 since 2004 as a reference point, it was completed in the mid to late 2010s. This raises the question of what was different about this transition. Did governments support a transition in particular ways that they did not apply in other transitions? What can we learn from this case for desirable future (sustainability) transitions? While our text does not answer these questions, it showcases that sustainability transitions literature can still benefit from studying socio-technical change without an immediate relation to sustainability issues (as opposed to e.g., renewable energy niches). Future research can further explore and learn from the connections between such different socio-technical transitions.

From a technology and innovation research perspective, this study highlights the benefits of applying the transition literature to emerging technologies related to Web3. Understanding the complex socio-technical systems underlying the societal transition towards a next generation of the internet, helps formulating concrete and actionable propositions on how to promote the Web3 niche.

Perhaps this study also provides a valuable starting point for further research towards a twin transition [37]—a societal transition that synergistically combines green and digital transitions to address grand societal challenges. This study has shown a practical way in which both disciplines can be integrated and there is great potential for further research exploring the interplay between digital innovation and socio-technical transitions.

References

1. G. Balbi, R.R. John, 2. Point-to-point: telecommunications networks from the optical telegraph to the mobile telephone, in *Communication and Technology*, ed. by L. Cantoni, J.A. Danowski (DE GRUYTER, 2015), pp. 35–56

2. P. Baran, On distributed communications networks. IEEE Trans. Commun. Syst. **12**(1), 1–9 (1964). https://doi.org/10.1109/TCOM.1964.1088883
3. A. Bauer, *Die effektive Einzel- und Gesamtvollstreckung von Blockchain-basierten Kryptowährungen* (Duncker & Humblot, Berlin, 2023)
4. V. Bellotti, M. Back, W.K. Edwards, R.E. Grinter, A. Henderson, C. Lopes, Making sense of sensing systems: five questions for designers and researchers, in *Proceedings of the SIGCHI Conference on Human Factors in Computing Systems*. (ACM, Minneapolis Minnesota USA, 2002), pp. 415–422
5. T. Berners-Lee, *Weaving the Web: The Original Design and Ultimate Destiny of the World Wide Web*, 1st edn. (HarperBusiness, San Francisco, 2000)
6. T. Berners-Lee, J. Hendler, O. Lassila, The semantic web. Sci. Am. **284**(5), 34–43 (2001)
7. B. Buchner, J. Kühling, Das Ende des Daten-Eldorados. Verfassungsblog (2023). https://doi.org/10.17176/20230710-111046-0
8. T. Burns, G.M. Stalker, *The Management of Innovation*, Rev. (Oxford University Press, Oxford, New York, 1994)
9. V. Buterin, A next-generation smart contract and decentralized application platform (2014)
10. E. Coiera, Information economics and the internet. J. Am. Med. Inform. Assoc. **7**(3), 215–221 (2000)
11. Court of Justice of the European Union (2015) Maximillian Schrems v. Data Protection Commissioner (Schrems I), Case C-362/14, Judgment of 6 October 2015. ECLI:EU:C:2015:650
12. Court of Justice of the European Union (2020) Data Protection Commissioner v. Facebook Ireland Ltd and Maximillian Schrems (Schrems II), Case C-311/18, Judgment of 16 July 2020. ECLI:EU:C:2020:559
13. Der Spiegel, Urteil: Verlag darf weiter Häuser fotografieren. Der Spiegel (1999). https://www.spiegel.de/netzwelt/web/urteil-verlag-darf-weiter-haeuser-fotografieren-a-56225.html. Accessed 30 Aug 2024
14. A. Dutra, A. Tumasjan, I.M. Welpe, Blockchain is changing how media and entertainment companies compete. MIT Sloan Manag. Rev.Manag. Rev. **60**(1), 39–45 (2018)
15. European Commission, antitrust: commission sends statement of objections to apple (2022). In: European Commission. https://ec.europa.eu/commission/presscorner/detail/en/ip_22_2764. Accessed 30 Aug 2024
16. European Commission (2019) EU-Funded Projects in Blockchain Technology. https://digital-strategy.ec.europa.eu/en/news/eu-funded-projects-blockchain-technology. Accessed 2 Sept 2024
17. European Parliament and Council of the European Union, Regulation (EU) 2023/1114 of the European Parliament and of the Council of 31 May 2023 on markets in crypto-assets, and amending Regulations (EU) No 1093/2010 and (EU) No 1095/2010 and Directives 2013/36/EU and (EU) 2019/1937 (2023)
18. L. Fuenfschilling, B. Truffer, The structuration of socio-technical regimes—conceptual foundations from institutional theory. Res. Policy **43**(4), 772–791 (2014). https://doi.org/10.1016/j.respol.2013.10.010
19. F. Geels, Technological transitions as evolutionary reconfiguration processes: a multi-level perspective and a case-study. Res. Policy **31**(8–9), 1257–1274 (2002). https://doi.org/10.1016/S0048-7333(02)00062-8
20. F. Geels, The multi-level perspective on sustainability transitions: responses to seven criticisms. Environ. Innov. Soc. Trans.Innov. Soc. Trans. **1**(1), 24–40 (2011). https://doi.org/10.1016/j.eist.2011.02.002
21. F. Geels, Socio-technical transitions to sustainability: a review of criticisms and elaborations of the multi-level perspective. Curr. Opin. Environ. Sustain.. Opin. Environ. Sustain. **39**, 187–201 (2019). https://doi.org/10.1016/j.cosust.2019.06.009
22. F. Geels, Ontologies, socio-technical transitions (to sustainability), and the multi-level perspective. Res. Policy **39**(4), 495–510 (2010). https://doi.org/10.1016/j.respol.2010.01.022

23. F. Geels, J. Schot, Typology of sociotechnical transition pathways. Res. Policy **36**(3), 399–417 (2007). https://doi.org/10.1016/j.respol.2007.01.003
24. F.W. Geels, Micro-foundations of the multi-level perspective on socio-technical transitions: developing a multi-dimensional model of agency through crossovers between social constructivism, evolutionary economics and neo-institutional theory. Technol. Forecast. Soc. Chang. **152**, 119894 (2020). https://doi.org/10.1016/j.techfore.2019.119894
25. E.C. Hanson, A history of international communication studies, in *Oxford Research Encyclopedia of International Studies* (2010)
26. M. Hossain, Grassroots innovation: the state of the art and future perspectives. Technol. Soc. **55**, 63–69 (2018). https://doi.org/10.1016/j.techsoc.2018.06.008
27. L. Kanger, F.W. Geels, B. Sovacool, J. Schot, Technological diffusion as a process of societal embedding: lessons from historical automobile transitions for future electric mobility. Transp. Res. Part D: Transp. Environ. **71**, 47–66 (2019). https://doi.org/10.1016/j.trd.2018.11.012
28. L. Kanger, J. Schot, B.K. Sovacool, E. van der Vleuten, B. Ghosh, M. Keller, P. Kivimaa, A.-K. Pahker, W.E. Steinmueller, Research frontiers for multi-system dynamics and deep transitions. Environ. Innov. Soc. Trans.Innov. Soc. Trans. **41**, 52–56 (2021). https://doi.org/10.1016/j.eist.2021.10.025
29. J. Köhler, F.W. Geels, F. Kern, J. Markard, E. Onsongo, A. Wieczorek, F. Alkemade, F. Avelino, A. Bergek, F. Boons, L. Fünfschilling, D. Hess, G. Holtz, S. Hyysalo, K. Jenkins, P. Kivimaa, M. Martiskainen, A. McMeekin, M.S. Mühlemeier, B. Nykvist, B. Pel, R. Raven, H. Rohracher, B. Sandén, J. Schot, B. Sovacool, B. Turnheim, D. Welch, P. Wells, An agenda for sustainability transitions research: state of the art and future directions. Environ. Innov. Soc. Trans.Innov. Soc. Trans. **31**, 1–32 (2019). https://doi.org/10.1016/j.eist.2019.01.004
30. M. Landwehr, A. Borning, V. Wulf, Problems with surveillance capitalism and possible alternatives for IT infrastructure. Inf. Commun. Soc.Commun. Soc. **26**(1), 70–85 (2023). https://doi.org/10.1080/1369118X.2021.2014548
31. J. Lanier, Who owns the future?—Du bist nicht der Kunde der Internetkonzerne. Du bist ihr Produkt. Penguin, London (2014)
32. J.Y. Lee, A decentralized token economy: how blockchain and cryptocurrency can revolutionize business. Bus. Horiz.Horiz. **62**(6), 773–784 (2019). https://doi.org/10.1016/j.bushor.2019.08.003
33. S. Lukasik, Why the Arpanet was built. IEEE Annals Hist. Comput. **33**(3), 4–21 (2011). https://doi.org/10.1109/MAHC.2010.11
34. J. Markard, R. Raven, B. Truffer, Sustainability transitions: an emerging field of research and its prospects. Res. Policy **41**(6), 955–967 (2012). https://doi.org/10.1016/j.respol.2012.02.013
35. Meta (2024) Facebook and Instagram to Offer Subscription for No Ads in Europe. In: Meta. https://about.fb.com/news/2024/11/facebook-and-instagram-to-offer-subscription-for-no-ads-in-europe/. Accessed 18 Nov 2024
36. P. Monge, S.A. Matei, The role of the global telecommunications network in bridging economic and political divides, 1989 to 1999. J. Commun.Commun. **54**(3), 511–531 (2004). https://doi.org/10.1111/j.1460-2466.2004.tb02642.x
37. S. Muench, E. Stoermer, K. Jensen, T. Asikainen, M. Salvi, F. Scapoo, *Towards a Green and Digital Future: Key Requirements for Successful Twin Transitions in the European Union* (Publications Office of the European Union, Luxembourg, 2022)
38. A. Murray, D. Kim, J. Combs, The promise of a decentralized internet: what is Web3 and how can firms prepare? Bus. Horiz.Horiz. **66**(2), 191–202 (2023). https://doi.org/10.1016/j.bushor.2022.06.002
39. J. Mylan, C. Morris, E. Beech, F.W. Geels, Rage against the regime: Niche-regime interactions in the societal embedding of plant-based milk. Environ. Innov. Soc. Trans.Innov. Soc. Trans. **31**, 233–247 (2019). https://doi.org/10.1016/j.eist.2018.11.001
40. S. Nakamoto, *Bitcoin: A Peer-to-Peer Electronic Cash System* (2008)
41. L. Oppermann, Web engineering, in *Lecture in Master-Course at Hochschule Bonn-Rhein-Sieg. Faculty of Computer Science* (2022)

42. L. Oppermann, Y. Uzun, F. Buchholz, U. Riedlinger, S. Fuchs, H. Stenzel, L. Odenthal, A. Altepost, M. Bau, Industrial metaverse? Human-centred design for collaborative remote maintenance and training using XR-technologies, in *XR Meets the Metaverse—Proceedings of the 8th International Augmented and Virtual Reality Conference 2023* (Las Vegas, USA, 2023)
43. T. O'Reilly, What is Web 2.0 (2005). https://www.oreilly.com/pub/a/web2/archive/what-is-web-20.html. Accessed 30 Aug 2024
44. A. Park, M. Wilson, K. Robson, D. Demetis, J. Kietzmann, Interoperability: our exciting and terrifying Web3 future. Bus. Horiz.Horiz. **66**(4), 529–541 (2023). https://doi.org/10.1016/j.bushor.2022.10.005
45. Peaq, What is the economy of things? (2021). https://www.peaq.network/blog/what-is-the-economy-of-things. Accessed 22 Sept 2024
46. J. Pohle, D. Voelsen, Das Netz und die Netze. Vom Wandel des Internets und der globalen digitalen Ordnung. Berlin J. Soziol. **32**(3), 455–487 (2022). https://doi.org/10.1007/s11609-022-00478-6
47. R. Raven, F. Kern, B. Verhees, A. Smith, Niche construction and empowerment through sociopolitical work. A meta-analysis of six low-carbon technology cases. Environ. Innov. Soc. Trans.Innov. Soc. Trans. **18**, 164–180 (2016). https://doi.org/10.1016/j.eist.2015.02.002
48. Reuters Institute for the Study of Journalism (2023) Reuters Institute Digital News Report 2023
49. D. Rosenbloom, Engaging with multi-system interactions in sustainability transitions: a comment on the transitions research agenda. Environ. Innov. Soc. Trans.Innov. Soc. Trans. **34**, 336–340 (2020). https://doi.org/10.1016/j.eist.2019.10.003
50. M. Rotenberg, Schrems II, from Snowden to China: toward a new alignment on transatlantic data protection. Eur. Law J. **26**(1–2), 141–152 (2020). https://doi.org/10.1111/eulj.12370
51. J. Schot, F.W. Geels, Strategic niche management and sustainable innovation journeys: theory, findings, research agenda, and policy. Technol. Anal. Strat. Manag. **20**(5), 537–554 (2008). https://doi.org/10.1080/09537320802292651
52. A. Smith, R. Raven, What is protective space? Reconsidering niches in transitions to sustainability. Res. Policy **41**(6), 1025–1036 (2012). https://doi.org/10.1016/j.respol.2011.12.012
53. E. Snowden, *Permanent Record*, 1st edn. (Metropolitan Books, New York, 2019)
54. S. Tardini, L. Cantoni, 6. Hypermedia, internet and the web, in *Communication and Technology*, ed. L. Cantoni, J.A. Danowski (DE GRUYTER, 2015), pp. 119–140
55. S. Zuboff, Big other: surveillance capitalism and the prospects of an information civilization. J. Inf. Technol. **30**(1), 75–89 (2015). https://doi.org/10.1057/jit.2015.5

Lisa Klug is a research associate at the Fraunhofer Institute for Applied Information Technology FIT and a PhD student in Innovation Science at Halmstad University, Sweden. Her work focuses on the value creation and governance of emerging digital technologies, with particular emphasis on digital platforms, blockchain, and Web3. With experience in both applied research and industry, she facilitates the use of digital solutions and supports companies in integrating innovation into their processes. The fusion of Lisa's research interests and those of her co-author Max Halbwachs, who studies sustainability transitions in urban mobility, leads to illuminating insights into the socio-technical transition to Web3. Together with Leif Oppermann, whose expertise in digital infrastructures complements this perspective, they apply transition theory to explore the systemic and institutional conditions under which Web3 technologies may evolve and diffuse.

Max Halbwachs is a Ph.D. student in Innovation Science at Halmstad University, Sweden. His research is located in the field of sustainability transitions, which aims at understanding and promoting socio-technical change towards sustainable societies. His PhD research focuses on how local authorities can foster the interaction and collaboration of actors in such processes through "intermediation". While the empirical focus of this research lies on the transitions towards sustainable urban mobility, he, together with Lisa Klug and Leif Oppermann, contributes to this book with insights from transition theory to understanding the broader socio-technical system in which the Web3 is emerging and developing. This can inform broader efforts to promote and diffuse the technology.

Dr. Leif Oppermann is head of the Mixed and Augmented Reality Solutions group in the Collaboration Systems research department at Fraunhofer FIT. He studied media informatics at the Hochschule Harz in Wernigerode and was a research fellow at the Mixed Reality Lab at the University of Nottingham, where he also earned his PhD in computer science. He is researching into applications of mobile Mixed Reality, web-based collaboration and ubiquitous computing for intelligence augmentation using a user-oriented cooperative design approach. Dr. Oppermann joined FIT in 2009 and led several national and international research projects. Most recently he led the German national project "5G Troisdorf IndustrieStadtpark" which produced a widely recognized Industrial Metaverse demonstrator, and influenced the positions of Fraunhofer and the German industry umbrella organisation BDI on the topic. He is co-author of the textbook "Virtual and Augmented Reality (VR/AR)—Foundations and Methods of Extended Realities (XR)" and Editorial Board Member of the Empathic Computing journal.

Trustworthy Tokenization in the Metaverse with Secure Digital Identities Through eIDAS 2.0?

Ignacio Alamillo and Steffen Schwalm

1 Dimensions of Digital Identities and the Limitations of eIDAS 1.0

Digital identities are the key for trustworthy digital transactions. Only if all actors in a process or ecosystem securely know with whom they are act digital trust will be ensured. Unique identification of legal or natural entities as well as their objects is the basement for a digital identity that allow the verification of companies (Do they really exist?), of the person acting on behalf of that company (Do they really exist?) and of their authorization (Is Alice authorized to act on behalf of company A?).

This means that digital identities comprise several dimensions [1] so:

- Natural entity: It's me.
- Legal entity; It's my company.
- Legal roles of a natural entity: It's my Power of Attorney.
- Credentials or attestations of a natural entity acting as natural or for a legal entity: it's my diploma or my driver license.
- Attestations of a virtual identity related to a natural/legal entity entity or virtual attribute related to natural or legal entity: It's my virtual me and/or the virtual car of my virtual me.
- Attestations allowing a natural entity to access something: It's what I'm allowed to access.
- Credentials for signing contracts or sealing documents: It's my signature or the seal of my company.

I. Alamillo (✉)
Facultad de Derecho, Universidad de Murcia, Murcia, Spain
e-mail: ignacio.alamillod@um.es

S. Schwalm
msg group, Ismaning, Germany

Currently, digital identities are typically issued by a centralized authority. Despite the widely used but privacy exposed social identities the main electronic identification means of natural entities are government eID issued by member states. The current [2] Regulation established a coherent and holistic legal framework on digital identities and trust services in the EU and EFTA but was mainly focused on governmental electronic identification schemes and means for natural entities. Those identities focused on identification of the person itself which is only small dimension of its identity as mentioned above.

In parallel decentralized digital ecosystems occurred in the context of emergence of distributed ledger technologies. DLT by its distributed design makes it easy to establish decentralized digital business models cross-industry and cross-country between. The technology gains it's biggest added value in transactions between > 3 parties which don't trust each other and so trust in a distributed network which is immutable by design [3, 4]. In the context of DLT and decentralized ecosystems also the new paradigm of self-sovereign identities has to be mentioned. SSI promise identity owner full control over its identity and attributes [Allen]. With the SSI principles some possible rules for decentralized digital identities were developed.

Security the identity information must be kept secure	Controllability the user must be in control of who can see and access their data	Portability the user must be able to use their identity data wherever they want and not be tied to a single provider
Protection	Existence	Interoperability
Persistence	Persistence	Transparency
Minimisation	Control	Access
	Consent	

All identity information is stored decentralized and only the holder should decide whom he'll give access or transmit identification information. One main postulate is that in SSI a trusted 3rd party might not be necessary anymore since DLT is used as decentralized PKI and immutable by design—so SSI may be trustworthy by itself [4, 5], [Yild22]. Currently SSI lacks the legal trust because current [2] mainly focused on government eID not integrating the new SSI-paradigm. There's no trust by default as expected in the fundamental idea of SSI and DLT—only by proof of a trusted 3rd party [5, 6]. Trustworthy digital transactions just require the proof of authenticity and integrity achieved e.g. with cryptographical measures such as qualified electronic signatures or seals according to eIDAS [5]. With the eIDAS Bridge the EU just developed possible legal and technical solution to bridge centralized approach of [eIDAS] referenced to government eID and (qualified) trust services with decentralized manner of DLT and possibly SSI [Al20]. Accelerated by success of DLT and developments like [EBSI] in Europe but also the limited utilization of existing (centralized) eID, the EU-Commission just revised eIDAS and proposed a re-engineered regulation in June 2021—recognizing decentralization on one hand and requirement of legal trust on the other one.

In parallel with Metaverse another dimension came up. The commingling of actual and virtual reality leads to combining of natural and legal entities on one hand but also virtual natural/legal entities (e.g. my avatar, my company in virtual reality) on the other hand. Combined with real and virtual machines and sensors as well as attestation of attributes for those real and virtual entities complex new business models are thinkable from augmented and virtual reality to digital twin or tokenization of real and virtual assets. Technically in many case DLT is used as infrastructure for transactions and self-created identities. The integration of legally compliant digital identities, attributes and trust services may enable metaverse to be use for legally compliant transactions too. This could transfer virtual realities, tokenization or digital twin from digital playground or grey zone into regulated and to wider useable ecosystems based on trustworthy digital identities but without revealing the actual legal entity if not needed—as long as combined with the possibilities of SSI and an enhancement of the scope on identities (not only identification of persons) and trust services (e.g. for DLT) of eIDAS in order to achieve legal trust on metaverse transactions [6].

Overlooking those developments and related regulative requirements on a secure and unique identification of natural and legal entities and their natural/virtual attributes or manifestations as well as the need to fulfilment of burden of proof against third parties or privacy requirements on one hand and decentralized ecosystems on the other hand there's the question on how digital identities and trust service may enable metaverse for legally compliant transactions by fulfilling those requirements[7]. The potentials of Metaverse are only useable if legally compliant and so traceable transactions based on trustworthy digital identities in all their dimensions possible. This would require the integration of virtual reality of Metaverse into the legal and technical reality defined by eIDAS ecosystem in Europe.

The paper describes based on introduction into the metaverse, the challenges within DLT and the changes through [8] how the new regulation on digital identities and trust services in Europe may ensure legal trust and proven security in metaverse applications. It also gives an overview on ongoing standardization and ends with an outlook for the future of comprehensive digital identities covering all their possible dimensions.

2 Dimensions of the Metaverse

2.1 Introduction

The term *Metaverse* is currently in use for a variety of digital contexts and often appears like a mixed bag of futuristic technology approaches and artefacts. Many consider themselves stakeholders, contributors, and beneficiaries in the upcoming Metaversian world, obviously a great lot of those does not understand the term in its entirety or merely has a vague, incomplete, or even wrong understanding of what it is and means. Even though young as a phenomenon, the Metaverse has already taken its place in digital cultures and needs to be interpreted as such, considering all implications on business, legal, and tech.

Historically the term *Metaverse* was coined by science fiction author Neal Stephenson in his renowned and awarded novel *Snow Crash*, published in 1992. In his story, the Metaverse resembles a massively multiplayer online game (MMOG) which is populated by both automated agents aka *system daemons* and user-controlled entities called *avatars*. With Snow Crash, Stephenson made avatar the de facto term for a graphical representation of a user or user character online. In his Metaverse, humans interact with each other and software agents via their avatars, in a 3D virtual space, which appears as an urban environment along a single 100 m-wide road that runs around the entire 65,536 km (2^{16} km) black spherical planet. Virtual real estate can be bought from a property management company and virtual buildings can be put on it. The Metaverse is accessed via terminals—personal goggles with high-quality or public low-quality ones—connected to a global, monopolized telecommunications network which evolved from the phone system. This fiction has fuelled and inspired today's understanding and incarnations of the Metaverse [7].

Contemporarily, the general idea of the Metaverse is a digital realm in which physical and virtual reality are combined to an extended, more immersive experience in a 3D virtual world, or many of those in a networked, interoperable state. It is perceived as the next evolutionary iteration of the internet, making social and economic connections in the virtual world feel more like physical reality.

2.2 Technology of the Metaverse

Metaverse' technological core is represented by a combination of modern IT concepts, hardware/software products and cryptographic procedures, further outlined in the following [9].

Virtual reality (VR). The stereotypic notion of VR is a person wearing a helmet-like headset bearing speakers and internal displays in front of the subject's eyes and sensory gloves. This provides an immersive experience for the VR user, who can hear and view as well as interact with the virtual world, a simulated 3D environment with objects and characters/avatars. The term was first used for science fiction in the 1982 novel *The Judas Mandala* by author Damien Broderick and popularized by VR pioneer Jaron Lanier by end of the same decade. The VR concept has sparked creativity of many authors and movie makers: From headset-less VR fictions like *Tron* (1982, *Tron: Legacy* 2010), *The Thirteenth Floor* (1999) and *The Matrix* (1999 plus later sequels) over *The Lawnmower Man* (1992) and *Johnny Mnemonic* (1995) to *Ready Player One* (2018) which became the instant classic and blueprint of what a great VR experience looks like. VR was assumed to make a commercial mainstream breakthrough already in the 1990s, which failed due to required gear being to clunky and expensive while delivering underwhelming experiences. The major technological advancements of the last two decades have changed the picture significantly, now combining affordability and impressive UX. Virtual reality also includes the composition of virtual, digital worlds with digital figures representing a real natural or legal entity, digital objects representing a real pendant or complete digital assets.

Mixed reality. In mixed reality configurations, the natural perception of a user is blended or mixed with artificial, computer-generated perceivable components. There is a further two-fold differentiation possible in mixed reality. One is *augmented virtuality* as a modification of a VR setup, in which objects or signal from physical reality. Examples are real furniture or walls being visible in the virtual environment or real external sounds being captured and inserted into the virtual audio. More prominent and already much more common is the second variant, *augmented reality* (AR). AR means a user is provided with additional computer-generated information within the real-world environment that enhances his perception of reality. Simple examples are *head-up displays* (HUD) in cars showing vehicle speed and speed limits or smartphones extending a video feed from internal cameras with e.g. names of stars or surrounding mountains. More sophisticated implementations are relatively lightweight eyeglasses which serve as minimally intrusive AR displays.

Digital Twin. Digital twins are on one hand digital representations of physical objects, such as industrial equipment and machinery, vehicles, and buildings. A digital twin is a virtual model designed to accurately reflect a physical object. Hence it is a data set of a real-world object which allows digital representation, modelling, and analyses. Digital twins offer a wide range of possibilities in industrial settings that benefit companies by saving enormous resources. Practical applications are e.g. in supply chain management, prototyping, manufacturing, predictive maintenance, or construction. Building Information Modeling (BIM) is already common practice in the construction industry, digital twins are considered the next evolutionary step there. On other hand in virtual reality the digital twin might also contain a digital figure or object representing a natural or legal entity resp. complete digital asset—assets typically manifested as Token.

Digital Asset, Token. In the economic sphere the Metaverse it is to some degree intertwined with *Web3*. The latter is a concept of a more decentralized internet in which digital assets can be traded without central intermediaries. This is achieved by cryptographically exchanging *tokens* which either represent the value of such an asset or the asset itself. Digital assets often occur as *Non-Fungible Tokens (NFT)*, which are unique and not divisible, hence can only be transferred as entire object, as opposed to cryptocurrencies, which can be obtained in fractions. NFTs usually represent or point to a digital artefact, like an image. This tokenization is attractive particularly in the Metaverse as virtual goods can be tagged, collected, and traded. Even though NFTs became popular in- and outside the Metaverse, besides a speculative hype they are subject to legal, economic, and ecological criticism. NFT are mainly used in DLT as decentralized infrastructure. Payment often done in cryptocurrencies also using DLT as infrastructure and (payment) wallets for holders to keep tokens, payments or (self-issued) identities etc.

A meticulous combination of the above makes a highly immersive, compelling, and interesting Metaverse experience possible today. But the Metaverse is far from being mainstream in 2023. The technological underpinnings have reached a quality level, though, which could invoke a broader interest. Early adopters are often from the gaming scene, as the required equipment is desirable and useful in their core realm as well. Certain VR/AR business applications could drive adoption and generate a pull effect in the personal domain [9].

2.3 Legal and Technical Challenges for Utilization in Regulated Environments

In case transaction within the metaverse shall be not only a game without legal consequences but legally secure in Europe same challenges must be solved. These include the unique and trustworthy identification of each legal or natural entity acting within the metaverse resp. their virtual twin or avatar as well as the digital objects or assets related to certain natural or legal entity. Currently only proprietary mostly self-issued identities are used in metaverse environments. Like SSI these identities lack off legal trust as a trusted 3rd party using proven identification procedures or eID schemes on certain LoA using relevant eID means according to relevant legal framework like eIDAS missed. This identification shall include not only the identification of the actual natural or legal entity but also their digital pendants, objects and assets—so all possible dimensions of digital identities as mentioned in Sect. 1. Regarding the fact that metaverse use technologies from Web3 such as decentralized infrastructures like DLT, digital assets and tokens etc. those techniques seem to be needed to be adopted in trustworthy manner into a then trustworthy metaverse in order to avoid complete change of the use case.

Beside the identities the non-repudiation of transactions has to be ensured with the utilization of (qualified) trust services also useable e.g. for the virtual twin or avatar of certain legal or natural entity while keeping the necessary security measures so e.g. a secure authentication for signature creation acc. ETSI EN 319 411[1] at QTSP. Currently platforms and infrastructures for metaverse applications are often provided by non-European market leading providers located outside Europe. This leads to GDPR issues because of foreseeable data transfer into third countries. Those platforms, tools and infrastructure in most case also not fulfil typical measures on privacy protection as privacy by design nor proven security by trusted 3rd parties as needed to achieve legal trust and established on within eIDAS trust framework in Europe. This means the challenges mentioned for SSI paradigm and DLT mentioned in Sect. 3 apply directly on metaverse applications too.

---

[1] ETSI EN 319 411–1. Electronic Signatures and Infrastructures (ESI). Policy and security requirements for Trust Service Providers issuing certificates. Part 1: General requirements. Version 1.3.1.

With the mentioned limitations of eIDAS 1.0 those challenges are not solvable as no virtual twin or digital asset can be identified legally compliant in this framework nor can a digital avatar sign a contract on behalf of its real natural or legal entity. Although the requirements on trustworthy digital transactions are clear, their implementation in metaverse require a new legal framework recognizing developments like decentralization of infrastructures, identities and transactions as well as virtualization and augmentation of reality and sophistication of identities on natural, legal and virtual entities, objects and assets in all their dimensions [9].

3 Distributed Ledger Technology

Basically, DLT is a decentralized distributed peer-to-peer network of technical nodes for data exchange and transaction execution. According to [10] a distributed ledger is in this case shared across a set of DLT nodes and synchronized between the DLT nodes using a consensus mechanism. The consensus mechanism ensures that all transactions are valid and unaltered. Its manner depends on the type of DLT so that the well-known prejudice that DLT implies unacceptable high energy need is only valid for some consensus mechanisms e.g. Proof of Work, other ones are much more efficient especially those ones in DLT with restricted access rights e.g. BFT, Proof of Authority, Proof of Stake. DLT networks allow the transfer of data or value from one party to another without having intermediates involved. Once written to the ledger the transactions are immutable, mainly based on hash protection of data stored on the chain. Any transaction can reliably be tracked on the chain. In case the DLT is organized in blocks it's called blockchain, so basically a blockchain is a special kind of DLT [3, 6]. If the factual distributed data set or transactions are bundled in sequential linked blocks it is called a blockchain—a special kind of DLT. The blocks can also include the hash of the previous block and so build the mentioned hash-protection and a so called "timestamp". This DLT- "timestamp" as well as DLT "signatures" have currently to be differentiated from timestamps defined in eIDAS and related standards due to its lack of a trustworthy source of time, missing creation and validation of digital signatures by trust service provider and missing Proof of Existence created by a third party instead of the system, here DLT, itself. The hash-based integrity protection of each block is based on Merkle-trees. This means that if authenticity or Proof of Existence within DLT needed they have to be added from (qualified) trust service providers acc. eIDAS. Similar challenges occur in case the parties participating in a transaction shall be made evident. In this case the DLT has to be combined with external systems to ensure unique and trustworthy identification of legal and/or natural entities [3, 6], [IDNTS31648].

If DLT should be used for trustworthy digital transactions, it is mandatory to fufill requirements on records management including long-term preservation of the evidence of authoritative records also against 3rd parties, until the end of the retention periods in force and to keep them provable—as it is required for any business IT-system. This means a valid records management ensuring integrity, authenticity, reliability, confidentiality and transferability of so authoritative records by trusted 3rd parties incl. evidence preservation for the whole retention period. Additionally proven security of a DLT network done by independent 3rd party based on international standards is an additional core requirement to use DLT in regulated environments with the need to fulfil burden of proof. Without additional measures like given in [11] DLT is currently not able to fulfil those as comprehensively described in [3, 5, 6, 12].

4 Fundamental Requirements on Trustworthy Digital Transactions

Trustworthiness of digital transactions and records means that the process and the records are really what they seem to be and that this is provable by independent 3rd parties. Trustworthy digital transactions ensure the unique and lossless evidence of authenticity, integrity, reliability of the electronic records which are created, received, stored and managed during the life-cycle of transaction against independent 3rd parties as long as they are needed. This means typically until the end of the defined retention periods based on and compliant to existing laws (between 2 and 110 years or permanent). Some main pre-condition are their availability as well as the protection of the confidentiality of records worthy of protection. The records contain content, metadata and transaction (process) data. The basic preconditions for this is the transferability [UN17] of the records. The evidence will be proven based on the records themselves so the named requirements and in consequence the evidence value of a record are significant properties of the electronic record itself ([WE18], [KHS14], [Ro07]). The utilization of cryptographic measures, e.g. qualified e-signatures, seals and time stamps acc. to eIDAS [Re14], enables users to preserve the evidence of their electronic records without losing the transferability of the records. The evidence value of a qualified electronic signature (e-signature) is the same as a handwritten signature, the seal makes the authenticity and integrity of the sealed record evident. These cryptographic measures are inherent and significant properties of the records. They require measures concerning long-term preservation focusing on the record itself not the storage, the software environment etc. to keep the trustworthiness of the records in the sense of preservation of the information of the data record and its evidence. Main precondition is the establishment of a valid records management according. This includes established policies, roles and responsibilities, processes as well as appropriate functionalities in business-IT to managing records properly during their whole life-cycle from the creation or receiving over utilisation and storage until archiving and disposition [We18].

These basic burdens of proofs and requirements on trustworthy digital records and transactions are independent from used IT-system, organization or process. Currently there is no regulation defining technology or institution as trustworthy by themselves. Trustworthiness always requires the evidence of the significant properties based on the records themselves as long as they are needed and without any losses. This requires especially the transferability of the records and so the utilisation of (qualified) electronic signatures, seals and timestamp acc. to eIDAS [We18], [3]. An evidence value of a record is an inherent property of the record itself. That is why records should only be archived in self-contained AIP which contain any necessary information (metadata, content, evidence relevant and technical evidence data) in a standardized container acc. to [13]. The proof is typically done by trustworthy 3rd parties such as courts, regulative authorities, auditors etc. depending on the legal requirements [We18].

This means trustworthiness can be achieved only by proof not by self-declaration. Essentially it is necessary to make compliance to legal requirements and prior art— so technical standards given and audited by trustworthy 3rd parties—evident [13], [We18].

5 Development Within eIDAS 2.0

5.1 Fundamentals

In December 2023 the ITRE Committee of European Parliament agreed to the final version of [8] as an amendment of [2]. The main goal of the update is not a replacement but further development of [2] in the context of decentralization and the upcoming SSI-paradigm but also further development on (qualified) trust service providers (QTSP). The technical framework of [8] is determined by the Architecture and Reference Framework developed in the eIDAS Toolbox through experts from Member States. As Reference [8] requires mandatory implementing acts for each component referencing European Standards from ETSI or CEN the regulation also creates a much more coherent technical framework than [2] where only fewer implementing acts were mandatory [6]. As the paper focus on Distributed Ledger the scope of the [8] considerations is on DLT related subjects.

5.2 EUDI Wallets and (Qualified) Trust Services Using DLT

The presumable biggest change in [8] is the requirements for every Member state to provide an EU-Digital Wallet to its natural entities. The Wallet could be published by member state, under authority of member state or recognized by member state. This makes also private wallet possible under the recognition of a Member State. Any EUDIW will contain a Personal Identification (so called PID for natural or legal entity as wallet holder) based on notified eID scheme on LoA "high" and has to achieve LoA "high" itself. Directly corresponding with the EU-Digital Wallet the new qualified attestation services acc. Art. 45a-e [8] have to be taken into account. (Qualified) Attestations (QEAA) are nothing more, nothing less than additional attributes so driver license, diplomas or vaccine passport of EUDI Wallet holder but with qualified seal from issuing QTSP. This means that EU-Digital Wallet will contain the core identity currently covered by government eID as well as additional attributes. The data to be attested in QEAA will be provided from so called authentic sources provided by Member States. Recognizing this close relationship between qualified attestation services and the wallet [8] contains the same requirements for mandatory implementing acts referring on European Standards for both—wallet and (qualified) attestation service. Both will be certified by independent Conformity Assessment Body which ensures the proven security. In the consequence [8] crosses digital identity means and (qualified) trust services—they determine each other and for both DLT as infrastructure is possible to use. The core requirements on QTSP like liability, periodical re-certification, reporting obligation on security issues etc. remain in [8]. Technically the EUDI Wallet as well as QTSP for QEAA can use DLT as decentralized infrastructure. The [ARF] as fundamental technical framework for [8] only defines protocols and formats as well as key management for the Personal Identification (PID) of natural and legal entities but no limitations on the infrastructure. Same applies to current standardization in this subject in CEN or ETSI [14–16]. Beside the EUDI Wallet and QEAA the [8] contains some changes on other (qualified) trust services and introduce new ones like QTSP for Electronic Ledger, (Art. 45 h), Management of secure signature creation devices (Art. 29a) or Archiving (Art. 45g). As [8] technology neutral DLT can be used as infrastructure for EUDIW as well as QTSP for QEAA but also any QTSP. Means on the other hand also that the term "Electronic Ledger" not necessarily applies for DLT only. One fundamental change is the binding of QTSP on NIS2 Directive. In the result any QTSP so also the one on electronic ledger or in context of the paper DLT become part of critical infrastructure and so have to fulfil foreseeable higher security requirements than under [2] The core requirements on QTSP like liability, periodical re-certification, reporting obligation on security issues etc. are applicable for all QTSP in [8] too. For each (qualified) trust service also mandatory implementing acts are required in [8] referencing European standards. As for each (qualified) trust service also DLT as infrastructure possible those standards will define the needed security and trust requirements and so close the gaps mentioned in Sect. 1.

5.3 QTSP for Ledger

With Section 11 [8] also introduces (qualified) trust services on Electronic Ledger (Art. 45h following). [8] defines that qualified ledgers "are created and managed by one or more qualified trust service provider or providers, establish the origin of data records in the ledger, ensure the unique sequential chronological ordering of data records in the ledger and record data in such a way that any subsequent change to the data is immediately detectable, ensuring their integrity over time". Although [8] is technology neutral the description in Art. 45i is in line with the definition of DLT in international standards [10] and contains core properties of DLT. As [8] contains the requirement of mandatory implementing acts referring to European standards it ensures coherent technical framework for DLT. Since the requirements on QTSP also apply for QTSP for Ledger these standards will also be the basement for certification by independent conformity assessment body and so ensure proven security and trust in DLT. It has to be stated that Section 11 focus on all use cases not covered by EUDI Wallet or all other (qualified) trust services so e.g. (qualified) signatures, seals, timestamps, attestations electronic delivery etc. Means that DLT can be used as infrastructure for any EUDI Wallet as well as any other QTSP too—the security will be proven within the conformity assessment of the CAB, but there's no need to use QTSP for Ledger as precondition to provide another (qualified) trust service nor an EUDIW [8, 9]. This differentiation is important as it lead to the core use cases for QTSP for Electronic Ledger as e.g. tokenization or digital assets, cryptocurrencies or traceability in supply chains and digital product pass.

5.4 Trustworthy Decentralization Through eIDAS 2.0

Reference [8] complements the eIDAS ecosystem (eID or PID in 14, existing QTSP) with recognition of developments on decentralized identities (EUDIW, QEAA), cybersecurity [17], [18] as well as (qualified) trust services (e.g. Archiving, Electronic Ledger) and the technology neutrality which allows utilization of existing technologies like PKI but also DLT for each component. Each element so also decentralized ones like EUDIW or Electronic Ledger/DLT is directly integrated into the eIDAS trust framework. There´s no trust by default in Europe. Trust only occurs based on European law, supervised by European and national supervisory bodies, accreditation of conformity assessment bodies under European standards, certification of trust services by CAB under supervision of national supervisory bodies and verifiable via European wide trusted lists [6, 19].

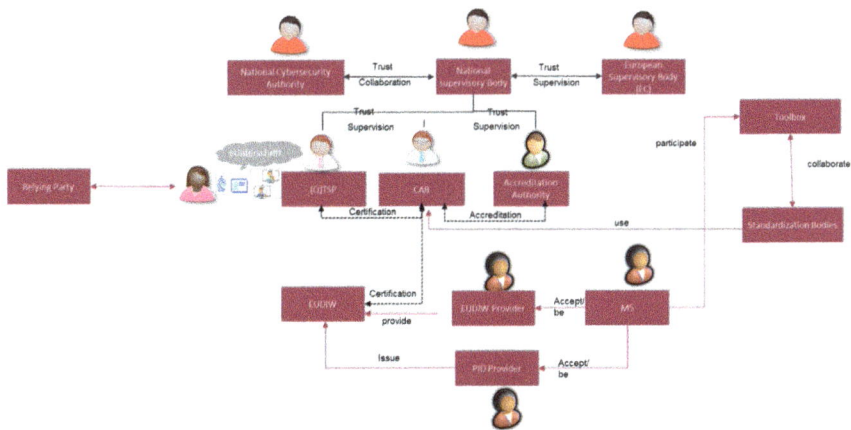

6 Trustworthy Tokenization in the Metaverse Through eIDAS 2.0

6.1 Trustworthy Digital Transactions Through EUDIW and Trust Services

As described in Sect. 5.2 the EUDI Wallet will contain the personal identification of its holder as well as related (qualified) attestations of attributes. This means that the relationship between a natural entity and its digital twin so e.g. avatar within the metaverse can be made evident, same with digital twin of machines or any other physical object in augmented or virtual reality. Practically the natural entity may identify itself against certain QTSP for QEAA just to get the digital twin e.g. avatar issued into the EUDI Wallet. In case of transactions in a virtual environment e.g. to purchase a virtual asset the EUDI Wallet holder use its EUDIW to present the QEAA to the certain relying party. The relying party may also be represented by a digital twin referenced to real legal entity (using PID for legal entities + QEAA). The virtual asset can be represented by a Non-Fungible Token which anchored on a Distributed Ledger provided by QTSP for Ledger according Sect. 11 [8] and referenced to the EUDI Wallet of holder purchasing e.g. a virtual house or another digital asset in virtual reality. Similar subject possible for any digital twin. A possibly necessary (qualified) signature to sign the purchase contract. In order to ensure the privacy of real legal entity the qualified certificate on which the qualified signature will be based may be issued by QTSP for qualified certificates with a pseudonym—so exactly the name of the digital twin of related natural entity, or better the QEAA of certain natural entity. In summary [8] provides all necessary tools and regulations to establish legal trust in transactions in the augmented and virtual reality or to use digital twin in a legally compliant manner.

As metaverse applications are often provided by market leading platforms the obligations for acceptance of EUDIW for those providers acc. to Art. 6db simplify the adoption of [8]. In summry [8]:

- Matches legal identity of natural entity with virtual equivalent for legal trust.
  - EUDIW + QEAA.
- Match virtual/real machine to legal/natural entity.
  - EUDIW incl. PID + QEAA.
- Ensures legally binding transactions.
  - EUDIW + PID + qualified signatures.
- Makes authenticity and integrity provable for trusted 3rd party.
  - EUDIW + qualified signatures.
- Avoids the disclosure of all PII due to privacy by design.

- Trustworthy DLT-infrastructure for Tokenization including traceability and evidence.
  - QTSP for Electronic Ledger resp. EUDI Wallet or QTSP using Ledger.

6.2 Privacy and Security Within Metaverse

As [8] requires privacy by design the [ARF] defines functionalities like Selective Disclosure or Zero Knowledge Proofs the natural or legal entity holding the EUDIW can decide in own sovereignty which data they want to provide to the relying party. This data sovereignty is only limited by the documentation requirements of the relying party which may require the provision of personal or other data to be able to use a certain service.

The conformity assessment of any EUDIW as well as any QTSP by independent CAB together with the supervision by National Supervisory Bodies and the obligations on liability etc. for QTSP and wallet providers ensure proven security and so legal trust in any wallet, personal identification but also QEAA representing digital twins, assets etc. in metaverse applications.

6.3 Trustworthy Tokenization Within Metaverse

None-Fungible Token often used for transaction on digital assets are typically anchored on DLT and in many cases in DLT using Ethereum protocol. The [EBSI] is using Hyperledger BESU so an Ethereum derivate. With its nodes provided by member states [EBSI] uses governmental trust anchor. The Digital Europe Programme currently supports dedicated European consortiums in the improvement of [EBSI] to be used for regulated use cases. This includes not only the further development of governance and specifications to achieve compliance on [8] but also the definition of requirements on QTSP for Ledger based on [EBSI]. In the result a European DLT-infrastructure will be created not only with governmental trust anchor but fully liable QTSP which provides a DLT infrastructure with proven security and trust through the conformity assessment by independent CAB acc. [8].

Practically the NFT can be issued by certain relying party for a certain natural or legal entity proven using its EUDIW including the PID as well as QEAA for the digital twin (e.g. avatar in virtual environments). The NFT itself will be anchored on a qualified ledger related to certain transactions and matched with EUDIW of a certain entity (e.g. via DID).

7 Conclusion and Necessary Standardization

Reference [8] defines the legal and through mandatory implementing acts for de facto all components also the technical framework for trustworthy decentralized ecosystem in Europe. As the regulation is technology neutral it also allows the utilization of DLT for each component from EUDI Wallet and all QTSP. With the QTSP for Electronic Ledger [8] establishes a dedicated (qualified) trust service for DLT. Due to the integration of DLT in the eIDAS trust framework all requirements on EUDI Wallet and QTSP like liability (EUDIW = member state), conformity assessment by independent CAB apply which ensures the proven security, legal trust and so solves the main gaps mentioned in Sect. 1 which limited a broad utilization of DLT in Europe. Beside EUDIW and other (qualified) trust services QTSP for ledger can be a game-changer not only in supply chain but also industry 4.0 or metaverse applications by matching legal and natural entities with their virtual twins, digital assets in transactions with real or virtual relying parties in real, augmented or virtual realities. Reference [8] makes it possible to use the identity of natural or legal entities in their complete variety in legally compliant manner. The regulation together with the de facto mandatory technical framework which applies also for EUDI Wallet and QTSP using ledger but especially the new QTSP for ledger [8] ensures trustworthy decentralization of real and virtual ecosystems using digital identities and shown in the picture below:

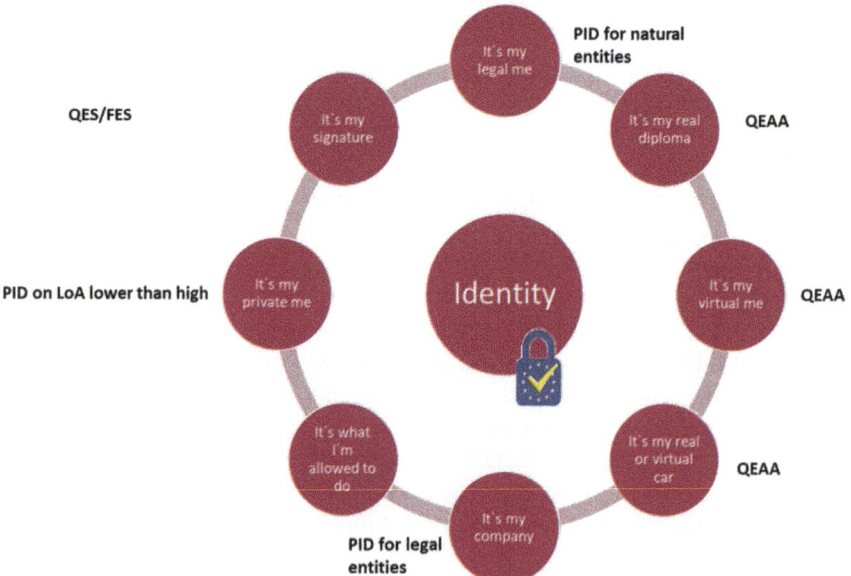

The focus in European standardization should be especially on the relationship between identities of natural/legal entities and their real and virtual characteristics. The harmonization or technical framework so existing Web 3 applications, payment wallets and the upcoming [8] ecosystem including the worldwide interoperability.

As regarding regulation, the implementing acts have to be published after 6 (EUDIW/QEAA) resp. 12 (all other subjects) month after publishment of [8] which is expected for March 2024, the research shall also focus on definition of concrete security and technical requirements. Especially the portfolio definition of QTSP for Ledger and in this context the adjustment of [EBSI] regarding [8] seem to be most important issues to be solved in order to create European trusted DLT infrastructure (not only) for the metaverse.

References

1. D. Huehnlein, Identitätsmanagement. Eine visualisierte Begriffsbestimmung. DuD • Datenschutz und Datensicherheit 3|2008, pp 163–165
2. Regulation (EU) No 910/2014 of the European Parliament and of the Council - of 23 July 2014 on electronic identification and trust services for electronic transactions in the internal market and repealing Directive 1999/93/EC. eIDAS, 2014
3. U. Korte et al., Records management and long-term preservation of evidence in DLT, in *Open Identity Summit 2021*, ed. by H. Roßnagel, C.H. Schunck, S. Mödersheim (Gesellschaft für Informatik e.V., Bonn, 2021), pp. 131–142
4. K. Werbach, The blockchain and the new architecture of trust. Massachusetts Institute of Technology (2018)
5. U. Korte et al., Criteria for trustworthy digital transactions—blockchain/ DLT between eIDAS, GDPR, data and evidence preservation, in *OpenIdentity Summit 2020. Lecture Notes in Informatics (LNI)* (Proceedings, Bonn, 2020), pp. 49–60
6. I. Alamillo, S. Schwalm, Self-sovereign-identity and eIDAS: a contradiction? Challenges and Chances of [14]. European Review of Digital Administration & Law—Erdal2021 **2**(2):89–108
7. D. Antin, The technology of the metaverse, it's not just VR, in *The Startup* (2020)
8. Proposal for a REGULATION OF THE EUROPEAN PARLIAMENT AND OF THE COUNCIL amending Regulation (EU) No 910/2014 as regards establishing a framework for a European Digital Identity. 2021/0136 (COD) (Version from 10.12.2023)
9. A. Kudra, S. Schwalm, Decentralised digital identity in the metaverse under eIDAS 2.0 chair for the responsible development of the Metaverse. Alicante (2024)
10. ISO 22739:2020: Blockchain and distributed ledger technologies—terminology (2020)
11. DIN TS 31648:2021. Criteria for trusted transaction. Records Management and Evidence Preservation in Distributed Ledger Technologies and Blockchain
12. ISO DTR 24332. Information and documentation—blockchain and DLT in relation to authoritative records, records systems, and records management
13. U. Korte, T. Kusber, S. Schwalm, Vertrauenswürdiges E-Government-Anforderungen und Lösungen zur beweiswerterhaltenden Langzeitspeicherung (2018)
14. European Digital Identity Wallets standards Gap Analysis
15. ETSI TS 119 471 Policy and Security requirements for Providers of Electronic Attestation of Attribute Services
16. ETSI TR 119 476 Analysis of selective disclosure and zero-knowledge proofs applied to Electronic Attestation of Attributes
17. DIRECTIVES DIRECTIVE (EU) 2022/2555 OF THE EUROPEAN PARLIAMENT AND OF THE COUNCIL of 14 December 2022 on measures for a high common level of cybersecurity across the Union, amending Regulation (EU) No 910/2014 and Directive (EU) 2018/1972, and repealing Directive (EU) 2016/1148 (NIS 2 Directive)
18. REGULATION (EU) 2019/881 OF THE EUROPEAN PARLIAMENT AND OF THE COUNCIL of 17 April 2019 on ENISA (the European Union Agency for Cybersecurity)

and on information and communications technology cybersecurity certification and repealing Regulation (EU) No 526/2013 (Cybersecurity Act)
19. S. Schwalm, Decentralised digital identity in the metaverse under eIDAS 2. Webinar of chair for the responsible development of the Metaverse. Alicante (2023)
20. I. Alamillo, S. Schwalm, C. Stoecker, R. Thiermann, Qualified ledgers: bridging the gap between blockchain technology and legal compliance (2024)
21. https://gitlab.opencode.de/bmi/eudi-wallet/eidas-2.0-architekturkonzept-v1
22. The Common Union Toolbox for a Coordinated Approach Towards a European Digital Identity Framework. The European Digital Identity Wallet Architecture and Reference Framework. December 2023. https://github.com/skounis/architecture-and-reference-framework/blob/80d00cf5ad1c3930235e4140b1fc8a975638f787/docs/arf.md
23. Federal Office for Information Security (BSI): Towards Secure Blockchains. Concepts, Requirements, Assessments (2019)
24. Eckpunktepapier für Self-sovereign Identities (SSI) unter besonderer Berücksichtigung der Distributed-Ledger-Technologie (DLT). Bundesamt für Sicherheit in der Informationstechnik. Bonn (2021)
25. DIN SPEC 4997: Privacy by blockchain design: a standardised model for processing personal data using blockchain technology (2020)
26. EBSI, European Blockchain Services Infrastructure, https://ec.europa.eu/cefdigital/wiki/display/CEFDIGITAL/EBSI. Accessed 30 Mar 2020
27. ETSI EN 319 411 Policy and security requirements for Trust Service Providers issuing certificates; Part 1: General requirements
28. ETSI TS 119 612 Electronic Signatures and Infrastructures (ESI); Trusted Lists
29. Regulation (EU) 2016/ 679 of the European Parliament and of the Council—of 27 April 2016—on the protection of natural persons with regard to the processing of personal data and on the free movement of such data and repealing Directive 95/46/EC (General Data Protection Regulation). GDPR (2016)
30. S. Schwalm, The (not only) social impact of the eIDAS 2.0 digital identity approach in Germany and Europe, in *CRYPTOASSETS, DEFI REGULATION AND DLT: Proceedings of the II Token World Conference* (2022), pp. 23–38
31. W3C VC Data Model v2.0. (2024)

Steffen Schwalm has more than 15 years experiences in design, implementing and rollout of trustworthy digital ecosystems based on digital identities and trust services in high regulated industries in Europe. He works as Senior Manager Digital Identity and Trust at msg. His main subjects are:

- Legal Compliance and Governance (e.g. eIDAS, eIDAS 2.0 etc.)
- Decentralized digital identities and Wallets
- Trust Services incl. Preservation
- Certification and Audit
- DLT and Blockchain

Steffen Schwalm takes part in international standardization at ISO, ETSI, CEN and is author of numerous scientific publications. Since long time he supports shaping European digital identity and trust framework in several European Consortiums and projects.

Modern Blockchain Wallets

Mikolaj Pawel Radlinski-Konas, Wolfgang Prinz, and Daniel Trauth

1 Introduction

The 2017 study entitled "Payments 4.0" deals with the digitization of payment processes [1]. It not only deals with the digitalization of cash, but also describes the growing interest in blockchain. In particular, the study emphasizes potential interaction innovations, new business models and new communication solutions that go hand in hand with blockchain technologies. Digital and flexible adaptation is also seen as a necessity.

Business models that are based on blockchain technologies and allow users to interact both actively and passively with the blockchain require the use of blockchain wallets. Such wallets are user-unfriendly, which makes user acceptance more difficult and inhibits their spread. The hurdle lies in the effort involved with a blockchain wallet [2].

Custodial wallets are an alternative to wallets [3]. Custodial wallets are operated by third parties who manage ‚the users' assets and are responsible for the security of the wallets. Analogous to these wallets are banks, which also manage the users' assets. A business model that has emerged through custodial wallets and also through multi-party billing wallets is called wallet-as-a-service. This study deals exclusively with wallet-as-a-service providers that promise to guarantee good user-friendliness.

M. P. Radlinski-Konas (✉) · W. Prinz · D. Trauth
Fraunhofer FIT, Sankt Augustin, Germany
e-mail: radlinski.mikolaj@gmail.com

M. P. Radlinski-Konas · W. Prinz
RWTH Aachen University, Aachen, Germany

Fig. 1 Non-custodial wallets

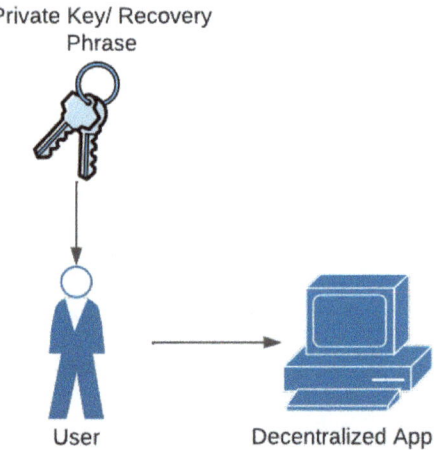

2 Blockchain Wallets

Blockchain wallets consist of a private and a public key. The public key serves as the address and the private key is used to sign and execute transactions. executing transactions. With blockchain wallets, the user receives the private key and a recovery record, which is used to restore the wallet on other devices devices if the original wallet is lost or deleted. is lost or deleted. In this case only the owner of the private key or the recovery of the recovery set can restore the wallet restore the wallet. If these keys including the wallet are lost, the entire assets are also entire assets are lost [4]. Cash works in the same way. If the cash is lost, the original owner loses the assets.

With custodial wallets, providers manage the respective private keys including the recovery set [3]. Figure 1 shows the management method of a non-custodial wallet, whereas Fig. 2 visualizes a custodial wallet.

In both cases, the user authenticates themselves in a decentralized app with a signature using the private key. In Fig. 2, however, the user does not directly possess the keys, but only has access to them. This access can be realized through Web2 authentication methods such as e-mail, Google, etc.

3 Wallet-as-a-Service

Wallet-as-a-Service (WaaS) is a business model that allows companies to integrate a wallet infrastructure into a system. WaaS can be based on the principles of both custodial and non-custodial wallets [5]. In the context of ease of use and mass adoption, WaaS companies are considered and evaluated that rely on Web2 authentication

Fig. 2 Custodial wallets (simplified)

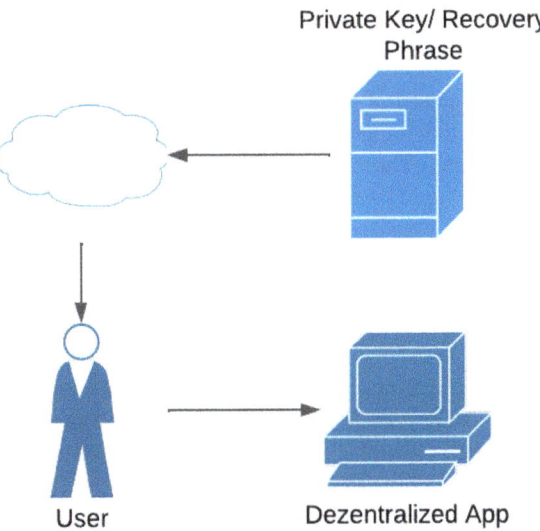

methods, relieving the user of the burden of managing the keys. In the case of these wallets, WaaS companies create different architectures to ensure the security of the wallets.

4 Criteria for Wallet-as-a-Service Wallets

In terms of user-friendliness, there are two users who are affected by WaaS wallets:

- Programmers who need to integrate the WaaS infrastructure into a project.
- Users who interact with the user interface of the implementation.

4.1 User

For the user, a particularly seamless interaction with the user interface is of the utmost importance. This includes a simple or familiar interaction method that does not require any additional or unusual effort. The associated costs must also be taken into account here. The programmer or the company decides which party will bear the costs.

4.2 Programmer

The integration of WaaS software is particularly important for programmers. This includes the implementation effort required and the supported frameworks and blockchains. The costs associated with WaaS software must also be taken into account.

4.3 Examined Wallets

This study provides a comparison of 10 different WaaS companies. In the following, the WaaS companies examined are presented and their respective promises outlined. The order is in alphabetical order and is not intended to reflect any kind of rating.

1. **Circle** promises a secure wallet with user-friendly interactions, fast implementation and scalability for billions of users [6].
2. **Coinbase-WaaS** promises seamless interaction of users with blockchain, allowing them not to worry about recovery records and trust the security of Coinbase [7].
3. **Crossmint** promises an interoperable NFT wallet with high security [8].
4. **Fireblocks** promises a scalable, flexible and secure wallet for thousands of businesses [9].
5. **Fordefi** promises a non-custodial wallet for seamless user interaction [10].
6. **Magic** promises a hassle-free and seamless Web3 experience [11].
7. **Privy** promises a secure WaaS wallet for high user integration in a short programming time [12].
8. **Self chain** promises a layer-1 blockchain including key-free WaaS wallets [13].
9. **walt.id** promises a simple infrastructure for digital wallets and identities [14].
10. **Web3Auth** promises a secure, scalable non-custodial WaaS through social logins [15].

5 Wallet Comparison

In this section, the 10 WaaS providers are examined in detail in order to evaluate the user-friendliness of their wallets. The chapter is divided into the individual wallet functionalities and special features, the supported blockchains, the associated costs, and finally developer and user friendliness.

5.1 Wallet Functionalities and Special Features

The WaaS companies presented have various special features and objectives that they pursue. The respective features are therefore listed in Table 1. It should be noted that user-friendliness was not explicitly noted here as a special feature because this feature is promised by many of the WaaS companies and will be examined later in the corresponding chapter.

Many of the WaaS companies use multi-party computation and support paymasters. Table 2 shows which of these companies support each of these technologies.

5.1.1 Multi-party Computatuion

Multi-party computation (MPC) is based on the principles of cryptography, whereby several participants are involved in a task [16]. Multi-party wallets use the technology

Table 1 WaaS special features

WaaS	Special features
Circle	Ful control over user keys
Coinbase-Waas	Wallets with coinbase integration
Crossmint	NFT-focues including NFT creation
Fireblocks	Different management types for private users
Fordefi	Control center
Magic	Patented key management
Privy	User-oriented dynamic customization
Self chain	Own blockchain
Walt.id	Digital identities
Web3Auth	Account abstraction, ownership transfer

Table 2 MPC and paymaster wallets

WaaS	MPC wallet	Paymaster
Circle	✓	✓
Coinbase-Waas	✓	✓
Crossmint	✓	✓
Fireblocks	✓	✗
Fordefi	✓	✗
Magic	✗	✓
Privy	✗	✓
Self Chain	✓	✗
Walt.id	✗	✗
Web3Auth	✓	✓

whereby several servers each store only part of the private key. Another approach would be for the servers to hold private information that only the user can decrypt [17]. With the help of this system, only the user can use the private keys without having to manage the keys themselves. WaaS companies use this technology for their wallets to create secure non-custodial wallets for users. The company itself has no access to the user's private keys in such wallets.

Figure 3 shows a visualization of this function using three keys. Depending on the implementation, *1* to *n* partial keys are required to calculate the correct private key.

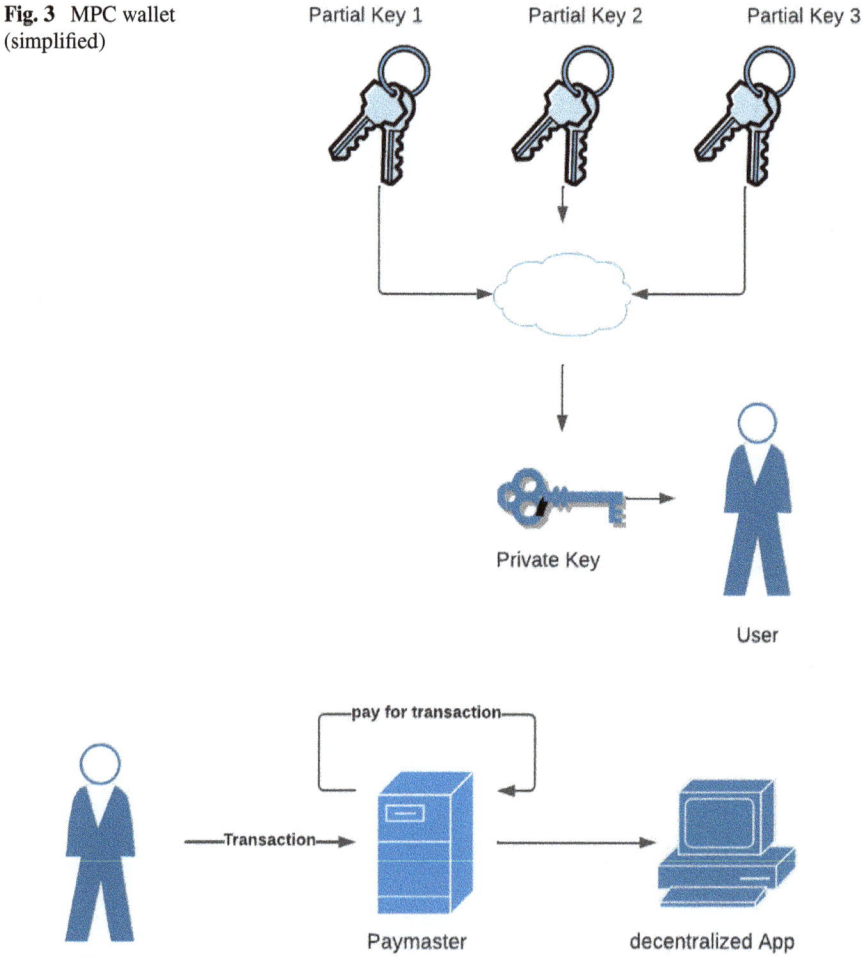

Fig. 3 MPC wallet (simplified)

Fig. 4 Paymaster (simplified)

5.1.2 Paymaster

The Ethereum standard ERC-4337 introduced the functionality of paymasters [18]. These are smart contracts that pay the costs of transactions without the user being liable for them. In the classic sense, account abstraction wallets are required for this. To simplify matters, this study equates the term paymaster with the functionality to pay for transactions of other participants. Thus, the term also includes the functionalities of gas stations [6]. Such gas stations contain blockchain-specific tokens that can be used to pay for transactions (Fig. 4).

Assume a company uses Ethereum as its blockchain. This company uses WaaS wallets for its users. As a loyalty program, the company gives away its own loyalty ERC-20 tokens to its customers. If a customer now wants to send the tokens to another wallet, transaction costs are charged here, which must be paid in ETH, the Ethereum native token. However, if the company has installed an automatic fee transfer system, the transaction costs are paid by this system. The customer therefore does not have to transfer ETH tokens to the new customer wallet and can send the loyalty ERC-20 tokens directly.

This functionality represents a seamless interaction between the user and the blockchain and promotes user-friendliness. However, it should be noted that the company using WaaS wallets must bear the costs for such applications.

5.2 Supported Blockchains

For companies, the blockchain on which they program is of particular importance, especially when it comes to the programming language and the functionalities of blockchains. An important aspect is therefore the supported blockchains that WaaS offers. Table 3 provides an overview of which blockchains are supported in each case.

It is striking that there is no blockchain at Walt.id. The reason for this is that IOTA was the only supported blockchain that no longer meets current requirements. The Walt.id Wallet Kit will also be discontinued in 2024 [19]. Walt.id is therefore not compared further in the following.

In addition, "all EVM-compatible blockchains" is listed under Privy. EVM stands for Ethereum Virtual Machine and defines the programming language and functionality of the Ethereum blockchain. EVM-compatible blockchains are blockchains that follow the principles of Ethereum and can read and process the same EVM code [20]. Such blockchains include Arbitrum, Polygon, Optimism and many other blockchains [21].

Table 3 Supported blockchains [(R) = restricted]

WaaS	Supported blockchains
Circle	Ethereum, Polygon, Avalanche (R)
Coinbase-WaaS	Arbitrum, Avalanche C-Chain, Base, Binance Smart Chain, Ethereum, Fantom Opera, Gnosis, Optimism, Polygon
Crossmint	Apex (R), Aptos (R), Arbitrum One, Arbitrum Nova, Astar zkEVM, Base, BSC, Etherum (R), Optimism, Polygon, Solana, Sui (R), Zora
Fireblocks	Arbitrum, Astar, Aurora_dev, Avalanche (C-Chain), Axelar, Base, Bitcoin, BSC, Canto, Celestia, Celo, Chiliz, Cosmos, dYdX, Etherum, Evmis, Fantom, HAT Chain, KAVA, Linea, Matic, Moonbeam, Miinriver, Oasys, Osmosis, Ronin, RSK, Shimmer, SmartBCH, Songbird, TokenX, TRON, Velas, XDC Network, zkEVM, Solana, Algorand
Fordefi	Bitcoin, Cosmos Hub, Akash, Archway, Axelar, Celestia, dYdX, Dymension, Noble, Osmosis, Sei, Stride, Solana, Arbitrum One, Avalanche, Base, BNB, Blast, Canto, Conflux, Dymension, Ethereum, Fantom, Gnosis, Kava, Linea, Manta Pacific, Mantle, Merlin, Optimism, Polygon, Polygon zkEvm, Scroll, Xai, zkLink Nova, zkSync Era
Magic	Polygon, Ethereum, Solana, Flow, Aptos, Algorand, Arbitrum, Avalanche, Base, BSC, Bitcoin, Celo, Chiliz, Cosmos, Cronos, Fantom, Harmony, Hedera, ICON, Loopring, Moonbeam, Neat, Optimism, Polkadot, Tezos, ZetaChain, Ziliqa
Privy	All EVM-compatible blockchains
Self chain	Self chain
Walt.id	–
Web3Auth	Alle secp256k1 & ed25519 curve Blockchains, Ethereum, Arbitrum, Avalanche, Base, BSC, Celo, Cronos, Flare, Harmony, Klaytyn, Moonbeam, Moonriver, Neon, Optimism, Polygon, SKALE, Songbird, zkEVM, zKyoto, Solana, XRPL, Algorand, Aptos, Cosmos, ImmuTabelleX, Near, Polkadot, Polymesh, StarkEx, StarkNet, Tezos

5.3 WaaS Costs

WaaS costs do not contribute to how the user interacts with the user interface and how user-friendly it is. However, these values are of great importance for a company that installs such infrastructures. Tables 4, 5 and 6 compare these costs, showing that different WaaS companies have different cost plans. In these tables it is noticeable that no data is given for Self Chain, this is due to the fact that the company is still in the initial phase and will carry out the implementations in the future. Walt.id is not included due to the closure of the wallet. These costs are for comparison purposes only and are not included in the user-friendliness factor.

5.4 Developer Friendliness

Developer friendliness plays an essential role when it comes to the success of the company. If a quick implementation is necessary, a simple developer interface is desirable. Such interfaces also help to reduce errors and make the system more robust. For WaaS wallets that offer free plans (see Table 4), demos were created to test developer friendliness.

Magic, Privy and Web3Auth were the quickest ways to create a working demo. During the implementation phase, Corssmint's servers were unavailable and therefore no demo could be created. A server failure creates unreliability and reflects negatively on the company.

Furthermore, a demo was developed with the Circle wallet, whereas the developer effort is higher than with the competitors, whereby the programmer has more control over the wallets.

5.5 Supported Frameworks

Depending on the experience, knowledge and specialization of programmers, the supported frameworks of the WaaS companies must be taken into account. Table 7 lists all supported frameworks:

5.6 Manageability

In order to bring blockchain applications to the user without further effort, seamless interaction is of great importance. Therefore, the easier a system is to use, the more likely it is that users will interact with it. All of the wallets mentioned in the study promise that the user can log in using familiar web2 login options. Especially when using the wallet, the functionality of an automatic fee transfer means that the user does not have to worry about blockchain-specific transactions.

For users who are more experienced with blockchain, the use of a frequently used or sustainable blockchain (Table 3) could contribute to users being more likely to interact with the system.

6 Conclusion

WaaS functionalities, costs, supported blockchains and frameworks, and user and developer friendliness were examined. Special emphasis was placed on user-friendliness, as this is crucial to the success of a company offering WaaS wallets.

Table 4 WaaS costs: Circle and Magic (costs per wallet)

WaaS	<1.000	<5.000	<10.000	<25.000	<50.000	<100.000	<250.000	>250.000
Circle	0$	0.05$	0.047$	0.04$	0.035$	0.03$	0.025$	0.02$
Magic	0$	0.05$	0.1$/custom	0.1$/custom	0.1$/custom	0.1$/custom	0.1$/custom	0.1$/custom

Table 5 WaaS costs: Privy, Web3Auth and Fireblocks (costs per month)

WaaS	< 1.000	< 2.500	< 10.000	> 10.000
Privy	0$	99$	299$	Custom
Web3Auth	0$	69$ (<3000)	399$	Custom
Fireblocks	500–550$	500–550$	500–550$ (<7.500)	Custom

Table 6 WaaS costs: Coibase-WaaS, Crossmint, Fordefi, Self chain and Walt.id

WaaS	Costs
Coibase-WaaS	39$ for 300.000 daily API interactions, otherwise custom
Crossmint	Custom
Fordefi	Custom
Self chain	–
Walt.id	–

Table 7 WaaS—supported frameworks and SDKs

WaaS	Frameworks and SDKs
Circle	REST APIs, iOS, Android, React Native, Node.js, web SDK (Javascript)
Coinbase-WaaS	Viem, React
Crossmint	React, iOS SDK, Android SDK
Fireblocks	Rest API, JacaScript, Python, Java, Rust
Fordefi	Ract Native, Android > = v7.0, iOS > = v13
Magic	Javascript, React Native, Android, iOS, Flutter, Unity, Server side SDK
Privy	NextJS, React, React Native
Self chain	–
Walt.id	–
Web3Auth	React, Next JS, React Native, Vue, Angular, Javasript, Web SDK, Android SDK, iOS SDK, Flutter, Unity

Seamless interaction can be improved with the help of automatic fee transfer, which is supported by the following wallets: Circle, Coinbase WaaS, Crossmint, Magic, Privy and Web3Auth. Of these six WaaS companies, a good implementation without server errors was possible with Circle, Magic, Privy and Web3Auth. Circle differs from these four WaaS wallets in that it requires more programming effort to ensure the security and interaction of the wallet. Greater programming effort can be seen as a negative, whereas greater control may be desirable for some companies. A negative aspect of Circle would be that only two blockchains are fully supported, whereas Magic, Privy and Web3Auth support over 10 blockchains.

Various scenarios are listed below and the corresponding fitting WaaS wallets:

A product is quickly developed that contains the basic functionalities of a blockchain wallet:

- Magic, Privy and Web3Auth

High control over wallet management and workflow is desired:

- Circle and Fireblocks

NFTs are an important aspect of the final product. Easy integration with various functionalities is required:

- Magic and Corssmint (although Crossmint has been affected by server outages in the past).

References

1. C. Bruck, "Payment transactions 4.0"—what impact does this have... Payment Transactions 4.0 (n.d.). https://www.bearingpoint.com/files/Studienergebnisse_BearingPoint_Studie_Zahlungsverkehr_4.0.pdf?download=0&itemId=386092. Accessed 10 May 2024
2. B.J. Hajjar, What's a DEFI wallet and how to choose the right one for my business. RIF (n.d.). https://rif.technology/content-hub/defi-wallet/. Accessed 09 May 2024
3. L. Merten,. What are custodial wallets?. Blockchainwelt (2023). https://blockchainwelt.de/Custodial-wallet/. Accessed 10 May 2024
4. S. Suratkar, M. Shirole, S. Bhirud, Cryptocurrency wallet: a review, in *2020 4th International Conference on Computer, Communication and Signal Processing (ICCCSP)* (IEEE, 2020), pp. 1–7
5. B. Editor, What is wallet-as-a-service (WAAS)?. Medium (2023). https://blog.bitgo.com/what-is-wallet-as-a-service-waas-a44f84fe8d70. Accessed 11 May 2024
6. Programmable wallets: Wallet as a Service. Circle (n.d.). https://www.circle.com/en/programmable-wallets. Accessed 12 May 2024
7. Embedded wallets—coinbase developer platform. embedded wallets—coinbase developer platform (n.d.). https://www.coinbase.com/de/developer-platform/products/embedded-wallets. Accessed 12 May 2024
8. Custodial NFT wallets: embedded wallets as a service. Custodial NFT Wallets|Embedded Wallets as a Service (n.d.). https://www.crossmint.com/products/Custodial-wallet-as-a-service. Accessed 12 May 2024
9. Wallets as a Service. Fireblocks (2024). https://www.fireblocks.com/platforms/wallets-as-a-service/. Accessed 12 May 2024
10. Wallet as a Service. Wallet as a service (n.d.). https://www.fordefi.com/wallet-as-a-service. Accessed 12 May 2024
11. The leading wallet-as-a-service plus essential NFT capabilities. Magic (n.d.). https://magic.link/. Accessed 12 May 2024
12. Onboard all your users to web3. Privy (n.d.). https://www.privy.io/. Accessed 12 May 2024
13. Self chain. Self Chain (n.d.). https://selfchain.xyz/. Accessed 12 May 2024
14. Powerful digital identity and wallet infrastructure. walt.id. (n.d.). https://walt.id/. Accessed 12 May 2024
15. Web3Auth. Key management SDKs with MPC and AA enabled (n.d.-b). https://web3auth.io/. Accessed 12 May 2024
16. O. Goldreich, Secure multi-party computation. Manuscript. Preliminary Version **78**(110), 1–108 (1998)

17. What is MPC (Multi-Party Computation)?. Fireblocks (2022). https://www.fireblocks.com/what-is-mpc/. Accessed 18 May 2024
18. Z. Lin, T. Wang, C. Zhao, S. Zhang, Q. Yang, L. Shi, A measurement investigation of ERC-4337 smart contracts on ethereum blockchain
19. Introduction: Wallet kit. walt.id. (n.d.-a). https://docs.walt.id/v/web-wallet. Accessed 19 May 2024
20. E. Hildenbrandt, M. Saxena, N. Rodrigues, X. Zhu, P. Daian, D. Guth, G. Rosu, KEVM: a complete formal semantics of the Ethereum virtual machine, in *2018 IEEE 31st Computer Security Foundations Symposium (CSF)* (IEEE, 2018), pp. 204–217
21. Top-Ethereum-ecosystem currencies by market capitalization. CoinGecko. (n.d.). https://www.coingecko.com/de/categories/ethereum-ecosystem. Accessed 19 May 2024

Mikolaj Pawel Radlinski-Konas is the President at the Aachen Blockchain Club, as well as the Developer Relations Lead at Solana Superteam Germany. Throughout his career, he has specialized in IT security and data communication—two fields that particularly interest him and that he believes are ideally complemented by blockchain technology. Early on, during his undergraduate studies, he developed his first trading bot on top of two blockchains and also wrote his bachelor's thesis on this topic.

Alongside his studies, Mikolaj gained over two years of hands-on experience in the Blockchain Reallabor project at Fraunhofer FIT as Blockchain Researcher, where he delved deeply into the applications of this technology and expanded his knowledge in a leading research environment.

He has a special interest in the scientific exploration of blockchain technology. To gain a deeper understanding of the technical fundamentals and potential of this technology, he decided to contribute two scientific articles to this book project. These articles examine the sustainability of blockchains and compare various Wallet-as-a-Service options, aiming to showcase the practical applications and challenges of the technology.

Sustainable Blockchains

Mikolaj Pawel Radlinski-Konas, Wolfgang Prinz, and Daniel Trauth

1 Introduction

In 2019, according to the study "Sustainability in the context of blockchain technology" [1], the German government was one of the first governments in the world to address the issue of the sustainability of blockchain technologies. The study emphasizes the need for the sustainable use of blockchain technologies in connection with the government's sustainability and climate protection goals. It emphasizes that not all blockchains are resource- and energy-intensive.

According to the white paper "Blockchain: fundamentals, applications and potential" [2], a trend is also emerging: the growing popularity of decentralized systems and the increasing attractiveness of programmable blockchains and smart contracts. These developments have not only sparked interest in digital currencies, but also the range of different blockchain platforms and their applications in industry and everyday life.

In its blockchain strategy, the German government describes blockchain technologies as a "building block for the Internet of the future" [3]. With this growing interest, new innovations have emerged in the areas of consensus algorithms and their energy consumption, which have a direct impact on the sustainability of these technologies.

In light of these developments, it is crucial to take a closer look at the sustainability of different blockchain technologies. This study aims to identify the most sustainable blockchain solution and analyze its potential impact on the environment and society. This includes aspects of sustainability, including energy consumption, scalability and ease of use.

M. P. Radlinski-Konas (✉) · W. Prinz · D. Trauth
Fraunhofer FIT, Sankt Augustin, Germany
e-mail: radlinski.mikolaj@gmail.com

M. P. Radlinski-Konas · W. Prinz
RWTH Aachen University, Aachen, Germany

2 Programmable Blockchains

The term programmable blockchain is used to describe a special type of blockchain that enables the creation and execution of programs (smart contracts) using its own programming language. An example of such a blockchain is Ethereum, the second largest blockchain by market capitalization [4]. Blockchains such as Bitcoin, which were primarily developed to store transactions with their native tokens, do not fall under this term.

Given the growing trend of integrating programmable blockchains into business structures and uploading smart contracts and data to different blockchains, it is of great interest to evaluate which of these blockchains are the most sustainable.

2.1 Sustainability and Its Significance for the Blockchain

According to the Bitcoin Energy Consumption Index [5], a single transaction on the Bitcoin blockchain consumes 1214.65 kWh of electricity. This index equates electricity consumption with 791,421 Visa transactions. With this consumption, 677.48 kg of CO_2 emissions are released by Bitcoin miners. Current transaction fees are also between 3\$ and 38\$ (US dollar) [6]. This data can be explained by the high level of activity, the size of the network and the Proof of Work (PoW) consensus mechanism. The architecture setup and the complex peer-to-peer communication require such amounts of energy.

Over the last few years, more and more blockchains have been developed that are based on other algorithms and consensus mechanisms and thus reduce power consumption, CO_2 emissions and transaction fees compared to Bitcoin [7]. A frequently used consensus mechanism alongside PoW is Proof of Stake (PoS). This is characterized by better scalability and energy and cost efficiency. Ethereum is an example of the implementation of Proof of Stake [7]. Due to the popularity of Ethereum, many other blockchains have been made Ethereum-compatible. This means that programs that work on Ethereum can also run on these so-called Etherem Virtual Machine (EVM) compatible blockchains [8]. This study primarily evaluates EVM-compatible and individual established blockchains. Before making a comparison, it is necessary to clarify the definition of sustainability and its relationship to blockchain technology.

2.2 What Does Sustainability Mean?

Sustainability means meeting the needs of the present generation without compromising the ability of future generations to meet their own needs. The aim here is to achieve a balance between the environment, society and the economy in order to ensure long-term quality of life [9]. This refers not only to a reduction in potential CO_2 emissions, but also to the scalability and interaction costs of blockchains.

2.3 Sustainability Criteria in Relation to Blockchain

Three main criteria are considered when examining the sustainability of programmable blockchains: Environment, society and economy. Blockchains are used by both companies and private individuals. One assumption of this study is that a tension triangle can be drawn between society and the economy, whereby only the two specified criteria can be fulfilled for each party (Fig. 1).

The most relevant aspects for the environment, society and the economy are listed below.

2.3.1 Environment

- Power consumption
- CO_2 Emissions.

The electricity consumption of a blockchain is determined either by quantifying the energy consumption per transaction or by the total consumption of electricity per year. As each blockchain processes a different number of transactions, it is more relevant to look at the consumption per transaction in comparison.

Blockchain servers, also known as nodes, are usually distributed over a large area and decentralized. Private individuals can be operators of such nodes and use any green electricity without specifying its origin. This means that CO_2 emissions from blockchains cannot be determined accurately [10]. For a scientific analysis, the

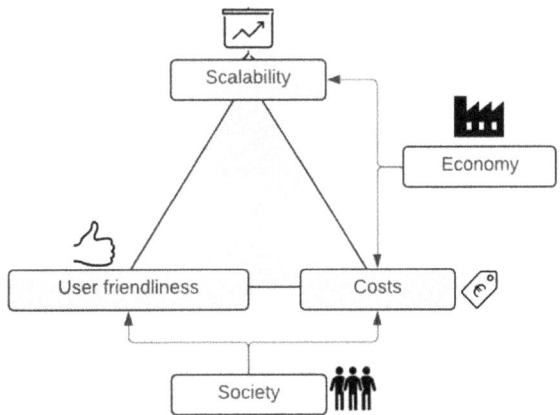

Fig. 1 Tension triangle—society and economy

electricity consumption of transactions is therefore considered in order to be able to draw a comparison.

2.3.2 Economy

- Scalability
- Cost.

Good scalability is an important issue in sustainability. The consensus mechanism plays a major role here, as it determines how nodes communicate with each other within a network and how the validation of a transaction is carried out [11]. The maximum number of transactions per second is used to quantify this.

Transaction costs have a high value for companies because they determine the costs of the internal work architecture and the final product. Transaction costs do not directly reflect the costs incurred in the creation of smart contracts of any size, but this data does provide a guideline value that indicates a trend for the final costs.

2.3.3 Society

- User friendliness
- Transaction costs.

The first programmable blockchain is Ethereum and introduced Solidity as a programming language [12]. Based on the concepts of Ethereum, many EVM-compatible blockchains were created with a large number of programmers in this programming language. Nevertheless, user-friendliness cannot be analyzed objectively, so this aspect is not given a strong weighting in the analysis.

Transaction costs also play a major role in society, as private individuals decide which blockchain to use based on the interaction costs of a blockchain.

In view of these criteria, blockchains will now be analyzed for their contribution to sustainability.

3 Blockchains Examined

This study provides a comprehensive comparison of 29 different blockchains. The blockchains examined are presented below and their respective promises are outlined. They are listed in alphabetical order and are not intended to reflect any kind of rating.

1. **Algorand** promises a dynamic, inclusive, borderless and scalable ecosystem [13]
2. **Aptos** promises a comprehensive platform for the realization of large projects with the help of AI in various areas such as gamimg, social media, etc. [14]

Sustainable Blockchains 69

3. **Arbitrum** promises to make Ethereum more sustainable and inclusive [15]
4. **Avalanche** promises confident and rapid development as well as a fast and scalable infrastructure [16]
5. **BNB Chain** promises a user-friendly blockchain with high reliability [17]
6. **Caradno** promises a secure, transparent and sustainable blockchain based on peer-reviewed research [18]
7. **Celo** promises to make Ethereum scalable and easily accessible [19]
8. **Cosmos** promises an affordable and scalable blockchain for entrepreneurs and users [20]
9. **Cronos** promises an EVM-compatible and scalable blockchain that also offers interoperability and financial support [21]
10. **EOS** promises an energy-efficient, user- and developer-friendly EVM-compatible blockchain [22]
11. **Ethereum** promises technologies that enable digital money and application software to be decentralized [23]
12. **Fantom** promises a developer-friendly environment with low transaction costs and fast transaction validations [24]
13. **Flow** promises a permissionless blockchain that instantly validates transactions and is mobile device optimized [25]
14. **Gnosis** promises a decentralized, cheap and neutral blockchain that ensures easy operation of nodes [26]
15. **Harmony** promises a scalable, secure and private blockchain specializing in microtransactions and market pricing [27]
16. **Hedera** promises good performance, security and compliance as well as simple tokenization of assets on a large scale [28]
17. **Immutable X** promises a scalable, secure and environmentally friendly gaming-focused blockchain [29]
18. **IOTA** promises secure and sustainable technologies that guarantee access for all people and digital applications and ensure free transactions [30]
19. **IOTA 2.0** promises one of the most environmentally friendly and sustainable distributed ledger technologies [31]
20. **Klaytyn** promises to be a sustainable, transparent and user-friendly blockchain [32]
21. **Near** promises developers a blockchain that enables apps to be scalable for billions of users and functional across all blockchains [33]
22. **Optimism** promises a superchain for interoperability between different blockchains and assures democratic and transparent governance [34]
23. **Polkadot** promises a scalable, efficient and secure overarching ecosystem that allows users to create their own blockchains [35]
24. **Polygon** promises a carbon–neutral, infinitely scalable infrastructure of blockchains based on zero knowledge technologies [36]
25. **Solana** promises a carbon–neutral, powerful developer platform and fast user experience [37]
26. **Stellar** promises a fast, cost-effective, energy-efficient and sustainable blockchain [38]

27. **TRON** promises high throughput rates, scalability and availability and claims to be the fastest growing blockchain [39]
28. **XRPL** promises a business-based blockchain that guarantees low transaction costs, high performance, sustainability and stability [40]
29. **Zilliqa** promises low transaction costs, high scalability, sustainability on the blockchain as well as developer friendliness and security in terms of development and interaction [41].

4 Blockchains Examined

In this section, 29 blockchains are examined in detail in order to assess their sustainability characteristics. The chapter is divided into energy consumption, transaction costs and user-friendliness.

4.1 Power Consumption

An essential aspect of the contribution to sustainability is reflected in the energy consumption of a blockchain. Here, attention was not paid to the total consumption of a blockchain, but to the energy consumption of an individual transaction. Some blockchains lack measurement data, which is either not provided by the respective blockchain company or cannot be analyzed due to missing data. In the following, 29 blockchains are compared in terms of energy consumption, the associated CO_2 emissions and the total consumption per year.

Table 1 (appendix) was sorted by energy consumption per transaction. Table 1 was not only sorted by energy consumption per transaction, but the respective blockchains were also grouped by color according to their energy consumption.

The increase in energy consumption in the first 11 blockchains is exponential, with an increase of up to 595 times between two levels. Figure 2 shows the four most energy-efficient blockchains as a bar chart:

Figure 3 lists the blockchains from Algorand to EOS. This figure illustrates how clear the differences between the blockchains are. In both Figs. 2 and 3, Algorand was shown to illustrate these differences.

It can be seen that IOTA 2.0 is in first place, a blockchain that is only available as a test network at the time of writing this study.

Furthermore, Table 1 (Appendix) contains a column for CO_2 emissions, which provides estimated values, as the exact origin of the energy sources is often unknown [10]. Some of the blockchains are taking initiatives to offset CO_2 emissions. These measures could include, for example, the planting of trees. Blockchains that offset more CO_2 emissions than they produce are Hedera, Polygon, Celo and Gnosis. Blockchains that offset as many CO_2 emissions as they produce are Algorand, Solana, EOS, Near, Immutable X and Avalanche. However, these factors alone do not provide

Sustainable Blockchains 71

Table 1 Power consumption and emissions

Blockchain	Power consumption per transaction [kWh]	CO_2-emissions per transaction [kg CO_2]	Power consumption per year [kWh]	CO_2-emissions per year [kg CO_2]	References
IOTA 2.0	< 0.01 [$1.883*10^{-9}$]	< 0.00475	18,923.25	–	[31, 45, 46]
IOTA	< 0.01 [$1.12*10^{-6}$]	< 0.00475	–	–	[31, 45, 47]
Hedera	< 0.01 [$3*10^{-6}$]	< 0.00475	–	–	[45, 48–50]
Algorand	< 0.01 [$8*10^{-6}$]	< 0.00475	650,181.40	298,433.30	[45, 48, 51, 52]
BNB	< 0.01 [$8*10^{-6}$]	< 0.00475	16,446.50	6,309.10	[45, 48, 53]
Fantom	< 0.01 [$2.4*10^{-5}$]	< 0.00475	8,200.00	–	[45, 54]
TRON	< 0.01 [$7*10^{-5}$]	< .00475	162,868.00	69,470.00	[45, 48]
Polygon	< 0.01 [$1.03*10^{-4}$]	< 0.00475	126,814.00	38,324.70	[45, 48, 55, 56]
Solana	< 0.01 [$1.83*10^{-4}$]	< 0.00475	22,660.900.00	9,162,400.00	[45, 48, 57–60]
Celo	< 0.01 [$5.5*10^{-4}$]	< 0.00475	46,893.73	17,210.00	[45, 61]
EOS	< 0.01 [$1.23*10^{-3}$]	< 0.00475	630.720	281.000	[45, 62, 63]
Stellar	< 0.01	< 0.00475	15,538.20	5,713.60	[45, 48]
XRPL	< 0.01	< 0.00475	57,382.20	23,378.60	[45, 48]
Near	< 0.01	< 0.00475	–	174,123.00	[45, 64]
Immutable X	< 0.01	< 0.00475	–	–	[45, 65]
Arbitrum	< 0.01	< 0.00475	–	–	[45]
Avalanche	< 0.01	< 0.00475	735,354.00	283,335.20	[45, 48, 66]
Flow	0.01 = < x < 1	0.00475 ≤ x < 0.475	180.000	56,281.626	[45, 67]
Harmony	0.01 = < x < 1	0.00475 ≤ x < 0.475	–	–	[45]
Klaytyn	0.01 = < x < 1	0.00475 ≤ x < 0.475	–	–	[45]
Aptos	0.01 = < x < 1	0.00475 ≤ x < 0.475	–	–	[45]

(continued)

Table 1 (continued)

Blockchain	Power consumption per transaction [kWh]	CO_2-emissions per transaction [kg CO_2]	Power consumption per year [kWh]	CO_2-emissions per year [kg CO_2]	References
Cronos	$0.01 =< x < 1$	$0.00475 \leq x < 0.475$	–	–	[45]
Zilliqa	$0.01 =< x < 1$	$0.00475 \leq x < 0.475$	–	–	[45]
Polkadot	$0.01 =< x < 1$	$0.00475 \leq x < 0.475$	172,394.80	48,627.80	[45, 48]
Optimism	$0.01 =< x < 1$	$0.00475 \leq x < 0.475$	–	–	[45]
Cosmos	$0.01 =< x < 1$	$0.00475 \leq x < 0.475$	632,577.50	290,353.10	[45, 48]
Ethereum	$0.01 =< x < 1$ [0.03]	$0.00475 \leq x < 0.475$	2,600,000.00	870,000.00	[45, 48, 68, 69]
Caradno	$0.01 =< x < 1$ [0.5479]	$0.00475 \leq x < 0.475$	941,338.30	340,162.90	[45, 47, 48]
Gnosis	–	–	–	–	[70]

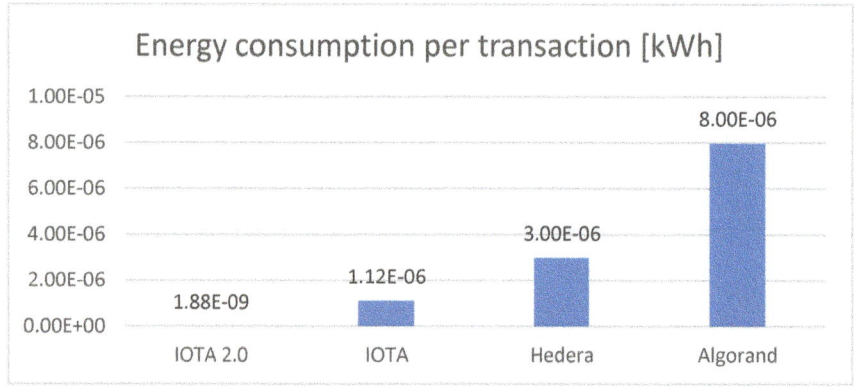

Fig. 2 Energy consumption per transaction of the four most efficient blockchains

any insight into the energy efficiency of the blockchain; they are more of an indication. This information is not heavily weighted in this study.

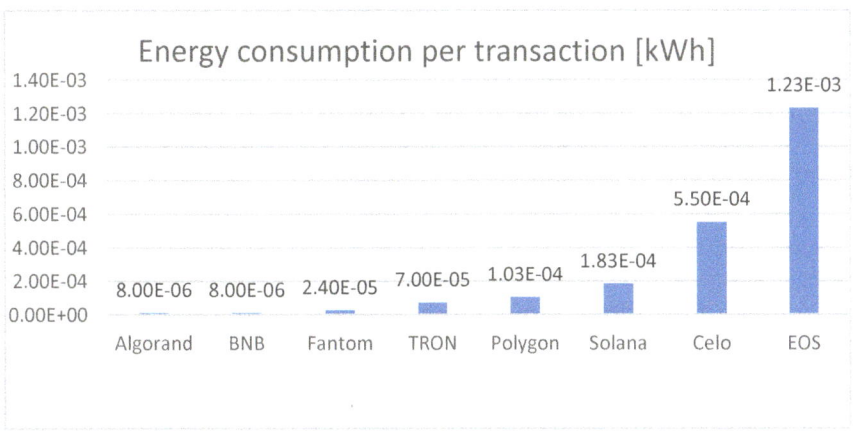

Fig. 3 Energy consumption per transaction (rank 4–11)

4.2 Transaction Costs

Transaction costs are an important aspect when it comes to assessing the sustainability of blockchains. They help to determine the economic efficiency of a blockchain.

As blockchains can be operated by any private individual and are not centralized, mechanisms have been developed to remunerate these private individuals. In the classic case, these private individuals receive a share of the transaction costs. This means that in most cases, every transaction is associated with additional costs. High transaction costs contribute to users being deterred and not interacting with the given network.

Table 2 (Appendix) shows the average transaction costs for 29 blockchains in US dollars. It is particularly striking that IOTA 2.0, IOTA and EOS do not incur any transaction costs.

The respective blockchains were also grouped by color. Due to the widely varying fees, three separate bar charts were created to visualize the cost differences (Figs. 4, 5 and 6).

4.3 User Friendliness

The area of user-friendliness is a challenge in the context of blockchain applications, as it is partly subjective. Nevertheless, an analysis of this area was undertaken. Aspects such as the time to finality, the theoretical maximum number of transactions per second (TPS), the virtual machine and the various consensus mechanisms were considered. The time to finality indicates how long it takes for a transaction to be finally stored on the blockchain and thus irrevocably built into the transaction chain [42]. This does not indicate the transaction confirmation time, i.e. the time it takes

Table 2 Transaction costs

Blockchain	Costs per transaction [$]	References
IOTA 2.0	0	[71]
IOTA	0	[71]
EOS	0	[72]
Harmony	0.000001	[73]
Stellar	0.00000124	[74]
Flow	0.00000185	[75]
TRON	0.000005	[76]
Hedera	0.0001	[77]
XRPL	0.0002	[78]
Algorand	0.00023	[79]
Celo	0.0006	[80]
Klaytyn	0.00386	[81]
Solana	0.004	[82]
Gnosis	0.005	[83]
Aptos	0.006	[84]
Near	0.01	[85]
Fantom	0.01	[24]
Cronos	0.01	[86]
Polygon	0.015	[87]
Zilliqa	0.03	[88]
Arbitrum	0.05	[89]
Caradno	0.125	[90]
Polkadot	0.18	[91]
BNB	0.2	[92]
Optimism	0.445	[93]
Cosmos	0.5	[94]
Ethereum	1.35	[95]
Avalanche	5.89	[96]
Immutable X	2.00%	[97]

for a transaction to be validated. The confirmation time can be manipulated by the user depending on the amount of gas fees.

The virtual machine (VM) instantiates a blockchain. It defines the functionality and process of a blockchain [43]. In the case of Ethereum, this is called EVM (Ethereum Virtual Machine). Although the consensus mechanism has an influence on the interactions of the respective nodes and their energy consumption, this is not considered a criterion for sustainability in the following, as only the true energy consumption is of particular importance. An overview of the data can be found in Table 3 (Appendix).

Fig. 4 Costs per transaction (position 1–5)

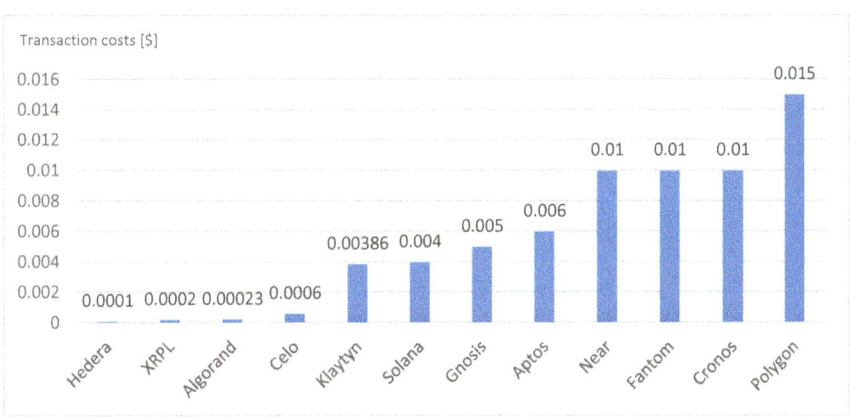

Fig. 5 Costs per transaction (position 5–16)

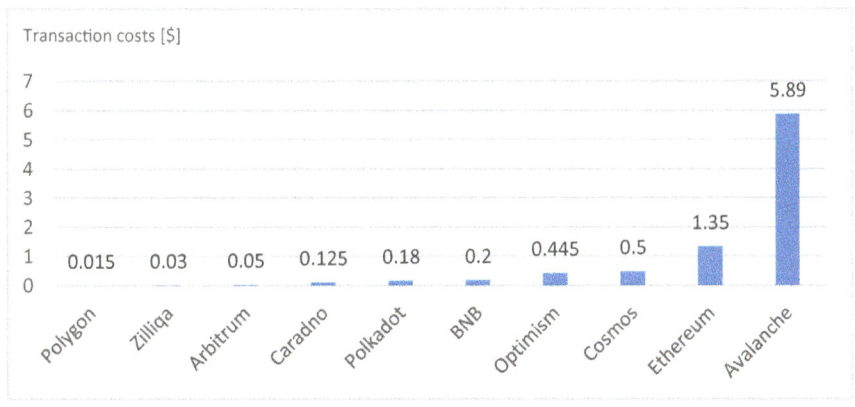

Fig. 6 Costs per transaction (position 16–25)

Table 3 Scalability and usability

Blockchain	Max. TPS	Time to finality [s]	Consensus mechanism	Engine type	References
Aptos	160,000	0.9	PoS	MoveVM	[98–100]
Polkadot	100,000	60	Nominated PoS	PolkaVM	[101, 102]
Near	100,000	1	Threshhold PoS	Custom/EVM	[85, 103, 104]
Solana	65,000	12.8	PoH	SolanaVM	[105]
Arbitrum	40,000	1.5	PoS	EVM	[106, 107]
Fantom	20,000	3	Fantom Opera	EVM	[24, 108]
Hedera	10,000	7	Hashgraph	EVM	[77, 109]
Cronos	10,000	5.5	PoA	EVM	[22, 86]
Cosmos	10,000	6	PoS	Evmos / EVM	[102, 105, 110]
Immutable X	9,000	300	ZK-Rollup	zkEVM / EVM	[111]
Polygon	7,500	3	PoS	EVM	[56, 112, 113]
Algorand	7,500	3	Pure PoS	AlogVM	[105, 114, 115]
Avalanche	4,500	0	PoS	EVM	[102, 105]
EOS	4,000	2	Delegated PoS	EVM	[44, 105, 116]
Klaytyn	4,000	0	PoS	EVM	[81, 117]
XRPL	3,000	–	XRP Ledger	EVM	[105, 118]
Zilliqa	2,828	–	PoW	EVM	[88, 119]
Stellar	2,500	5	Stellar Consensus Protokoll	Custom	[105, 120, 121]
BNB	2,222	1	PoS Auhority	EVM	[44, 122, 123]
TRON	2,000	57	Delegated PoS	EVM	[105]
Harmony	2,000	1	PoS	EVM	[124]
Optimism	2,000	2	PoS	EVM	[106]
Flow	1,000	1	PoS	EVM	[125–127]
Celo	1,000	5	PoS	EVM	[128]
IOTA 2.0	1,000	10	DAGs	EVM	[31, 105]
IOTA	1,000	10	DAGs	EVM	[31]
Caradno	1,000	35	PoS	EVM	[105, 129]
Gnosis	156	240	Delegated PoS	EVM	[122, 130, 131]
Ethereum	119	960	PoS	EVM	[132]

4.4 Scalability

Furthermore, the theoretical maximum number of transactions per second (TPS) plays a relevant role when it comes to scalability. Assuming a blockchain can process 10 transactions per second and 1000 transactions are requested at the same time, 990 users would have to wait until their transactions are processed. These latencies also have an impact on user-friendliness.

Another important factor is the applicability and user interaction as well as the scope of the application area. Although these tend to be observable, they are difficult to quantify. All scalability and usability-related data are listed below.

As scalability is the main focus, Table 3 (Appendix) has been sorted by maximum TPS. In last place is the Ethereum blockchain, which can process 119 transactions per second. In contrast, blockchains such as Flow, Polkadot, Aptos and Near achieve up to over 100,000 TPS.

In principle, it is not possible to say how many transactions per second are necessary to ensure a sustainable future, but a higher number of transactions per second indicates better scalability. Figure 5 lists all blockchains in descending order of maximum achievable TPS (Fig. 7).

It can also be seen that many blockchains have a time to finality of less than 10 s, which is significantly less than the time to finality of Bitcoin (approx. 1 h) [44]. These values show the final storage of transactions on the blockchain, but they do not contribute significantly to user-friendliness.

There is a clear trend in the engine types, as many of them are compatible with the Ethereum Virtual Machine (EVM), which also reflects the largest share of the market [4].

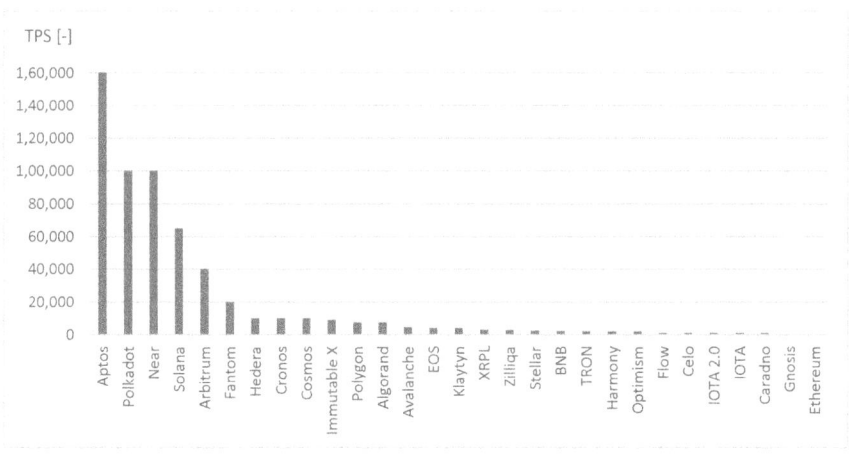

Fig. 7 Transactions per second

5 Conclusion

Environmental as well as social and economic aspects that are particularly relevant to sustainability were examined. Particular emphasis was placed on the electricity consumption of transactions, which can be quantifiably compared and allow conclusions to be drawn about real CO_2 emissions. Scalability and transaction costs were also taken into account.

In the power consumption category, the clear winner was IOTA 2.0. With a power consumption of $1.883 * 10^{-9}$ kWh per transaction, this is 595 times more efficient than IOTA, the second most efficient blockchain in terms of power consumption (Table 1). It should now be noted that this blockchain is currently a test blockchain. In terms of transaction costs, IOTA, IOTA 2.0 and EOS share first place with $0 costs per transaction, although it should be noted that blockchains with transaction costs of a few to only fractions of a cent are also almost negligible.

Finally, scalability played a decisive role, with the number of theoretically achievable maximum transactions per second being analyzed. Aptos, Polkadot, Near, Solana, Arbitrum and Fantom, which can process well over 10,000 transactions per second, stand out in particular.

If we now consider all the aspects of sustainability mentioned above, it is clear that IOTA 2.0 is the clear winner in terms of ecology. If we now consider scalability, 1,000 TPS could represent a bottleneck. Good alternatives would be Hedera, Algorand, Fantom, Polygon, Solana and EOS. The ecological features of Hedera and Algorand should be emphasized in the context of their blockchain infrastructures. Hedera aims to achieve a carbon negative footprint by offsetting more carbon dioxide than it produces. Hedera also achieves a transaction speed of up to 10,000 transactions per second (TPS) with transaction costs of around $0.0001 and an electricity consumption of $3 * 10^{-6}$ kilowatt hours (kWh). Algorand declares itself to be CO_2—neutral and has transaction costs of around 0.00023$ and an electricity consumption of $8 * 10^{-6}$ kWh. It should be noted that Algorand's transaction costs are 23 times higher than Hedera's and it also consumes 2.67 times less electricity per transaction.

In addition, Solana is characterized by a remarkable scalability of up to 65,000 TPS and also pursues the goal of being CO_2—neutral. However, the reliability of the network has suffered due to previous outages, which also has a negative impact on sustainability efforts. It can therefore be seen that different blockchains have different advantages and disadvantages.

The original assumption from Fig. 1 stated that user-friendliness was not particularly relevant for the economy. The same applies to society and scalability. However, the blockchain comparison shows that scalability and user-friendliness are closely linked in relation to blockchain.

Ultimately, it is therefore not possible to crown THE best blockchain in the area of sustainability. Different blockchains have different aspects that speak for themselves. When selecting the appropriate blockchain technology, it is crucial to carefully consider the specific use case and the associated requirements. In some cases, it may be advantageous to use a blockchain with lower TPS if this is accompanied by a corresponding reduction in emissions.

Appendix

See Tables 1, 2 and 3.

References

1. C. Culotta, S. Brüning, A.T.Schulte, D. Gesmann-Nuissl, C. Märkel, R. Beck, Sustainability in the context of blockchain technology: application examples, challenges, and action fields (2022)
2. Blockchain—Fraunhofer Fit (n.d.). https://www.fit.fraunhofer.de/content/dam/fit/de/documents/Blockchain_WhitePaper_Fundamentals-Applications-Potentials.pdf. Accessed 27 Apr 2024
3. Blockchain strategy of the federal government. The Federal Government informs | Homepage (n.d.-b). https://www.bundesregierung.de/breg-de/service/publikationen/blockchain-strategy-of-the-federal-government-1672384 (Last accessed: 2024, April 27)
4. Kryptowährung: Kurse, Charts und Marktkapitalisierung. CoinGecko (n.d.). https://www.coingecko.com/de (Zuletzt aufgerufen am: 2024, April 04)
5. Bitcoin-Energieverbrauchsindex. Digiconomist (2024, Januar 10). https://digiconomist.net/bitcoin-energy-consumption (Zuletzt aufgerufen am: 2024, April 04)
6. Durchschnittliche Bitcoin-Transaktionsgebühr (I:BATF). YCharts (n.d.). https://ycharts.com/indicators/bitcoin_average_transaction_fee (Zuletzt aufgerufen am: 2024, April 04)
7. Blockchain energy consumption an exploratory study—aramis (n.d.-b). https://www.aramis.admin.ch/Default?DocumentID=68053&Load=true. Accessed 27 Apr 2024
8. E. Hildenbrandt, M. Saxena, N. Rodrigues, X. Zhu, P. Daian, D. Guth, G. Rosu, Kevm: a complete formal semantics of the ethereum virtual machine, in *2018 IEEE 31st Computer Security Foundations Symposium (CSF)* (IEEE, 2018), pp. 204–217
9. J.H. Franz, What is sustainability? in *Sustainable Development of Technical Products and Systems: The Engineering Profession in Transition* (Springer Fachmedien Wiesbaden, Wiesbaden, 2022), pp. 7–14
10. C. Mora, R.L. Rollins, K. Taladay, M.B. Kantar, M.K. Chock, M. Shimada, E.C. Franklin, Bitcoin emissions alone could push global warming above 2 C. Nat. Clim. Chang. **8**(11), 931–933 (2018)
11. B. Lashkari, P. Musilek, A comprehensive review of blockchain consensus mechanisms. IEEE Access **9**, 43620–43652 (2021)
12. V. Buterin, Ethereum white paper. GitHub Repository **1**, 22–23 (2013)
13. Algorand Foundation (n.d.). https://www.algorand.foundation/. Accessed 04 May 2024
14. The world's most production-ready blockchain. Aptos (n.d.). https://aptosfoundation.org/. Accessed 04 May 2024
15. The future of ethereum. Arbitrum (n.d.). https://arbitrum.io/. Accessed 04 May 2024
16. Avalanche: create without limits: Dapp platform. Avalanche: Create Without LimitsldApp Platform (n.d.). https://www.avax.network/. Accessed 04 May 2024
17. BNB Smart Chain (BSC): bring smart contracts to BNB chain. BNB Chain (n.d.). https://www.bnbchain.org/en/bnb-smart-chain. Accessed 04 May 2024
18. Cardano—making the world work better for all: Cardano. Cardano RSS (n.d.). https://cardano.org/. Accessed 04 May 2024

19. Homepage. Celo (n.d.). https://celo.org/. Accessed 04 May 2024
20. The internet of Blockchains. Cosmos (n.d.). https://cosmos.network/. Accessed 04 May 2024
21. Cronos (n.d.) Cronos Chain: Defi and Metaverse for the next billion users. https://cronos.org/. Accessed 04 May 2024
22. EOS EVM. EOS Network (2024, January 9). https://eosnetwork.com/eos-evm/. Accessed 04 May 2024
23. Ethereum.org (n.d.). Homepage. ethereum.org. https://ethereum.org/de/. Accessed 04 May 2024
24. Baue deine beste App. Fantom (n.d.). https://fantom.foundation/ (Zuletzt aufgerufen am: 2024, Mai 04)
25. Flow is building the future of culture and community in WEB3. Flow is building the future of culture and community in Web3 (n.d.). https://flow.com/. Accessed 04 May 2024
26. Gnosis chain. Gnosis Chain (n.d.). https://www.gnosis.io/. Accessed 04 May 2024
27. Harmony ONE (n.d.). https://www.harmony.one/. Accessed 04 May 2024
28. Hello future. Hedera (n.d.). https://hedera.com/. Accessed 04 May 2024
29. Powering the next generation of web3 games. Immutable (n.d.). https://www.immutable.com/. Accessed 04 May 2024
30. Home. IOTA (n.d.). https://www.iota.org/. Accessed 04 May 2024
31. I. Foundation, Energieverbrauch von IOTA 2.0. IOTA Foundation Blog (2023, Juni 9). https://blog.iota.org/energy-consumption-of-iota-2-0/ (Zuletzt aufgerufen am: 2024, Mai 04)
32. A sustainable and verifiable blockchain built for all. Klaytn Foundation (n.d.). https://klaytn.foundation/. Accessed 04 May 2024
33. Blockchains, abstracted. NEAR (n.d.). https://near.org/. Accessed 04 May 2024
34. Home. Optimism (n.d.). https://optimism.io/. Accessed 04 May 2024
35. Polkadot: Web3 interoperability: decentralized blockchain. Polkadot Network (n.d.). https://polkadot.network/. Accessed 04 May 2024
36. Web3, aggregated. The value layer of the internet (n.d.). https://polygon.technology/. Accessed 04 May 2024
37. Web3 infrastructure for everyone. Solana (n.d.). https://solana.com/de. Accessed 04 May 2024
38. A blockchain network for payments and tokenization. Stellar (n.d.). https://stellar.org/. Accessed 04 May 2024
39. Tron network: decentralize the web. TRON (n.d.). https://tron.network/. Accessed 04 May 2024
40. XRP Ledger Home. XRP LEDGER (n.d.). https://xrpl.org/. Accessed 04 May 2024
41. Zilliqa (n.d.). https://www.zilliqa.com/. Accessed 04 May 2024
42. Chainspect, What is time to finality (TTF)? [real TTF in 2024] (2024, April 10). https://chainspect.app/blog/time-to-finality-ttf. Accessed 28 Apr 2024
43. Virtual Machines. Avalanche Dev Docs (n.d.). https://docs.avax.network/learn/avalanche/virtual-machines. Accessed 28 Apr 2024
44. Binance Academy, Finality (n.d.). https://academy.binance.com/en/glossary/finality. Accessed 28 Apr 2024
45. Wie grün ist Ihre Kryptowährung? Liste der Top 100+Kryptos (2024). Cryptowisser (n.d.). https://www.cryptowisser.com/crypto-carbon-footprint/ (Zuletzt aufgerufen am: 2024, April 04)
46. I. Stiftung, Iota 2.0 ama. IOTA Foundation Blog (2024, Februar 13). https://blog.iota.org/iota-20-ama/ (Zuletzt aufgerufen am: 2024, April 05)
47. Blockchain-Protokolle und ihr Energie-Fußabdruck. Adan. (2023, Januar 12). https://www.adan.eu/en/publication/blockchain-protocols-and-their-energy-footprint/ (Zuletzt aufgerufen am: 2024, April 05)
48. CCRI-Indizes. CCRI Indizes (n.d.). https://indices.carbon-ratings.com/ (Zuletzt aufgerufen am: 2024, April 04)
49. UCL Centre for blockchain technologies discussion paper. Hedera (n.d.-b). https://hedera.com/ucl-blockchain-energy. Accessed 15 May 2024

50. Going carbon negative at Hedera hashgraph|Hedera (n.d.-b). https://hedera.com/blog/going-carbon-negative-at-hedera-hashgraph. Accessed 05 Apr 2024
51. J. Martin, M.D, Grün ist gut: Algorand verbraucht nur 80 kW Energie und kompensiert den Kohlenstoffausstoß wie zuvor versprochen. AlgoDaddy (2023, Oktober 6). https://www.algodaddy.org/2023/02/algorand-uses-only-80-kw-energy-carbon-offsetting.html (Zuletzt aufgerufen am: 2024, April 05)
52. How algorand offsets its carbon footprint (n.d.-b). https://algorandtechnologies.com/news/how-algorand-offsets-carbon-footprint. Accessed 05 Apr 2024
53. Admin, Grüne Kodierung und Energieverbrauch: Blockchain und Kryptowährungen. Exove (2024, Januar 18). https://www.exove.com/blogs/green-coding-and-it-energy-consumption-blockchain-and-cryptocurrencies/ (Zuletzt aufgerufen am: 2024, April 05)
54. S. Pomposi, Fantom, die umweltfreundliche Blockchain. Fantom Insights (2021, November 23). https://blog.fantom.foundation/fantom-the-eco-friendly-blockchain/ (Zuletzt aufgerufen am: 2024, April 05)
55. KlimaDAO, Polygon wird klimaneutral durch KLIMADAO: Das Grüne Manifest in Aktion (2022, Juni 19). https://www.klimadao.finance/resources/polygon-goes-carbon-neutral-via-klimadao-the-green-manifesto-in-action (Zuletzt aufgerufen am: 2024, April 05)
56. Polygon pos: Das effizienteste Blockchain-Protokoll. Polygon PoS|Das effizienteste Blockchain-Protokoll (n.d.). https://polygon.technology/polygon-pos (Zuletzt aufgerufen am: 2024, April 05)
57. K. Wright, Bericht behauptet, dass jeder Solana TX weniger Energie verbraucht als 2 Google-Suchen. Cointelegraph (2021, November 26). https://cointelegraph.com/news/report-claims-each-solana-tx-uses-less-energy-than-2-google-searches (Zuletzt aufgerufen am: 2024, April 05)
58. Solana's Energy Use Report: Dezember 2023. Solana (2023, Dezember 6). https://solana.com/news/solana-energy-use-report-december-2023 (Zuletzt aufgerufen am: 2024, April 05)
59. Solana Klima-Dashboard: Trycarbonara Übersicht. Solana Climate Dashboard|Trycarbonara Übersicht (n.d.). https://solanaclimate.com/ (Zuletzt aufgerufen am: 2024, April 05)
60. Solana versus Ethereum: Erneute Debatte bricht über Effizienz, Entwicklung und Skalierbarkeit aus. FXStreet (n.d.). https://www.fxstreet.com/cryptocurrencies/news/solana-vs-ethereum-renewed-debate-erupts-over-efficiency-development-and-scalability-202310191136 (Zuletzt aufgerufen am: 2024, April 05)
61. C. Foundation, Eine kohlenstoffnegative Blockchain? sie ist da und sie heißt Celo. Medium (2021, Mai 28). https://blog.celo.org/a-carbon-negative-blockchain-its-here-and-it-s-celo-60228de36490 (Zuletzt aufgerufen am: 2024, April 05)
62. E.N. Foundation, Climate positivity: Eos Network's commitment to sustainability strengthened through ENF and upland collaboration (2023, October 12). https://eosnetwork.com/blog/eos-blockchain-climate-positive-2023/. Accessed 05 Apr 2024
63. E. Authority, The future of blockchain is sustainable. Making EOS the first major carbon neutral blockchain (n.d.). https://eosauthority.com/green/. Accessed 05 Apr 2024
64. N. Team, Near Foundation schließt sich der Ethereum Climate Platform an. NEAR Protocol (2023, Juli 17). https://near.org/blog/near-foundation-signs-up-to-the-ethereum-climate-platform (Zuletzt aufgerufen am: 2024, April 05)
65. Immutable has measured their emissions and started their net zero journey. has measured their emissions and started their net zero journey (n.d.). https://www.our-trace.com/brand/immutable. Accessed 22 Apr 2024
66. Mattereum, Wie @avalancheavax @nori @mattereum heute die Welt verändert hat. Im Frühjahr hat Vinay Gupta (CEO von Mattereum) eine grobe Berechnung der CO2-Emissionen von @avalancheavax durchgeführt, um zu sehen, ob es sich um eine grüne Kette handeln könnte. es war niedrig. wirklich niedrig. ich habe es mit @Nori überprüft und sie haben die Schätzung bestätigt. 1/5. Twitter (2021, November 4). https://twitter.com/mattereum/status/1456315959141076993 (Zuletzt aufgerufen am: 2024, April 05)
67. Flow ist wegweisend für nachhaltige Technologie in WEB3. Flow ist der Wegbereiter für nachhaltige Technologie im Web3 (n.d.). https://flow.com/sustainability (Zuletzt aufgerufen am: 2024, April 05)

68. Ethereum.org. (n.d.). Ethereum Energy Consumption. ethereum.org. https://ethereum.org/en/energy-consumption/ (Zuletzt aufgerufen am: 2024, April 05)
69. A. Sarkar, Der Merge senkt den Stromverbrauch des Ethereum-Netzwerks um über 99,9 %. Cointelegraph (2022, October 29). https://cointelegraph.com/news/the-merge-brings-down-ethereum-s-network-power-consumption-by-over-99-9 (Zuletzt aufgerufen am: 2024, April 05)
70. Gnosis chain is carbon neutral with Offsetra. Offsetra (n.d.). https://offsetra.com/profile/gnosischain. Accessed 22 Apr 2024
71. Introduction. IOTA Wiki (2024, February 2). https://wiki.iota.org/get-started/introduction/iota/introduction/. Accessed 23 Apr 2024
72. Eos Eos Price, live charts, and news in United States (n.d.-a). https://www.coinbase.com/price/eos. Accessed 05 Apr 2024
73. Harmony. Transactions (n.d.). https://docs.harmony.one/home/general/technology/transactions. Accessed 23 Apr 2024
74. Stellar-Netzwerk: Grenzüberschreitende Zahlungen mit XLM. Gemini (n.d.). https://www.gemini.com/cryptopedia/stellar-blockchain-payments-xlm-coin#section-enhanced-by-stellar-networks-built-in-features (Zuletzt aufgerufen am: 2024, April 05)
75. Gebühren. Flow Developer Portal (n.d.). https://developers.flow.com/build/basics/fees (Zuletzt aufgerufen am: 2024, April 023)
76. BitPowr, How to send fee-less tron transactions. Bitpowr (n.d.). https://bitpowr.com/blog/how-to-send-fee-less-tron-transactions (Zuletzt aufgerufen am: 2024, April 05)
77. What is the Ethereum Virtual Machine & How Does It Work?|Hedera (n.d.-e). https://hedera.com/learning/smart-contracts/ethereum-virtual-machine. Accessed 05 Apr 2024
78. Q. AK, Krypto-Transaktionsgebühren: XRPL dominiert mit niedrigsten Gebühren, lässt BTC, eth, Ada, Matic hinter sich. Coinpedia Fintech News (2023, November 26). https://coinpedia.org/news/ripple-news-xrp-shatters-traditional-transaction-fee-norms-outperforming-btc-eth-ada-and-matic/ (Zuletzt aufgerufen am: 2024, April 05)
79. Struktur. Algorand-Entwickler-Portal (n.d.-a). https://developer.algorand.org/docs/get-details/transactions/ (Zuletzt aufgerufen am: 2024, April 05)
80. Messari Crypto News (n.d.-a). https://messari.io/report/state-of-celo-q4-2023 (Zuletzt aufgerufen am: 2024, April 05)
81. Klaytn, Ein Vergleich der Latenzzeiten von Blockchain-Netzwerken. Medium (2022, Juli 20). https://medium.com/klaytn/a-comparison-of-blockchain-network-latencies-7508509b8460 (Zuletzt aufgerufen am: 2024, April 05)
82. T. Spät, Wie hoch ist die Solana Gas Fee? CoinCodex (2024, Februar 19). https://coincodex.com/article/24933/solana-gas-fees/ (Zuletzt aufgerufen am: 2024, April 05)
83. Gnosis Kettenstatistik. Blockscout (n.d.). https://gnosis.blockscout.com/stats (Zuletzt aufgerufen am: 2024, April 05)
84. Messari Crypto News (n.d.-a). https://messari.io/report/state-of-aptos-q3-2023 (Zuletzt aufgerufen am: 2024, April 05)
85. R. Nambiampurath, Was ist ein Nahprotokoll? The Defiant (2022, September 6). https://thedefiant.io/what-is-near-protocol-2 (Zuletzt aufgerufen am: 2024, April 05)
86. Z. Attar, Cronos blockchain: high speed, low fees, and energy efficiency. Medium (2023, November 24). https://med. Accessed 05 Apr 2024
87. Messari Crypto News (n.d.-a). https://messari.io/report/state-of-polygon-q3-2023 (Zuletzt aufgerufen am: 2024, April 05)
88. Zilliqa (n.d.). https://www.zilliqa.com/what-is-zil. Accessed 05 Apr 2024
89. Was sind Arbitrum-Gas-Gebühren?. Arbitrum-Gas-Gebühren - Leitfaden für Einsteiger & Vergleich (n.d.). https://www.hord.fi/blog/arbitrum-gas-fees (Zuletzt aufgerufen am: 2024, April 05)
90. Messari Crypto News (n.d.-a). https://messari.io/report/state-of-cardano-q4-2023 (Zuletzt aufgerufen am: 2024, April 05)
91. Polkadot zeigt branchenführende Skalierbarkeit im positiven Ende bis 2023. Polkadot Network (n.d.). https://polkadot.network/blog/polkadot_q4_update_data (Zuletzt aufgerufen am: 2024, April 05)

92. Binance Smart Chain Durchschnittliche Transaktionsgebühr (I:BSCATFND). YCharts (n.d.-a). https://ycharts.com/indicators/binance_smart_chain_average_transaction_fee_es (Zuletzt aufgerufen am: 2024, April 05)
93. Messari Crypto News (n.d.). https://messari.io/report/optimism-q3-2023-brief (Zuletzt aufgerufen am: 2024, April 05)
94. Atom die gebühren: Wie viel Kostet es, Kosmos zu senden (Atom). ATOM Die Gebühren: Wie viel kostet es, Cosmos zu senden (ATOM) (n.d.). https://www.cropty.io/de/fees/atom (Zuletzt aufgerufen am: 2024, April 05)
95. Ethereum durchschnittliche Transaktionsgebühr (I:EATFND). YCharts (n.d.-b). https://ycharts.com/indicators/ethereum_average_transaction_fee (Zuletzt aufgerufen am: 2024, April 05)
96. Ein Anstieg der Transaktionsgebühren bei Avalanche: A closer look. Coinlive (2023, Dezember 18). https://www.coinlive.com/news/a-surge-in-avalanche-s-transaction-fees-a-closer-look (Zuletzt aufgerufen am: 2024, April 05)
97. Fees. Immutable Documentation (n.d.). https://docs.immutable.com/docs/x/fees/. Accessed 05 Apr 2024
98. Crypto.com. Was ist Aptos (APT)? (n.d.) https://crypto.com/university/what-is-aptos-token-apt (Zuletzt aufgerufen am: 2024, April 05)
99. Aptos wird zum Spitzenreiter bei der schnellsten Zeit bis zur Finalität: Messari. RSS (n.d.). https://www.bsc.news/post/aptos-emerges-as-frontrunners-for-fastest-time-to-finality-messari
100. Move - Eine Web3-Sprache und -Laufzeit. Aptos Docs (2024, März 16). https://aptos.dev/concepts/move-on-aptos/ (Zuletzt aufgerufen am: 2024, April 05)
101. fvnmGEUzSW2RapB8a3XRYF, Autor, Schultz, L., 8544927f-ba8f-41ea-b2b5-d2c5f9a2d338, & 2022-01-07T08:42:50Z. Was ist Polkadot (DOT) und wie funktioniert es? Firi (2023, April 13). https://firi.com/cryptocurrency/polkadot-dot/what-is-polkadot (Zuletzt aufgerufen am: 2024, April 05)
102. Seq, Vergleich zwischen Avalanche, cosmos und polkadot. Medium (2021, November 17). https://medium.com/avalanche-hub/comparison-between-avalanche-cosmos-and-polkadot-a2a98f46c03b (Zuletzt aufgerufen am: 2024, April 05)
103. S. Lucas, Near-Protokoll: Skalierbarkeit durch Sharding - Krypto: Sygnum Bank - in Krypto investieren mit einer regulierten Schweizer Bank. Sygnum Bank (2023, Oktober 6). https://www.sygnum.com/future-finance/crypto/near-protocol-scalability-through-sharding/ (Zuletzt aufgerufen am: 2024, April 05)
104. Was ist near & wie funktioniert es? wer hat near geschaffen?. Kriptomat (2022, September 2). https://kriptomat.io/cryptocurrencies/near/what-is-near-protocnear-protocol/ (Zuletzt aufgerufen am: 2024, April 05)
105. A. Bhalla, Top-Kryptowährungen mit ihren hohen Transaktionsgeschwindigkeiten [aktualisiert]. Blockchain, AI & Web3 Certifications (2024, Februar 29). https://www.blockchain-council.org/cryptocurrency/top-cryptocurrencies-with-their-high-transaction-speeds/ (Zuletzt aufgerufen am: 2024, April 04)
106. Person, Alles, was Sie über Layer 1 (und 2) Transaktionsfinalität wissen müssen - curvegrid. RSS (2023). https://www.curvegrid.com/blog/2023-06-28-all-you-need-to-know-about-layer-1-and-2-transaction-finality (Zuletzt aufgerufen am: 2024, April 04)
107. R. Stevens, Was ist Arbitrum? Beschleunigung von Ethereum durch optimistische Rollups. Decrypt (2023, März 21). https://decrypt.co/resources/what-is-arbitrum-speeding-up-ethereum-using-optimistic-rollups (Zuletzt aufgerufen am: 2024, April 05)
108. Fantom Network (FTM) (n.d.). https://support.bitso.com/hc/en-us/articles/6844980235412-Fantom-Network-FTM (Zuletzt aufgerufen am: 2024, April 05)
109. HBAR (ℏ). Hedera (n.d.). https://hedera.com/hbar (Zuletzt aufgerufen am: 2024, April 05)
110. J. Kubinec, Evmos hofft, Ethereum-Entwickler zur IBC zu locken, indem es Cosmos-Transaktionen abschafft. Blockworks (n.d.). https://blockworks.co/news/evmos-cosmos-ethereum-evm (Zuletzt aufgerufen am: 2024, April 05)
111. Pintu, Was ist Immutable X (IMX)? Erklärt - Pintu Academy. Explained - Pintu Academy (2023, März 17). https://pintu.co.id/en/academy/post/what-is-immutable-x (Zuletzt aufgerufen am: 2024, April 05)

112. Was ist Ihnen lieber - maximale Sicherheit oder billigere Transaktionen? Die Wertschöpfungsschicht des Internets (n.d.). https://polygon.technology/blog/what-do-you-prefer-maximum-security-or-cheaper-transactions (Zuletzt aufgerufen am: 2024, April 04)
113. Die Polygon-Blockchain ist da. zondacrypto (n.d.). https://zondacrypto.com/de/aktuelles/the-polygon-blockchain-is-here (Zuletzt aufgerufen am: 2024, April 05)
114. Warum Algorand?. Algorand Developer Portal (n.d.). https://developer.algorand.org/docs/get-started/basics/why_algorand/ (Zuletzt aufgerufen am: 2024, April 05)
115. Solving the "Blockchain trilemma." (n.d.-d). https://algorandtechnologies.com/technology/solving-the-blockchain-trilemma. Accessed 05 Apr 2024
116. I. Zia, Eos Leap 6: How it introduces fast finality, flexibility, and a new consensus. DailyCoin (2024, February 22). https://dailycoin.com/what-eos-leap-6-hard-fork-brings/. Accessed 05 Apr 2024
117. Overview. Klaytn Docs. (2024, März 27). https://docs.klaytn.foundation/docs/learn/ (Zuletzt aufgerufen am: 2024, April 05)
118. Münzstätten, XRP's turbocharged blockchain: 12,24 Millionen Transaktionen pro Stunde. Binance (2024, 8. Januar). https://www.binance.com/en/square/post/2449496484593 (Zuletzt aufgerufen am: 2024, April 05)
119. Stakin, Zilliqa and EVM—expanding the ecosystem (2023, May 19). https://blog.stakin.com/zilliqa-and-evm-expanding-the-ecosystem/. Accessed 05 Apr 2024
120. Wie Stellar funktioniert: eine schnelle, nicht-technische Anleitung (n.d.-b). https://resources.stellar.org/hubfs/Proof%20of%20Agreement%20explainer.pdf (Zuletzt aufgerufen am: 2024, April 05)
121. A. Takyar, Stellar-vs-EVM-basierte-Blockchains. LeewayHertz (2023, Februar 22). https://www.leewayhertz.com/stellar-vs-evm-based-blockchains/ (Zuletzt aufgerufen am: 2024, April 05)
122. Chainspect, TPS Dashboard [echte Metriken]. (n.d.). https://chainspect.app/dashboard (Zuletzt aufgerufen am: 2024, April 04)
123. T. Zimwara, BNB's tech roadmap for Opbnb targets 10,000 transactions per second - blockchain bitcoin news. Bitcoin News (2023b, November 29). https://news.bitcoin.com/bnbs-tech-roadmap-for-opbnb-targets-10000-transactions-per-second/ (Zuletzt aufgerufen am: 2024, April 05)
124. G. Roy, Investieren in Harmonie (eins) - alles, was Sie wissen müssen. Securities.io (2024, März 8). https://www.securities.io/investing-in-harmony/ (Zuletzt aufgerufen am: 2024, April 05)
125. C. Beat, Flow Krypto: NBA top shot. Medium (2023, Januar 24). https://medium.com/coinmonks/flow-crypto-nba-top-shot-bee813c3af94 (Zuletzt aufgerufen am: 2024, April 05)
126. A. V. Khatibi, Touring sonic - was ist Zeit bis zur Endgültigkeit? Fantom Einblicke (2023, November 25). https://blog.fantom.foundation/touring-sonic-what-is-time-to-finality/ (Zuletzt aufgerufen am: 2024, April 05)
127. EVM über Flow. Flow baut die Zukunft von Kultur und Gemeinschaft im Web3 (n.d.). https://flow.com/upgrade/crescendo/evm (Zuletzt aufgerufen am: 2024, April 05)
128. M. Bilušić, Celo - Blockchain mit schnellen, günstigen und sicheren Transaktionen Altcoin Buzz. Altcoin Buzz (2022, Juni 3). https://www.altcoinbuzz.io/reviews/altcoin-projects/celo-blockchain-with-fast-cheap-and-secure-transactions/ (Zuletzt aufgerufen am: 2024, April 05)
129. Layer-1 Performance: Comparing 6 leading blockchains. Weltnachrichten über Kryptowährung und Blockchain-Technologie aus verschiedenen Quellen (2023, Juli 28). https://cryptonews.net/news/analytics/21373397/ (Zuletzt aufgerufen am: 2024, April 05)
130. R. Neiheiser, G. Inácio, L. Rech, C. Montez, M. Matos, L. Rodrigues, Praktische Beschränkungen von Ethereums Layer-2. IEEE Access **11**, 8651–8662 (2023)
131. Die Dao-Kette. Gnosis-Kette (n.d.-a). https://www.gnosis.io/blog/ps13aapsjp0kdex7bnylbms2ui1-2inqwczug3zbjg0 (Zuletzt aufgerufen am: 2024, April 05)
132. Chainspect, Was sind Transaktionen pro Sekunde (TPS)? [echte TPS im Jahr 2024] (2024, April 4). https://chainspect.app/blog/transactions-per-second-tps (Zuletzt aufgerufen am: 2024, April 05)

The Purpose and Difficulties of Decentralization

Arpad Djuraki

1 Introduction

The following text delves into the multifaceted realm of decentralization within the context of blockchain technology. It navigates through key concepts, challenges, and trade-offs associated with decentralization, drawing insights from pioneers like Satoshi Nakamoto. The exploration extends to diverse applications, ranging from decentralized applications (DApps) to Decentralized Autonomous Organizations (DAOs) and the innovative realm of Decentralized Science (DeSci). Additionally, the text addresses the inherent vulnerabilities in decentralized blockchains, dissecting potential threats and attacks while spotlighting the evolving landscape of security measures. This comprehensive overview captures the essence of decentralization's impact on technology, applications, and the intricate balance between security and innovation.

2 Decentralization

Decentralization is a key concept in blockchain technologies. It is the distribution of power and control to a larger number of actors, in contrast to centralized systems where ownership and control are concentrated in the hands of a few. Decentralization offers a number of benefits, including security, transparency, and participation. However, it also comes at a cost. Decentralized systems can be slow and difficult to make changes, and throughput can be limited. The blockchain trilemma states that

A. Djuraki (✉)
Environmental Engineering, Aachen Blockchain Club, RWTH Aachen University, Aachen, Germany
e-mail: adjuraki@aachen-blockchain.de

it is impossible to have a blockchain that is perfectly scalable, secure, and decentralized. Any improvement in one of these properties will come at the expense of one or both of the other two properties.

This chapter will discuss the concept of decentralization in more detail, as well as the challenges and trade-offs involved.

2.1 General Definition

Decentralization is about spreading ownership and control among many entities. This is in contrast to centralized systems. So, decentralized systems have two key attributes: security and stability. This is because no single entity can make changes to the system, so it is less likely to be compromised. However, decentralization also comes at a cost. It can be slow and difficult to make changes to a decentralized system, and throughput can be limited [1].

The decentralized protocols are for example useful for foundational infrastructure, such as money, payment systems, and smart contracts. For applications built on top, more centralized solutions may be better, as they allow for more rapid innovation. However, if these products and services can be built on decentralized infrastructure, they can overcome a lot of the traditional overhead due to the safety and openness of blockchains. Decentralization is not a black or white state. There are some key thresholds that matter to achieve certain properties, with their advantages as well as their disadvantages. For example, if one entity has full control over a system, it is easy to make effective decisions, manage it, and drive improvements. Nevertheless, the controlling entity may have different motivations than other stakeholders, which could lead to exploitation [1]. A good example for decentralization beyond the blockchain is NASA. They designed their space flights computers to be redundant by allowing multiple systems to vote on the output of a computation, so if one makes a mistake, the other computers will override it by majority [2].

Against that, centralized systems need to be regulated to work, such as through property rights. The rules create trust and safety, which allows people to build things in peace and for the long-term benefit of society, which then leads to productivity and prosperity. In Table 1 the differences between these two systems could be shown.

In summary, decentralization is a trade-off. It offers security and stability, but it can also be slow and difficult to make changes. Centralized systems can be more efficient, but they need to be carefully regulated to avoid exploitation [1].

2.2 Decentralization According to Satoshi Nakamoto

Decentralization is a key concept in blockchain technologies. In his Bitcoin whitepaper, Satoshi Nakamoto defines decentralization as the distribution of power and control to a larger number of actors. This stands in contrast to a centralized

Table 1 Comparison centralization—decentralization [1]

	Centralization	Decentralization
Ownership	One in control	No Controlling entity
Decision-making	Top-down	Consensus
Decision time	Fast	Slow
Requirements	Trust in decision-maker	Distributed operation and control
Risks	Corruption, exploitation	Capture, concentrated ownership
Works well with.	Competition	Public goods, open protocols
Works badly with.	Monopolies	Individual products and services

structure, in which power and control lie in the hands of a small group of people. Therefore, it can contribute to a fairer and more democratic world.

Nakamoto argues that decentralization is important for a number of reasons [3], including:

- Security:

A decentralized structure is less vulnerable to attacks. In a decentralized blockchain environment, all participants have access to the same data and can track and confirm it. This makes it very difficult to manipulate or change the data in the blockchain.

- Transparency:

In a decentralized environment, it is more difficult to censor or manipulate information. This can help to build trust and trustworthiness in the blockchain.

- Participation:

Decentralization can help to ensure that more people are involved in decision-making processes. This can help to ensure that the blockchain is used and managed by a wider population.

2.2.1 Challenges for Decentralized Systems

Satoshi named also the challenges of a decentralized blockchain. So, there are Costs, Complexity and Security issues.

In connection with the costs, decentralized blockchains can be expensive to implement and maintain. This is because all participants must bear a portion of the infrastructure that supports the blockchain. In point of complexity, decentralized blockchains can be complex and difficult to manage. This is because a decentralized blockchain requires a number of different components to work together. The security aspect is decentralized blockchains can be vulnerable to certain types of attacks. Since the data in a decentralized blockchain is stored in a decentralized manner, it is more difficult for attackers to manipulate it. However, decentralized blockchains are still vulnerable to certain types of attacks, such as denial-of-service (DoS) attacks or 51% attacks, which will be discussed in Chap. 4 [3].

These challenges can make the adoption and deployment of decentralized blockchains more difficult. Still, it is important to be aware of these challenges in order to address them and maximize the benefits of decentralized blockchains.

2.3 The Blockchain Trilemma

Decentralization, security, and scalability are the three most important properties of blockchains (Fig. 1). In order to achieve scalability, either security, decentralization, or both must be sacrificed. In other words, decentralization in combination with security leads to slow transaction throughput. For example, Bitcoin and Ethereum are decentralized and secure, but not very scalable. Other blockchains are more scalable, but less decentralized and secure. However, no protocol has yet been able to achieve all three properties simultaneously. This dilemma is known as the blockchain trilemma. Other blockchains are trying to solve this dilemma of decentralization to the benefit of higher-level user functionality, while Bitcoin and Ethereum tend to value decentralization the most [1].

Further examples are EOS and Tron, they are more scalable, but less decentralized and secure. And still other blockchains, such as the GHOST protocol, attempt to combine decentralization and scalability, but are not secure [4]. The blockchain trilemma states that it is impossible to have a blockchain that is perfectly scalable, secure, and decentralized. Any improvement in one of these properties will come at the expense of one or both of the other two properties [5].

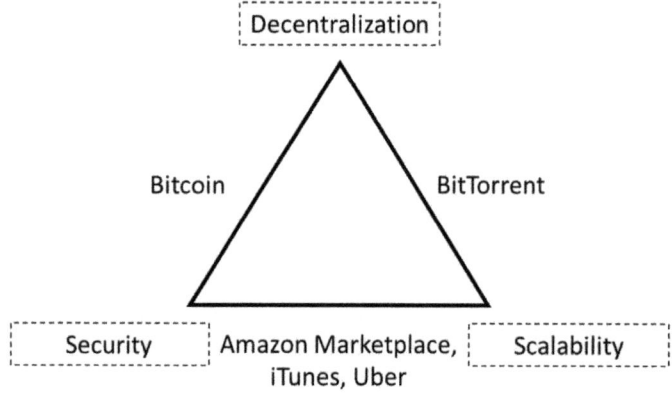

Fig. 1 The blockchain trilemma [4]

3 Decentralization in Application

Decentralized applications were proposed much earlier than blockchain technology [6]. An example would be BitTorrent, which is a decentralized application.

Decentralization should be used where it makes sense. Just because a blockchain application is decentralized does not mean that it has to be 100% decentralized. The goal of any blockchain solution is to meet the needs of its users, which may or may not include certain levels of decentralization. So, when it comes to blockchain technology, the term "decentralized network" is often mentioned. Many people still find it difficult to explain what a decentralized network is, whether there is a difference between decentralized and distributed networks, and what advantages these network structures have over centralized networks. Each different type of network architecture (Fig. 2) comes with its own advantages and disadvantages. Below, the fundamental differences between centralized, decentralized, and distributed networks are discussed in the appendix in Table 3 [7].

Each network architecture has its benefits and tradeoffs. For example, decentralized blockchain systems, unlike distributed systems, typically prioritize security over performance. So, when a blockchain network scales up or out, the network becomes more secure, but performance slows down because each member node must validate all data being added to the ledger. Adding members to a decentralized network can make it safer, but not necessarily faster.

3.1 Decentralized Applications

Blockchain-based decentralized applications, known as DApps, increase trustworthiness while lowering the cost of the trusted authority. The application possibilities

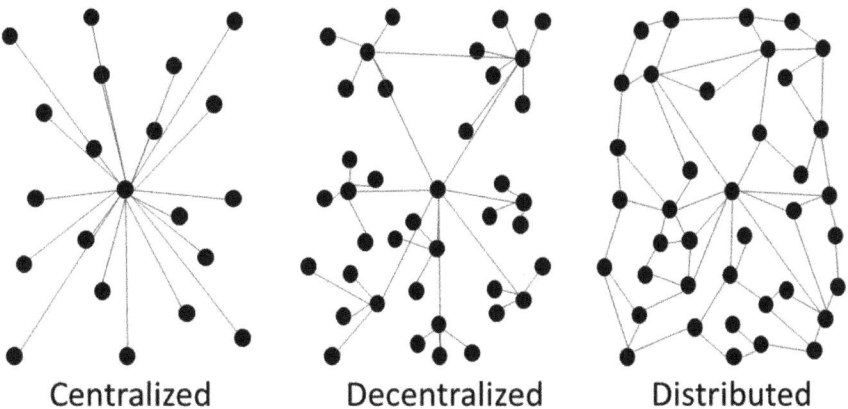

Fig. 2 Network types by Touron [8]

Table 2 Architectures and its examples [6]

Architecture	DApp
Native client as a DApp	Bitcoin, Zcash, Monero
Smart contract as a DApp	DanKu, EurocupBet, The DAO
Web and contract as a DApp	MakerDAO, Uniswap, Curve, Compound, Aave
Fully-decentralized DApp	TornadoCash

are therefore very interesting for the area of finance and the Internet of Things. They have attracted considerable attention in both industry and science in recent years.

However, there is a significant gray area between these two types. Therefore, the definition of blockchain-based decentralized applications is still pending, according to Zehng et al. [6] unclear. This article describes the architectures of blockchain-based decentralized applications and are summarized into four types:

These are Native Client as DApp, Smart Contract as DApp, Web and Contract as DApp and fully decentralized DApp, based on their different architectures (Table 2).

3.2 Decentralized Autonomous Organization

A DAO is a blockchain-based system that allows people to coordinate and govern themselves. They are governed by a set of smart contracts implemented in a blockchain, and their governance is decentralized [9].

The modern meaning of the term DAO (Decentralized Autonomous Organization) can be traced back to the earlier concept of a Decentralized Autonomous Corporation (DAC), which was mentioned a few years after the emergence of Bitcoin [3]. Today, the term DAO is much more common, and therefore this term is used.

3.2.1 Distinctive Characteristics

DAOs allow people to coordinate and self-govern online. While there is no minimum group size specified, the term "organization" is a general term that refers to a unit that includes: multiple people acting towards a common goal and are not legally registered as an organization.

The following characteristics have been defined for DAOs by Hassan and De Filipi [9]:

- A DAO source code is deployed on a blockchain with smart contracts. Functions like Ethereum probably always a public blockchain.
- The source code of a DAO is implemented within a blockchain equipped with smart contract functionalities, such as Ethereum, which is arguably perpetually a public blockchain.

- Because these rules are defined using smart contracts, they are self-executing regardless of the will of the parties.
- DAO governance should remain independent of central control.

3.2.2 DAO

True DAOs are companies that are fully decentralized and have no physical ties. The shareholders of a true DAO are defined by tokens issued on a blockchain.

The challenges for true DAOs arise from the fact that token holders are only identifiable as shareholders by a pseudonymized blockchain address. This makes it difficult to determine the legal points of reference for these companies. Accordingly, it is still difficult to define legal points [10].

Here are some examples of the challenges:

- Unclear legal system: The legal system that applies to a true DAO is unclear. This is because token holders are only identifiable as shareholders by a pseudonymized blockchain address.
- Securities law questions: If the DAO issues tokens, further securities law and regulatory questions arise.
- Unclear legal capacity: A true DAO itself is not capable of acting in these structures and will be considered to have little legal capacity due to the unclear corporate law classification.

3.2.3 DINO

Unlike DAOs, DINOs (Decentralized in Name Only) are not completely decentralized, having centralized authority. This can be defined by a person, a group of people or an organization. DINOs are often embedded in existing legal forms to ensure legal certainty for their members [10].

The following types of DINOs can be cited as examples:

- DAOs embedded in a limited liability company (GmbH)
- DAOs embedded in a foundation
- DAOs embedded in a cooperative
- DAOs that are integrated into an association.

In the example of Germany, DAOs are not fully regulated and therefore also mean that there is no uniform legal situation.

3.2.4 Limits to DAOs

DAOs enable transparent and democratic decision-making, but they can also lead to coordination problems. For example, voting on every decision by the members of the DAO can be more time-consuming than traditional top-down decision-making by leaders. Additionally, DAOs can be vulnerable to hacks and exploitation if their smart contract code is flawed [11].

So far, the majority of regulators are trying to fit DAOs into existing laws. In the United States, for example, the Securities and Exchange Commission (SEC) has clarified that DAO tokens can be treated as securities and therefore fall under securities laws [11].

However, there are also a few examples of tailored regulatory frameworks. In the United States, the state of Wyoming passed a law in 2021 that allows DAOs to attain legal status and register as a Limited Liability Company (LLC). This would give DAOs the ability to act like a traditional business, enter into contracts (e.g., hiring employees), clarify liability, and lend legitimacy to DAOs. Outside of the United States, there are also a few countries that are DAO-friendly. In Switzerland, for example, DAOs have opted to register as Swiss associations or DAAs (Decentralized Autonomous Associations). These legal forms bind DAOs into an existing legal framework, allowing them to better interact with the offline world and obtain limited liability. Another example is the Cayman Islands, which have introduced a new legal form called a "Foundation Company". These companies can act like a registered trust while retaining the separate legal personality and limited liability of a corporation. The passage of the Virtual Asset Service Provider Act allows DAOs to use a Foundation Company as a legal basis for their operations [11].

3.3 Decentralized Science

DeSci, short for Decentralized Science, is an initiative to improve the scientific landscape by integrating Web3 technologies. These technologies include elements such as blockchains, smart contracts and decentralized storage protocols. As described by Ethereum [12], Decentralized Science is a dynamic movement that seeks to create a shared infrastructure for activities such as funding, verification, accreditation, storage and dissemination of scientific knowledge. The basis of this movement is the pursuit of fairness and equality, which takes advantage of the possibilities of Web3. Essentially, DeSci is the fusion of open source science and blockchain [13].

3.3.1 DeSci and TradSci

The traditional academic system is inefficient and slows down research. There are six main problems that can be distinguished according to Shilina [14]. Research funding, peer review and research publication, intellectual property ownership, access to research and awareness, reproducibility and reproducibility of research results, and communication and collaboration between researchers. This description is shown in Fig. 3.

Below in the appendix is Table 5, which shows the key problems of traditional science and differences to decentralized science.

The Purpose and Difficulties of Decentralization

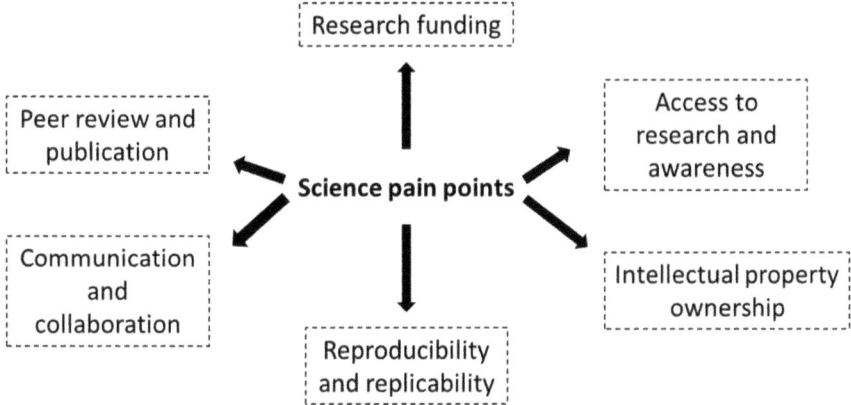

Fig. 3 Six main science pain points—the main DeSci drivers [14]

3.3.2 Token-Economy-Attempts

According to [15], four token economy approaches to research financing are shown. What opportunities and risks these contain and what connections exist:

- Token markets incentivize the early discovery and proper evaluation of talent and ideas. By linking individual stakes to tokens, the market value of these would be reflected and therefore represent an indicator of how the respective research is collectively assessed. Token markets can fulfill the function of prediction markets for researchers and research projects, they can bind token holders to a project, finance projects, and incentivize the opening of the research cycle.
- Governed token markets combine token markets in connection with a central authority. For example, a funder could set up a redistribution market, but instead of a free market for financing, only "value vouchers" of the funder would be traded. This would lose the specific advantages of open, borderless token markets, but the funder could define the target group of potential market participants and their roles, as well as meet the future regulation of such markets in its country.
- Research DAOs could fulfill open and cross-border tasks of funders, but also of research institutes. Such virtual organizations could negotiate research programs smoothly and transparently, set up smart contracts, and pay out budgets. A DAO can optionally issue tokens to trade itself on the market and it can partially or fully pre-define certain roles internally.
- Remodeling of Transactions Within the Research Community: This includes in particular the linking of credentials or monetary remuneration to the outcome of certain prescribed peer reviews via smart contracts. For example, to implement a "money-back" function for non-reproducible results.

3.4 Further Possible Applications

Example of further decentralized applications are the trains of the East Japan Railway Company [16]. In this case, these can be treated like patterns that have remained in nature. In biology manner, many transport processes are controlled in a decentralized manner. In nature, a central nervous system only emerged very late and is only present in a few living beings. The whole central nervous system is also strongly decentralized in a deeper way. There is no strong central point, but many decentralized nodes, which makes it less vulnerable to failures of parts. This means that most processes are controlled in a decentralized manner and function through self-organization. A school of fish has a few rules that each member of the school adheres to. Through these, harmonious and uniform actions can be achieved without a central leadership [17].

Furthermore, it would be conceivable to use decentralized solutions in the field of electromobility and the energy sector. On the one hand, blockchain can enable one of the biggest obstacles to the widespread use of electromobility to be solved by a simple billing model based on blockchain. The widespread use of electromobility is only possible if there are corresponding charging stations for the users of the vehicles available across the board. A difficulty today is the uncomplicated billing at charging terminals, which are often set up in public places and can therefore be used by anyone. Through blockchain technology, a model could be realized in which the driver parks the vehicle to shop in a store, for example, and the car automatically registers with the charging station and charges during the parking process. As soon as the driver leaves the parking lot, the charging station automatically charges the electricity used via the blockchain [18].

Another application area would be the integration of blockchain in the field of smart devices. The future communication of smart devices with each other and with third-party devices within and outside households and companies requires a communication carrier over which information and transactions can be moved and stored [18].

In addition, the function of decentralized documentation of transactions can be used for the widespread archiving of all billing data for electricity consumption. In conjunction with the Smart Meter Rollout, blockchain can be used for consumers to read and bill their digital electricity meters. In principle, it is also possible to think about further related applications that are not related to the electricity market, such as the billing of heating costs and hot water, which are currently carried out by professional meter reading services [18].

There is also DeAI, or Decentralized Artificial Intelligence. This is an approach aimed at decentralizing AI systems and services. Unlike traditional AI models that are centralized and static, DeAI focuses on the individualization, personalization, security and distribution of intelligence in intelligent systems [19].

The advantages of DeAI are as follows:

- The distribution of intelligence across different devices and nodes, resulting in improved flexibility, reliability and adaptability.

- In addition, maximizing local resources and minimizing dependence on centralized servers can improve the efficiency and performance.

4 Attacks on Decentralized Blockchains

Decentralized blockchains are increasingly becoming relevant in business and society. However, they are not immune to threats. Attacks on blockchains can be motivated by financial gain, sabotage, or simply to demonstrate technical prowess.

This chapter will discuss the different types of attacks that can be launched against decentralized blockchains, as well as the limitations of these attacks and mitigation strategies that can be employed.

4.1 Blockchain Threats

Blockchains systems become increasingly relevant in business and society [20]. Therefore, they should be secure and transparent, but they are not immune to threats. Hacking, scalability, interoperability, regulation, privacy, and energy consumption are some of the key risks to blockchain technology. If the network is not secure enough, blockchains can be hacked. Hackers can attempt to manipulate the network by altering the data in blocks, obtaining private keys, or mounting a 51% attack to gain control of the network [21].

4.1.1 Types of Attacks

Cybersecurity attacks are intentional and unauthorized access to systems that pose a threat to the security objectives of information systems. Attackers attempt to achieve an unauthorized goal through a series of planned steps. They use various tools to exploit vulnerabilities in a system. The motivations of attackers are diverse and depend on various factors, such as the type of attack, the application attacked, the layer of the information system attacked, and other aspects [20].

Phishing and Social Engineering Attacks

One way to attack crypto users is through phishing and social engineering attacks. The goal of the attacks is to trick victims into disclosing important information or finances, which can have catastrophic consequences for the victim. For example, the attacker creates a fake website or email that appears to be from a legitimate cryptocurrency exchange, wallet, or other service [22]. The victim is then asked to provide their data, such as login credentials, seed phrase or private key, so that the attacker can then use the data to steal the victim's assets. In order to minimize the risk, it is recommended to use reputable exchanges, wallets or services and in addition, the user should not pass on any information to third parties or use a service without two-factor authentication [23].

51% Attacks

A 51% attack on a blockchain network occurs when an attacker or group controls more than 50% of the network. This allows them to manipulate the network by controlling the majority of nodes. With this power, the attacker can roll back transactions, double spend, and block verification of new transaction [24]. This type of attack is extremely dangerous to the security and integrity of a blockchain network [21].

Dusting

Dusting is an attack in which a small amount of, for example, Bitcoin which is transferred to an address with the intent to confuse, deceive the recipient, or disrupt the regular operations of a blockchain network. The word "dust" therefore refers to the fact that these small transactions often result in a small amount or tiny fractions of Bitcoin that are too small to be economically spent or traded. Furthermore, dusting attacks can also be used to overload transaction processing capacity with a large number of tiny transactions [21].

Double-Spend Attack

A double-spend attack refers to an attempt by a hostile actor to spend digital money more than once. This can happen in a decentralized system where many servers store identical copies of a public transaction in a ledger whereby the money reaches at different times. The victims in this scenario were the exchanges rather than the end consumers or private wallet users [21].

DDos Attacks

One of the most common attacks on internet services is the Distributed Denial-of-Service (DDoS) attack [25]. Even though blockchain technology is defined by a peer-to-peer system, it is vulnerable to this kind of attacks. A DDoS attack is another category of cryptographic attack that can disrupt the normal operation of blockchain networks. Its involves sending numerous requests to a network or server, overloading the blockchain with traffic and making it inaccessible to legitimate users.

Most of them are launched against cryptocurrency exchanges, mining pools, and other blockchain-related services [21].

4.2 Limitation of Attacks

It is important to note that attacks also have limitations, including the cost and difficulty of execution, the size of the network, the consensus mechanism, the effectiveness of mitigation strategies, and the potential for human error. As blockchain technology advances, it is likely that the risks associated with attacks will continue to decrease [26]. To reduce the risks of blockchain attacks, it is important that blockchain developers incorporate resilient security measures and conduct ongoing testing to identify vulnerabilities and improve network security.

4.2.1 Cost and Difficulty

Blockchain attacks are often expensive and difficult to carry out as they require a significant amount of computing power and resources. A 51% attack, for example, requires a large amount of computing power that is not always achievable for an individual or a small group of attackers [26].

4.2.2 Network Size

The size of the blockchain network also influences the effectiveness of attacks. A larger network is typically more resilient to attacks because it requires more nodes and more computing power to control the network [26].

4.2.3 Consensus Mechanisms

Blockchain technology uses a consensus mechanism to ensure the accuracy and security of transactions. Different consensus mechanisms have different levels of security and vulnerability to attacks [26].

4.2.4 Mitigation Strategies

Blockchain developers are constantly working to develop new security measures to prevent or combat attacks. However, these measures are constantly being developed further by attackers, so that the security of blockchain networks is constantly being put to the test [26].

4.2.5 Human Error

Although attacks are often attributed to malicious actors, human error can also play a significant role in blockchain security breaches. Mistakes like mismanaging private keys or failing to update software can leave networks vulnerable to attacks [26].

5 Conclusion

The examination of decentralization, considering its benefits and challenges, underscores its transformative potential and the intricate difficulties it brings across various domains.

Decentralization, particularly in the realm of blockchain technology, proves to be a concept that distributes power among numerous entities, thereby promoting security, transparency, and increased participation. However, this paradigm shift comes with compromises, including potential slow processes and the impossibility of achieving perfect scalability, security, and decentralization simultaneously, as illustrated by the Blockchain Trilemma. The concept of decentralization extends beyond blockchain, with decentralized applications (DApps), Decentralized Autonomous Organizations (DAOs), and even Decentralized Science (DeSci) making substantial progress. The development of token economies in research financing and the application of decentralized principles in various sectors such as electromobility and smart devices illustrate the versatility of decentralization. Nevertheless, challenges loom. The different chapters delve into the vulnerabilities of decentralized blockchains, emphasizing the constant threat of various attacks such as phishing, 51% attacks, dusting, double-spending, and DDoS attacks. The limitations of these attacks are considered, taking into account factors such as cost, difficulty, network size, consensus mechanisms, defense strategies, and the potential for human error.

In summary, the future shaped by decentralization is promising, offering increased security, transparency, and inclusivity. An example from Prof. Strücker's research in the field of Decarbonization using regenerate Finance to address the existing Trust Gap can be inserted here. However, these visions are not without hurdles. The decentralized world must navigate challenges like security risks, potential exploitation, and the need to master evolving and adaptive governance models. Striking the right balance between the advantages of decentralization and the associated complexities will be crucial. On the path to a decentralized future, the continuous development of robust security measures, regulatory frameworks, and educational initiatives will be paramount to ensure a sustainable and resilient decentralized world.

Appendix

See Tables 3, 4 and 5.

Table 3 Fundamental differences between networks [27, 28]

	Centralized	Distributed	Decentralized
Network/hardware resources	Maintained and controlled by single entity in a centralized location	Spread across multiple data centers and geographies; owned by network provider	Resources are owned and shared by network members; difficult to maintain since no one owns it
Solution components	Maintained and controlled by central entity	Maintained and controlled by solution provider	Each member has exact same copy of distributed ledger
Data	Maintained and controlled by central entity	Typically owned and managed by customer	Only added through group consensus
Control	Controlled by central entity	Typically, a shared responsibility between network provider, solution provider and customer	No one owns the data and everyone owns the data
Single point of failure	Yes	No	No
Fault tolerance	Low	High	Extremely high
Security	Maintained and controlled by central entity	Typically, a shared responsibility between network provider, solution provider and customer	Increases with increasing number of network members
Performance	Maintained and controlled by central entity	Increases as network/hardware resources scale up and out	Decreases with increasing number of network members
Example	ERP system or national currencies	Cloud computing	Blockchain technology like Bitcoin

Table 4 Blockchain attack group and its assaults [21]

Blockchain attack group	Types of attacks
Blockchain network attacks	• Distributed denial of service (DDos) • Sybil attacks • Transaction malleability attacks Timejacking • Routing attacks • Long range attacks on proof of stake networks • Eclipse attacks
User wallet attacks	• Phishing • Flawed key generation • Attacks on cold wallets • Vulnerable signatures • Dictionary attacks • Attacks on hot wallets
Smart contract attacks	• Vulnerabilities in contract source code • Vulnerabilities in virtual machines
Transaction verification mechanism attacks	• Finney attacks • Race attacks • Vector76 • Alternative history attacks • 51% or majority attacks
Mining pool attacks	• Selfish mining • Fork after withholding

Table 5 Comparison between DeSci and TradSci [12]

Decentralized science	Traditional science
Distribution of funds is determined by the public using mechanisms such as quadratic donations or DAOs	Small, closed, centralized groups control the distribution of funds
You collaborate with peers from all over the globe in dynamic teams	Funding organizations and home institutions limit your collaborations
Funding decisions are made online and transparently. New funding mechanisms are explored	Funding decisions are made with a long turnaround time and limited transparency. Few funding mechanisms exist
Sharing laboratory services is made easier and more transparent using Web3 primitives	Sharing laboratory resources is often slow and opaque
New models for publishing can be developed that use Web3 primitives for trust, transparency and universal access	You publish through established pathways frequently acknowledged as inefficient, biased and exploitative
You can earn tokens and reputation for peer-reviewing work	Your peer-review work is unpaid, benefiting for-profit publishers
You own the intellectual property (IP) you generate and distribute it according to transparent terms	Your home institution owns the IP you generate. Access to the IP is not transparent
Sharing all of the research, including the data from unsuccessful efforts, by having all steps on-chain	Publication bias means that researchers are more likely to share experiments that had successful results

References

1. P. Knirck, M. Zollinger, Decentralization—does it actually matter? (2022)
2. W. Metcalfe, Ethereum, smart contracts, DApps, in *Blockchain and Crypto Currency* (Singapore, Springer, 2020), p. 141
3. S. Nakamoto, Bitcoin: a peer-to-peer electronic cash system (2008)
4. Trifecta, Trifecta: the blockchain TriLemma solved (2019)
5. G.D. Monte, D. Pennino, M. Pizzonia, Scaling blockchains—without giving up decentralization and security (2020)
6. P. Zheng, Z. Jiang, Z. Zheng, J. Wu, Blockchain-based decentralized application: a survey. IEEE Open J. Comput. Soc. (2023)
7. C. Staff, „gemini.com, 12 07 2021. [Online]. Available: https://www.gemini.com/cryptopedia/blockchain-network-decentralized-distributed-centralized. [Zugriff am 22 12 2023]
8. M. Touron, berty.tech, 20 06 2019. [Online]. Available: https://berty.tech/blog/decentralized-distributed-centralized. [Zugriff am 22 12 2023]
9. S. Hassan, P.D. Filippi, Decentralized autonomous organization. Internet Policy Rev. (2021)
10. Lukas, G. Hemmelmayr, J.-G.A. Hannemann, M. Marz, D.R. Müller, Z. Vig, Working group-DAO – Whitepaper. (Blockchain Bundesverband, Berlin, 2023)
11. Fisch, P.P. Momtaz, The rise of decentralized autonomous organizations (DAOs): a first empirical glimpse, in *Venture Capital* (2022)
12. Ethereum, ethereum.org, 15 12 2023. [Online]. Available: https://ethereum.org/de/desci/. [Zugriff am 22 12 2023]
13. R. Sanchez, blockzeit.com, 16 08 2023. [Online]. Available: https://www.blockzeit.com/de/decentralized-science-nachste-thema-krypto-web3/. [Zugriff am 22 12 2023]
14. S. Shilina, Decentralized science (DeSci): Web3-mediated future of science, 01 2023. [Online]. Available: https://medium.com/paradigm-research/decentralized-science-desci-web3-mediated-future-of-science-2547f9a88c40. [Zugriff am 22 12 2023]
15. L. Heller, I. Blümel, Disruption der Forschungsförderung—Mit Dezentralisierung zu einer offeneren Wissenschaft, in *Laborjournal* (2019), pp. 16–21
16. F. Kitahara, K. Kera, K. Bekki, *Autonomous Decentralized Traffic Management System* (IEEE, Chengdu, 2000)
17. H. Baumgarten, Das Beste der Logistik—Innovationen, Strategien, Umsetzungen (Springer Berlin, Heidelber, 2008)
18. F. Hasse, D.A.V. Perfall, T. Hillebrand, E. Smole, L. Lay, M. Charlet, Blockchain—Chance für Energieverbraucher? (2016)
19. L. Cao, Decentralized AI: edge intelligence and smart blockchain, metaverse, Web3, and DeSci. IEEE Intell. Syst. (2022)
20. T. Guggenberger, V. Schlatt, J. Schmid, N. Urbach, *A Structured Overview of Attacks on Blockchain Systems*. Dubai (2021)
21. M.B. Malik, M.F.B. Zolkipli, Blockchain threats: a look into the most common forms of cryptocurrency attacks. Borneo Int. J. 20–32 (2023)
22. S. Campbell, F.M. Moghaddam, Social engineering and its discontents, in *The Psychology of Radical Social Change* (2018), pp. 103–121
23. K. Weber, A.E. Schütz, T. Fertig, N.H. Müller, Exploiting the human factor: social engineering attacks on cryptocurrency users, in *Lecture Notes in Computer Science,* Nr. 12206 (2020), pp. 650–668
24. S. Sayeed, H. Marco-Gisbert, Assessing blockchain consensus and security mechanisms against the 51% attack. Appl. Sci. (2019)
25. M. Saad, J. Spaulding, L. Njilla, C. Kamhoua, S. Shetty, D.H. Nyang, D. Mohaisen, Exploring the attack surface of blockchain: a comprehensive survey, in *Communications Surveys and Tutorials*, Bd. 22 (2020), pp. 1977–2008
26. K. Borge, A. Kokane, S. Prof. Sumant, A study of blockchain attacks, in *NCRD´s Technical Review,* Bd. 1, Nr. 7 (2022)

27. Amazon Web Services, Inc., amazon.com, 01 10 2025. [Online] Available: https://aws.amazon.com/de/web3/decentralization-in-blockchain/. [Zugriff: 01 10 2025]
28. H. Graf, blog.novatrend.ch, 25 10 2017. [Online] Available: https://blog.novatrend.ch/2017/12/25/zentrale-dezentrale-undverteilte-systeme/. [Zugriff: 01 10 2025]

Arpad Djuraki is a successful graduate of RWTH Aachen University with a degree in Environmental Engineering, specializing in Urban Water, and a background in industrial mechatronics. As a co-founder of the Aachen Blockchain Club e.V. (ABC) and team leader of the industry team, he established valuable connections within the blockchain realm and fostered partnerships between companies and the ABC. His expertise was further refined in the DeFi Talent Program at the Frankfurt School Blockchain Center under Prof. Sandner.

Arpad Djuraki is also distinguished by his keen interest in technology and investments. This passion has led him to invest not only in stocks but also in security and utility tokens within the blockchain domain.

His motivation for contributing to the Web.3 Compendium is multifaceted. Through his involvement in this renowned compendium, Arpad Djuraki aspires to contribute to knowledge dissemination on a specific topic, empowering interested individuals to delve deeper into the intricacies of blockchain technology. Additionally, he views collaboration with diverse authors sharing similar interests as an excellent opportunity to refine his skills in research, writing, and critical thinking.

Decentralization Levels of L2 Scaling Solutions on Ethereum

Lorenz Raphael Lehmann

1 Fundamentals of Ethereum

Ethereum is a blockchain primarily recognized as a smart contract platform [1]. Before we can dive into the decentralisation of Layer 2 (L2) scaling solutions, we first need to understand the foundations of Ethereum. Ethereum falls into the category of a Layer 1 blockchain, which means it operates in an isolated environment, which forms the bedrock upon which other layers and technologies are built and integrated [1].

Ethereum's architecture revolves around three core operations: distributed consensus, data availability and transaction execution [1, 2]. Each of these plays a critical role in how Ethereum functions and maintains its integrity [2].

Distributed consensus is a central operation Ethereum prises itself in [3]. Consensus is secured through a proof of stake mechanism, where validators stake Ethereum tokens (ETH) and participate in a democratic process to establish the network truth [3]. This system is characterised by probabilistic finality, meaning the likelihood of a transaction being reversed diminishes as more blocks are added to the blockchain [4]. Anyone can contribute to this decentralized network by setting up a node. These nodes are crucial for maintaining Ethereum's blockchain state, processing transactions, and engaging in the peer-to-peer consensus process. The Ethereum Virtual Machine (EVM) is encapsulated within these nodes and various node clients exist, with Geth being the most popular [5].

The second key operation is data availability [3]. The network functions as a universal source of truth. It hosts all necessary data open to anyone to be able to verify executed transactions. The total data is also known as a ledger. Data availability ensures transparency and traceability of all activities on Ethereum [6].

L. R. Lehmann (✉)
RWTH Aachen University, Aachen, Germany
e-mail: lorenz.lehmann@rwth-aachen.de

Transaction execution is the third operational pillar of Ethereum's architecture [5]. The EVM, which serves as the runtime environment for smart contracts on Ethereum, operates in a sandboxed and isolated manner. This isolation ensures that the EVM cannot interact with any other network, filesystem or external processes [5]. Isolation is key in maintaining a high level of security and integrity [5]. Smart contracts, which are written in a high-level programming language (e.g. Solidity), are converted into bytecode, which allows encoding and decoding with an Application Binary Interface (ABI) [5, 6]. The ABI enables interactions with the smart contracts [6].

The goal of a L2 scaling solutions is to enhance Ethereum's performance and capabilities while maintaining these three operational components [2].

2 The Scaling Problem for Ethereum

2.1 Development of Ethereum

Ethereum's development was marked by significant technological advancements and milestones. Conceived by Vitalik Buterin in 2013, Ethereum emerged from a vision to create a more versatile and functional blockchain compared to the then-dominant Bitcoin [7]. This vision materialized into active development in 2014, leading to the Ethereum network going live on July 30, 2015. Initially, Ethereum operated on a proof-of-work (PoW) consensus mechanism, mirroring the security model of Bitcoin but offering more functionalities through smart contracts [5].

One import upgrade to the Ethereum protocol was the switch to a proof-of-stake (PoS) consensus mechanism [8]. This transition, which took over a year and was fully realized on September 15, 2022, marked a significant leap forward in addressing environmental concerns associated with blockchain technology. The move to PoS dramatically reduced Ethereum's energy consumption by approximately 99% [8]. Despite this advancement, Ethereum continues to face challenges in terms of scalability. Through a large amount of technical debt, milestones first conceptualised in the whitepaper are still to be fully developed. Scalability in particular remains a critical focus, given the growing demand for blockspace [5].

2.2 How the Scaling Problem Emerged

The Ethereum blockchain has faced persistent scaling challenges. Originally build to process a maximum of approximately 15 transactions per second [7], this limitation has remained unchanged even after the transition to the PoS consensus mechanism. This throughput limitation has become increasingly apparent as the blockchains popularity has grown. One event in particularly stands out, the launch of the application Cryptokitties by Dapper Labs in November 2017 [9]. The surge in transaction demand due to the game's popularity was so significant, that is caused huge network congestion [9].

As Ethereum continues to attract new use cases, its inability to accommodate more transactions becomes more pronounced [5]. The network has been operating at its capacity limits for several years [5]. Efforts to increase the gas limit per block are tempered by the need to maintain manageable computational requirements for hardware and node operators [10]. This is crucial for ensuring the networks decentralised operation. In 2021 alone, just 10 applications consumed about 40% of the blockspace [11]. Limited blockspace leads to an auction system, where users bid up their transaction fee to be included onchain, invariably driving up transaction costs. There are instances where even simple transactions like transferring Ethereum between wallets can cost upwards of 5$ in ETH [12]. This cost requires high-value transactions to justify the fees, which is a significant barrier to broader adoption. In comparison to traditional transaction systems like Visa, who process around 45,000 payments per second, Ethereum is far behind [5].

Three ways have been conceptualised on how to scale Ethereum: sharding, sidechains and L2 scaling solutions [5, 13].

Sharding, an upgrade within the inner workings of the Ethereum blockchain, involves partitioning the global state and computation into shard chains [14]. Through this concept, the network throughput can be increased by executing transactions in parallel [15]. Sharding would enable each node to store and validate only a subset of the total network, thus reducing the individual workload and increase throughput [14].

Sidechains represent another approach to scale Ethereum. These operate as separate blockchains linked only to Ethereum [16]. The chains run completely offchain and don't require any changes to the Ethereum protocol. These new chains require two functions: a bridge to transfer assets between the L1 and the sidechain and a mechanism for posting snapshots of their state roots to Ethereum. However, sidechains introduce certain risks. They are not as decentralized and transactions executed on sidechains are not verified by Ethereum consensus network. Additionally, sidechain bridge contracts operate on the assumption that transactions executed on the sidechain are valid according to its rules, which could be a point of potential exploit [16].

L2 scaling solutions take offer a third way to scale the throughput of transaction on Ethereum and will be discussed in more details in the following chapters.

2.3 L2 Development

L2 scaling solutions are designed to enhance the scalability and efficiency of L1 blockchains [5]. The fundamental principle behind L2 scaling solutions is to leverage the existing security of the L1 (in this case Ethereum) while offloading certain other computational tasks [5]. Therefore, the development of a L2 scaling solutions does not involve around creating a new blockchain from scratch. The focus of L2 solutions is to execute transactions offchain, which are later finalised on the L1 blockchain [3]. This mechanism allows for increased scalability and throughput, as the L1 is relieved from processing a large volume of transactions directly. This can increase throughput without compromising on the decentralised security [3].

In the case of Ethereums, as the consensus mechanism is highly effective, but scalability hindered by limitations in data availability and execution, a L2 scaling solutions seems to be the ideal solution. However, the development and implementation of L2 solutions, particularly by shifting execution and data availability offchain, are complex tasks [3]. Challenges revolve around maintaining the delicate balance between security and decentralisation. While offchain processing offers significant performance improvements, it requires careful design to ensure that the decentralised and secure nature of blockchain technology is not compromised [6].

2.4 Type of L2s

In the evolving landscape of Ethereum L2 scaling solutions, a variety of types have emerged, each with distinct characteristics and mechanisms. Among the primary types of L2 scaling solutions are: Channels, Plasma, Rollups and Validiums, each addressing the scalability challenge in a unique way [5].

Channels were one of the first L2 scaling solutions developed. A typical example are payment channels, where two or more parties who frequently transacting create a contract and deposit a set amount of funds [15]. This approach involves recording numerous transaction signatures, with funds being transferred only when the channel is closed. Such channels require upfront capital and participants must diligently maintain records to safeguard against potential fraud. Also, there has been limited success in extending this concept beyond payment channels. An example of a Channel scaling solution is Bitcoin's Lightning Network [17].

Plasma was first introduced in 2017, which utilizes smart contracts to facilitate the creation of "child chains" [18]. These child chains are essentially scaled-down versions of the Ethereum blockchain, organised within a hierarchical structure. Plasma chains execute transactions off-chain while maintaining exit mechanisms for users to withdraw assets [18]. However, the requirement for transaction verification by users themselves to prevent fraud has significantly hindered widespread adoption. Notably, projects like OMG Network and Boba Network initially focused on Plasma but have since shifted their development efforts towards Rollups [11].

Rollups encompass both optimistic and zero-knowledge rollups and have emerged as the most prominent L2 scaling solution at the time of writing [19]. They shift the execution of transactions offchain while preserving data availability and consensus on the Ethereum L1. This approach ensures the security benefits of Ethereum. Rollups provide an EVM-compatible experience for users on L2, focusing on removing execution complexity and compressing transaction data before posting them onchain. Still the major cost factor for Rollups is the expense of settling transaction data on L1 [6, 19].

Validiums share similarities with Rollups in that they move execution off-chain while keeping consensus on Ethereum [20]. However, Validiums go one step further by only posting highly compressed state root hashes to L1. The data availability aspects are moved offchain and overseen by a Data Availability Committee (DAC). This mechanism necessitates a higher degree of trust from users regarding the availability of data on demand. Validiums are inherently more cost-effective compared to Rollups, as they don't have to endure to high cost in posting transaction data to L1 [6, 20].

Each L2 solutions represents a unique approach to scaling Ethereum, balancing factors like trust, cost, and user experience. As the Ethereum ecosystem continues to grow, these L2 solutions play a crucial role in enhancing its scalability and broader adoption [5].

2.5 Architecture of Rollups

To enable offchain execution, Rollups introduce three new entities—Sequencer, Prover and Verifier—who collectively contribute to the operational integrity and security of the Rollup [5].

The first of these entities is the *Sequencer*. A Sequencer is a specialized node with the responsibility of receiving transactions from users on the L2 network. The primary function of the Sequencer is to execute these transactions in an orderly and efficient manner. The Sequencer plays an important role in maintaining the integrity of the L2, ensuring that user transactions are processed [21].

Another significant entity in the L2 framework is the *Prover*. The Prover is a node that generates validity proofs, particularly for zero-knowledge based Rollups. These proofs are generated based on all L2 transactions recorded by the Sequencer [5].

The third critical new entity is the *Verifier*, whose task is in the attesting to the validity of transactions that have been processed on the L1 network [5]. In the context of Optimistic Rollups, the Verifier operates on the principle of "innocent until proven guilty" [22]. This approach implies that transactions are presumed valid unless a specific invalid transaction is identified, at which point a fraud proof is submitted [21]. This mode of operation allows for an efficient verification process, as it reduces the need for constant verification of each transaction. Contrary to optimistic Rollups, zero-knowledge Rollups work after the principle of "guilty until proven innocent" [22]. This means that transactions are not considered valid on the L1 network until a proof is submitted by the Prover. This approach ensures a high level of security, as each transaction must be explicitly validated before being accepted [21].

3 Decentralization Levels of Rollups

While projects have been focusing on scaling Ethereum, the issue of security concerns in regards to centralisation of the three new entities have gained attention in the Ethereum community. This has brought the concept of decentralisation levels to wider attention, serving as a framework for assessing the degree of decentralisation of Rollups [23].

A decentralisation levels concept was first discussed in November 2022 by Ethereum co-founder Vitalik Buterin, to represents a structured approach to understand how platforms evolve from centralised control towards a decentralised system. The following framework of decentralisation levels for Rollups was introduced by Luca Donno from L2beat in June 2023, presenting a three-stage concept. This framework categorises the decentralisation status of Rollups, ranging from Level 0, indicating high centralization, to Level 2, which signifies a state of high decentralization. The numbering starts from 0, following the convention used in programming and coding [23].

Level 0 represents the initial phase of a L2, where entities are primarily centralized and operated by the developers. In this stage, the criteria include: posting of L1 state roots for emergency withdrawals to L1, ensuring data availability on L1 and the ability to reconstruct the state of the L2 with only data from L1. This level is akin to the developmental phase, where centralised control is good for rapid deployment and quick bug fixes, that can pose a high risk to user fund [23].

Level 1 is the first-time decentralisation is mandatory for certain aspects. This stage requires the establishment of a council to democratically and transparently manage changes. Criteria for achieving Level 1 status include: a correct proof system to confirm state roots posted on L1, the capability of at least five external actors to submit a fraud proof, independent user exit mechanisms without operator permission and a minimum 7-day delay for upgrades to allow users to exit in case of unwanted changes [23].

Level 2 epitomizes the ultimate goal of decentralisation for L2s, where the Rollup is fully managed by smart contracts and every single one of the three entities operates in a permissionless manner. To reach this stage, criteria include: a sufficient time window (30 days) for users to exit during an unwanted upgrade, protection of users from governance attacks and a security council that acts only in response to on-chain detected errors. The ethos of Level 2 is that the code itself becomes the ultimate authority, embodying the decentralized ethos of Ethereum [23].

As of the time of writing, L2beat records indicate a diverse landscape of 34 different L2 solutions. Among these, 2 have achieved Level 2 status, 3 are at Level 1, and 17 are at Level 0. This leaves 12 projects that have not yet reached any of the defined levels of decentralisation [24, 25].

4 Conclusion

Decentralising Layer 2 scaling solutions on Ethereum is a still ongoing process and full of challenges. Centralisation risk remain prevalent in many L2 projects, as evidenced by the current distribution of projects across the decentralisation levels [24, 25]. However, the potential for discovering effective and scalable solutions appears promising, with the framework of decentralization levels providing a clear roadmap for development and progression.

One of the critical concerns is the accumulation of technical debt. As projects strive to innovate and scale, they often make compromises that could lead to complex challenges in the future. However, there is an underlying optimism in this scenario. The collaborative and open-source nature of the blockchain community allows projects to learn from each other's experiences. This collective knowledge means that once a project successfully navigates the path to full decentralisation and scalability, others can follow suit.

Ultimately the confidence and determination of the community remains strong: Ethereum will be scaled [5]. The progress made so far and the continued efforts of developers and the community at large indicate a strong commitment to overcoming said obstacles [10, 12, 20]. The decentralization levels framework [23], while highlighting the current state of centralization, also underscores the community's dedication to the ethos of decentralization and security.

References

1. W. Metcalfe, Ethereum, smart contracts, DApps. Blockchain Crypt Curr. **77**, 77–93 (2020)
2. C. Sguanci, R. Spatafora, A. Mario Vergani, Layer 2 blockchain scaling: a survey. arXiv preprint arXiv:2107.10881 (2021)
3. Z. Xu, L. Chen, L2chain: towards high-performance, confidential and secure layer-2 blockchain solution for decentralized applications. Proc. VLDB Endow. **16**(4), 986–999 (2022)
4. K. John, et al.: Economics of Ethereum. Available at SSRN 4783695 (2024)
5. W. Zhang, T. Anand, Ethereum architecture and overview, in *Blockchain and Ethereum Smart Contract Solution Development: Dapp Programming with Solidity*. (Apress, Berkeley, CA, 2022), pp. 209–244
6. Y. Faqir-Rhazoui, J. Arroyo, S. Hassan, A comparative analysis of the platforms for decentralized autonomous organizations in the Ethereum blockchain. J. Internet Serv. Appl. **12**, 1–20 (2021)
7. V. Buterin, A next-generation smart contract and decentralized application platform. White Paper **3**(37), 2–1 (2014)
8. K. Arshi, A.K. Goharshady, Congesting Ethereum after EIP-1559. In: 2024 IEEE International Conference on Blockchain and Cryptocurrency (ICBC). IEEE (2024)
9. X.J. Jiang, X.F. Fan Liu, Cryptokitties transaction network analysis: the rise and fall of the first blockchain game mania. Front. Phys. **9**:631665 (2021)
10. L. Marchesi, et al., Design patterns for gas optimization in Ethereum, in *2020 IEEE International Workshop on Blockchain Oriented Software Engineering (IWBOSE)* (IEEE, 2020)
11. DefiLlama. DefiLlama, www.defillama.com. Accessed 27 June 2024

12. A. Laurent, L. Brotcorne, B. Fortz, Transaction fees optimization in the Ethereum blockchain. Blockchain: Res. Appl. **3**(3), 100074 (2022)
13. S. Woo, et al., GARET: improving throughput using gas consumption-aware relocation in Ethereum sharding environments. Cluster Comput. **23**, 2235–2247 (2020)
14. M. Schäffer, M. Di Angelo, G. Salzer, Performance and scalability of private Ethereum blockchains, in *Business Process Management: Blockchain and Central and Eastern Europe Forum: BPM 2019 Blockchain and CEE Forum*, Vienna, Austria, September 1–6, 2019, Proceedings 17. (Springer International Publishing, 2019)
15. M. Spain, S. Foley, V. Gramoli, The impact of Ethereum throughput and fees on transaction latency during icos, in *International Conference on Blockchain Economics, Security and Protocols (Tokenomics 2019)*. (Schloss-Dagstuhl-Leibniz Zentrum für Informatik, 2020)
16. A. Singh, et al., Sidechain technologies in blockchain networks: An examination and state-of-the-art review. J. Netw. Comput. Appl. **149**, 102471 (2020)
17. A.M. Antonopoulos, O. Osuntokun, R. Pickhardt, Mastering the lightning network. (O'Reilly Media, Inc. 2021)
18. J. Poon, V. Buterin, Plasma: scalable autonomous smart contracts. White paper 1–47 (2017)
19. M. Armstrong, *Ethereum, Smart Contracts and the Optimistic Roll-Up* (2021)
20. O. Belz, Layer 2 ecosystem of the Ethereum blockchain. Науковий журнал «Економіка i регіон» **2**(93), 129–133 (2024)
21. M. Derka, et al., Sequencer level security. arXiv preprint arXiv:2405.01819 (2024)
22. M. Moosavi, et al., Fast and furious withdrawals from optimistic rollups. In: *5th Conference on Advances in Financial Technologies (AFT 2023)*. (Schloss-Dagstuhl-Leibniz Zentrum für Informatik, 2023)
23. L. Donno, Introducing stages—a framework to evaluate rollups maturity. *L2BEAT*, 19 June (2023)
24. Scaling Solutions Summary. *L2BEAT*. https://l2beat.com/scaling/summary. Accessed June 2024
25. I. Roşca, Alexandra-Ina Butnaru, and Emil Simion.Security of Ethereum Layer 2s. Cryptology ePrint Archive (2023)

Lorenz Raphael Lehmann Serves as the Research Lead at growthepie, a role that underscores his expertise in the web3 space and his commitment to tokenizing the future. He completed dual master's degrees in Chemical and Energy and Power Engineering from RWTH Aachen University in Germany and Tsinghua University in China, providing him with a solid educational foundation that spans two continents. Driven by a vision to level the playing field and enable equal access to financial technology for everyone, Lorenz is predominantly passionate about infrastructure through Layer 2 scaling solutions and prediction market designs. His commitment to web3 extends beyond building; he is also dedicated to educate the public on blockchain technology, believing that knowledge empowers community transformation. Lorenz's interest in web3 began during his university years when he founded the Aachen Blockchain Club. This student initiative at RWTH Aachen focuses on broadening the understanding of blockchain technology among students. Under his leadership as president, the club thrived, helping to establish a prominent web3 scene in Aachen.

Smoothing Blockchain Adoption with Abstracted Smart Accounts

Felix Hildebrandt

As blockchain technology steadily spreads into various sectors of society, its broader adoption barriers are becoming more present. Apart from all its benefits, limitations on the user experience include backups, protection, and payments, as well as the prevalent need for embedded user information and digital goods. This paper analyzes existing obstacles for decentralized economies and outlines solutions using abstracted smart contract standardizations for Ethereum-based blockchains. A derived development curve illustrates how a secure, efficient, and user-friendly experience evolves across different tech stacks. These new standardizations facilitate smoother onboarding and herald a new social and identity era of enhanced interoperability for decentralized ledger technology.

1 Barriers to Blockchain Adoption

Managing identity on a large scale requires recovery, the assignment of rights, and references to public data points. While offering self-sovereignty and censorship-resistant base layers, the security and convenience of wallets are often outsourced to tokens, external registries, or centrally governed environments. This phenomenon leads to a mixed nest of different account systems for each service, similar to Web 2.0. Before diving into the prevalent account standardizations, navigating the challenges of blockchain accounts is essential to understanding desired changes.

F. Hildebrandt (✉)
Chemnitz, Germany
e-mail: felix.hildebrandt@gmx.net

© The Author(s), under exclusive license to Springer Nature Switzerland AG 2025
W. Prinz and D. Trauth (eds.), *Tokenizing the Future*,
https://doi.org/10.1007/978-3-031-91405-8_8

1.1 Permissions and Backups

Regular accounts do not come with granular permission. Losing the wallet's seed phrase of a regular Externally Owned Account (EOA) means that stored assets become irretrievable, excluding further actions from the user account. This dilemma hinders user-friendly onboarding, known from regular web services, as individuals need extensive research on backup solutions.

Furthermore, each private key is endowed with immediate administrative privileges. To mitigate risks, users often distribute their assets across multiple EOAs. However, this strategy poses a significant challenge when it comes to reconnecting values and maintaining social interactions. Depending solely on a single, unchanging passphrase to safeguard an entire account or identity is a high-stakes gamble and is widely regarded as an anti-pattern in the industry.

Additional multi-signature applications like Safe [1] offer upgraded security when managing valuable assets by requiring multiple EOAs to execute an action. However, reducing the key's significance is primarily designed for having multiple groups or instances rather than individuals, as accepting everyday interactions around a single identity with varying personas feels out of place for user accounts.

1.2 Paid Services and User Protection

As a community of node operators run blockchain networks, there are costs associated with their use each time new data is written to the block. Immediately after setting up a wallet, users must connect to a crypto exchange to acquire coins and execute operations on the network. The verification on crypto exchanges can take days, among other things, until users are in a position to act. Here, indirect or token payments would foster onboarding. Service providers could pay for their customers' transactions and consider other compensation options. Making transactions "gas-less" and allowing optional payment methods is a new trend seen across crypto. However, it often correlates with the providers' custom server setups or secondary wallets, as its tech has yet to be integrated natively for accounts.

Even after the users receive their currency to power their transactions, there are no safeguards for incorrect data entries. Mistakes are irreversible, such as transfers to incorrect recipient addresses or wrongly typed amounts. On top of that, a significant concern is the lack of consent and spam protection. As everyone in the network can send transactions to any possible address, the openness is exploited for fraud. Due to transactions often being presented in a very technical way, inexperienced users might only sometimes understand the actions they sign. This opportunity resulted in a whole wave of social-engineering wallet exploits, where marketplaces like Opensea [2] had to implement ways to hide assets within the front end to reduce their direct exposure. The topic of protection is crucial when digital goods or certificates are account-bound, i.e., they cannot be transferred after their initial receipt. Counterparts should always be able to approve actions affecting their account to establish a healthy data economy that complies with the presented legal frameworks.

1.3 The Gap of User Information

The EOA structure, limited to a plain key pair without storage, starkly contrasts modern web platforms that enable shared user interactions. Unlike these platforms, the EOA cannot store or manage user-based information directly on the account. This limitation underscores the need for additional concepts beyond the protocol layer to establish an identity beyond the assets an account holds.

The limitation of the EOA structure is a reflection of the asset-focused development that sparked the NFT hype in 2021. While this approach fostered community building [3], it fell short of meeting people's integrational needs. The underlying technology, centered around assets, could not provide the desired level of engagement, leading many to revert to traditional web services.

The most widespread project aiming to add more context to addresses is the Ethereum Name Service (ENS) [4]. The idea of ENS is similar to today's web domains. Instead of typing the raw address of a website's server, browsers handle human-readable domains that resolve to the correct address underneath. As long as the domain is owned, the operator can refer it to any webpage. ENS provides similar functionality for blockchain addresses, linking names to various crypto addresses as long as the domain is owned via a subscription model. On top of that, the name can house text records like references to social media platforms, email addresses, profile pictures, or custom text entries.

However, once access is lost, a domain can not be restored. People can only acquire their names if their old subscriptions run out. The approach of renewal is what complicates identity management. If the subscription expires, all historical actions on the network dissociate from the original name, erasing the collected reputation and references. Individuals could acquire expired domains and inherit a user's historical identity associated with the name. Reputation and properties are most important, especially in social collectives or creative environments. Therefore, it is critical to link data to the active account directly.

1.4 Embedment of Digital Goods

Since the asset economy is not integrated into the accounts of the protocol, it is impossible to make data queries from them. Its associated contract has to be invoked individually to verify the ownership of a particular token. Even if a project's platform is open source and deployed on the blockchain, many services predominantly rely on centralized backends to grasp account information. To comprehensively view an account's assets, services must go through complex and continuous retrieval operations using node machines or centralized block explorers and data services like Etherscan [5].

While smaller decentralized applications can avoid external data points by statically integrating their associated or allowlisted addresses, larger platforms face a dilemma. They need comprehensive asset information without overburdening their technical infrastructure or making demands on their users. This often leads to reliance on external providers, which in turn intensifies the dependency on centralized entities and introduces vulnerabilities, including potential system outages or errors. This situation calls for a more robust and secure solution.

Having an integrated data point makes reading broad user information manageable. Because an account can own assets without having them in their transaction history, the above steps also involve storing and maintaining the results, as services could not retrieve the data ad hoc. A service has to perform the following steps to fetch owned assets:

1. Get the account addresses of an identity.
2. Check asset balances by performing a data analysis against the accounts.
3. Pass all owned assets back to the application interface.
4. Call every resulting contract address for their data.
5. Embed contract data into the service's front end.

Figure 1 illustrates the complex flow of broad asset data fetches. To gain shared ownership insights or fetch NFT issuers, further external checks are necessary. However, if accounts seamlessly integrate into the on-chain asset economy, operations can be kept within the application layer, enhancing security through a chain of smart contract calls. Such integration could also facilitate streamlined consent for asset transfers, offering a more efficient and secure way of managing assets.

Another lack of embedment can be seen in the assets, relating to how the valuable data is attached to the smart contract. Typically, the token's data is not stored within the smart contract. Instead, it refers to an off-chain file often saved in a JSON format. This full off-chain referencing comes with significant limitations. Notably, smart contracts cannot read these references, making them unsuitable for any other contract built to work with those datasets.

Additionally, data references use URLs rather than direct cryptographic hash references, which ensures the data remains unaltered. Future asset standards should incorporate hash references in the contract storage, enabling consistent data verification, irrespective of where it is stored. The compound would allow updating the asset's storage solution while verifying the original data, making decentralized assets more sustainable.

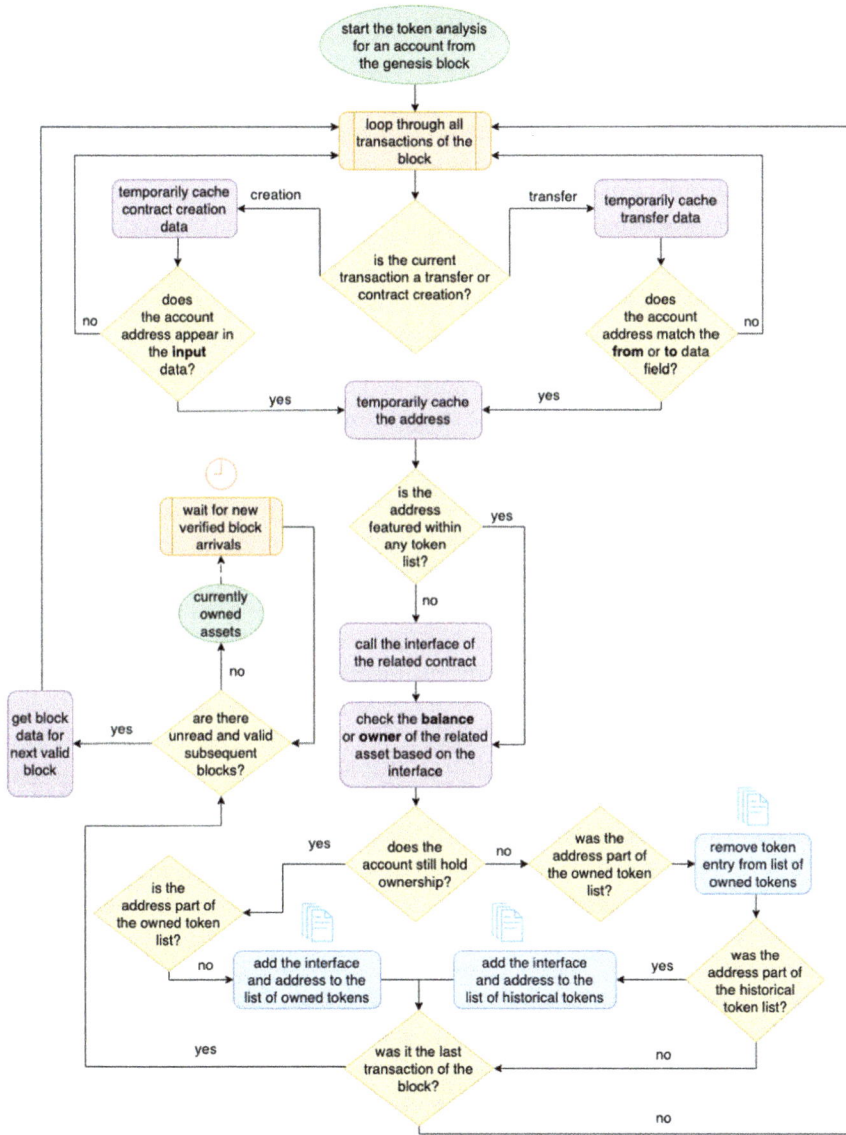

Fig. 1 Verifying ownership of unknown assets

1.5 Navigation of User Interaction

Accounts serve as a pivotal anchor for digital communication, much like they have always been in traditional Web 2.0. However, while transitioning into the decentralized domain, creating modern identities using regular EOAs becomes a challenge. Table 1 classifies the core data economy obstacles.

These challenges could be addressed entirely at the application layer using smart contracts. However, the associated costs and possible scalability must be considered. If something is exclusively on the application layer, it fosters custom implementations without direction while increasing complexity over contracts and off-chain server infrastructure. The topics of payment, integration, and permission should be embedded in the protocol long term to improve the network's base functionality. On top of that, specific accounts could be designed with the needs of the user base in mind.

It's worth noting that every transaction processed through a smart contract is more expensive than a direct EOA transfer. The more functionality a transaction triggers, the higher the execution cost. To keep future cost increases within acceptable limits, it's crucial for open-source smart contract development to adopt generic standardizations. In the future, platforms could potentially mitigate costs by implementing indirect payment methods, a concept familiar to the current web, where the user indirectly pays for services through advertising or other means.

Table 1 Analyzed data economy obstacles

Category	Challenges	Affected blockchain layer
Permissions	No granular roles	Application
	Immediate administrative privileges	Protocol/Application
Payment	No indirect payment	Protocol/Application
Backups	Static backup phrase	Protocol/Application
Integration	Heavily complex data analytics	Protocol/Application
	Assets are separated from the account	Protocol/Application
	Asset verification relies on storage	Application
Protection	No safeguard for on-chain actions	Application
	Lack of spam protection	Application
	No consent on the transaction	Application
Description	No way to attach account data	Application
	Sole asset-focused economy	Application

2 Abstraction of EVM Accounts

As blockchain accounts have faced various barriers for nearly a decade, multiple solutions have been outlaid. Vitalik Buterin, one of the Ethereum founders, states that EOAs are a main constraint for mass adoption, and the abstraction of accounts remains one of the industry's most important topics [6].

Account abstraction is a continually evolving area in the Ethereum ecosystem aimed at making accounts more flexible, secure, and feature-rich. While not a single standard, it is a process that has been in active development since 2016, involving multiple Ethereum Improvement Proposals (EIPs) and Ethereum Request for Comments (ERCs), covering both network or contract-related standardizations. Each submission has features that contribute to a more robust account system.

In summary, account abstraction tries to standardize how the body of future blockchain accounts should be set up and what transaction flow they follow. The fundamental challenge of it is the verification of transactions by network operators. Traditional EOAs have the advantage that they allow for straightforward verification because a transaction is guaranteed valid if parameters like balances and computation prices are met and signed by the address key. In addition, EOAs offer a unique nonce value that can be tracked to determine the transaction order. However, this verification mechanism gets even more complex if the account can no longer be assigned to a single operator. Within the field of account abstraction, the goal is to make this verification more flexible, allowing for multiple signing keys and validation schemes for a single account, while maintaining safety standards across the existing setup.

2.1 Historical Standardizations

Introduced in early 2016, EIP-86 was one of the first glimpses at account abstraction on Ethereum. The protocol and EOAs regularly use a static verification method, ECDSA, for every action of its whole network. The proposal made it possible to abstract away signature verification and nonce checking from EOAs by forwarding them to the custom logic of smart contracts. The most significant take was that these contracts could become accounts directly handling such operations, making transaction processing obsolete and paving the way for greater functionality to bundle actions, allowing for indirect payment, or upgrading the signature algorithm for security without being tied to the network [7]. After a lack of attention and demand, it stayed inactive while the concept was revisited in subsequent standardizations.

ERC-725, proposed in late 2017, provided an interface for smart contract-based accounts, allowing users greater control via exchangeable EOA keys while keeping the regular account unchanged. The initial proposal included a preview of a manager to regulate which key could do what and how claims could be attached. Later on, these functionalities split into separate standards: ERC-734 [8] and ERC-735 [9]. The vision was that smart accounts could include features like social recovery and greater security for assets while abstraction does not have to be solidified within the protocol. The concept allows developers to work on feature-rich account ecosystems to sense possibilities and transaction verification in parallel. One of the proposal's highlights was the option to attach storage, providing a potential basis for identity and profile management on-chain. While the EOA controllers can rotate, these identities could have set backups on hardware wallets or instances for limited access [10]. This flexibility allows for improved onboarding, where a service provider initially creates a profile, but the user can take control afterward. The standard resulted in the ERC-725 Alliance [11] that officially developed and prototyped it. Due to the lack of features and structure at the time being, the majority of identity moved on to verifiable credential solutions. However, development beyond ERC-725 continued to build on social economies.

Midway through 2018, ERC-1271 provided a standard method for verifying smart contract signatures. While signing could regularly be done on EOAs due to their private key, contract accounts could not interact with applications utilizing these signatures. Usually, smart contracts had to rely on cumbersome and potentially less secure methods to validate signatures, often requiring off-chain verification or complicated on-chain logic. The standard ECDSA signatures often resulted in multiple callbacks to the EOA, breaking the validation chain on the backend. The complexity especially presented a barrier to seamless interactions for multi-signature wallets when granting rights for asset transfers. The proposed improvement with the standard signature validation allowed smart contracts to authenticate signatures directly through a single function, simplifying the architecture of decentralized applications and keeping the security measures within contract-to-contract interactions. The verification method has been particularly beneficial for multi-signature wallets and identity development [12]. The standard was the first within the account abstraction field to become final and greatly expanded contract communication for various projects.

EIP-2938 later sought to redefine user and contract interactions by introducing a new category of abstracted transactions in mid-2020. The proposal would allow contracts to initiate and cover the cost of transactions, much like EOAs, paving the way for more versatile and user-friendly account functionality. This standardization was the first abstraction concept focusing on transactions instead of the account itself. While it was a big step forward for accounts with relatively small protocol interventions, the standard remained inactive due to the unclear direction of accounts [13]. However, the concept continued to remain as a prototype idea that was incorporated into later standards. One of the key benefits of going with a new transaction type is that the protocol ensures they meet validity criteria before even sending it off to the network's transaction pool. In an ideal solution, abstraction must tackle both transactions and account specifications. Otherwise, transactions could spam the pool while waiting for a smart contract to validate them.

As one of the final milestones in the history of abstraction, EIP-3074 proposed another protocol operation, allowing users to delegate control of their EOA to smart contracts. The outlaid structure allowed regular accounts to behave like contracts without changing the EOA framework. The benefit of introducing a new mechanism, rather than creating a new transaction type, as in EIP-2938, is that it does not require any changes to existing wallets and decentralized applications. However, both solutions have their negative points. The idea of EIP-3074 was similar to ERC-725, but instead of keeping EOA controllers, they could envoy rights to contracts themselves. These could then initiate transactions on behalf of the user. Here, transaction fees can be paid from an account different from the initiator. On top of that, it would come with greater flexibility, as sponsored transactions with custom tokens could be used as refunds [14]. Because the original EOA framework remains unchanged, the initial key retains immediate administrative privileges even if unused, potentially posing a safety risk. This issue was tackled by combining the proposal with the extension of EIP-5003 [15], which makes it possible to revoke any previous signing key of an EIP-3074-based account directly in the protocol. Both proposals have already become part of the core standards track of Ethereum and are likely to be used within new account changes.

2.2 Account Imbalances

Based on the history of standards, it is apparent that the issues surrounding signing and account features represent two distinct challenges for account abstraction. On the one hand, EOAs are the exclusive account entities that can sign. On the other hand, smart contracts bring all the functionality. So far, EOA controllers for smart contract accounts, as described by ERC-725 or used within multi-signature applications, still represent the state-of-the-art functionality but are often seen as second-class citizens within operational networks that mainly onboard users through EOAs. When updating the network's account system, the network's entry point should be consistent across all users, while backward compatibility with the previous EOA system is guaranteed. Such considerations are necessary for the risk of dividing user groups. In this regard, abstraction is also a matter of equality.

Due to EOAs serving dual roles of account management combined with signing, it is not easy to introduce new features without altering the protocol. The signing functionality should be separated from the account frame to tackle both problems modularly. Such a split allows for independent development of account features while keeping ownership rights from the cryptographic keys that are already widespread.

2.3 ERC-4337 Abstraction

Based on the smart contract limits outlined within account imbalances, the Ethereum network is on the verge of a pivotal transformation in account management, brought about by the standardization of ERC-4337. The proposal aims to revolutionize how accounts are treated in the network to spread the adoption of smart contract wallets.

Hence, it is positioned as the definitive framework for realizing account abstraction and is already part of Ethereum's official standards track [16].

The concept outlines how the previously described separation of signer functionality and account can be integrated into the application and protocol layer. As the split acts as an ideal foundation, the proposal also includes features from previously stagnated standards such as token payment, bundling, or custom signature schemes.

On the technical side, ERC-4337 introduces a new architecture where every account becomes a smart contract deployed via a factory contract during its initial transaction. Unlike traditional Ethereum accounts, which derive their addresses from a single private key offline, these new abstract accounts have addresses generated at the application level using wallet-specific data [16]. This update paves the way for greater functionality and security.

The proposal replicates the behavior of existing EOAs to ensure backward compatibility. Instead of immediately generating an address on account creation, a signer key pair is handed out without any on-chain deployment. Afterward, the public key can be checked against a smart contract interface to obtain the on-chain account address. This call can be done immediately after the initial setup to receive funds at the address. The account, however, can be fully deployed later on whenever their first transaction is sent using this original key.

One of the most noteworthy features is the account flexibility once deployed. Since the signature scheme is no longer hard-coded into the account, transactions can carry multiple or custom types of signatures. The enhancement allows an account to become a native multi-signature wallet, where multiple keys can be added or removed while keeping the same address. The static anchor point mainly solves the hurdles of identity management and reputation. However, freedom of choice regarding signature schemes also enables multi-calls in applications, where various user interactions from different smart contracts can be executed within a single transaction. It is a powerful feature for saving network fees and facilitates the development of more feature-rich but efficient blockchain-based applications. These new signatures will be validated on-chain, adding an extra layer of security. Consequently, signatures must also be validated on-chain as described within ERC-1271 rather than checking the ECDSA key offline.

To handle the enhanced functionalities, ERC-4337 introduces a new type of transaction. The so-called user operations have a separate, higher-level transaction pool within the network. The separate pool ensures that the new transaction type does not interfere with traditional transactions and the block building. The separation is particularly needed as the new signature schemes will bundle multiple user operations into a single transaction [17], which is later sent to the main pool [18]. The bundling reduces overall network traffic and costs, thereby enhancing scalability.

Within the separate network pool, a new entity of a bundler plays a crucial role. The bundler collects user operations, packages them into standard Ethereum transactions, and covers the transaction fees. The bundled transactions are sent to a single entry point smart contract for validation. The entry point extracts each user operation from the transaction, validates them separately by calling the abstracted user accounts, and checks their signature, nonce, and ownership. Upon success, the entry point executes the user operations and initiates refunding the bundler fees from the user accounts based on their used computation.

The bundler can be upgraded with a separate smart contract paymaster module for more versatile payment options. The module allows the bundler to specify a particular token that can be used for the refunding process. Before aggregating user operations into a single transaction, the bundler's system verifies that the user accounts have sufficient balances and funds the paymaster contract with the network's native coin. Instead of using its wallet to cover the transaction costs, the bundler then points to the funded paymaster contract to handle the network fee for the transaction. In return, the paymaster contract demands its refunds as ERC-20 [19] compatible tokens previously allowlisted in the transaction metadata. Within the verification phase, the entry point checks the balance on the paymaster to ensure it has enough funds to complete the transaction. It then waits for the paymaster to finish its internal token validation process again. If everything is correct, the entry point regularly executes the user operations. It then triggers the paymaster to deduct the appropriate token amount from the user accounts to cover the computation costs incurred. This payment mechanism is designed so the paymaster cannot act in bad faith. The entry point contract controls the initiation of the payment or refund, ensuring a secure and trustworthy transaction process.

ERC-4337 is designed to operate both at the application and protocol levels. Initially, off-chain bundlers or paymasters will be standalone servers facilitating transactions to nodes. In this case, every server maintains its separate pool of user operations. While the application layer approach functions correctly, it can come with multiple downsides. The largest providers would have the most extensive user operations to choose from, bundling more efficiently with lower refund fees and effectively marginalizing others.

On top of that, setting up off-chain servers can be complex and carry the risk of centralization if a few prominent bundlers dominate the market. Therefore, these roles are imagined to be integrated into the protocol in later stages. The embedment would mean nodes becoming bundlers or paymasters while running the regular network. The protocol integration would help keep the network fair and allow for a bundling economy inside the network, improving speed and operation costs.

Alongside optimization, another reason for the protocol embedment is the unification of accounts on Ethereum, aiming that every account becomes abstracted. Without the change, they will only be used for a network subsection, remaining as second-class members in the ecosystem designed for EOAs. As ERC-4337 can emulate similar functionality more modularly, the standard eliminates the need for a separate account and address generation. With the protocol integration and simultaneous removal of EOAs, every single user on the network could be onboarded in a more user-friendly way, only requiring the plain cryptographic keys on the back. The freedom of choice makes it suitable to test and further specify the concept within live environments before statically embedding it into the EVM protocol when adoption is present.

3 Insights into Proxy Ecosystems

Like the development of account validation mechanisms, management solutions operating on top of them have also developed steeply. Where account upgrades focus on token payment and rotatable keys, smart contract ecosystems help to add functionality using interconnected standardizations.

As mentioned in the history of standards, the concept of ERC-725 impacted its developer alliance, which has been researching rotatable EOAs and identity management for years. Instead of being heavily involved within ERC-4337, the focus was mainly on outlining what can go beyond and solve additional missing features. Over the years, Fabian Vogelsteller, creator of ERC-725, further fine-tuned modular components to enhance the base entity model. The primary standard was divided into two parts: One for executing program code or creating new smart contracts and the other one for defining an expandable storage list. This list exists as a parameter and can be filled with data elements like links or VCs, making the handling extremely flexible without changing the contract itself.

Many projects within the ERC-725 Alliance focused on core identity solutions for finance regarding uncollateralized loans, investments, KYC, or better custody. For instance, ERC-3643 outlaid a solution for permissioned issuance. Nevertheless, Fabian Vogelsteller and the LUKSO project built a fully standardized ecosystem for social and creative applications, utilizing the vast potential for a drastically new economy beyond DeFi. The project and related proposals then became the main driving force for proxy smart accounts and actively fostered the development of developer libraries to interact with and decode information about such accounts.

3.1 Standardized LSP Ecosystem

The history of account abstraction reveals complexities in synchronizing developer community efforts, especially when multiple interdependent standards emerge simultaneously. LUKSO's approach, with its unique standard proposals called LSPs, compatible with the EVM, aimed to circumvent these issues by developing several standards concurrently, separate from Ethereum's ERCs.

The standard of ERC-725, translated into LPS0, represents the foundation of a smart account [20]. Beyond others, it utilizes ERC-1271 signature validation for abstraction. By combining the account with the LSP1 Universal Receiver standard, this account can receive notifications about incoming and outgoing actions. Developers can attach custom flows and behaviors to these on-chain events. An example would be rejecting or approving certain digital goods or currency transactions, solving the issue of consent. If the recipient does not accept certain payments, they could be sent back to the original address [21]. Redirects or blocklists to reduce spam are also conceivable, as actions can be delegated.

The Universal Receiver standard can then be bundled with LSP5 Received Assets [22] and LSP12 Issued Assets [23], a storage framework to directly see the account's owned or issued digital goods. When the asset is entirely spent or sent, the received assets list is directly updated within the call. This feature is a milestone for the infrastructure of data economies in decentralized networks and a practical solution to a common problem. As described within the main adoption barriers, the collective viewing of currencies was previously only possible for centralized services that scan the blockchain and display transactions in a readable format. Now, smart contracts can act completely decentralized by directly querying the addresses of accounts with a simple call of the ERC-725 developer library instead of running a node or connecting to complex and centralized off-chain systems, providing practical benefits for networks and their data economies.

Another critical point is the LSP6 Key Manager, which evolved from ERC-734, giving the EOAs different roles and rights. Before, simple accounts could only cover full access and offered little security for managing content. Even ERC-4337 only provides a base concept for admin key rotation, leaving additional permission for developers. The key manager standardization is a suitable answer. By default, the proposal gradually offers nine permissions across signing, transfers, and ownership. When acting through the account, the key manager in front of it is called first to check if the controller has the correct permissions. If access is granted, the transaction can be executed [24]. As everything is modular, the Key Manager interfaces are exchangeable and can even be attached to other smart contracts to enable tokens with governance.

One of the LSP standards also specifies the memory register for ERC-725 accounts, as the storage is only useful if one knows how to access and interpret it. The generic LSP2 JSON Schema ensures that metadata within the account is both readable and writable in an automated way [25]. It sets the guidelines for attaching information, such as asset ownership or file references. Another criticism analyzed within the blockchain adoption problems was that current goods on a blockchain can be verified, but often not their attached data. To this end, the attached data has a hash key indicating whether the original information is still unaltered for any storage key.

The storage concept can also add LSP3 Profile Metadata to the smart account. This enhancement turns a basic anonymous account into a Universal Profile that attaches publicly viewable information directly to the account, similar to how people are used to within the current social media landscape [26]. Since this metadata link is stored directly in the account's storage, the profile can be used for any action on the blockchain without creating a new profile for each application. Similarly, external services can write custom information for storage as well.

The approach of the Universal Profiles can be cited as "public first, private second," as the account can already have general information as in regular social media apps. Later, services will utilize the convenient structure underneath to dock private claims onto them or even stay fully anonymized. As David Silverman said in one of his talks: "If you build a project for private purposes first, it is locked in, and going public would not be an option. So being public and flexible is better for placing directions and power into developers' hands, not making decisions for them" [27].

A similar LSP4 Digital Asset Metadata standard brings equal features for exchangeable or unique assets on the blockchain, allowing for updatable metadata. Here, multiple sources of information, like media files, names, or descriptions, can be attached [28]. Even profiles of numerous artists can be linked directly to solve the current reputation problem of EOAs. Added profiles could then quickly enter the data economy of the digital good by utilizing royalties or reputation. If there is a link with the key manager, rights assignments for subsequent content modification can also be implemented for assets. The modularity of the ecosystem comes in handy, as the new asset standards, namely LSP7 Digital Asset [29] and LSP8 Identifiable Digital Asset [30], can all combine previous concepts to enhance regular Tokens and NFTs. Among other things, these include notifications, the batch transfer of goods, links to the creators, and updatable datasets. Figure 2 demonstrates how the standardized features come together to an identity architecture with the interwoven lifecycle of digital values.

In the future, a lot of data traffic will likely happen through such new smart accounts, adding a layer of complexity to asset navigation. Especially active blockchain users need help managing many digital goods. The issue was tackled as separate standards, LSP9 Vault [31] and LSP10 Received Vaults [32]. Here, accounts can create multiple different smart contract instances that behave like subfolders of a profile to organize data and assets. Not only can they be used to sort a person's possessions or provide additional security, but they also allow trusted applications to open or write into specific profile subfolders. While doing so, users can enjoy applications without signing each action individually.

The concept of shared management but self-sovereign ownership can be bundled with a separate standardization for relay services. In these, transactions are not sent directly from the account to the blockchain but are passed to an external service

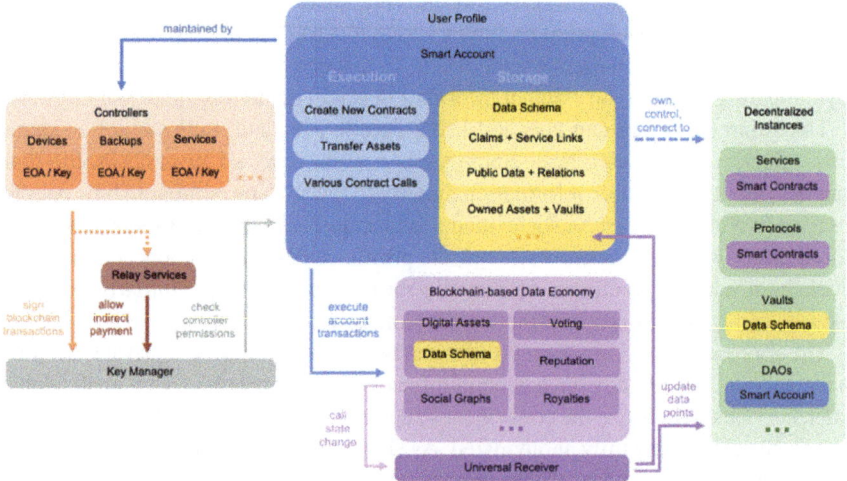

Fig. 2 Abstract LSP ecosystem flow

that executes them on behalf of the user. Bundlers and paymasters make a similar approach without their protocol integration. Here, LUKSO has brought its own RPC API to let services remotely pay for the transactions while remaining protected by cryptographic signatures. Users can then utilize indirect payments until fully integrated into the EVM. There will likely be a new market that will allow users to participate in the network entirely without their crypto assets. These concepts are similar to mobile phone providers' monthly data volume. In return, users can use advertising, free-to-play concepts, or subscription models.

3.2 Lightweight Abstraction Approaches

While LUKSO is focused on developing interoperable standards for the core accounts of the future, numerous other initiatives focus on protocol or wallet-specific features regarding proxy execution. For instance, Argent X [33] is actively developing a 4337-compatible web wallet that includes proxy fraud detection with double signatures, definable spending limits for held tokens, and cookie-based keys for web pages where a wallet is created and can already sponsor their first transactions. After a certain amount of time or asset value, the user can be asked to overtake the created wallet. The adaptive design could be a significant breakthrough for in-browser sessions to manage users on the go, which may take their belongings into self-custody after some value generation. Games could profit from session keys to manage in-game items within a wallet. On top of that, it has a guardian functionality to create backup keys with a time gate to restore the account.

Another on-chain approach for profiles was laid out by the Lens Protocol in 2022, using an NFT as a social media profile. The protocol focuses on managing modular and decentralized social graphs where users can take their followers, reputation, and interactions across any app building [34]. While a single EOA still holds everything, the NFT can reference a profile manager smart contract to delegate social actions to a different wallet. The composite approach would mean a profile could safely be stored using browser wallets with less security that can only execute certain operations. However, every functionality is specifically for its social media protocol. As an indirect payment, a second proxy wallet, called a dispatcher, can be sponsored with the network's native coin to pay for every transaction. The intermediate wallet then acts as the signer for every transaction going through the Lens Hub contract. With the delegated signing privileges to the dispatcher, every transaction can be forwarded without additional signing on the user's front end [35]. Still, operators must acquire platform-independent values funds first. Compared to Universal Profiles, there would be a separate profile and token payments for each protocol instead of a global account solution. The concepts could even be combined, having a universal identity but distinct profiles for every central area of digital life.

3.3 Cross-Layer Network Evolution

Beyond the described proxy solutions, there has been a trend toward using subordinate or distinct networks. Argent X is building on StarkNet [36], which is focused on scalable and privacy-preserving transactions. Lens Protocol runs on Polygon [37] and has its separate Momoka network [38] to map all social interactions. As outlined in user interaction barriers, smart contract operations are more expensive than regular network transactions. Here, speed and price are the main reasons for using L2 solutions [39] that dock onto Ethereum instead of using it directly. Costs are the reason why Ethereum is focusing on embedding bundling for user operations, as regular costs related to abstraction would not be eligible by default.

In addition to the costs, there is another reason why LUKSO has established its standalone network. If abstracted contracts interact with regular EOA wallets, the counterpart will hinder convenience as it can not adapt to functionality. However, the problem of the second user class does not arise if each user is given an abstracted account directly upon entry. By instantiating a new network, the entry can be done directly via the abstracted LSP0 account to use all features. For existing EOA networks, the counterpart would be the protocol integration of account abstraction.

StarkNet, for instance, is a more scalable bundling network that uses a direct protocol-based abstraction model. While still able to dock onto Ethereum, they already got rid of EOAs, meaning that contracts are derived from the user's keys and created implicitly. After users compute a contract address from their keys, they can already receive funds that can be accessed after the related smart contract is deployed by using the received coins. Here, every developer can directly deploy various custom account features into the account's core while complying with the protocol standardization [40]. Users can later sign transactions using their private key, whose signature is checked against the generated smart contract address and its protocol-based nonce [41]. The network presents a semi-integrated abstraction that is less extensive than Ethereum's ideas.

Compared to LUKSO's Universal Profiles, developers working with StarkNet have the flexibility to embed features directly into the core and determine their structure and appearance. However, this also means that the user experience and security can heavily vary from wallet to wallet, and interoperability will not exist across the board. In the future, networks will try to enrich their abstraction design, similar to the features of ERC-4337.

With such separate standalone networks for different economies in the blockchain field, general scalability for mainstream adoption can be securely raised. Additionally, early users can be subsidized through lower network participation. Subsidization and entry point convenience could not work out on an already settled network.

However, L2 solutions have drawbacks. Governed by multi-signatures, these subnetworks employ a relatively centralized way of constructing blocks, compromising asset security. They can also act as data silos that trap users' assets due to high transfer costs or bottlenecking during high-demand periods. Furthermore, only strict assets can be easily transferred, while social graphs and interactions remain largely unportable [42]. Where L2s provide an excellent solution for purpose-driven environments and first adoption, they do not seem fitting for generic accounts that act as the center of a user's decentralized identity. Here, abstraction has to be awaited on more robust, primary networks.

3.4 Curve of Smart Account Adoption

According to Vitalik Buterin's remarks at the 2022 Ethereum conference in Bogota, the integration of ERC-4337 could be a complex process, potentially taking more than half a decade to implement fully [43]. It has yet to be decided how exactly EOAs will be removed. Two solutions would be modifying the transaction, a concept leaning on EIP-2938, or the account ruleset outlined in EIP-3074 and EIP-5003.

The entry point contract of ERC-4337 was already audited and deployed on the Ethereum mainnet in March 2023, validating bundled user operations. However, the adoption of ERC-4337 is expected first on subordinate L2 or unoccupied networks. However, there already has been another proposal, EIP-7702, similar to EIP-3074, to offer similar functionality for the time the recent ERC-4337 is not fully implemented within L1, due to its complexity.

While broader account solutions have been gaining traction, utilizing EOAs on their backend, it's important to understand the limitations of plain account abstraction. While it's a step in the right direction for the blockchain space, it's unable to compete with the contract systems being built today, with LSPs being the figurehead. This realization paves the way for a seamless transition to pure signing keys, a move that could enable optimization in the future.

A similar combined approach can be applied to today's relay solutions for indirectly paying transactions. Such servers could switch to become bundlers or paymasters directly in a much more convenient way. Here, both network and application development act in a supportive manner by already rudimentary showing what would be possible to implement and simplify account management later on strategically.

Previous trends can be directly transferred into Table 1 to see delimitations. Table 2 shows how backend infrastructure like payment, backups, and scalability are separated from the individual data exchange problems directly facing the users. The bottom line, however, is that these run parallel to overcome analyzed data-economy problems.

The integration graph shown in Fig. 3 elaborates on the path taken by different account systems. Currently, EVM users start with a pure protocol and key-based account with low costs and functionality. Over time, these accounts can be enriched with additional features via smart contracts, elevating their mainstream functionality while moving to the application layer. These smart contract-based accounts can be further optimized and integrated with transaction bundlers, reducing transaction costs to enhance scalability. The goal is to effortlessly manage all activities involving

Table 2 Leading decentralized data economical solutions

Category	Challenge	Standardized solution
Payment	No indirect payment	Relay service APIs, Account abstraction (Mainly ERC-4337)
Scalability	Static Signature Schemes	
Permissions	No granular roles	
	Immediate administrative privileges	Smart contract ecosystems (Mainly LSPs), Asset and security features (EIPs/ERCs)
Protection	No safeguard for on-chain actions	
	Lack of spam protection	
	No consent on the transaction	
Description	No way to attach account data	
	Sole asset-focused economy	
Integration	Heavily complex data analytics	
	Assets are separated from the account	
	Asset verification relies on storage	

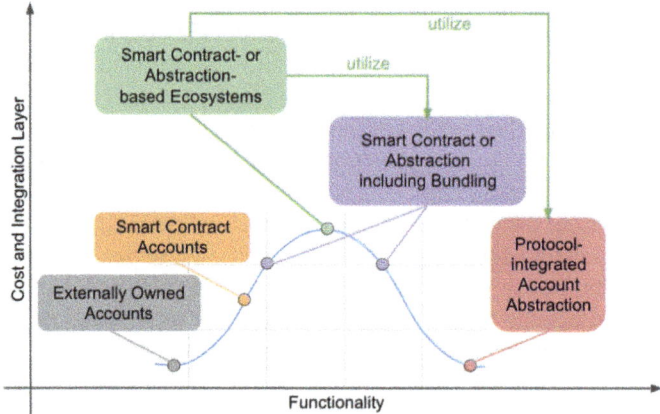

Fig. 3 Evolution of abstraction levels and integration layers

accounts directly within the protocol while maintaining previously acquired functionality. The unification would result in a fair and decentralized network with unified user onboarding.

3.5 Summative Insights

Account abstraction standardizes the body of future blockchain accounts and their validation principles. The almost decade-long evolution of abstract smart accounts is trying to unleash the full potential of blockchain accounts, fostering the next wave of adoption. Here, the focus points are modular transaction verification, indirect payment, and updatable permissions, making accounts operation centers for entities of any kind.

Account Abstraction is divided into two main areas: updating the account's frame and managing systems on top of the account. Several prototype wallets have already attempted to integrate rotatable keys, bundling, and token payments into extensions, focusing mainly on the frame and existing DeFi world. More interesting, however, will be upcoming social environments utilizing functionality for profiles and reputation. Here, the most diverse management standardization ecosystem within the Ethereum ecosystem is called LSPs, where on-chain accounts can become universal profiles with rich context, embedded asset interactions, and convenience features known from regular social media platforms.

References

1. Safe Wallet, Web3 Account abstraction developer stack. Gnosis. Accessed 30 May 2024. [EVM Smart Contracts]. Available: https://safe.global/
2. A. Fauvre-Willis, Hiding suspicious NFT transfers on OpenSea. OpenSea. Accessed 30 May 2024. [Online]. Available: https://opensea.io/blog/articles/hiding-suspicious-nft-transfers-on-opensea
3. S. Casale-Brunet, M. Zichichi, L. Hutchinson, M. Mattavelli, S. Ferretti, The impact of NFT profile pictures within social network communities, in *GoodIT: Proceedings of the ACM Conference on Information Technology for Social Good*. (Association for Computing Machinery, New York 2020), pp. 283–291. Accessed 30 May 2024. [Online]. Available: https://doi.org/10.1145/3524458.3547230
4. "Ethereum Name Service." ENS. Accessed 30 May 2024. [EVM Smart Contracts]. Available: https://ens.domains/
5. Etherscan. Etherscan. Accessed 30 May 2024. [Ethereum Block Explorer]. Available: https://etherscan.io/
6. V. Buterin, A history of account abstraction, in *EthCC* [6], Paris, July 18, 2023. Accessed 30 May 2024. [Online]. Available: https://www.youtube.com/live/iLf8qpOmxQc?si=_C5fOLPH84DZe5qo
7. V. Buterin, EIP-86: abstraction of transaction origin and signature (2017). Accessed 30 May 2024. [Online]. Available: https://eips.ethereum.org/EIPS/eip-86
8. F. Vogelsteller, ERC-734: key manager (2017). Accessed 30 May 2024. [Online]. Available: https://github.com/ethereum/eips/issues/734
9. F. Vogelsteller, ERC-735: claim holder (2017). Accessed 30 May 2024. [Online]. Available: https://github.com/ethereum/eips/issues/735
10. F. Vogelsteller, T. Yasaka, ERC-725: General data key/value store and execution (2017). Accessed 30 May 2024. [Online]. Available: https://github.com/ethereum/eips/issues/735
11. ERC-725 Alliance, Ethereum Identity Standard. Accessed 30 May 2024. [Online]. Available: https://erc725alliance.org/
12. F. Giordano, M. Condon, P. Castonguay, A. Bandeali, J. Izquierdo, B. Masius, ERC-1271: standard signature validation method for contracts (2018). Accessed 30 May 2024. [Online]. Available: https://eips.ethereum.org/EIPS/eip-1271
13. V. Buterin, A. Dietrichs, M. Garnett, W. Villanueva, S. Wilson, EIP-2938: account abstraction (2020). Accessed 30 May 2024. [Online]. Available: https://eips.ethereum.org/EIPS/eip-2938
14. S. Wilson, A. Dietrichs, M. Garnett, M. Zoltu, EIP-3074: AUTH and AUTHCALL opcodes (2020). Accessed 30 May 2024. [Online]. Available: https://eips.ethereum.org/EIPS/eip-3074

15. D. Finlay, S. Wilson, EIP-5003: Insert Code into EOAs with AUTHUSURP (2022). Accessed 30 May 2024. [Online]. Available: https://eips.ethereum.org/EIPS/eip-5003
16. V. Buterin et al., ERC-4337: account abstraction using alt mempool (2021). Accessed 30 May 2024. [Online]. Available: https://eips.ethereum.org/EIPS/eip-4337
17. Privacy and Scaling Explorations, BLS Wallet: Bundling up data, Medium. Accessed 30 May 2024. [Online]. Available: https://medium.com/privacy-scaling-explorations/bls-wallet-bundling-up-data-fb5424d3bdd3
18. Ethereum Foundation, "ERC-4337 Documentation." Accessed 30 May 2024. [Online]. Available: https://www.erc4337.io/docs
19. F. Vogelsteller, V. Buterin, ERC-20: token standard (2015). Accessed 30 May 2024. [Online]. Available: https://eips.ethereum.org/EIPS/eip-20
20. F. Vogelsteller, LSP0 ERC725 account (2021) Accessed 30 May 2024. [Online]. Available: https://github.com/lukso-network/LIPs/blob/main/LSPs/LSP-0-ERC725Account.md
21. J. Carvalho, F. Vogelsteller, LSP1 universal receiver (2019). Accessed 30 May 2024. [Online]. Available: https://github.com/lukso-network/LIPs/blob/main/LSPs/LSP-1-UniversalReceiver.md
22. F. Vogelsteller, LSP5 received assets (2021). Accessed 30 May 2024. [Online]. Available: https://github.com/lukso-network/LIPs/blob/main/LSPs/LSP-5-ReceivedAssets.md
23. F. Vogelsteller, LSP12 issued assets (2022). Accessed 30 May 2024. [Online]. Available: https://github.com/lukso-network/LIPs/blob/main/LSPs/LSP-12-IssuedAssets.md
24. F. Vogelsteller, J. Cavallera, LSP6 key manager (2021). Accessed 30 May 2024. [Online]. Available: https://github.com/lukso-network/LIPs/blob/main/LSPs/LSP-6-KeyManager.md
25. F. Vogelsteller, LSP2 ERC725Y JSON schema (2020). Accessed 30 May 2024. [Online]. Available: https://github.com/lukso-network/LIPs/blob/main/LSPs/LSP-2-ERC725YJSONSchema.md
26. F. Vogelsteller, LSP3 profile metadata (2020). Accessed 30 May 2024. [Online]. Available: https://github.com/lukso-network/LIPs/blob/main/LSPs/LSP-3-Profile-Metadata.md
27. D. Silverman, Building on lens protocol, Amsterdam, Mar 26, 2022. Accessed 30 May 2024. [Online]. Available: https://youtu.be/usgYL-KYd7I
28. F. Vogelsteller, LSP4 digital asset metadata (2020). Accessed 30 May 2024. [Online]. Available: https://github.com/lukso-network/LIPs/blob/main/LSPs/LSP-4-DigitalAsset-Metadata.md
29. F. Vogelsteller, C. Weck, M. Stevens, A. Kumar, LSP7 digital asset (2021). Accessed 30 May 2024. [Online]. Available: https://github.com/lukso-network/LIPs/blob/main/LSPs/LSP-7-DigitalAsset.md
30. F. Vogelsteller, C. Weck, M. Stevens, A. Kumar, LSP8 identifiable digital asset (2021). Accessed 30 May 2024. [Online]. Available: https://github.com/lukso-network/LIPs/blob/main/LSPs/LSP-8-IdentifiableDigitalAsset.md
31. LSP9 Vault (2021). Accessed 30 May 2024. [Online]. Available: https://github.com/lukso-network/LIPs/blob/main/LSPs/LSP-9-Vault.md
32. LSP10 Received Vaults (2021). Accessed 30 May 2024. [Online]. Available: https://github.com/lukso-network/LIPs/blob/main/LSPs/LSP-10-ReceivedVaults.md
33. Argent X Browser Wallet. StarkNet, Oct 17, 2023. Accessed 30 May 2024. [Online]. Available: https://github.com/argentlabs/argent-x
34. Lens Protocol. Aave. Accessed 30 May 2024. [Social Network Protocol]. Available: https://docs.lens.xyz
35. Lens Dispatcher. Aave. Accessed 30 May 2024. [Social Network Protocol]. Available: https://docs.lens.xyz/docs/dispatcher
36. "StarkNet." StarkNet. Accessed 30 May 2024. [L2 Blockchain]. Available: https://docs.starknet.io/documentation/
37. Polygon PoS Network. Polygon (2017). Accessed 30 May 2024. [L2 Blockchain]. Available: https://polygon.technology/polygon-pos
38. Lens Protocol, "Introducing Lens V2." Accessed 30 May 2024. [Online]. Available: https://mirror.xyz/lensprotocol.eth/-hJH-2IYSe56rK7IEdwSI17hUWt-paTyAs1r4Zes0uQ

39. Ethereum Organization, "Layer 2: Ethereum for Everyone." Accessed 30 May 2024. [Online]. Available: https://ethereum.org/en/layer-2/
40. StarkNet JS Account Creation. StarkNet. Accessed 30 May 2024. [Online]. Available: https://www.starknetjs.com/docs/guides/create_account/
41. "Starknet Account Interface." StarkNet. Accessed 30 May 2024. [Online]. Available: https://docs.starknet.io/documentation/architecture_and_concepts/Accounts/approach/
42. M. Köppelmann, Ethereum's L2 limitations: need for more!, Denver (2023). Accessed 30 May 2024. [Online]. Available: https://youtu.be/dimUOYzhkmk
43. V. Buterin, Ethereum in 30 minutes, Bogota (2023). Accessed 30 May 2024. [Online]. Available: https://youtu.be/UihMqcj-cqc

Felix Hildebrandt is a software developer and researcher. He studied computer science for application development at the University of Applied Sciences in Mittweida, where he completed his Master's degree in Distributed Ledger and Blockchain Technology. Following his interests, he became active in the Ethereum ecosystem and worked on well-known projects such as slock.it (TheDAO), Blockchains LLC (Incubed), and LUKSO.

Dedicated to the research and development of decentralized data economies and account systems, he builds decentralized apps for new on-chain accounts and social solutions. By writing articles, designing prototypes, and proposing standards, he promotes the importance of giving internet users more rights over their data, building fair relationships, and creating provider-independent data networks.

Decentralized Account Management: A Web3 Solution

Felix Hildebrandt

Within the Internet, user accounts became essential anchor points of data. In traditional account management, these user entities are mainly controlled by centralized platforms, often leading to numerous issues, including data breaches and privacy violations. As a counterpart, decentralized Web3 accounts have transformed how information can be stored, shared, and monetized, leading to user-centric control and security within the last decade. This paper analyzes the evolution of identity handling on the Internet and the pressing need for a dedicated identity protocol. It critically examines the influence of centralized platform policies and legal challenges regarding user interactions. In conjunction with the findings, the article portrays the fundamental architecture for secure and privacy-conserving management used to embed fair relationship models and shows future opportunities.

1 Traditional Web Service Management

In a technical sense, the Internet is a global server network that uses protocols to transfer data between specific device addresses [1]. Users can store data on servers, which can later be accessed and retrieved individually. When displaying plain page content, the data exchange through device addresses is sufficient for basic web browsing without user interaction. However, in the current era of online connectivity, defined as Web 2.0, most pages feature social interactions or act as services for big user bases. Therefore, individual providers also opt to handle user accounts for their customers.

From a neutral standpoint, the goal is to offer people new tools to simplify or enjoy everyday life, making communities grow. In return, companies gain access

F. Hildebrandt (✉)
Chemnitz, Germany
e-mail: felix.hildebrandt@gmx.net

to marketing tools to understand user needs better, improve their infrastructure, and drive profits. Nevertheless, by building and providing such platforms, companies also access personal data from customers and create analytics on user profiles to optimize value creation [2]. Data has become the leading business model and the most critical asset for web services.

The online identities mainly consist of multiple user accounts that must be created for almost every individual platform. Upon registration or login, the user is then granted access to the account's information on the operator's part. In this respect, the service provider is the sole administrator with complete control over the data management. This centralized control offers convenience and security, but it also means that users are dependent on the provider for access to their data.

Customers may register by linking already existing accounts of larger providers they support to increase convenience. However, joint accounts, while offering the benefit of consolidated access, carry the risk of losing access to any connected product if access is either lost, attacked, the account is compromised, or the intermediate service becomes unavailable at some point. This potential loss of access underscores the importance of considering the risks before opting for interwoven account structures.

1.1 Foreign Identity Handling

From the perspective of identity custody, multiple problems arise. Not only is one's data held externally, and users acquire the right to manage it, but linked providers can also monitor users' interactions related to their services at any time and create a business out of it, as the account data is managed in a centralized and concealed way [3]. As service providers create bespoke software designed for their specific requirements, they also have individual account architectures, meaning users can not easily take data to other services, fueling data duplication, lock-ins, and management overhead.

The challenges of developing digital identities can be traced back Fig. 1. At the core, there is a corporate dependency, managing the server architecture and access to the service while the user only registers for it. While convenient, the user needs to catch up in an imbalance regarding the ownership of content created throughout the given service.

This architecture originated from the early internet servers and the mere display of content. The Internet protocols were designed for machines with unique device addresses but not for humans acting and needing to authenticate through them. No embedded systems verify individuals directly. There are only measurements to check access to devices or linked accounts, like two-factor authentication, session keys, or intermediate biometric steps. However, the lack of an actual identity protocol is one of the leading causes of cybercrime and identity theft, causing enormous financial and personal damage [4].

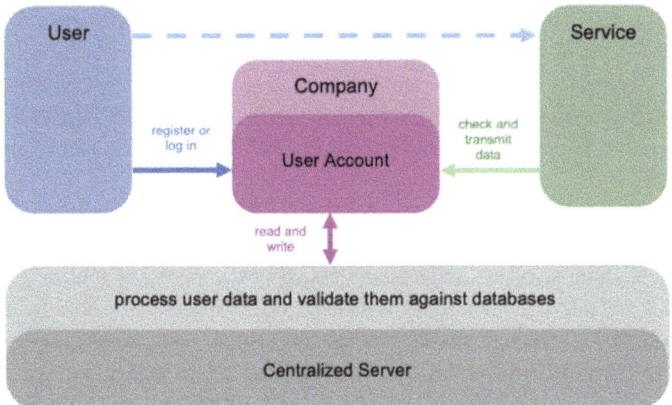

Fig. 1 Simplified Web2 login model

Besides the user logins, there is also a considerable need for unique digital assets. In the real world, our signature represents our identity. However, the files we transmit online are only copies of data held and archived through service providers. Typically, we scan verified documents to be able to use them digitally. With dozens of services, each associated with separate accounts, it is easy to lose track of ownership, when it was processed, and whether the stored data is up-to-date. In this situation, users need to trust platforms that store and secure their data. All this leads to the point where the dilemmas of security, privacy, and property cannot be addressed with the prevailing computer network architecture [5].

1.2 Chatting and Social Media Exchange

On top of the plain infrastructure, an identity mainly forms from interactions. At the beginning of the Internet, e-mail was a fundamental way of online communication and has always been a cornerstone of web technology. As protocols and platforms evolved, so did the ways of interaction between users. Within Web 2.0, conversations expanded to forums and social networks. As embedded within the platforms, chatting allowed sharing experiences and emotions more fluidly in a public or private manner. Services also enabled real-time reactions, knitting together global communities with shared interests and values. Chatting nowadays has blended with social media, as even ordinary services support status messages, handles, video-sharing, and audio functionality. However, there are several points to criticize that are closely related to the accounting itself.

Widespread chatting and social media platforms have unique features and user interfaces. While fragmentation aids in distributing weight and diversifying the landscape, it leads to interoperability issues as data is kept in separate platform silos, creating barriers in the architecture, protocols, and development processes. Users cannot transfer or back up their chats to different platforms, as they want to optimize accounts for their product specifically. If users lose access to a provider, their messages, memories, and digital history on that service are often lost forever. Even if users are friends on multiple services, they only remain with parts of their conversations.

The non-existence of transfers of accounts or parts of their belongings hardens network effects, and migrating users to newer or safer platforms is a significant challenge [6]. If most of an individual's social circle operates within a particular platform, the incentive to switch diminishes, even if there are valid concerns. This phenomenon ensures that encapsulated accounts remain dominant as they are strictly embedded in the infrastructure of user interactions- and relate to the ties of friends and family.

Secondly, as online communication becomes deeply ingrained in our daily lives, the security and privacy of these conversations gain the utmost importance, especially for data protection. Many chat platforms do not encrypt messages by default, leaving them vulnerable to unauthorized access. Web services can regularly be found in data breaches and leaks [7], all raising concerns over the untouchability of personal space in the digital world. Even outages can become critical. Within the last few years, multiple instances have arisen where major platforms were unavailable, leaving millions unable to communicate, underscoring the heavy reliance on platform-centered identity roots.

The behavioral restriction of social media platforms tied with identity management has become particularly evident with whistleblower reports from 2018 highlighting Facebook's Cambridge Analytica scandal [8]. Users who choose an account with a provider are also bound to their algorithms and changes. Suppose users want to stay in touch with friends or maintain their public page. In that case, they are often interwoven with feeds and news directed by the platforms' algorithms, impacting relationships and public image. The operator's closed source algorithms quickly lead to perception manipulation.

Digital systems and their underlying algorithms can prioritize and amplify radical or divisive content and connect those directly to someone's profile [9]. They frequently gravitate towards monopoly statuses to maximize advertising revenues fueled by disconnected networks. The reach-focused goals of services have tremendous implications for any campaign. Within Facebook's data breach, users were tracked for years, and their data, interactions, search queries, and visited places were analyzed to serve targeted ads and influence posts [10]. The bundled access over identities is threatening democratic processes. From a neutral standpoint, services should be separated from identity management, allowing users to choose communication channels without losing their identity, data, and reputation.

1.3 Legal Challenges for Platforms

Combining account infrastructure and generated content also raises the question of ownership and overall user rights within the relationship with service providers. In 2018, the General Data Protection Regulation (GDPR) concluded that "anything that helps identify an individual, whether it relates to an individual's professional, private or public life" is considered a private data set and counts as an individual's property [11], even if it resides on the company's server [12]. Accordingly, the digital identity rules refer to almost all information about such digital identity or accounts and their associated communication.

However, even if users are given the right to view, manage, and delete the data collected about them transparently, companies can still process the data beforehand. How quickly companies can analyze data to their desired advantage is simply a matter of computing power. While the GDPR restricts how this data can be obtained, companies can still obscure parts of data sets to exploit legal loopholes. In addition, innovations in data processing allow more and more information to be extracted from procured data, which thwarts the effectiveness of regulations [13].

As seen from the rapid development of terms and conditions over the last decade, companies are pursuing increasingly specific access rights to remain precisely at the limits of what is legally possible or undefined [14]. Since the account exists on the platform directly, users often have little choice: Either agree to the new rules or close the account and give up the built social life on this platform, which can not be withdrawn. Updates become psychological overhead for their customers, as the content they create on the platform is governed and owned by the same platform.

As the rules of the GDPR and service acts progress within legislative periods, companies constantly adapt to regulations to give users extended rights to track or delete data [15]. However, implementing systems to verify these rules is a huge task and could lead to the tedious restructuring of digital ecosystems. While user rights and corporate transparency increase, the need for follow-up and oversight will have to grow equally. As the variety of attacks and enforced requirements are tremendously raising the maintenance cost of central data centers, it is almost only possible to safeguard users comprehensively at scale with a dedicated identity layer [4, 5].

Current Web 2.0 security models are a top-to-bottom approach, relying on the company's ability to protect its infrastructure and user accounts. Any lack of coverage puts data at risk, especially in smaller businesses. As a result, users must comply with the company's standards. Ideally, data sovereignty and security should be directly embedded into user rights as digital creations and ownership have become native elements of society, also heavily pushed within the European Commission to protect citizens [16].

1.4 Measurement of Interaction Barriers

It is foreseeable that enterprises, government entities, platforms, and devices will need more robust identity management to counteract highlighted traffic imbalances. Rethinking how data is stored and managed is of enormous importance for society. Table 1 systematically categorizes the challenges associated with traditional web services from the previous chapters, evaluating the common issues related to identity handling, social interactions, and legal hurdles across various aspects.

In summary, the impact of centralized control and commercial interests on user autonomy and data protection has become a barrier. Legal challenges such as ownership of the value created and compliance with protection regulations further complicate the landscape. The outlined findings highlight the need for holistic solutions considering the technical, social, and legal aspects of creating more secure and user-centric frameworks.

As this enormous criticism has not gone unnoticed, federated social media such as Mastodon and Bluesky are also experiencing a dramatic upswing. In contrast, an account is created on servers that follow a specific open-source protocol instead of coming directly from a single company or app [17]. Different implementations and platforms based on this can then decide how to proceed separately for updates and implementations. Users can transfer their data as they follow the same protocol and can be imported. However, despite the improvement in the joint operation of these

Table 1 Traditional obstacles of web services

	Identity handling	Social interactions	Legal challenges
Lock-in	Account setup and duplicated data for every service in use	No portability for posts, reputation, and chats	Created value is owned property of the platform
Backend	User data extraction and private value optimization done by corporations	Behavioral restriction for users based on algorithms and feeds	Race conditions from companies while processing data
Maintenance	Account owned by centralized companies and only managed through user logins	Reliability issues, closed source data formats, and backup issues	Massive effort to convert or update systems with user rights
Security	Risks through intermediates or device and password logins	Extended tracking and risk of external data breaches or leaks	Top-to-bottom safety model from company to customer
Values	No personal ownership for digital goods and profiles	Commercialized relationship models	Inverted identity relationship model based on real-world
Protection	Mandatory to trust company for data custody, active preservation	Privacy concerns based on platforms individual features and terms	Companies applying to the minimum of protection rules, interfering business

platforms and data withdrawal, the criticism remains that there needs to be direct integration for new identity technology to map real ownership for digital goods.

2 Data Economy on Global Ledgers

Based on the analysis of traditional web services, it should become a principle of digital identities, allowing users to retain complete control and consent over their data.

Web3, the third evolution of internet technologies, introduces blockchain networks to address the critical need for data ownership. Blockchains aim to create fair and equal relationships between users and services by defining a decentralized way of data exchange. Due to its architecture, such networks enable the creation of unique digital values maintained based on an utterly user-centered account structure.

In detail, public blockchains work without the need for any central actors. Complex cryptography makes the network of independently working servers secure to create virtually immutable and extending data storage built out of datasets called blocks. The manifestation of committed blocks then acts as a security measurement, as it would be uneconomical to attack the network to alter previously generated states due to cryptographic barriers. In contrast to previous iterations of the internet, which primarily focused on enhancing data collection and browser-based functionalities without radically altering server communication, blockchains represent a fundamental redesign of the digital economy's backbone by forming global computing units [3].

The standout advantage: blockchains introduce the desired and requested foundation for a global identity layer. All interactions and datasets added within blockchain constructs must be digitally signed and permitted by users. The signature and its manifestation within blocks make it possible to exchange and own data without creating duplicates or double spending of values. All those actions relate to a network entity, not just a device connected to a service provider. Through the novel data economy, multiple parties can request and verify the same information about an individual without application-specific storage solutions. The concept defines a new starting point for data protection rights, authenticity, and value creation. While blockchain information is public and should not be used to store private manners directly, this identity layer can be utilized in subnetworks, protocols, and platforms that dock onto it, signing and referencing data proofs of users.

2.1 Decentralized Architectures

With built-in cryptography on blockchain networks, digital signatures can be used to verify accounts instead of usernames and passwords from Web 2.0. The cryptographic mechanism uses a key pair of private and public parts. All actions on a blockchain are linked to a public key and can be compared to a reference of a person or one of his instances. The private key represents a handwritten signature or password. Only the user decides when, where, and what they sign with it, automatically tying it to the public counterpart of the identifier on the chain.

While using cryptographic keys is central to ensuring security and privacy in digital communications, their role extends beyond safeguarding individual users and their credentials. In the broader ecosystem of interactions, the relationships established through these keys are not owned by a single entity or platform, reflecting a collective and equal existence similar to human interactions in the real world. This balance marks a significant milestone for the Internet. In traditional web architectures, companies still have sovereignty over connections, and users are merely given more rights to access specific data points. In this regard, blockchain technology is helping to enable an unprecedented level of independence within the digital realm.

Even the enormous effort of complying with the law and verifying the integrity of personal data in the centric model for companies and states can now be solved more efficiently and user-oriented using unique key signatures of users. Adding advanced cryptographic methods such as zero-knowledge proofs allows even private datasets to be verifiably attached to these accounts without revealing them directly to third-party services [18].

Blockchains also stand out with exceptional security and resilience. Users' sole access to their data reduces companies' system management and IT security costs. Instead of a central server, geographically distributed computers run the same network software in parallel and verify information independently. In correlation, there is a strong trend toward publishing the source code, as management depends entirely on the accepted consensus of the protocol and functions autonomously. All operators have the right to review their exact specifications and security measures [19].

Despite all its benefits, the common side effect of blockchain technology requires the user to take on more responsibility for their accounts and belongings, similar to dealing with valuable assets in the real world. On top of that, there needs to be active incentives for the network's operating community. Table 2 describes the central development needs to enable a seamless integration of decentralized economies and their users' transition from regular accounts seen today into a more blockchain-based future.

The primary focus points of new decentralized architectures are mainly structured into three common categories: blockchain accounts themselves, their data embedment, and the network's scalability. From a data economic standpoint, lowering the responsibility and required expertise from a user standpoint to create and maintain a digital identity on the global ledger is relevant for the users' onboarding process. In several areas, developers try to enhance the account functionalities related to the key signatures to add user data and descriptions. How claims and data issuing will take place is closely connected to more functionality. On top of that, scalability is essential to speed up ledger interactions and make them available for the masses, mainly when people are used to the responsiveness and maturity of regular internet infrastructure used daily. From the viewpoint of current identity providers, blockchain accounts are

Table 2 Development subjects for the decentralized economy schemas

Area of concern	Description and challenges	Solutions and development	Tech department
Asset embedment	How data can be verifiably incorporated into user-centric accounts	• SSI claims and token • Issuing protocols • Recovery mechanisms	Data economy
Blockchain accounts	How blockchain accounts can become feature-rich centers for identities	• Abstracted smart accounts • Permission integration • Data maintenance flows	
Network scalability	How distributed ledgers can handle costs and speed for extensive user bases	• Sidechains • Rollups and bundling • Network layering	Infrastructure

still in a relatively early stage of development or adoption. The following chapters elaborate on the distinct principles and challenges of the data economy.

2.2 Data Relationships

When establishing user behavior to verify and manage data globally, actors must use a public and decentralized registry, as every participant needs unrestricted read access to verify credentials or track their history. Operators then put data or related references onto a publicly viewable blockchain ledger to check data connections independently. The handling of such an account scheme is called self-sovereign identity (SSI), as the operator is responsible for the possession and management of his account and its data [4]. Even though the main objective of SSI is to detach authentication processes from web platforms, identity facilitators, and certification agencies, the SSI principle can be extended to a wide range of shared or self-issued data.

As shown in Fig. 2, there are three critical roles for self-sovereign verification and data management: the issuer, the verifier, and the actual user, each having a public address or key as a decentralized identifier (DID). Similar to how individuals in the real world need official documents to validate their identity, users in the digital space start with an empty digital identity account [20] and request certificates from issuers. After the request is fulfilled, the issuer can sign the certificate with its private key on the blockchain, referencing the user's account to enrich the user's identity. After its handout, the certificate becomes a signed proof in the global record, called a verifiable credential (VC) [21]. The holder can then continuously use services that check these credentials independently.

DIDs and VC can be applied flexibly, including private companies and service providers that are not necessarily connected to a blockchain. Nevertheless, compared to regular web infrastructure, the user benefits from only associating one account

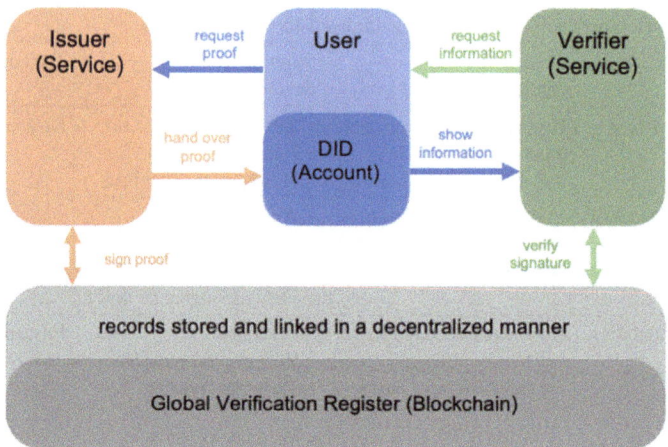

Fig. 2 Decentralized data verification model

with a ledger, which can then be used for many different services and requests. If the blockchain network is involved, credentials can even have a value connected to them. Through the user-centered methodology, network effects can be challenged without losing the roots of confidential or public account data.

The concept of DIDs and VCs is not limited to a specific domain. It can be extended to core elements in almost all areas of life where verification of arbitrary credentials, transcripts, or IDs is required. For instance, in e-commerce, the verification of users before payment could be streamlined through SSI integration. Banking services could leverage VCs and DIDs for claims, simplifying bureaucracy and enabling the native issuance of digital documents. Health documents could also be instantly approved digitally, or a comprehensive history of illnesses regarding medical interventions could be easily displayed.

As data is stored indefinitely and remains immutable in the formed chain, blockchain storage is enormously cost-intensive. Although blockchains offer the perfect environment for SSI, decentralized ledgers should never be used for storage, especially personal information. External solutions and the previously mentioned zero-knowledge proofs will become fundamental building blocks to comply with European law. Regarding the GDPR, however, SSI is not meant to restrict Web 2.0's big data but rather to own and manage data flows and identity claims in a user-centered manner. Services can still measure behaviors or collect data. The primary benefit is the opportunity for built-in consent for owned private information. Users must sign the collection of verifiable data referenced on a ledger that always captures the current content or status. Reconfirmation for queries may be necessary for long-term storage if account information does not remain with the service provider or was updated, significantly reducing duplicates known from Web 2.0 [4].

As the space evolves, more development is expected to explore different ways of implementing storage and claims based on the schema, pushing the boundaries of privacy, security, and interoperability. Ultimately, the methodology only needs the underlying encryption keys, some identity interface, and accessible storage on both ends.

2.3 Integration of Blockchain Accounts

When sharing data is distributed through multiple actors, the private key is used for verification. As analyzed by decentralized architectures and their relations, it is evident that integrating new solutions depends on leveraging the signatures of the underlying private keys. The challenge is to structure the data infrastructure and flow to address scalability and functionality effectively. A basic framework is needed within the blockchain so these cryptographic keys can become accounts and digital identities owning datasets.

On Ethereum's EVM, a protocol that stands as a beacon of programmable blockchain technology, a unique type of user account exists: the Externally Owned Account (EOA). Unlike traditional accounts, an EOA is derived from the public part of the key, allowing the user to be reached and manage their digital assets. This address serves as a distinctive identifier across the network. Importantly, each private key corresponds to a single account, a crucial requirement for participation [22].

To adjust to the traditional and feature-rich accounts known from Web 2.0, modern blockchains like the EVM also feature a programmable application layer operated by smart contracts. They allow developers to execute custom logic on the blockchain using signatures from EOAs. However, smart contracts are final in their functionality once deployed, included within a block, and come with computational costs, representing a limited good [23]. In order to give blockchain accounts more long-term and upgradable functionality, a combination of EOA signatures and smart contract functionality has to be considered to retain Web 2.0-like flexibility. The demand correlates to the previous subjects of the decentralized economy, creating generic standardizations and abstracting contract functionality into blockchain protocols.

2.4 Economic Opportunities and Challenges

At its core, blockchain technology can strengthen democratic principles, addressing shortcomings often seen in closed, platform-specific accounts of traditional web services. When building on global collaborative and interoperable networks, Web3 and its drive to open source create possibilities to mitigate common network effects seen within the internet and prioritize the protection of individuals. Outlaid frameworks empower users and allow them to select desired software services at more fine-tuned levels while freely retaining and transporting their data, social interactions, and reputation. Therefore, open standardization plays a significant role in more transparency and collaboration on code, ensuring broad consensus and remaining secure.

However, despite its promise, the technology remains in early adoption, and barriers to broader adoption are becoming more evident. Compared to the current internet, huge limitations impact user experience, such as challenges with backups, security, and scalability related to interacting with blockchain technology. Additionally, there needs to be more integrated user information to make accounts more social [24]. Creating and implementing new standards will be vital to enable general use and abstract the core principles of self-sovereign accounts and use blockchain beyond finance.

References

1. P.B. Nath, Md.M. Uddin, TCP-IP model in data communication and networking. Am. J. Eng. Res. **4**(10), 102–107 (2015)
2. A. Kaushik, Web analytics 2.0: the art of online accountability and science of customer centricity. Sybex 1ff, 241ff (2009)
3. S. Voshmgir, *Token Economy: How the Web3 reinvents the Internet*, 2nd edn. (BlockchainHub Berlin, 2020), pp. 30f, 39f
4. A. Preukschat, D. Reed, D. Searls, *Self-Sovereign Identity: Decentralized Digital Identity and Verifiable Credentials*. (Manning Publications, 2021), pp. 3–9., 101ff
5. G. Gilder, *Life After Google: The Fall of Big Data and the Rise of the Blockchain Economy*. (Simon and Schuster, 2018), p. 7
6. J. Gans, Enhancing competition with data and identity portability. Hamilt. Proj. **Policy Proposal**(10), 7–10 (2018)
7. Proxyrack, Social media security report. Accessed 30 May 2024. [Online]. Available: https://www.proxyrack.com/blog/social-media-security-report/
8. I. Atik, *Investigation of Facebook-Cambridge Analytica Data Privacy Scandal*. (Instituto Politécnico do Cávado, 2020). Accessed 30 May 2024. [Online]. Available: https://www.academia.edu/41701131
9. H. Berghel, Malice domestic: the Cambridge analytica dystopia. Inst. Electr. Electron. Eng. **51**(5), 84–89 (2018). https://doi.org/10.1109/MC.2018.2381135
10. W. Christl, S. Spiekermann, networks of control: a report on corporate surveillance, digital tracking, big data and privacy. Facultas 15–21 (2016). Accessed 30 May 2024. [Online]. Available: https://crackedlabs.org/dl/Christl_Spiekermann_Networks_Of_Control.pdf
11. General Data Protection Regulation, GDPR-EU. 2023, p. Article 4. Accessed 30 May 2024. [Online]. Available: https://www.privacy-regulation.eu/en/article-4-definitions-GDPR.htm
12. General Data Protection Regulation, GDPR-EU. 2023, p. Article 13. Accessed 30 May 2024. [Online]. Available: https://www.privacy-regulation.eu/en/article-13-information-to-be-provided-where-personal-data-are-collected-from-the-data-subject-GDPR.htm
13. A. Gandomi, M. Haider, Beyond the hype: Big data concepts, methods, and analytics. Int. J. Inf. Manag. **35**(2), 137–144 (2015). https://doi.org/10.1016/j.ijinfomgt.2014.10.007
14. M. Wasserman, V. Chidambaram, J. Mohan, Analyzing GDPR compliance through the lens of privacy policy, in *Heterogeneous Data Management, Polystores, and Analytics for Healthcare, in Lecture Notes in Computer Science*, vol. 11721 (Springer, Cham, 2019), pp. 82–95. https://doi.org/10.1007/978-3-030-33752-0_6
15. A. Voss, Fixing the GDPR: Towards Version 2.0. EPP—European People's Party, 2021. Accessed 30 May 2024. [Online]. Available: https://www.axel-voss-europa.de/wp-content/uploads/2021/05/GDPR-2.0-ENG.pdf

16. European Commision, "European Digital Identity Act." Accessed 30 May 2024. [Online]. Available: https://commission.europa.eu/strategy-and-policy/priorities-2019-2024/europe-fit-digital-age/european-digital-identity_en
17. Y. Roth, S. Lai, Securing federated platforms: collective risks and responses. JOTS **2**(2) 2024. Accessed 30 May 2024. [Online]. Available: https://doi.org/10.54501/jots.v2i2.171
18. U. Feige, A. Shamir, Zero-knowledge proofs of identity. J. Cryptol. **1**, 77–94 (1988)
19. M.J. Casey, P. Vigna, In blockchain we trust. MIT Technol. Rev. **121**(3), 10–16 (2018)
20. "Decentralized Identifiers DIDs v1.0." July 19, 2022. Accessed 30 May 2024. [Online]. Available: https://www.w3.org/TR/did-core/
21. "Verifiable Credentials Data Model v1.1." Mar. 03, 2022. Accessed 30 May 2024. [Online]. Available: https://www.w3.org/TR/vc-data-model/
22. A. Antonopoulos, Wood, G.: *Mastering Ethereum: Building Smart Contracts and Dapps*, 1st edn. (O'Reilly Media, 2018), pp. 61–76
23. A. Antonopoulos, G. Wood, *Mastering Ethereum: Building Smart Contracts and Dapps*, 1st edn. (O'Reilly Media, 2018), pp. 127–129
24. G.E. Weyl, P. Ohlhaver, V. Buterin, Decentralized Society: Finding Web3's Soul, 2022, pp. 15–19. Accessed 30 May 2024. [Online]. Available: https://papers.ssrn.com/sol3/papers.cfm?abstract_id=4105763

Felix Hildebrandt is a software developer and researcher. He studied computer science for application development at the University of Applied Sciences in Mittweida, where he completed his Master's degree in Distributed Ledger and Blockchain Technology. Following his interests, he became active in the Ethereum ecosystem and worked on well-known projects such as slock.it (TheDAO), Blockchains LLC (Incubed), and LUKSO.

Dedicated to the research and development of decentralized data economies and account systems, he builds decentralized apps for new on-chain accounts and social solutions. By writing articles, designing prototypes, and proposing standards, he promotes the importance of giving internet users more rights over their data, building fair relationships, and creating provider-independent data networks.

Web3—arsNFT Badges as an Application Option

Alexander Robert Skurka

1 Introduction

In the context of a digital era characterized by innovation and technological disruption [1, p 9, 19], tokenization represents a significant milestone that has a central influence on the transformation of our economy and society. The transfer of assets, identities and authentication methods to the digital sphere not only opens up new perspectives, but also presents challenges and opportunities. In this exciting paradigm shift, I intend to make a substantial contribution to a seminal chapter that takes an in-depth look at the current developments and applications of tokenization [2].

This chapter focuses in particular on two key elements that represent the innovative current of tokenization: Non-Fungible Tokens (NFTs) and digital product certificates [3] or product passports [4]. The uniqueness of NFTs as individual digital assets, coupled with the authentication and traceability provided by digital product certificates, creates a fascinating synergy that goes far beyond the traditional boundaries of digital exchange.

This research goes beyond the scope of mere technology exploration; it addresses the profound significance of these developments for our society and economy. The integration of NFTs not only enables unprecedented digital representation of assets, but also fosters new forms of creativity and cultural expression. At the same time, digital product certificates help to ensure the authenticity and quality of products in a globalized economy.

In the rest of this chapter, we will explore the depths of NFTs and digital product certificates, from their basic definitions to their impact on various industries and the evolving digital landscape. But the insights go beyond simple tokenization. Our journey takes us to the sophisticated mechanisms of 3-factor protection (3FP) and token proof, which serve as cornerstones of security in the emerging Web3 world.

A. R. Skurka (✉)
Munich, Germany
e-mail: app.tricard.info@arsnet.de; arsdev@arsnet.de

The Web3 world [5], characterized by decentralized technologies and blockchain, provides a platform for trustworthy and secure interactions. In this context, we will analyze how 3-factor protection and token proof not only act as security measures but also strengthen trust in this innovative technology landscape.

This expedition through the world of tokenization and security aims not only to under-stand current developments, but also to look to the future. How will tokenization continue to transform the way we create, transfer and secure value? What role will NFTs and digital product certificates play in the evolutionary landscape of digital assets? These questions will be addressed throughout this chapter and explored in greater depth through precise analysis and practical application examples.

Welcome to a journey through the spheres of tokenization, where innovation and security go hand in hand to shape a forward-looking digital reality.

2 From Theory to Practice

The 'arsNFT Badge' represents a pioneering fusion of open badge and blockchain in the specific context of non-fungible tokens (NFTs). This innovative fusion manifests itself as a paradigmatic advance in digital recognition and combines the proven principles of Open Badge with the security features of blockchain technology.

The structure of the 'arsNFT Badge' is based on a sophisticated architecture that goes beyond the purely visual and carries a deeper meaning. Each badge, represented as a non-fungible token on the blockchain, undergoes a tokenization process that makes it a unique and immutable digital entity. This process not only guarantees the authenticity of the badge, but also enables transparent and secure tracking of its origin and use.

The integration of metadata enhances the intelligence of the 'arsNFT Badge' by adding additional information about the exhibitor, the criteria for the award and a unique identifier. This metadata makes the badge not only visually appealing, but also machine-readable and interoperable.

The tokenization of the 'arsNFT Badge' not only provides aesthetic enrichment, but also creates a bridge between the digital and physical worlds. The issuance of arsNFT badges as physical representations of digital recognitions underlines the union of analog and blockchain, creating a unique symbiosis.

The areas of application of the 'arsNFT Badge' extend across various sectors, including the promotion of a strong digital identity, the recognition of cultural contributions in the arts sector and the transparent documentation of skills and certificates in the education sector.

The 'arsNFT Badge' thus proves to be not only an aesthetically pleasing digital award, but an innovation that combines the principles of open badge and blockchain in a harmonious unison. This trailblazer reflects not only the technological excellence but also the transformative power in shaping future paradigms of digital recognition in the NFT context.

3 Open Badges: An Evolution of Digital Recognition

In the era of digital transformation, the concept of "open badges" has taken on a significant role in digital recognition. These digital awards not only represent symbols, but are an innovative method of making skills and achievements visible in the digital world. This section is dedicated to a comprehensive analysis and appreciation of open badges, their emergence, significance and diverse areas of application.

The creation of the Open Badges

The origins of Open Badges [6] date back to 2011, when the "Mozilla Open Badges" project was launched. Mozilla, known for the Firefox web browser, aimed to create an open and standardized way to digitally verify skills and achievements. The result was the Open Badge Framework [7], a system that enables the creation, collection and verification of digital badges (Image 1).

The structure of Open Badges

Open badges are much more complex than simple graphics. They are structured data sets that contain essential information on the skills or achievements acquired. Each badge contains metadata such as the issuer of the badge, the acquisition criteria, the date of issue and a unique identifier. This structured information makes Open Badges not only visually appealing, but also machine-readable and interoperable.

Significance and recognition

The relevance of Open Badges extends beyond traditional certificates. These digital awards offer a flexible and easily verifiable way to showcase individual skills and achievements. Companies, educational institutions and organizations are increasingly recognizing the value of open badges as a complement to traditional credentials.

Areas of application for Open Badges

1. **Education and training**: In educational institutions, Open Badges serve as digital rewards for acquired skills, allowing students to enrich their digital portfolios with compelling evidence of their skills and achievements.
2. **Professional development**: In a professional context, Open Badges act as verifiable signs of specific skills or certifications. Employees can transparently present their professional development and make it visible to potential employers.
3. **Online communities**: In digital communities, open badges are awarded as a means of recognition for special contributions or achievements, which promotes motivation and cohesion in virtual communities.
4. **Corporate sector**: Companies use Open Badges to document internal training or to recognize special employee achievements. This not only promotes employee development, but also increases motivation and commitment.

Challenges and future prospects

Despite their success, open badges face challenges such as standardization and widespread recognition. Nevertheless, current developments suggest that open badges will play an increasingly important role in the future of digital recognition. For example, the integration of blockchain technology in conjunction with open badges could further strengthen security and verifiability.

Conclusion

Open Badges have fundamentally changed the way skills and achievements are presented and recognized. Their digital nature, coupled with structured metadata, creates a transparent and interoperable platform for recognizing individual achievements. Despite existing challenges, open badges are undoubtedly a key element in shaping a future-oriented education and recognition landscape.

4 Open Badges 2.0 in the Context of 3-factor Protection (3FP) for the Web3 World

Reasons for the selection

The decision to use Open Badges 2.0 as the preferred version in the context of 3-factor protection (3FP) for the Web3 world is based on an in-depth analysis of the various versions of this standard [9]. Open Badges, as information-rich visual tokens of verifiable achievements, have gone through several stages of development. Nevertheless, it is clear that version 2.0 is particularly suitable for meeting the requirements of 3-factor protection and for seamless integration into the Web3 world.

Furthermore, Open Badges 2.0 enables external validation through the integration of validation resources on the web. This means that the authenticity of an Open Badge can be verified not only by the issuer, but also by independent parties. This external validation functionality significantly strengthens the credibility of the information provided.

Validation portals

IMS Global Open Badges 2.0 Validator [10].
   https://openbadgesvalidator.imsglobal.org/.
   See Image 2.
   badgr—Open Badges 2.0 Validator [11].
   https://badgecheck.io/.
   See Image 3.

Open Badges v2.0

IMS Final Release

Date issued	April 12, 2018
Status	IMS final release [12]
Latest version	https://www.imsglobal.org/spec/ob/v2p0

https://www.imsglobal.org/sites/default/files/Badges/OBv2p0Final/index.html
This specification is free for anyone to use or implement (Image 4).

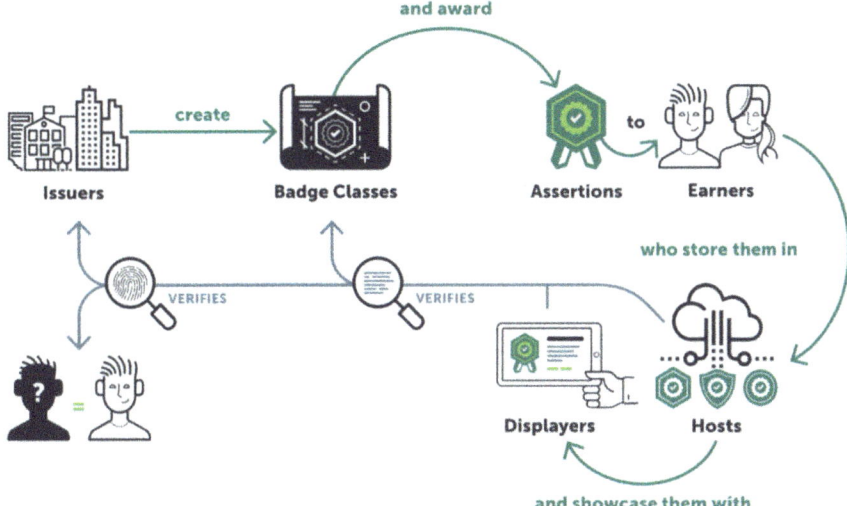

Image 1 Create and award openBadges—https://openbadges.org/build [8]. This image is licensed under a **Creative Commons Attribution 4.0 International License**

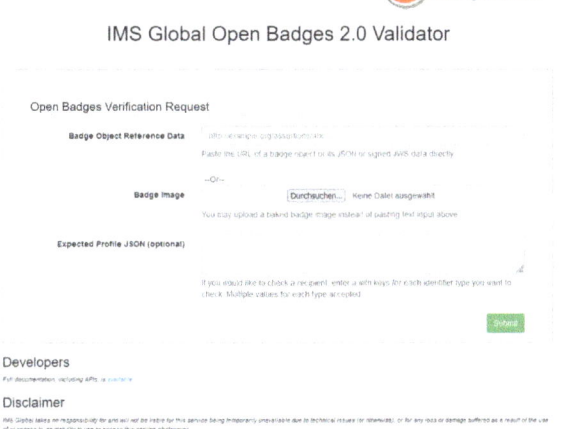

Image 2 IMS global open Badges 2.0 validator—(Screenshot of the website [online] © 1EdTech Consortium, Inc.) [10]

Image 3 badgr—open Badges 2.0 validator—(Screenshot of the website [online] Powered by Badgr & By Instructure) [11]

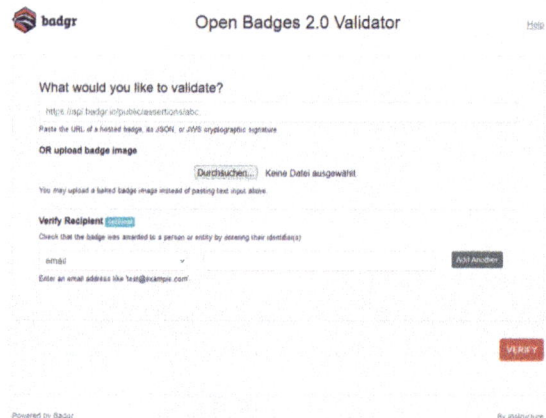

Image 4 Open badges 2.0 diagram—Open Badges Specification 3.0 [online] © 1EdTech Consortium Inc. [13] https://www.imsglobal.org/spec/ob/v3p0/#fig-open-badges-2-0-diagram

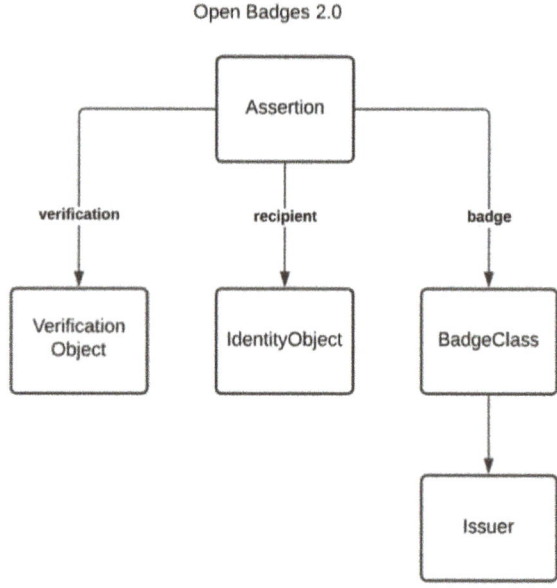

Versions 2.x and above, but especially Open Badges 2.0, have been deliberately designed for compatibility and interoperability with other digital standards in the field of digital credentials. This includes the Comprehensive Learner Record and the Competencies and Academic Standards Exchange (CASE)®. The integration of Open Badges 2.0 into this comprehensive framework enables a holistic view and adaptation to different digital ecosystems.

Open Badges 2.0 is characterized by powerful features including endorsements, internationalization, multilingual capabilities, version control, accessibility improvements and full adoption of JSON-LD. These features make Open Badges 2.0 a stable and advanced basis for the implementation of 3-factor protection in the Web3 world.

Furthermore, the integration of a REST-based API was introduced in Open Badges 2.1, known as BadgeConnect® API. This feature, which fully inherits the existing data model of Open Badges 2.0, supports the mobility of learning by enabling learners to transfer their credentials between different systems. This aspect is crucial for interoperability and learner empowerment.

The introduction of Open Badges 3.0 (OBv3) as the latest version of the standard offers further benefits, including enhanced security and privacy features through layers of cryptographic proof. This version is not only aligned with the W3C Verifiable Credential (VC) Data Standard, but also with the Comprehensive Learning Record 2.0 standard, enabling extensive integration and application possibilities. In addition, Open Badges 3.0 supports a native API that further facilitates learning mobility.

Overall, the use of Open Badges 2.0 in the context of 3-factor protection provides a solid foundation for security, interoperability and adaptability in the dynamic and innovative Web3 world. This choice is based on a careful consideration of the different versions and their specific features to meet the security and functionality requirements of the digital landscape (Image 5).

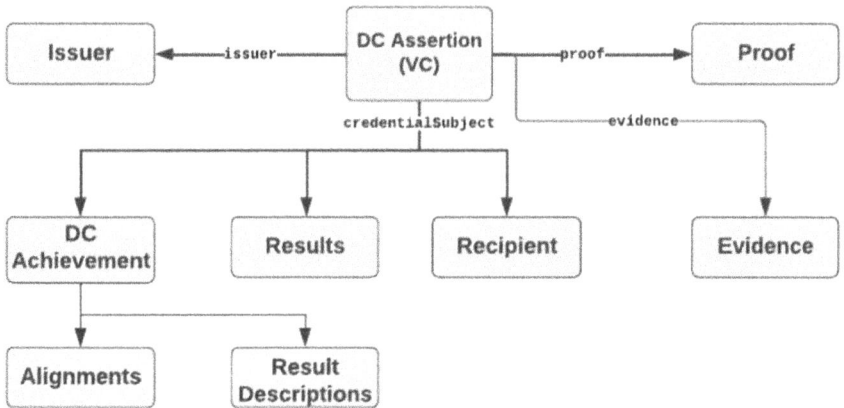

Image 5 Shows the major conceptual components of an open badge verifiable credential—Open Badges Specification 3.0 [online] © 1EdTech Consortium Inc. https://www.imsglobal.org/spec/ob/v3p0/#fig-diagram-show-the-major-conceptual-components-of-an-open-badge-verifiable-credential [13]

Open Badges Specification

Candidate Final Public.
    Spec Version 3.0.
    Candidate Final Public.

Document version	1.0.10
Date issued	September 22, 2023
Status	This document is for review and adoption by the 1EdTech membership [13]
This version	https://www.imsglobal.org/spec/ob/v3p0/main/

5   3-Factor Protection (3FP)

The symbiosis of analog, blockchain and digital in the security architecture of the Web3 world

The concept of 3-Factor Protection (3FP) represents a groundbreaking security architecture that establishes an unprecedented security assurance in the Web3 world through the unique integration of analog, blockchain and digital. This chapter deepens the under-standing of the fundamental elements of 3FP and explores their practical applications.

The basics of 3-factor protection

Analog: Physical security in the digital era

The analog component of the 3FP focuses on physical security by integrating physical elements such as arsNFT badges. This creates a bridge between the real and digital worlds and ensures that security not only exists virtually, but can also be experienced physically.

Blockchain: the immutability of data

The blockchain forms the core of the 3FP. Due to its decentralized and unchangeable nature, it ensures reliable verification and traceability of information. Every step and every interaction is recorded in the blockchain, which generates an audit trail and makes manipulation virtually impossible.

Digital: The intelligence of the virtual world

The digital factor of the 3FP represents the intelligence of the virtual world. Advanced encryption technologies, authentication mechanisms and AI-supported security protocols are used here. This digital layer not only enables secure access to information, but also dynamic adaptation to changing security requirements.

The symbiosis in practice

Now that we have a solid overview of the terminology and definitions of various elements, we will focus on the Proof of Concept (PoC) of **TRIcard** and **arsNFT Badge**. In this context, we will conduct an in-depth analysis.

arsNFT Badges: The connection between analog and blockchain

The use of arsNFT badges as a physical representation of digital recognitions illustrates the seamless connection between analog and blockchain in the 3FP. These badges, represented as NFTs on the blockchain, offer not only aesthetic appeal, but also a secure, unambiguous assignment in the digital world (Image 6).

Example (Image 7):
After an arsNFT badge has been created, the required data is consolidated by Open Badge and calculated in a new "CurrentBlockData" using previous hash values. It is then integrated into the DominoChain as a new element (Image 9).

In specific scenarios, the arsNFTs created are canceled, which is displayed as "REVOKED" in the web application. This measure in accordance with Open Badge 2.0 ensures that the certificates lose their validity and are no longer considered valid (Image 10).

Token proof: The pinnacle of security

Token proof in 3FP ensures that every access or transaction is protected by a token-based verification system. The combination of NFTs, POAPs or digital product certificates, securely stored and verified via the blockchain, creates an unparalleled level of security.

TokenProof has been added to the TRIcard arsNFT badges to unlock the possibilities of a secure token (3-factor protection ~ 3FP) (Image 11).

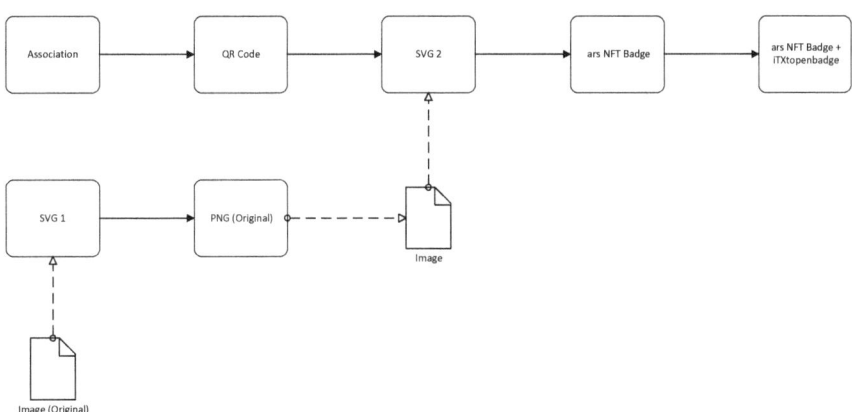

Image 6 Creation of arsNFT badge | TRIcard and arsNFT badge—Specification 1.0 [© ARSnet]

Image 7 Example #1 arsNFT badge | created by TRIcard Application [© ARSnet]

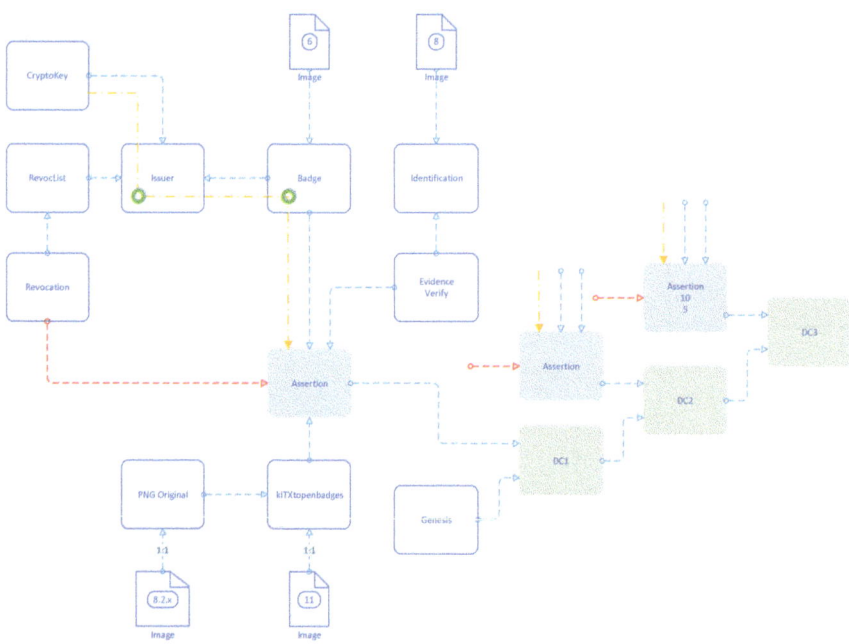

Image 8 Entry in domino chain | TRIcard and arsNFT badge—Specification 1.0 [© ARSnet]

Web3—arsNFT Badges as an Application Option

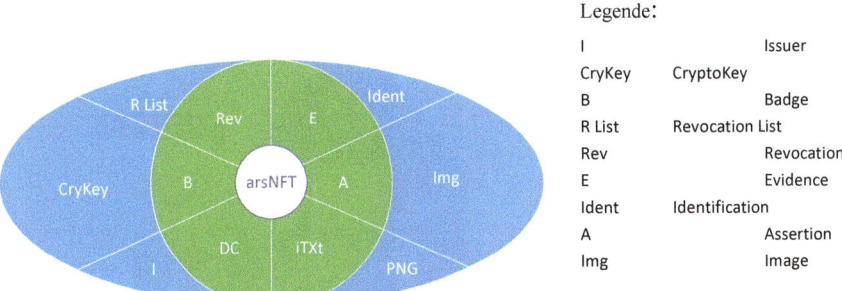

Image 9 arsNFT badge components | TRIcard and arsNFT badge—Specification 1.0 [© ARSnet]

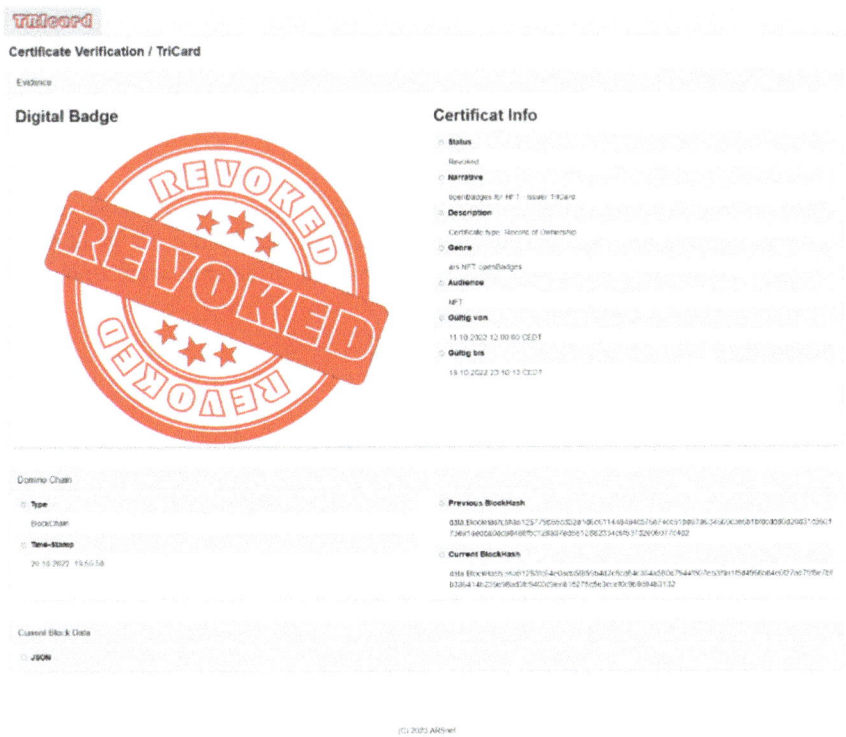

Image 10 Example #2 arsNFT badge—revoked | created by TRIcard Application [© ARSnet]

Integration into the Web3 world

The 3FP integrates seamlessly into the Web3 world through the use of open standards and interoperable protocols, which enables broad acceptance and adaptability in various digital ecosystems.

Image 11 Tokenproof with TRIcard arsNFT Badge | TRIcard and arsNFT badge—Tokenproof Livecycle Specification 1.0 [© ARSnet]

The importance of 3-factor protection

3-factor protection surpasses traditional security concepts by not only ensuring the security of information, but also creating trust in the digital landscape. In a time of in-creasing cyber threats, the 3FP acts as a robust line of defense against potential threats.

Conclusion

Three-factor protection (3FP) represents an evolutionary development in the field of security architecture. Through the unique integration of analog, blockchain and digital, it not only creates a secure environment, but also sets new standards for trust and integrity in the Web3 world. The harmonious interplay of these factors offers a comprehensive security solution that successfully overcomes the challenges of the digital era (Images 12 and 13).

Visit https://app.tricard.info for more information.
Example—Certificate Verification/TriCard:

https://app.tricard.info/nft/TriCard.nsf/openBadges_v2_verifyUI.xsp?documentId=195FD77A4C722ADBC125896F00462012

Example—Tokenproof mit TRIcard arsNFT Badge:

https://app.tricard.info/nft/TriCard.nsf/QR_Check_in.xsp?documentId=195FD77A4C722ADBC125896F00462012

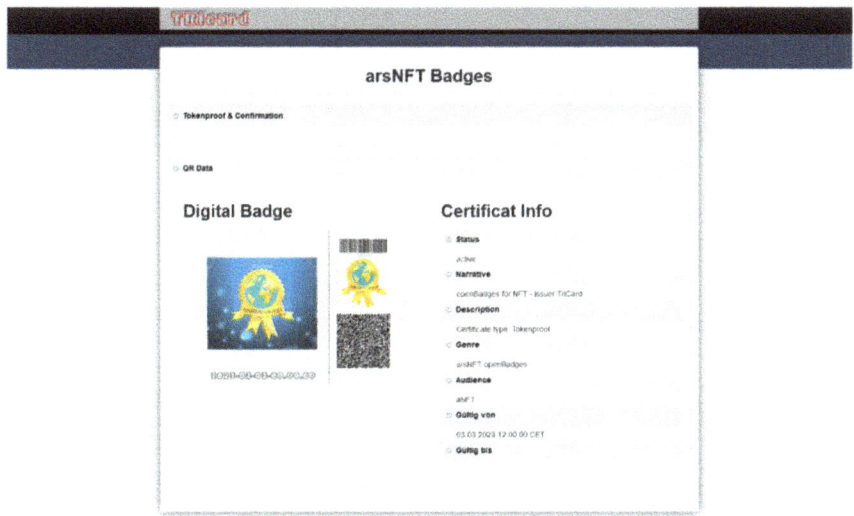

Image 12 Example #1 Tokenproof with TRIcard arsNFT badge | created by TRIcard Application [© ARSnet]

Image 13 Example #2 Tokenproof with TRIcard arsNFT badge | created by TRIcard Application [© ARSnet]

References

1. Fraunhofer-Gesellschaft, Blockchain and Smart Contracts (online) (2023, Nov 13). Available: https://www.fraunhofer.de/content/dam/zv/de/forschung/artikel/2017/Fraunhofer-Positionspapier_Blockchain-und-Smart-Contracts_v151.pdf
2. BTC-ECHO GmbH, NFT (Non Fungible Token) (online) (2023, Nov 13). Available: https://www.btc-echo.de/academy/bibliothek/non-fungible-token-nft/
3. Haufe-Lexware GmbH & Co KG, Der Digitale Produktpass (online) (2023, Nov 13). Available: https://www.haufe.de/sustainability/umwelt/der-digitale-produktpass_575774_593538.html

4. Federal Ministry for the Environment, Nature Conservation, Nuclear Safety and Consumer Protection (BMUV), What is a digital product passport? (online) (2023, Nov 13). Available: https://www.bmuv.de/faq/was-iWasisweb3?st-a-digital-product-passport, https://www.haufe.de/sustainability/umwelt/der-digitale-produktpass_575774_593538.html
5. IDG Tech Media GmbH, What is Web3? (online) (2023, Nov 13). Available: https://www.computerwoche.de/a/was-ist-web3,3552475
6. The Mozilla Foundation and Peer 2 Peer University, in collaboration with The MacArthur Foundation, Open Badges for Lifelong Learning (online) (2023, Nov 13). Available: https://wiki.mozilla.org/images/5/59/OpenBadges-Working-Paper_012312.pdf
7. Peer 2 Peer University and The Mozilla Foundation, in collaboration with The MacArthur Foundation an Open Badge System Framework (online) (2023, Nov 13). Available: https://wiki.mozilla.org/images/f/f3/OpenBadges_-_Working_Badge_Paper.pdf
8. 1EdTech Consortium Inc, IMS-Open Badges Home-Build (online) (2023, Nov 13). Available: https://openbadges.org/build
9. 1EdTech Consortium Inc, OpenBadges Summary (online) (2023, Nov 13). Available: https://www.imsglobal.org/activity/openbadges
10. 1EdTech Consortium Inc, IMS Global Open Badges 2.0 Validator (online) (2023, Nov 13). Available: https://openbadgesvalidator.imsglobal.org/
11. Badgr, Open Badges 2.0 Validator (online) (2023, Nov 13). Available: https://badgecheck.io/
12. 1EdTech Consortium Inc, Open Badges v2.0 IMS Final Release (online) (2023, Nov 13). Available: https://www.imsglobal.org/spec/ob/v2p0
13. 1EdTech Consortium Inc, Open Badges Specification 3.0 (online) (2023, Nov 13). Available: https://www.imsglobal.org/spec/ob/v3p0/

Alexander Robert Skurka In over 35 years of programming experience and over 25 years of practical experience, I have been able to expand and deepen my knowledge of various programming languages, database development (with and without workflow) and IT landscapes in the areas of communication, interfaces, networks and security. As a speaker at national IT conferences, I was able to pass on my extensive expert knowledge to numerous participants. My overall knowledge was also underpinned by the latest scientific studies in the field of business informatics.

In addition to my professional activities, I regularly deal with the integration of business applications and communication systems such as Unified Communication and Collaboration—UC2 or Office programs with Internet Business Office Services—iBOS2.0. My area of responsibility includes web architectures and XML application development as well as the project management of corresponding implementations in this area.

In the context of Web 3.0, I am working intensively on the decentralization of systems, the use of blockchain technology to increase security and transparency (with TRIcard) as well as interoperability between different applications. These technologies offer the potential to revolutionize existing IT infrastructures and create innovative solutions for integration and communication, such as 3-factor protection (3FP) with TRIcard, a symbiosis of analog, blockchain and digital in the security architecture of the Web3 world.

For my beloved wife and our children, who always provide love and inspiration.
To my parents, who paved the way.
And to my teachers, who ignited the spark of knowledge.
Thank you for your support on this journey.
Headquarters Munich|Germany.

Web3 Security

José Carlos Ramírez and Isaac Agudo

1 Web3 Security

Any given week Web3 security incidents land in news sites or social media. This is not just a matter of the total amount of "hacks" happening but of a trademark of public blockchain technologies: the information is open to everyone. When a project's funds get drained anyone can watch it in real-time, if a convoluted multi-stage attack is carried out anyone can analyze it, when a wallet tries to scatter funds they can be tracked (with some exceptions). The blockchain's transparency made it clear for every player in the ecosystem: security should be the baseline, not a last-minute patch.

It should be noted that not every security incident is rooted in a Decentralized Application (DApp) vulnerability. This chapter focuses on the security risks of smart contracts, but most thefts of funds out there are related to scams, social engineering, phishing, and the like. These techniques are well-known by traditional Web2 cyber-criminals, they need some additional twists but the main target is still the same: the end user. Although security awareness training is out of the scope of this chapter, it is a cornerstone on the road to mass adoption.

There is one key trait of the current DApp and smart contracts development culture that directly affects the number of security bugs reaching production: innovation and experimentality. Being the fastest player in rolling out a new feature, trying a new concept or getting the first batch of users are sometimes primed over following proper Software Development Lifecycle patterns. A direct consequence of this is deploying less polished software, which also means more security bugs. Web3 software should

---

J. C. Ramírez (✉)
Zaragoza, Spain
e-mail: jc@jcsec.io

I. Agudo
Málaga, Spain
e-mail: isaac@uma.es

look less like the small mobile games industry and more like traditional financial and banking software, having a slower pace and undergoing several quality and assurance steps from the beginning of the development.

1.1 Threat Actors and Attack Surface

To start looking into the risk of DApps and smart contracts we will first look into a list of threat actors and analyze the attack surface of this kind of software. Let's first get a quick definition of these terms: A **threat actor** is anyone who has the potential to cause damage to your system, either intentionally or unintentionally. The **attack surface** refers to the set of elements, components or areas in the system that could potentially be misconfigured or vulnerable, each of which is referred to as an **attack vector**.

Let's start with a quick analysis of the overall actors of a smart contract of any type to see if they could be considered threat actors or not. No one will question that anonymous and participant addresses - the users - should not be trusted as those are usually the source of external attacks. Administrators or privileged roles of the protocol are sometimes considered trusted members, the same with the developers who craft the code. However, this last assumptions must be taken with a grain of salt. Keys can be compromised, or individuals may have personal interests in subverting the behavior of the system, as in the case of rug pulls. In a zero-trust environment, every one of the listed actors should be considered a threat actor, and the impact of their actions considered when assessing the risks of a DApp.

There are many more elements in a DApp that are exposed to risk than just the smart contracts alone. First, there is the organization in charge of the DApp. Any piece of software in their personal laptops or internal servers could be subject to compromise. As mentioned below, the people that are part of the organization too. Unauthorized access to any of those could result in key compromise of a privileged address that could be used to modify the contract, its configuration or any wallet of importance.

Then there are the Web2 components. Many DApps have some of their components in a traditional infrastructure such as the frontend, DNS, corporate email, and others. These components could also be used to attack the users of the DApp directly or as part of a multi-stage attack that causes damage in some way.

Finally, we should look at the blockchain components, which include, among others, node software, wallets, consensus algorithms, and, of course, smart contracts. Any of these pieces of software could be compromised or affected by security flaws at any point in its supply chain, which could ultimately affect the user or their funds in one way or another.

In summary, overall security is difficult because of the large number of vectors that can ultimately cause a security incident. For the rest of this chapter, we will focus specifically on smart contracts, but it is important to realize that we should consider many more elements when we want to assess the risk of our DApp.

2 Common Vulnerabilities in Smart Contracts

There is a large list of potential security issues that could affect our smart contract written in the Solidity language. Here we will discuss a subset of the more common or more relevant of these vulnerabilities in no particular order.

2.1 Traditional Software Vulnerabilities

Several vulnerabilities can be found in smart contracts that have largely been studied in traditional software and are not exclusive to blockchain technologies. The following security issues can be found in web, mobile, desktop, or any other type of software.

Access controls: access control issues arise when a user can access a function or feature that was not intended for them. For example, modifying some core configuration or being able to transfer funds that do not belong to the caller.

It is worth highlighting one particular issue of solidity smart contracts when implementing access controls that have caused security issues in the past. The variable `msg.sender` holds the address that issued the actual call to the function, on the other hand, `tx.origin` stores the address that issued the transaction that resulted in the current call. This distinction can open the door to access control issues, as a phishing attack against a privileged user could result in the arbitrary execution of sensitive functions by the attacker.

Arithmetic issues: Mathematical operations and handling of numeric values can be affected by several long-known vulnerabilities. Arithmetic overflows and underflows are probably the best known, where a piece of data goes beyond the maximum value that a type can store and reaches zero, or goes below zero and is set to the maximum value. Aside from overflows, rounding issues when performing divisions could affect our calculations, as the default is to round to the lower bound. Casting integers could result in unintended truncation of values, which could affect any subsequent calculation that uses the variable

Logic bugs: Various bugs fall into this category because it doesn't define a specific incorrect behavior or effect. Instead, we consider logic bugs to be any kind of problem where the code causes a detrimental situation for the core logic of the product, behaves differently than specified, or is based on incorrect assumptions. An example of a logic bug in a Web2 e-commerce would be choosing the price you pay for an item instead of the price set by the store.

Getting a little deeper into DeFi, some examples of logic errors could be: incorrect calculations of asset distribution, tricking the system to receive additional rewards, funds locked in the contract, or positions not subject to liquidation under any circumstances, to name a few.

2.2 Weak Randomness

Randomness in Ethereum is a challenge by definition, as Ethereum is deterministic. The EVM will always yield the same results for a given set of inputs plus an initial state, meaning there is no variation in the outputs and no randomness.

Even so, randomness—pseudo-randomness to be exact—is still desirable for multiple applications and products that could be built using blockchain technology. At the moment the recommended solution is to leverage specialized Oracles that bring off-chain data into the blockchain. However, many developers have tried to incorrectly create randomness, resulting in predictable results that defeat the purpose of trying to implement randomization.

In-chain information such as the block hash and timestamp among others has been used as part of custom algorithms to try to craft a random value. Such pieces of information are available to anyone, so it is possible to pre-calculate the results of the (non-) random number by replicating the algorithm. Therefore, anonymous users could rig the system to always submit the correct value at the correct time to profit from the vulnerable smart contract.

The following sources of data should not be used as the base to create a pseudo-random number, no matter the amount or complexity of the operations that they would undergo:

- `block.coinbase`
- `block.difficulty`
- `block.number`
- `block.timestamp/now`
- `block.blockhash`

2.3 Publicly Accessible Secrets

Public blockchain technologies like Ethereum keep all historic data publicly stored on the ledger. That is one of their perks after all. Anyone can easily use a block explorer like https://etherscan.io/ to look into a smart contract's transaction history, storage or creation time.

Anything that gets forged into a block anywhere will be publicly available. When a new smart contract is created, the initial transaction includes the source address that is deploying it, any arguments that the constructor received and the contract's bytecode—code in a non human-readable format that can be interpreted given enough time and skills. From that on, any other transaction call that is targeted to our contract or from our contract will be equally accessible, including any information that is included in the call.

Therefore, if any secret information is stored on chain without being encrypted, it will not be secret anymore. Look into the following short scenarios to showcase the dangers of not being aware of the above:

- A smart contract that implements clear text password-based access controls will reveal its initial password either hardcoded in the code, as part of the constructor's arguments or in any of the future transactions that attempt to be authorized.
- A smart contract that implements a hashed password-based access controls will not reveal the secret initially as it stores a hash instead, but the first successful authentication will record forever which was the correct secret on chain. In addition, some hashes can be directly brute-forced in a short amount of time.
  - Note that access controls should be enforced through address allowlisting, password-based variations should never be used in smart contracts.
- A smart contract implementing the well-known "rock-paper-scissors" game, which requires each player's selection to be revealed at once, could be rigged as any of the player's submissions will be public and the smart contract will process them sequentially i.e. The second player will be able to learn the choice of the first player before submitting his choice.
- Every bid of an auction will be public after submission, restricting some of the more common approaches where this information is considered a secret until the final stage.

If a secret is required to be stored on-chain, it should never be done in plain text. A secret that gets encrypted and decrypted off-chain, could be considered safe as long as the off-chain systems don't get compromised.

Additionally, mechanisms such as Commit and Reveal schemes allow users to temporarily share secret information. For instance, a rock-paper-scissors smart contract that implements such a mechanism can not reveal each of the users' selections until all of them have submitted theirs.

2.4 Front-running

Front-running is the action of sending a transaction with the intention of getting it included in a block before one that has been observed in the mempool, benefitting the user in any way.

Validators usually prioritize transactions that pay a higher "priority fee" or tip for the gas when they choose which transactions to include in a given block. Therefore, if two similar transactions are found in the mempool at the same time, the one that paid a higher gas fee will most likely be included first.

Front-running is not a vulnerability itself but more of a feature. However, not considering the impact of front-running transactions could create a security risk in a smart contract.

Anyone can look into a public mempool looking for transactions that could benefit them if a specifically crafted one gets executed before the observed one. How can front-running create a negative impact on the users?

- Features where the transaction order matters.
  - E.g: if the solution of a "guess the number" game is announced by submitting a call to the contract, an attacker could front-run the announcement by submitting the winning number themselves.
- When there is a competition between users to submit information.
  - E.g: If a user submits their answer in competition to solve some mathematical problem, an attacker could front-run them by copying the target solution and submitting it.
- Arbitraging: operating with assets, for example, swapping or selling, in one or more markets with the intention of taking advantage of additional knowledge. In this case, knowing which transactions are waiting to be executed in the mempool.

To protect users from being affected by front-running the developers should design the externally facing entry-points to not depend on the time and order of execution. As this is not feasible in some scenarios, private mempool are also a good option, where the transactions are only visible to the validators. Finally, Commit and Reveal schemes could also be applied in some scenarios as temporarily protecting the submitted information would also prevent a valid front-run.

2.5 Oracle Manipulation

Oracles provide external data to smart contracts that allow the implementation of advanced features. Although an oracle is expected to provide off-chain data, on-chain data providers are also referred to as oracles in some scenarios.

As oracles are a central piece of some DeFi products, there is no doubt that an alteration in the provided data hugely affects the user and its protocol. Oracle manipulation refers to forcing oracles to temporarily report incorrect information that can be leveraged by the attacker for profit.

To better understand the concept, look at the below example on how this kind of attack can be carried out against a Lending protocol that uses a DEX's pool as Oracle. The victim protocol requests the spot price of the pool to calculate if the user-provided collateral is enough to grant the request or to decide if the loan has been repaid.

1. Pool manipulation: the attacker first has to manipulate the balance of the pool. For example, by requesting a Flash Loan and swapping it for a different asset in the pool.

2. Further exploitation: while the price is still artificially low, the attacker can either repay any current debt or ask for a loan for a fraction of the price that this operation will normally require.
3. Closing: the attacker would swap back into the victim pool and repay the Flash Loan.

This is just a quick illustrative example of how oracles could be manipulated, that is part or the wider group of oracle-related security issues. Oracle manipulation-related incidents are one of the most common types of high-value smart contract security incidents.

Properly consuming oracle data is not trivial and should include several steps such as selecting pools with high liquidity, using multiple trusted oracles, configuring advanced features like TWAP or checking the data's freshness among others.

3 Secure Programming in Solidity

3.1 Classifications: SWC Registry and EEA EthTrust Security Levels

One of the first efforts to provide a common understanding of security vulnerabilities in smart contracts is the Smart Contract Weakness Classification (SWC).[1] This registry was developed having in mind the popular Common Weakness Enumeration (CWE)[2] in order to provide a common language for smart contract vulnerabilities that could enable a more systematic analysis of current vulnerabilities similar to the OWASP TOP 10[3] or SNAS TOP 25.[4]

All the entries in the SWC include a description of the vulnerability, related CWE, information on how to avoid this vulnerability, related references and sample code that can help understand how the vulnerability works.

The SWC is no longer actively maintained, it is incomplete and outdated, its last update was in 2020. However, it is still a good starting point when learning about most common smart contract vulnerabilities. For guidance on smart contract vulnerabilities, current options are the Enterprise Ethereum Alliance (EEA)[5] EthTrust Security Levels specification,[6] the Smart Contract Security Verification Standard (SCSVS) and the Smart Contract Security Field Guide.[7]

---

[1] https://swcregistry.io/.
[2] https://cwe.mitre.org/.
[3] https://owasp.org/www-project-top-ten/.
[4] https://www.sans.org/top25-software-errors/.
[5] https://entethalliance.org/.
[6] https://entethalliance.org/specs/ethtrust-sl/.
[7] https://scsfg.io/.

The EEA EthTrust Security Levels is developed by the Enterprise Ethereum Alliance (EEA) as part of its EEA EthTrust Certification. Apart from introducing some general security considerations and recommended good practices, it defines three Security Levels:

- Security Level [S] is the most basic level of security, easily checked using automated tools. It avoids complications that would require manual checking, leaving them for higher security levels.
- Security Level [M] is a medium level of security that requires the involvement of an experienced human auditor and some manual code review.
- Security Level [Q] is the highest level, including a comprehensive security assessment that covers all aspects of the smart contract: business logic, coding quality, and documentation.

This is an example of a basic security requirement in Level[S].

It is fairly easy to check using automated tools if the code includes the CREATE2 opcode, in case it does the auditor would need to check manually whether this could lead to a security risk under the considerations of security level[M].

Another interesting initiative is the Smart Contract Security Verification Standard,[8] whose current version is 2.0, that aims to maintain a checklist to standardize the security of smart contracts for developers, architects, security reviewers, and vendors. The controls are divided in three categories:

- General and well known security vulnerabilities related with the design of the project, its code, upgrade policies, etc.
- Vulnerabilities related to external dependencies such as oracles, bridges or liquidity pools.
- Vulnerabilities related to the Integration with external projects.

3.2 Battle Tested Contracts

The Solidity language is relatively new but is heavily influenced by other popular programming languages such as Javascript, Python or C++. As with any modern programming language, it is always better to look for available libraries before trying to reinvent the wheel. This is especially true when it comes to cryptographic and security-related functionality.

---

[8] https://github.com/ComposableSecurity/SCSVS.

One of the main sources for trusted solidity code is the OpenZeppelin Contracts Libraries.[9] In these libraries we can find contracts for:

- Access Control: Everything related to how to restrict access to your solidity methods
- Tokens: Covering most popular token definitions, e.g. ERC20 and ERC721
- Utilities: Generic tools such as signature verification, new data types and modifiers to make your code safer.

The latest version of the library at the time of writing is v5 and offers 11 APIs. This is just a short summary of the available modules:

- `Pausable`: Module to implement an emergency stop mechanism to halt activity in your contracts.
- `ReentrancyGuard`: Module to protect from reentrancy attacks.
- `AccessControl`: Module to implement role-based access control in your smart contract.
- `PullPayment`: Module to implement a pull-payment strategy where the recipient must withdraw the funds from the contract.
- `SafeMath`: Module that implements safer math checks in arithmetic operations.
- `*Proxy`: Modules to implement the proxy upgrade pattern in your smart contracts.

José Carlos Ramírez is a security engineer and lecturer specializing in blockchain technologies and smart contract security. Throughout his career, he has been focused on offensive security techniques; first, as a pen-testing consultant working on web applications, mobile applications and cloud environments, and later as a smart contract auditor in both the Cosmos and Ethereum ecosystems.

Since 2022 he has been involved in teaching smart contract security for Spanish organizations such as the University of Malaga. In addition, he has been working on public resources for students that can be found on his Github, such as the mock audit environment "Failapop" depicting a vulnerable multi-contract protocol written in Solidity.

Nowadays he has left the consultancy work to be part of the internal security team at Matter Labs. Leveraging his expertise in smart contracts and blockchain security he helps build the most secure ZK-based technology possible in the zkSync stack.

Isaac Agudo is the Director of the program on blockchain technologies at the University of Málaga, where he also holds the position of Associate Professor. His research focuses on applied cryptography, with an extensive record of publications in the field. He is the inventor of several patents and has participated in numerous research projects in different roles, contributing his

---

[9] https://www.openzeppelin.com/contracts.

expertise across both academic and technical domains. He is also the founder of Decentralized Security, a university spin-off operating at the intersection of blockchain and cybersecurity. His work supports the development of secure and innovative solutions in blockchain and cryptographic technologies.

Decentralised Finance (DeFi)

DeFi and Its Implications on Enterprise Software and Business Processes

Simon Engel

1 Introduction to Decentralized Finance (DeFi)

In recent years, Decentralized Finance (DeFi) has emerged as a transformative and disruptive force within the financial industry. DeFi concepts find adoption in other industries like gaming or enterprise software. The decentralized nature of DeFi leverages blockchain technology, allowing for trustless and permissionless financial transactions, thereby challenging traditional financial intermediaries.

1.1 Definition and Overview of DeFi

Decentralized Finance (DeFi) is a revolutionary concept that utilizes blockchain technology to establish a financial ecosystem free from traditional intermediaries like central banks and other common financial institutions. It enables open and permissionless access to various financial services and eliminates the need for centralized authorities. DeFi's defining features include trustlessness, interoperability, transparency, and programmability, all made possible through the use of smart contracts and decentralized applications (DApps) on blockchain platforms like Ethereum [1].

S. Engel (✉)
Heidelberg, Germany
e-mail: simon.engel19@gmail.com

1.2 Evolution and Growth of DeFi

The evolution of DeFi can be traced back to the inception of Bitcoin in 2009 and the subsequent development of Ethereum in 2015, which introduced the concept of smart contracts. This laid the foundation for DeFi and led to several notable milestones:

- The Initial Coin Offering (ICO) boom in 2017 fueled the development of DeFi projects.
- The rise of decentralized exchanges (DEXs) like Uniswap and SushiSwap provided alternatives to centralized exchanges [2].

DeFi's growth has been driven by liquidity provision, yield farming, institutional interest, and regulatory challenges. It has attracted substantial investments and institutional attention, which has contributed to its rapid expansion. With a global crypto market cap of $1.42 trillion in November 2023 [3].

1.3 Key Concept and Components of DeFi

DeFi encompasses a wide array of key concepts and components that work synergistically to create a decentralized financial ecosystem. Understanding these elements is essential for navigating the DeFi landscape effectively. Following is a list of the major concepts. It's not exhaustive due to the constant innovation and change happening.

Smart Contracts

A foundational element of DeFi, are self-executing contracts with coded terms. They automate financial transactions, removing the need for intermediaries. Ethereum serves as the primary platform for DeFi smart contracts.

Decentralized Applications (DApps)

DeFi DApps are software applications that run on blockchains without central control. These include lending platforms (e.g., Compound), DEXs (e.g., Uniswap), and yield farming protocols (e.g., Yearn Finance).

Decentralized Exchanges (DEXs)

DEXs enable direct cryptocurrency trading without intermediaries. They rely on smart contracts and automated market-making algorithms. Notable DEXs include Uniswap and SushiSwap.

Liquidity Pools

Consist of assets used for trading on DEXs. Users contribute assets to these pools, earning fees and rewards. Platforms like Balancer and Curve offer liquidity pool services.

Yield Farming

Involves providing liquidity to DeFi protocols in return for rewards, often in the form of governance tokens. Users optimize their returns by strategically moving assets between different protocols.

Decentralized Lending and Borrowing

DeFi lending platforms (e.g., Aave and Compound) enable users to lend their cryptocurrencies and earn interest. Borrowers can secure loans by using their assets as collateral, bypassing traditional banking systems.

Decentralized Stablecoins

Stablecoins like DAI and USDC are cryptocurrencies designed to maintain a stable value. They play a pivotal role in DeFi by offering a stable medium of exchange and store of value.

2 Understanding Enterprise Software and Business Processes

2.1 Enterprise Software: Definition and Importance

Enterprise software, often referred to as Enterprise Resource Planning (ERP) software, is a class of software applications designed to streamline and manage various aspects of an organization's operations. It plays a crucial role in modern businesses, enhancing efficiency, data integration, and decision-making processes. Enterprise software typically includes modules for accounting (Finance), human resources (HR), supply chain management (SCM), customer relationship management (CRM), and more. It integrates various functions within an organization, allowing for a seamless flow of information and resources. The importance of enterprise software lies in its ability to centralize data, automate tasks, and optimize business processes, ultimately leading to increased productivity and cost savings [4].

2.2 Business Processes: Overview and Significance

Business processes are the structured, systematic activities that organizations undertake to achieve specific goals and deliver value to customers. They encompass a wide range of operations, from procurement and production to sales and customer support. Understanding and optimizing these processes are vital for achieving operational efficiency and maintaining a competitive edge. Business processes are not static; they evolve and adapt to changing circumstances, which makes their management and optimization an ongoing endeavor. Significantly, aligning business processes with enterprise software systems can lead to streamlined operations, improved data management, and enhanced customer service. This alignment is crucial for organizations aiming to remain agile and responsive to market changes [5].

2.3 Benefits of Traditional Enterprise Software and Business Processes

Process and Workflow Optimization

Enterprise software is able to provide great transparency on common business processes like accounting, finance, bookkeeping, customer service, warehouse management, supply chain management and other functions. It increases the predictability, plannability, and flexibility to deal with the usual business operations. In addition, manual work and the potential to automate process steps with software and new technologies can provide a basis for future improvements.

Single-Source-of-Truth

Having an integrated enterprise software landscape with managed databases, master data management, and master data governance can help to provide a single-source-of-truth to all business functions in the enterprise. This prevents confusion, double-work, and makes analysis on important business objects easier.

Real-Time Reporting and Analytics

With enterprise software in place, analytics and real time reporting becomes much easier. Establishing KPIs and tracking them with the right system provides stakeholders at all function with the right data at the right time.

2.4 Challenges and Limitations of Traditional Enterprise Software and Business Processes

Legacy Systems and Integration Challenges

Many organizations still rely on legacy and highly customized enterprise software systems, which are often outdated and inflexible. Integrating these systems with modern technologies and evolving business needs can be a complex and costly endeavor. Data silos may arise, hindering information flow across the organization and impeding data-driven decision-making. Modernizing these systems without disrupting ongoing operations is a significant challenge that organizations face.

Scalability and Adaptability

Traditional enterprise software may lack the scalability and adaptability required to accommodate rapid growth or shifting market demands. As businesses expand, they must ensure that their software and processes can evolve in tandem, or they risk inefficiency and lost opportunities. The challenge is to maintain a delicate balance between stable, reliable processes and the ability to adapt to change [6].

Technological Advances and Emerging Trends

The pace of technological change is relentless. Organizations must grapple with the challenge of adopting emerging technologies like Artificial Intelligence (AI), Internet of Things (IoT), and blockchain while maintaining their existing enterprise software and processes. Balancing innovation with stability is a persistent challenge in enterprise software.

3 Exploring the Implications of DeFi on Enterprise Software

DeFi provides exciting opportunities to rethink enterprise software and business processes. Following chapter will provide an overview on the integration, advantages, and disadvantages.

3.1 Integration of DeFi with Enterprise Software

The integration of Decentralized Finance (DeFi) with traditional enterprise software systems presents a compelling opportunity to enhance various aspects of business operations. DeFi protocols and technologies can be seamlessly integrated with enterprise software to provide new functionalities, streamline processes, and unlock novel possibilities for businesses. The integration process involves the creation of secure interfaces and data connectors that allow DeFi applications to interact with existing enterprise software, including Enterprise Resource Planning (ERP), Customer Relationship Management (CRM), and supply chain management systems (SCM).

3.2 Advantages and Benefits of DeFi for Enterprise Software

Enhanced Security

DeFi introduces enhanced security measures through blockchain technology and smart contracts. Data immutability, cryptographic verification, and decentralized control mechanisms contribute to a more secure environment for financial transactions and data management within enterprise software.

Cost Reduction

DeFi's automation capabilities and elimination of intermediaries can lead to significant cost reductions within enterprise software systems. Smart contracts, for instance, automate payment processes, reducing administrative overhead and the potential for errors. Additionally, DeFi's decentralized nature minimizes the need for costly intermediaries, resulting in cost savings across various operational facets.

Transparency and Audibility

DeFi's transparency and immutable record-keeping on the blockchain enhance the traceability and auditability of financial transactions. This feature is invaluable in regulatory compliance, financial reporting, and internal control processes within enterprise software systems. It ensures that all transactions are transparent and accountable, fostering trust and compliance with industry regulations.

3.3 Disadvantages and Drawbacks for DeFi in Enterprise Software

Regulatory Uncertainty

One of the significant challenges of integrating DeFi into enterprise software is navigating the evolving regulatory landscape. DeFi operates across borders, making it subject to varying and sometimes conflicting regulatory requirements. Enterprises need to carefully assess and adapt to changing regulations, which can be time-consuming and complex.

Smart Contract Vulnerabilities

While smart contracts are a core element of DeFi, they are not immune to vulnerabilities. Coding errors or vulnerabilities in smart contracts can lead to financial losses and security breaches. Enterprises must implement rigorous testing and security measures to mitigate these risks effectively.

Market Volatility

DeFi systems often rely on cryptocurrency assets, which are known for their price volatility. Enterprises that integrate DeFi solutions may be exposed to market fluctuations, impacting their financial stability. This aspect requires careful risk management and hedging strategies to mitigate potential losses.

Complex Integration

Integrating DeFi with existing enterprise software can be complex and resource-intensive. It may require substantial changes to existing systems, potentially causing operational disruptions. Organizations must carefully plan and execute the integration process to minimize potential drawbacks.

3.4 Enhancing Efficiency and Transparency in Business Operations

DeFi's impact on business processes extends to enhancing efficiency and transparency. By utilizing blockchain technology, DeFi offers an immutable and transparent ledger where all transactions and data are recorded. This feature can be leveraged to improve the traceability and auditability of business operations, crucial for regulatory compliance and internal control processes.

Furthermore, DeFi's transparency extends to financial transactions, providing stakeholders with a real-time view of transactions, payments, and financial data. This transparency is vital for building trust among stakeholders and fostering accountability within the organization. Enhanced visibility into financial processes can lead to better decision-making, as it enables organizations to identify and address bottlenecks, inefficiencies, and discrepancies more quickly.

The use of DeFi can also provide a competitive advantage, as it enables organizations to offer more transparent and efficient services to their customers. For instance, financial institutions can provide real-time visibility into the status of transactions, reducing customer inquiries and improving overall satisfaction. The transparency offered by DeFi can be a key differentiator in a market where customers increasingly value accountability and visibility.

4 Case Studies: Examples of DeFi Integration in Enterprise Software

In this chapter, we will explore real-world case studies of organizations that have successfully integrated Decentralized Finance (DeFi) into their enterprise software systems. These examples highlight the practical applications and benefits of DeFi in diverse business contexts.

4.1 Case Study 1: DeFi Implementation in Supply Chain Management Software

Organizations: SAP and the U.S. Pharma Industry [7].

Background: The U.S. Drug Supply Chain Security Act (DSCSA) got signed in 2013 with a 10-year implementation plan. Objective is to make the pharmaceutical supply chain more transparent. It included one important key feature to make it possible to authenticate all pharmaceutical products returned to the wholesaler.

Solution: Each package is verified using a scanner and code to uniquely identify each product. One use case includes where the wholesaler is scanning the product to verify the authenticity. Through blockchain it's possible to track the journey of each product and make sure it's no counterfeit. The real world asset and it's financial can be traced across the whole value chain.

Benefits

- **Data protection**: Only product data is used. No personal data like patient data.
- **Effort**: It takes just 6 months to put the solution in place.
- **Transparency**: The blockchain-based system enhanced the transparency of transactions, allowing all stakeholders to view the status of the drug.
- **Security**: The blockchain cannot be altered and increases protection from counterfeit and traceability across multiple parties.

4.2 Case Study 2: DeFi Integration in Financial Management Systems

Organization: MakerDAO and Request Finance [8]. (Note from the authoer: MakderDAO has rebranded to Sky.money after submission of the article.)

Background: MakerDAO is a decentralized autonomous organization which consists of all member holding Maker-Tokens (MKR). Maker started to issue DAI which is an over-collateralized stablecoin pegged 1:1 to USD. It was one of the first stable coins enabling more finance use cases in DeFi without the volatility of regular crypto tokens like Bitcoin and Ethereum.

DeFi Integration: MakerDAO decided to move to their financial flow to request finance to provide Maker grantees and freelances an easier way to receive payments.

Benefits

- **Less manual task**: Community members expressed satisfaction with an invoicing solution that enabled them to "configure and disregard," featuring a notification system for updates and functionalities designed to reduce the time invested in billing management.

- **Reduced risk and fraud**: Through the payment request mechanism, funds are no longer sent to a manually entered address but are automatically directed using Request's intelligent payment technology. Upon completion of the payment, the invoice undergoes automatic reconciliation, saving time and ensuring accurate bookkeeping.
- **Decentralized invoicing**: Request Finance plays a pivotal role in supporting the decentralized finance (DeFi) community, advocating for a paradigm that empowers individuals to assert authority over their financial affairs and data. User ownership extends to all pieces of information, which are securely stored on the Ethereum blockchain.

4.3 Case Study 3: DeFi-Enabled Customer Relationship Management (CRM) and Loyalty Software

Organization: Lufthansa and Polygon [9].

Background: Flight guests of Lufthansa Group Airlines are able to scan their boarding card and get an NFT for it. The center piece is the integration of blockchain technology in the loyalty program and gamification.

DeFi Integration: Uptrip (a loyalty mobile app provided by Lufthansa) uses Polygon to mint the NFTs. Polygon is a Ethereum Layer-scaling solutions which makes transactions cheaper, faster, while using the EVM as execution engine.

Benefits

- **Openness**: Airline guests are able to transfer the NFT to their own wallet.
- **Interoperability**: With the NFT transferred to an own wallet the users are able to sell or trade it in the Web3 ecosystem like OpenSea or other marketplaces.
- **Gamification**: Through gamification the uses stay longer on the product and come back to use the offer.
- **Education**: Having a simple MVP complemented with an additional loyalty program with great benefits proofed to be a great way to educate non-Web3 users about the possibility.

5 Future Trends and Opportunities in DeFi and Enterprise Software

This chapter provides an overview on emerging technologies in DeFi and its implications with opportunities and challenges for enterprise software.

5.1 Emerging Technologies and Innovations in DeFi

The future of Decentralized Finance (DeFi) holds promise with a range of emerging technologies and innovations set to shape the landscape. These developments not only enhance the capabilities of DeFi but also present opportunities for integration with enterprise software systems.

Layer 2 Solutions

Layer 2 solutions, such as rollups and sidechains, are emerging to address the scalability and congestion issues faced by the Ethereum network. These solutions enable faster and more cost-effective transactions, making DeFi more accessible for enterprises. Integrating Layer 2 solutions with enterprise software can further streamline financial operations [10].

Non-fungible Tokens (NFTs)

NFTs, which represent unique digital assets, have gained popularity in the art, gaming, and entertainment industries. The integration of NFTs with DeFi can provide new opportunities for enterprises, such as collateralizing NFT assets for loans or creating innovative financial products based on digital collectibles.

Decentralized Autonomous Organizations (DAOs)

DAOs are gaining prominence as a form of decentralized governance and decision-making in DeFi. Enterprises can leverage DAOs to enable decentralized voting and consensus mechanisms for important business decisions. This not only enhances transparency but also fosters community engagement and trust.

Central-Bank Digital Currency (CDBC)

CDBC represent a public issued digital money usually distributed by the central banks. There are several projects by leading central banks to explore how and what benefits they can create.

5.2 Potential Disruptions and Transformations in Enterprise Software

Financial Operations

DeFi integration can revolutionize financial operations within enterprises. Traditional banking processes, such as payment approvals, payroll, and expense management, can be automated and executed more efficiently through smart contracts. DeFi's transparency also enhances auditability and financial reporting.

Supply Chain Management

Supply chain management can benefit from DeFi's transparency and automation. The use of smart contracts in supply chain finance can streamline the financing of transactions, optimize inventory management, and reduce risks related to delayed payments.

Cross-Border Trade

DeFi solutions for cross-border payments can significantly impact international trade. Enterprises can reduce transaction costs and settlement times, making global trade more efficient. This transformation can lead to increased international business opportunities and trade partnerships.

5.3 Opportunities for Enterprises to Leverage DeFi for Competitive Advantage

Financial Efficiency

By automating financial processes and reducing intermediary costs, enterprises can enhance their financial efficiency. This cost-saving advantage can be reinvested in business growth or passed on to customers, making the organization more competitive.

Innovation and Product Development

Enterprises can explore innovative financial products and services based on DeFi principles. This includes offering DeFi-based loans, decentralized finance products, and customized financial solutions for customers. Such innovations can set enterprises apart from competitors and attract a broader customer base.

Global Expansion

DeFi-enabled cross-border payments and international financial operations can open doors for enterprises to expand globally. Enterprises can tap into new markets, establish international partnerships, and engage with a broader customer base, increasing their global presence and competitiveness.

6 Conclusion and Future Outlook

In conclusion, DeFi offers exciting opportunities for businesses to adopt DeFi. In addition, it gives enterprise software vendors and their ecosystem new innovative technologies to make business processes at scale better.

6.1 Implications for Enterprises and Business Leaders

The implications of DeFi in enterprise software are profound and offer significant benefits for organizations and business leaders.

- **Embrace Innovation**: Business leaders should embrace DeFi as an innovative and transformative technology that can enhance financial processes and provide a competitive edge. By exploring DeFi solutions and integrating them into enterprise software, organizations can position themselves as industry leaders.
- **Prioritize Security**: Security is paramount when adopting DeFi. Enterprises must invest in rigorous security measures, conduct regular audits, and stay vigilant against smart contract vulnerabilities and phishing attempts. Protecting user data and financial assets is essential.
- **Navigate Regulatory Landscape**: The evolving regulatory landscape poses challenges for DeFi adoption. Business leaders should stay informed about changing regulations and engage with authorities to ensure compliance. Collaboration with legal experts is essential for navigating the complex regulatory environment.
- **Educate Employees and Users**: Proper education and training are crucial for employees and users interacting with DeFi platforms. Providing guidance on recognizing phishing attempts, secure wallet management, and best practices for DeFi security can mitigate risks and build user trust.
- **Plan for Integration**: Integration of DeFi with enterprise software is a complex process that requires careful planning and execution. Business leaders should collaborate with IT and finance teams to create a roadmap for integration and manage potential disruptions.

6.2 Predictions for the Future of DeFi in Enterprise Software

Predictions

- DeFi Integration Growth: The adoption of DeFi in enterprise software will continue to grow as organizations recognize the benefits of enhanced financial efficiency, transparency, and automation.
- Regulatory Evolution: Regulatory frameworks will evolve to accommodate DeFi, providing clearer guidelines and compliance standards. Enterprises should anticipate regulatory changes and adjust their operations accordingly.
- Increased Security Measures: Enterprises will place a strong emphasis on security, investing in advanced cybersecurity technologies, audit processes, and user education to protect against evolving threats.

References

1. V. Buterin, Ethereum: A Next-Generation Smart Contract and Decentralized Application Platform (Ethereum White Paper, 2013)
2. W. Mougayar, The Business Blockchain: Promise, Practice, and Application of the Next Internet Technology (Wiley, USA, 2016)
3. Coinmarketcap, Cryptocurrency market. https://coinmarketcap.com/charts/. Accessed: 11 Nov 2023
4. T. Davenport, Putting the enterprise into the enterprise system. Harv. Bus. Rev. **76**(4), 121–131 (1998)
5. R.S. Kaplan, D.P. Norton, The balanced scorecard—measures that drive performance. Harv. Bus. Rev. **70**(1), 71–79 (1992)
6. C. Eckes, The Six Sigma Revolution: How General Electric and Others Turned Process into Profits (Wiley, USA, 2002)
7. A. Schmitz, Pharmaceutical Blockchain Verifies Returns. SAP, https://news.sap.com/2019/11/pharmaceutical-blockchain-return-verification/. Accessed 11 Nov 2023
8. Request, MakerDAO—Unlocking compliant cryptocurrency flows. Request Finance https://www.request.finance/use-cases/makerdao. Accessed: 19 Nov 2023
9. Miles & More, NFT-Smmelkarten-App geht an den Start. Miles & More: https://www.miles-and-more.com/de/de/program/news/nft-trading-card-app-goes-live-news-2023.html. Accessed: 11 Nov 2023
10. M. Teutsch et al., TrueBit: a scalable verification solution for blockchains, in *Proceedings of the 2017 ACM SIGSAC Conference on Computer and Communications Security*, pp. 703–720 (2017)

Simon Engel is Director, Product Management and Product Strategyt at SAP. In this role, he is responsible for driving cross-product management and product strategy across SAP's solution areas and with strategic customers and partners. Previous to his role in the Cross Product Engineering and Experience Area, he was part of the SAP Product Engineering Board Office and Customer Officer to the Executive Board Member. Responsible for driving and directing strategic customer and partner engagements. Simon started his career as part of the vocational training at SAP and holds a bachelor's degree in business information systems from the Baden-Wuerttemberg Cooperative State University (DHBW).

Currently he's conducting his MBA at the University of St. Gallen (HSG). His interest in blockchain and Web3 started with a rotation and leveraging the technology for reconciliation efforts between bank payments and enterprise processes. Realizing the advantages of public-permission blockchains led him to dive deeper into the technology. Always with the need to solve business problems and enable business outcomes. Since, then he's been actively working in the space e.g. writing blobs, contributing to DAOs, or bringing enterprise processes to blockchains. He believes in the tremendous opportunity of blockchain and privacy-enhancing technologies giving people and organizations more freedom, self-determination, and sovereignty across the globe. https://www.linkedin.com/in/simonengel/, https://x.com/Engel_Simon.

Programmable Money: Aligning Your Money with Your Values

Selin Sezer

1 Introduction

Optimal well-being, according to values theory exemplified by [1] and [2], is believed to be achieved through the alignment of one's behavior and environment with their personally held value system [3]. This notion is also reflected in our everyday financial decisions, where we are constantly confronted with the question of how to use our money to maximize happiness. The concept of individuals making rational economic decisions solely based on self-interest has been debunked, as financial choices are influenced by emotional, historical, familial, and personal factors, encompassing a broad spectrum of considerations beyond purely financial criteria [4]. Various studies [1, 2, 5] highlight the important role personal values play in shaping spending behaviors. Examples include but not limited to sustainable practices [6], ethical considerations [7], conscious consumption [8], and commitment to fair-trade principles [9]. The extensive literature further establishes this connection by illustrating the influence of values on diverse consumer behaviors, ranging from preferences in food choices and leisure travel to pro-environmental attitudes [9–12]. Moreover, the exploration of the relationship between the self and money [13, 14] underscores the multifaceted role of money as both an extension of the self and a symbol embodying aspects of power, success, gender identity, and overall well-being. Furthermore, the significance of these individual decisions reaches the global market, with fair-trade consumers serving as a compelling illustration of their substantial impact [15].

The convergence of globalization and digitalization, culminating in the emergence of innovative payment solutions from Fintech and Big Tech players, has not only driven the evolution of financial landscapes but also compelled central banks to assess potential upgrades to payment systems and currency frameworks over the past decades [16]. The introduction of Bitcoin in 2009 played a pivotal role in popularizing

S. Sezer (✉)
Fraunhofer FIT, Sankt Augustin, Germany
e-mail: selin.sezer@fit.fraunhofer.de

digital money, paving the way for different variations like cryptocurrencies, stable coins, and central bank digital currencies (CBDCs) [17]. According to a survey by the Bank for International Settlements in 2022, 93% of central banks actively engaged in CBDC initiatives, showcasing a growing trend towards the practical implementation of CBDCs globally, with a particular emphasis on the advancement of retail CBDC projects, and the survey predicts the potential existence of 15 retail and nine wholesale CBDCs in circulation by 2030 [18]. These actions provide the potential for incorporating innovative features, among which programmability stands out, contributing heightened value when compared to existing systems. In this paper, our focus centers on a specific form of programmability, specifically programmable money augmented with self-sovereign identity—an emerging digital identity solution. We explore how this combination can be effectively utilized to pave the way for innovative approaches to integrating personal values into individual spending habits.

The organization of the paper is as follows. Section 2 provides an introduction to programmable money. Section 3 navigates through SSI concepts, shedding light on their implications for programmable money. Section 4 further explores the potential of programmable money on reflecting personal values into our financial choices, and finally, Sect. 5 concludes the chapter by presenting challenges that need to be further investigated.

2 Programmable Money

When digital money integrates specific attribute information and inherent logic for self-control, it assumes the identity of programmable money [19]. Unlike traditional forms of currency, programmable money is designed to have programmable features, allowing for customization of its functionality based on predefined conditions or criteria, creating a dynamic and intelligent form of money.

The programmability of such currency enables it to be spent or earned under predefined circumstances [20]. This means that transactions involving programmable money can be subject to specific conditions, such as who can spend it, what it can be spent on, when it can be spent, and other criteria set by the party funding the payment. The peer-to-peer transferable ability for these conditions to spread through the currency's life cycle is attributed to the direct embedding of behavior into the money, distinguishing it from systems that manage the currency and allowing for dynamic updates tailored to specific use cases [21–23].

The goal of programmable money is to introduce a level of automation and customization to financial transactions, reducing friction, increasing efficiency and transparency, and providing new possibilities for innovative financial applications. The adoption of programmable money presents numerous advantages, notably decreasing the risk of misappropriation in diverse funding scenarios like donations, grants, loans, and social welfare programs. Greater financial oversight is another merit, demonstrated in instances such as personal conditional transfers (e.g., pocket money [24]) and specified spending directives in legal documents (e.g., last will).

Project budgets stand to gain from automated auditing through predefined expenditure criteria, and applications like customer incentives benefit from refined data analytics. Furthermore, programmable money allows earmarking funds for targeted purposes, such as green energy endorsement by governments to address climate change [21, 22].

The programmable money concept is often associated with the use of smart contracts and blockchain technology, allowing for decentralized and automated execution of transactions. Despite existing programmable payment methods like standing orders and direct debits, the challenge of implementing intricate logic continues, limiting flexibility. Smart contracts provide a solution, enhancing flexibility and streamlining complex business process integration with payments. Additionally, efficiency is heightened by automating critical exchange processes, ensuring faster and cost-effective transactions, eliminating the need for intermediaries [25]. Finally, the smart contracts make it possible to dynamically check the conditions in a transparent way before the transaction occurs, contributing to the efficiency once more in comparison to traditional systems that handle this typically manually and post-transaction [23].

Achieving the desired level of complexity necessitates a seamless synchronization of data pertaining to the external environment, particularly the transaction dynamics, including details about participating parties, satisfaction of arbitrary conditions, transaction locations, and timing—an information flow challenging to attain within the confines of a Blockchain setting. In theory, the essential data can be sourced from diverse third-party APIs; however, concerns arise regarding their reliability, the intricate process of information gathering, and the speed of retrieval, highlighting the need for a standardized method to represent diverse conditions and assess their fulfillment in a reliable and transparent manner. Existing Blockchain applications [23, 26, 27] tackle this issue by simplifying conditions to values easily verified within a smart contract (e.g., price limits, expiration dates) or by manually curating eligibility lists for programmable money transactions.

While these methods may suffice for illustrative purposes, a scalable implementation of programmable money, capable of accommodating complex conditions, calls for an automated standard approach that isn't contingent on actors requiring excessive trust. Here, the use of verifiable credentials has the capability to enrich a large-scale programmable money ecosystem by dispersing identity management responsibilities among ecosystem participants, incorporating extra trust through collaboration with reliable institutions, securing user data sovereignty, establishing a resilient decentralized structure free from third-party dependencies, and providing a cohesive representation for various spending conditions associated with identities [21, 22].

3 Self-Sovereign Identity (SSI)

The Internet originated without inherent identification methods for connections [28]. Since then, developing a universal, uncomplicated means to identify individuals, organizations, or entities online has proved challenging and prompted diverse approaches and a continuous evolution in design [29]. In the contemporary digital landscape, major corporations like Google and Facebook serve as Identity Providers (IdP), issuing electronic identities that streamline user identification, authentication, and authorization for both their internal services and external service providers. While the convenience of a single identity for multiple services is a benefit for users who would otherwise be overwhelmed by managing numerous username-password combinations, the concentration of digital identities within a few IdPs poses security concerns, making them lucrative targets for hackers and raising issues of unauthorized data use and trade [30, 31]. Also, the validity of data provided by customers becomes a critical factor for service providers, posing risks of compromised data quality and sometimes generating costs related to fraud that lack traceability back to a specific individual [32].

The landscape of digital identity is undergoing a transformative shift with the emergence of a decentralized identity paradigm namely Self-Sovereign Identity, presenting users with unprecedented control over their identity data independent of centralized authorities and intermediaries. Recent advancements, particularly in Blockchain technology, have sparked motivation to explore feasibility of such a self-sovereign identity management system in various use-cases [33, 34]. In the SSI system, key actors play distinct roles in the lifecycle of verifiable credentials (VCs) and presentations (VPs). The holder, represented by entities like students, employees, and customers, possesses verifiable credentials and generates verifiable presentations. Issuers, which can be corporations, non-profits, governments, or individuals, assert claims about subjects, create verifiable credentials, and transmit them to holders. Subjects, such as individuals, animals, or objects, are entities about which claims are made. Verifiers, encompassing employers, security personnel, and websites, receive and process verifiable credentials. Additionally, a verifiable data registry acts as a system mediating the creation and verification of relevant data, including decentralized identifiers (DID), keys, credential schemas, and more. DIDs serve to distinguish entities such as individuals, objects, organizations, or abstract concepts, employing cryptographic algorithms for proof of information ownership. This decentralized model eliminates the necessity for a central registry or authority, mitigating the risks associated with a single point of failure or control [35]. This ecosystem establishes a triangle of trust, where holders, issuers, and verifiers interact within a framework facilitated by verifiable data registries [36] (See Fig. 1). It is also noteworthy that verifiable credentials are only as trustworthy as the verifier's trust in the issuer, therefore the trust assurance in business decisions are crucial [37].

Verifiable credentials function as digital statements containing claims about a subject, attributed by an issuer. Holders, manage them in their digital wallet, encapsulating details about the subject such as attributes, relationships, or entitlements.

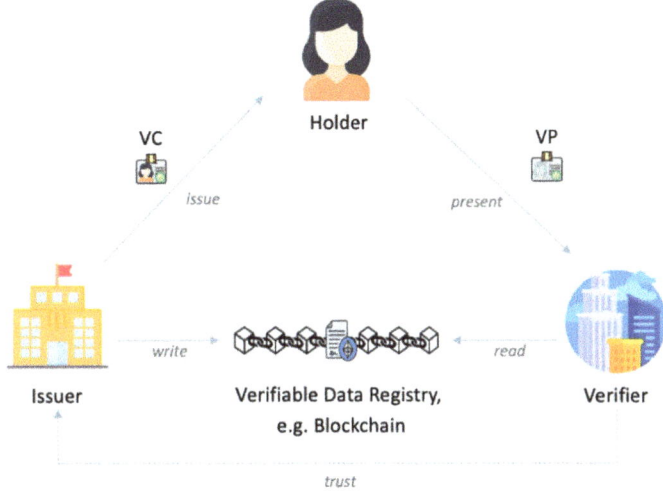

Fig. 1 SSI trust triangle [38]

The efficiency of verifiable credentials lies in their rapid digital verification facilitated by cryptographic protocols [37].

The versatility of verifiable credentials is showcased through various compelling use cases, each demonstrating a different transformative potential. Berg et al. [39] presents a use case outlining the implementation of an ecosystem for digital product passports, a standardized digital record designed to enhance product circularity and sustainability by providing detailed insights into a product's origin, composition, and lifecycle aspects, specifically focusing on the lifecycle of plastic products. A sample product passport provided by the manufacturer typically contains key information like product name, description, waste classification code, recycling information, product barcode, trade item number, and material property standards. Various organizations can enhance the material/product registry by linking information to either the manufacturer's DID or the product's DID, incorporating for example, sustainability certificates like the German eco label "Blauer Engel" or Product Ethics Passports such as the "Fairtrade" certificate. Tan et al. [40] outlines the execution of a pilot initiative that utilizes the European Blockchain Services Infrastructure (EBSI) to authenticate education credentials across borders. The key focus is on verifying student ID details and transcripts of records in a pilot case between KU Leuven and Università di Bologna, highlighting the distinctive digital transformation of the exchange process, particularly for exchange students. Another study that focuses on healthcare is MediLinker [41]. MediLinker functions as a blockchain-based identity wallet, enabling patients to issue and share verifiable credentials with healthcare providers. These VCs, aligned with the FHIR v4.0.1 standard, encompass demographic details, profile photos, and medication history for enhanced data interoperability across diverse Electronic Health Record (EHR) systems. These examples only scratch the surface, demonstrating that the range of potential applications is vast and extends beyond the instances mentioned.

Throughout history, the relationship between identity and currency has been deeply entwined [29], and the potential collaboration between digital money and digital identity, facilitated by blockchain infrastructure, has a transformative potential in the context of programmable money. Prompt validation of transaction conditions is vital for Programmable Money, and DIDs assure continuous accessibility for credential verification. They also enable users to selectively disclose data, preventing transaction profiling. The direct verification of credentials, avoiding third-party reliance, coupled with secure data exchange channels, establishes a swift, secure, and reliable system for sharing personal information. Trusted institutions, like government authorities or certified identity providers, issuing tamper-proof verifiable credentials, enhance system trust and minimize the risk of misuse, instilling confidence and in some cases, regulatory compliance. This involvement allows the issuer to extend trust to the spender, creating an additional layer of trust alongside the "code is law" principle. Portable VCs, not confined to the issuer, promote reuse and advanced cross-references across various services, particularly in scenarios involving conditional money exchange [21, 22].

In Fig. 2, we depict an instance of information flow in a blockchain-based programmable money system. The controller sets spending parameters, such as restricting money usage to individuals under 18 for book purchases. These conditions are then linked to a designated amount of digital money and sent to the sender. When the sender is at the bookstore, the decentralized application prompts the request for necessary credentials. The recipient provides identity credentials issued by the municipality to prove her age is below 18, while the bookstore presents product credentials confirming the items as books. The transaction is only completed if these conditions are satisfied, and subsequently in this scenario, the conditions are disassociated from the currency. One can observe that the trust triangle, as previously mentioned, is also present in this context, involving the controller and various issuers.

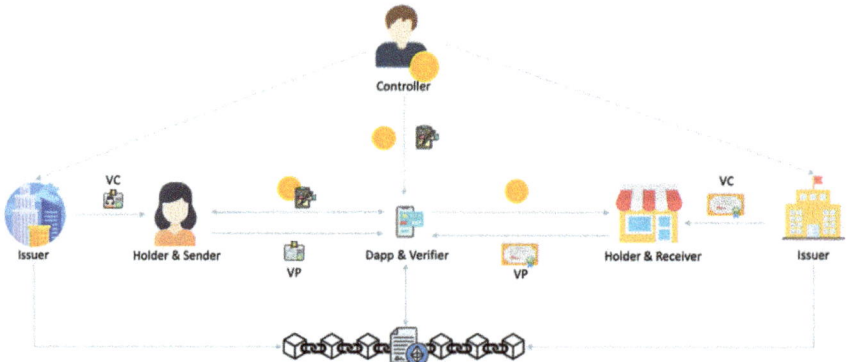

Fig. 2 Programmable money information flow [38]

4 Value-Embedded Money

Programmable money can act as a potential catalyst for reshaping the dynamics of spending behaviors of individuals by aligning their financial decisions with personal values and cultivate responsible financial habits. Simultaneously, from a prosocial standpoint, where individuals allocate funds to others rather than exclusively to oneself [3], the impact of programmable money becomes evident in encouraging behavioral changes that contribute to a broader societal as well as personal well-being.

For advocates of environmental sustainability, programmable money could be configured to only allow transactions for products with "eco-friendly" credentials, thereby enforcing a commitment to green initiatives. Ethical consumers may use programmable money to restrict spending to brands adhering to fair labor practices, indeed verified by credentials issued by accrediting organizations promoting a more equitable fashion industry. Supporting local businesses can be facilitated by setting spending conditions that favor transactions within a certain radius. Utilizing programmable money with verifiable credentials provides individuals with the means to solidify these aspirations beyond mere wishes.

For individuals seeking more responsible financial decision-making, programmable money offers additional advantages by assisting in measures against impulsive buying. Engaging in reckless financial decisions without careful consideration, especially in the era of consumerism and pervasive marketing, can lead to a series of consequences, encompassing not only financial and legal troubles but also inflicting psychological distress such as depression, guilt, and strained interpersonal relationships [42]. The underlying causes of impulsive buying, especially noticeable among the youth, are diverse and include several factors [43]. Pillai et al. [44] argues that the issue often arises not from a lack of financial literacy but from challenges associated with applying this knowledge practically to real-life situations. Taking this into account, programmable money can bridge the gap between financial knowledge and practical application. By enabling the creation of budgets for specific spending types, programmable money can effectively curb excessive consumption and introduce time-bound constraints, fostering more equitable money distribution throughout the month and mitigating impulsive financial behaviors. An additional example of its practical application is addressing the compelling need for savings, such as for college funds or retirement, where programmable money can provide a structured approach and verifiable credentials can assure that these allocations are freed on a timely manner.

The alignment of spending choices with individual values can seamlessly translate into prosocial spending. Taking the example of someone who prioritizes the environment and sustainable choices from the personal spending, this individual may choose to gift or donate programmable money specifically designated for purchasing green products, secured by verifiable credentials issued by trusted institutions in the form of product passes. The insights from [45] highlight that providing financial compensation and incentives, especially for the acquisition of "green" products such as eco-labeled or organic food, exerts a more substantial motivating influence on subsequent environmentally conscious purchases when compared to verbal encouragement and praise. By directing programmable money towards environmentally friendly options, this individual not only aligns their spending choices with their values but also utilizes incentivization through programmable money to encourage and support sustainable consumer behavior, thereby contributing significantly to their overall well-being [3]. An additional example targeting the formative years of financial attitudes (ages 17–21) [46] involves pocket money structured with programmable money [24], encouraging sustainable and responsible practices among emerging adults. Jorgensen et al. [47] emphasizes the positive impact of focusing on changing financial attitudes during this crucial period. In this scenario, not only can specific product groups be targeted, but also different rewarding schemes can be unlocked with the help of verifiable credentials. For instance, parents might allocate a certain amount of money that can only be used if their child present an education credential that shows that she has passed the Math class with grade over 7.5, thereby generating financial incentives with academic achievement.

5 Conclusion

Contrary to traditional notions, the evolving perception of money emphasizes its role as a tool that, when used skillfully, can contribute to increased happiness by nurturing positive relationships, instilling a sense of mastery, and enhancing overall well-being at both individual and societal levels, with the risk of negative consequences if used inappropriately [3]. Therefore, any application of programmable money should take ethical concerns into consideration to avoid misuse and discrimination. Although the integration of SSI concepts into programmable money opens vast possibilities and presents a transformative vision for the future of financial transactions, a thorough investigation into technical requirements and risks is highly necessary. Balancing privacy concerns and regulatory compliance when handling identity-related personal data poses a significant challenge. The fragmented state of the SSI infrastructure poses additional complications for the deployment of programmable money, as expounded in this paper, requiring its extensive adoption and standardized use for optimal functionality. Also, the utilization of a stable digital currency, backed by trusted institutions, emerges as a foundational element to enhance user acceptance, and facilitate seamless connections across use cases [21, 22]. Despite the considerable challenges, the possibilities for tailoring programmable money are vast. The scenarios outlined in this paper offer a glimpse into how these customizations can scale to organizational or governmental levels, highlighting the adaptability of programmable money in meeting various needs with the help of verifiable credentials.

References

1. M. Rokeach, *The Nature of Human Values* (Free Press, New York, NY, US, 1973)
2. S.H. Schwartz, Are there universal aspects in the structure and contents of human values? J. Soc. Iss. **50**, 19–45 (1994). https://doi.org/10.1111/j.1540-4560.1994.tb01196.x
3. G. Hill, R.T. Howell, Moderators and mediators of pro-social spending and well-being: the influence of values and psychological need satisfaction. Pers. Individ. Differ. **69**, 69–74 (2014). https://doi.org/10.1016/j.paid.2014.05.013
4. J.J. Kaye, M. McCuistion, R. Gulotta, D.A. Shamma, Money talks: tracking personal finances, in *Proceedings of the SIGCHI Conference on Human Factors in Computing Systems* (ACM, Toronto Ontario Canada, 2014), pp. 521–530
5. S.H. Schwartz, W. Bilsky, Toward a universal psychological structure of human values. J. Pers. Soc. Psychol. **53**, 550–562 (1987)
6. J. Thøgersen, F. Ölander, Human values and the emergence of a sustainable consumption pattern: a panel study. J. Econ. Psychol. **23**, 605–630 (2002). https://doi.org/10.1016/S0167-4870(02)00120-4
7. D. Shaw, E. Grehan, E. Shiu, L. Hassan, J. Thomson, An exploration of values in ethical consumer decision making. J. Consum. Behav. **4**, 185–200 (2005). https://doi.org/10.1002/cb.3
8. M. Pepper, T. Jackson, D. Uzzell, An examination of the values that motivate socially conscious and frugal consumer behaviours. Int. J. Consum. Stud. **33**, 126–136 (2009). https://doi.org/10.1111/j.1470-6431.2009.00753.x
9. C.J. Doran, The role of personal values in fair trade consumption. J. Bus. Ethics **84**, 549–563 (2009). https://doi.org/10.1007/s10551-008-9724-1
10. T. Dietz, L. Kalof, P.C. Stern, Gender, values, and environmentalism. Soc. Sci. Q. **83**, 353–364 (2002). https://doi.org/10.1111/1540-6237.00088
11. P. Honkanen, B. Verplanken, Understanding attitudes towards genetically modified food: the role of values and attitude strength. J. Consum. Policy **27**, 401–420 (2004). https://doi.org/10.1007/s10603-004-2524-9
12. R. Madrigal, Personal values, traveler personality type, and leisure travel style. J. Leis. Res. **27**, 125–142 (1995). https://doi.org/10.1080/00222216.1995.11949738
13. R.W. Belk, Possessions and the extended self. J. Consum. Res. **15**, 139–168 (1988). https://doi.org/10.1086/209154
14. M. Prince, Self-concept, money beliefs and values. J. Econ. Psychol. **14**, 161–173 (1993). https://doi.org/10.1016/0167-4870(93)90044-L
15. P.L. Taylor, D.L. Murray, L.T. Raynolds, Keeping trade fair: governance challenges in the fair trade coffee initiative. Sustain. Dev. **13**, 199–208 (2005). https://doi.org/10.1002/sd.278
16. M. Ferrari Minesso, A. Mehl, L. Stracca, Central bank digital currency in an open economy. J. Monet. Econ. **127**, 54–68 (2022). https://doi.org/10.1016/j.jmoneco.2022.02.001
17. J. Xu, Developments and implications of central bank digital currency: the case of China e-CNY. Asian Econ. Policy Rev. **17**, 235–250 (2022). https://doi.org/10.1111/aepr.12396
18. Making headway—results of the 2022 BIS survey on central bank digital currencies and crypto. BIS, Bank for International Settlements, Monetary and Economic Department, Basel (2023)
19. H. Masashi, H. Junichiro, Realizing Programmability in Payment and Settlement Systems (Bank of Japan, 2022)
20. J. Granados, A. Schlüter, Blockchain and payments for environmental services: tools and oppertunities for environmental protection (Bremen, 2023)
21. S. Sezer, W. Prinz, Combining self-sovereign identity with digital currencies to enable programmable money, in *Proceedings of the 1st Blockchain and Cryptocurrency Conference (B2C'2022)* (IFSA Publishing, Barcelona, 2022), pp. 23–28
22. S. Sezer, W. Prinz, Unlocking the future of money: programmable money enabled by self-sovereign identity. Blockchain Cryptocurrency **1**, 56–62 (2023)
23. I. Weber, M. Staples, Programmable money: next-generation blockchain-based conditional payments. Digital Finan.**18** (2022). https://doi.org/10.1007/s42521-022-00059-5

24. M. Avital, 3.1 Smart money: blockchain-based customizable payments system. Dagstuhl Rep. **7**, 104–106 (2017)
25. A. Bechtel, A. Ferreira, J. Gross, P. Sandner, The future of payments in a DLT-based European economy: a roadmap, in *The Future of Financial Systems in the Digital Age*, ed. by M. Heckel, F. Waldenberger (Springer, Singapore, 2022), pp. 89–116
26. T. Kolehmainen, G. Laatikainen, J. Kultanen, E. Kazan, P. Abrahamsson, Using blockchain in digitalizing enterprise legacy systems: an experience report, in *Software Business*, ed. by E. Klotins, K. Wnuk (Springer, Cham, 2021), pp. 70–85
27. Monetary Authority of Singapore, Purpose Bound Money (PBM) Technical Whitepaper (Monetary Authority of Singapore (MAS), 2023)
28. K. Cameron, The Laws of Identity(2005)
29. A. Preukschat, D. Reed, Why the internet is missing an identity layer—and why SSI can finally provide one, in *Self-Sovereign Identity: Decentralized Digital Identity and Verifiable Credentials* (Manning Publications, 2021)
30. M. Mollik, N. Pohlmann, Trust as a Service—Vertrauen als Dienstleistung—Validierung digitaler Nachweise mit der Blockchain (IT Sicherheit, 2019)
31. F. Schardong, R. Custódio, Self-sovereign identity: a systematic review, Mapping and Taxonomy. Sensors **22**, 5641 (2022). https://doi.org/10.3390/s22155641
32. V. Schlatt, J. Sedlmeir, S. Feulner, N. Urbach, Designing a framework for digital KYC processes built on blockchain-based self-sovereign identity. Inf. Manag. **59**, 103553 (2022). https://doi.org/10.1016/j.im.2021.103553
33. M.S. Ferdous, F. Chowdhury, M.O. Alassafi, In search of self-sovereign identity leveraging blockchain technology. IEEE Access **7**, 103059–103079 (2019). https://doi.org/10.1109/ACCESS.2019.2931173
34. C.D. Nassar Kyriakidou, A.M. Papathanasiou, G.C. Polyzos, Decentralized identity with applications to security and privacy for the internet of things. Comput. Netw. Commun. (2023). https://doi.org/10.37256/cnc.1220233048
35. M. Mollik, R. Klemens, Analysing the SSI World—Whitepaper. eco Association of the Internet Industry e.V (2023)
36. M. Sporny, D. Longley, D. Chadwick, Verifiable credentials data model v1.1, in *W3C Recommendation* (2022). https://www.w3.org/TR/vc-data-model/
37. A. Preukschat, R. Joosten, O. van Deventer, The basic building blocks of SSI, in *Self-Sovereign Identity: Decentralized Digital Identity and Verifiable Credentials* (Manning Publications, 2021)
38. Icons in Figures: Flaticon.com
39. H. Berg, R. Kulinna, C. Stöcker, S. Guth-Orlowski, R. Thiermann, N. Porepp, *Overcoming Information Asymmetry in the Plastics Value Chain with Digital Product Passports*, Wuppertal Paper no. 197 (2022)
40. E. Tan, E. Lerouge, J. Du Caju, D. Du Seuil, Verification of education credentials on european blockchain services infrastructure (EBSI): action research in a cross-border use case between Belgium and Italy. BDCC **7**, 79 (2023). https://doi.org/10.3390/bdcc7020079
41. J.R. Bautista, D.T. Harrell, L. Hanson, E. de Oliveira, M. Abdul-Moheeth, E.T. Meyer, A. Khurshid, MediLinker: a blockchain-based decentralized health information management platform for patient-centric healthcare. Front. Big Data **6**, 1146023 (2023). https://doi.org/10.3389/fdata.2023.1146023
42. M. Lejoyeux, A. Weinstein, Compulsive buying. Am. J. Drug Alcohol Abuse **36**, 248–253 (2010). https://doi.org/10.3109/00952990.2010.493590
43. R. Qin, Analysis on the causes and solutions of teenagers' excessive and impulse consumption. Adv. Soc. Sci. Educ. Human. Res. **554** (2021)
44. K.R. Pillai, R. Carlo, R. D'souza, Financial Prudence among Youth (2010)
45. P. Lanzini, J. Thøgersen, Behavioural spillover in the environmental domain: an intervention study. J. Environ. Psychol. **40**, 381–390 (2014). https://doi.org/10.1016/j.jenvp.2014.09.006
46. G. Meredith, C. Schewe, The power of cohorts: Americans who shared powerful experiences as young adults, such as the hardship of the great depression, fall into six cohorts. Am. Demogr. **16**, 22–22 (1994)

47. B.L. Jorgensen, D. Foster, J.F. Jensen, E. Vieira, Financial attitudes and responsible spending behavior of emerging adults: does geographic location matter? J. Fam. Econ. Iss. **38**, 70–83 (2017). https://doi.org/10.1007/s10834-016-9512-5

Selin Sezer is a Web3 developer and research associate at Fraunhofer FIT, pursuing a PhD in Computer Science at RWTH Aachen University. With a background in Web3 applications for supply chains, circular economy, and education, she focuses on digitalization and blockchain adoption. Her current research explores novel conditional payments with programmable money and explores the impacts of decentralized digital identity solutions in this context.

Selin holds a Bachelor's degree in Computer Engineering from Istanbul Technical University and a Master's in Media Informatics from RWTH Aachen. Passionate about integrating Web3 into daily tools, she aims to make a significant impact on the digital landscape through practical applications of these technologies.

A Short History of Decentralized Finance (DeFi)

Marcelo Emmerich

1 Introduction

DeFi, or decentralized finance, is inextricably linked to the rise of cryptocurrencies and blockchain technology. It is both a large-scale vision for a new way of conducting financial transactions and an umbrella term for various decentralized financial products and services known as protocols. While DeFi owes its existence to the launch of the Ethereum blockchain and smart contracts in 2015, it saw its major period of growth, known as DeFi Summer, in 2020.

The core values of the mother of all cryptocurrencies, Bitcoin, which include decentralization, transparency, security, and financial inclusivity, are deeply embedded in the ethos and operational mechanisms of DeFi. Here's how these values are reflected in DeFi:

Decentralization: The foundational principle of Bitcoin is decentralization, which means removing the control of a central authority over the currency. DeFi takes this principle further by applying it to various financial services. In DeFi, financial applications and services operate on decentralized networks, typically blockchain platforms. This decentralization allows for peer-to-peer financial transactions without the need for intermediaries like banks or financial institutions, echoing Bitcoin's vision of a decentralized financial system.

Transparency: Bitcoin operates on a public ledger, allowing anyone to verify transactions. DeFi platforms also emphasize transparency, as they are mostly built on public blockchains. This transparency ensures that all transactions are visible and auditable by anyone, fostering trust and reliability in the system. It aligns with the ethos of Bitcoin, where openness and transparency are paramount.

M. Emmerich (✉)
Conventic GmbH, Bonn, Germany
e-mail: m.emmerich@conventic.com

Security: Bitcoin introduced the concept of cryptographic security in financial transactions. DeFi inherits and extends this security feature. Smart contracts, which are self-executing contracts with the terms of the agreement directly written into code, are used extensively in DeFi platforms. These smart contracts are secured by the same cryptographic principles as Bitcoin, ensuring robust security measures against fraud and unauthorized access.

Financial Inclusivity: One of Bitcoin's key goals is to provide financial services to the unbanked or underbanked populations. DeFi aligns with this goal by offering financial services that are accessible to anyone with an internet connection. It removes barriers like the need for a traditional bank account or credit history, promoting financial inclusivity and equality. This democratization of financial services mirrors Bitcoin's vision of making financial transactions accessible to everyone, regardless of their geographic location or economic status.

Permissionless and Borderless: Bitcoin allows anyone to participate without permission from a central authority. Similarly, DeFi is permissionless; users can access financial services without needing approval from a centralized body. Furthermore, both Bitcoin and DeFi are borderless, meaning they are accessible globally, not confined to the rules or regulations of any single country.

Innovation and Adaptability: Bitcoin's emergence encouraged continuous innovation in the financial and technological sectors. DeFi reflects this by constantly evolving, with new platforms and services emerging regularly. This continuous innovation ensures that DeFi stays adaptable and relevant, similar to how Bitcoin has evolved over time to meet changing user needs and technological advancements.

In summary, DeFi not only embodies the core values of Bitcoin but also extends them to a wider range of financial applications, pushing the boundaries of what is possible in the realm of finance and technology.

2 The Evolution of DeFi

Decentralized Finance has evolved over time, leading to the informal categorization into DeFi 1.0 and DeFi 2.0. These phases represent different stages in the development and complexity of DeFi ecosystems. Here's an overview of each:

2.1 DeFi 1.0

DeFi 1.0, the first phase in the evolution of decentralized finance, roughly began around 2018 and gained significant momentum in 2019. This initial phase was characterized by the emergence and growth of foundational DeFi applications and platforms, primarily built on the Ethereum blockchain. Key milestones and developments during this period include:

MakerDAO's Launch (2017–2018): Although MakerDAO, one of the first DeFi projects, was launched in late 2017, it became significantly influential in 2018, laying the groundwork for DeFi 1.0. MakerDAO introduced the DAI stablecoin, a decentralized and collateral-backed cryptocurrency, which became a fundamental component of the DeFi ecosystem (Fig. 1).

Rise of Decentralized Exchanges (2018–2019): The emergence and growth of decentralized exchanges (DEXs) like Uniswap and Kyber Network were pivotal during this period. These platforms enabled peer-to-peer trading of cryptocurrencies without the need for traditional, centralized intermediaries (Fig. 2).

Growth in Lending Platforms (2019): The rise of decentralized lending platforms, such as Compound and Aave, marked a significant development in DeFi 1.0. These platforms allowed users to lend and borrow cryptocurrencies in a trustless manner.

Initial Liquidity Pools (2019): The concept of liquidity pools, where users could provide liquidity to DEXs and earn fees or rewards, started to gain popularity, setting the stage for the yield farming and liquidity mining trends that would define the later stages of DeFi.

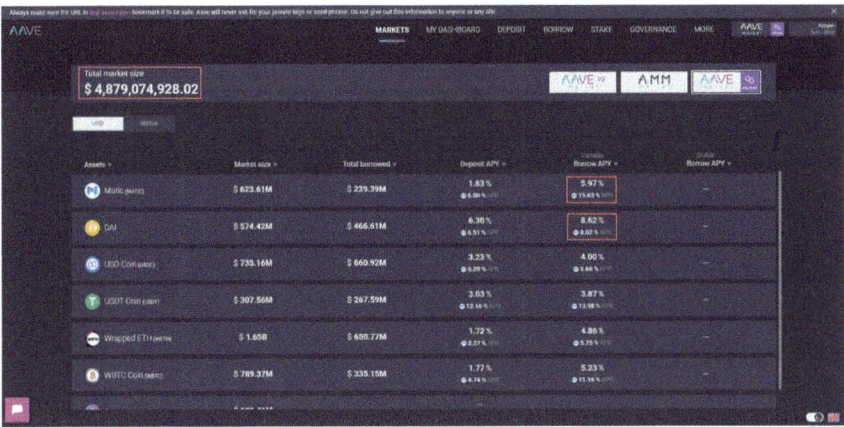

Fig. 1 Shows the AAVE markets dashboard page in 2021 [1]

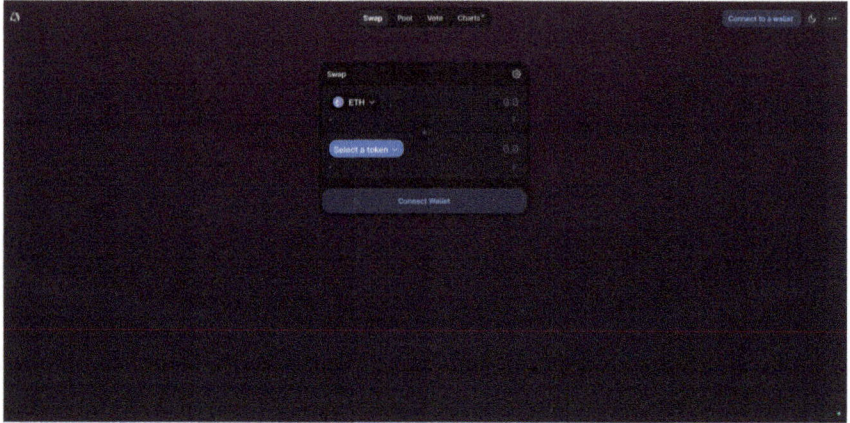

Fig. 2 Shows the web interface of Uniswap in 2022 [2]

2.2 DeFi 2.0

DeFi 2.0 generally refers to the next evolution of decentralized finance, which began to take shape around late 2020 and into 2021. This phase was characterized by the introduction of more sophisticated protocols and mechanisms, addressing some of the limitations and challenges encountered in DeFi 1.0. Key aspects and developments of DeFi 2.0 include:

Advanced Liquidity Solutions: Projects like SushiSwap and Balancer brought innovative approaches to liquidity provision and management, aiming to address issues like impermanent loss and liquidity fragmentation.

Enhanced Yield Farming: DeFi 2.0 saw the emergence of more complex and diverse yield farming strategies, offering users various ways to optimize their returns.

Improved Governance Models: There was a shift towards more sophisticated governance frameworks, giving community members greater control over the direction and decisions of DeFi projects.

Focus on Sustainability: Protocols in this phase started focusing on long-term sustainability, including mechanisms to ensure more stable and sustainable yield farming and liquidity provision models.

Risk Management and Insurance: Projects began to focus more on risk management and introduced various forms of insurance to protect users against smart contract failures or other DeFi risks.

Scalability Solutions: With the high transaction fees and network congestion on Ethereum, DeFi 2.0 also saw initiatives to improve scalability, including layer 2 solutions and the integration with other blockchain networks.

DeFi 2.0 marked a maturation of the DeFi space, with protocols and platforms striving for greater efficiency, user-friendliness, and integration with the broader financial ecosystem. This phase represents a period of consolidation, optimization, and expansion of the initial concepts and practices introduced in DeFi 1.0.

2.3　The DeFi Summer of 2020

The term "DeFi Summer" refers to a period in 2020, particularly centered around the middle of the year, when the decentralized finance (DeFi) sector experienced explosive growth and widespread attention in the cryptocurrency community. This period is marked by several key characteristics:

Surge in Total Value Locked (TVL): There was a dramatic increase in the total value locked in DeFi protocols. This metric became a crucial indicator of the sector's growth and popularity.

Yield Farming and Liquidity Mining: DeFi Summer saw the rise of yield farming and liquidity mining as major trends. Users were incentivized to provide liquidity to DeFi platforms in exchange for governance tokens, leading to substantial returns for many participants.

Launch of High-Profile Protocols: Several high-profile DeFi projects, including Compound and Uniswap, launched or introduced significant updates during this period. The Compound protocol's COMP token distribution event in June 2020 is often cited as a kickoff for the DeFi Summer.

Increased Mainstream Attention: The rapid growth of the DeFi space caught the attention of both the mainstream financial press and a broader array of investors, drawing more participants into the ecosystem.

Network Congestion and High Gas Fees: The Ethereum network, which hosted most of these DeFi applications, experienced significant congestion and high transaction fees due to the increased activity.

Rise in Token Prices: Many DeFi-related tokens saw substantial increases in price, reflecting the heightened interest and speculative activity in the sector.

DeFi Summer represents a pivotal moment in the evolution of decentralized finance, marking its transition from a niche interest within the crypto space to a major area of focus and innovation that continues to influence the development of blockchain technology and financial applications.

The history of Decentralized Finance (DeFi) marks a transformative chapter in the financial sector, showcasing a paradigm shift from traditional centralized financial systems to a more open, inclusive, and decentralized approach. This journey, though relatively short, has been marked by rapid innovation, significant challenges, and a profound impact on how financial services are perceived and utilized.

2.4 Key Milestones and Innovations

DeFi's roots can be traced back to the creation of Bitcoin in 2009, which laid the groundwork for decentralized currencies. However, the true genesis of DeFi began with Ethereum's smart contract capabilities, enabling more complex financial functions beyond mere transactions. Pioneering projects like MakerDAO introduced decentralized lending and borrowing, while protocols such as Uniswap and Compound revolutionized asset exchange and yield generation with automated market maker models and liquidity pools.

2.5 Challenges and Resilience

The journey of DeFi has not been without challenges. Issues like scalability, security vulnerabilities, regulatory hurdles, and market volatility have tested the resilience and adaptability of DeFi platforms. Despite these challenges, the sector has shown remarkable growth and innovation, addressing issues through technological advancements like layer-2 scaling solutions, improved security protocols, and more robust governance models.

2.6 Impact and Future Potential

DeFi has democratized access to financial services, allowing individuals worldwide to participate in financial activities without the need for traditional banking institutions. It has also spurred the development of new financial products and services, challenging conventional financial models and offering users more autonomy and control over their assets.

The future of DeFi holds immense potential. As the technology matures and integrates further with traditional finance, it promises to enhance financial inclusivity, transparency, and efficiency. However, this future also hinges on addressing existing challenges, particularly in terms of regulatory compliance, user security, and platform stability.

2.7 Conclusion

In conclusion, the history of DeFi is a testament to the power of blockchain technology in reshaping the financial landscape. It stands as a beacon of innovation and change, redefining the boundaries of what is possible in finance. As DeFi continues to evolve, it carries the potential to fundamentally alter how we interact with financial systems, making them more accessible, equitable, and aligned with the digital age. The journey of DeFi is far from over; it is continually evolving, promising a future where finance is more open, decentralized, and in the hands of the many rather than the few.

3 Case Studies

3.1 Case Study: The Olympus DAO Model

Olympus DAO introduced a novel approach in the DeFi (Decentralized Finance) ecosystem, aiming to create a stable yet decentralized currency system. This case study explores the concepts behind Olympus DAO, its unique mechanism, and the implications it has on the broader DeFi landscape.

Olympus DAO was launched in early 2021, amidst the burgeoning growth of the DeFi sector. Its primary goal was to create a decentralized reserve currency, OHM, that is not pegged to any traditional fiat currency but maintains stability and value. Unlike typical stablecoins, OHM's value is derived from a basket of assets held in the DAO's treasury, making it a reserve currency akin to how gold used to back some fiat currencies.

Olympus DAO operates as a DAO, meaning it is fully autonomous and governed by its community of OHM holders. These holders can propose and vote on changes to the protocol, ensuring a democratic and decentralized decision-making process.

3.1.1 The OHM Token

OHM is not pegged to any specific asset but is instead backed by a diverse range of assets in the treasury. This backing gives OHM intrinsic value, differentiating it from purely algorithmic stablecoins. The treasury backs every OHM with a basket of assets (e.g., stablecoins, other cryptocurrencies, and potentially other real-world assets). The value of these assets provides a floor value for OHM, offering stability and trust in the inherent value of the token.

3.1.2 Protocol Owned Liquidity (POL)

Olympus DAO introduced the concept of POL, where the protocol itself owns and controls its liquidity. This approach reduces reliance on external liquidity providers, minimizing the risk of liquidity crises.

3.1.3 Staking and Bonding

OHM holders can stake their tokens in the protocol to receive more OHM, derived from the protocol's expansion. Additionally, users can purchase OHM at a discount by bonding other assets with the protocol, which are then added to the treasury.

3.1.4 Implications and Challenges

Economic Model

Olympus DAO's model represents a shift from traditional fiat-pegged stablecoins, offering a crypto-native currency with inherent value. It combines elements of fiat-backed and algorithmic stablecoins but introduces a unique asset-backed approach.

Challenges

Volatility: Despite the aim for stability, OHM has experienced significant price volatility, raising questions about its effectiveness as a reserve currency.

Sustainability: The long-term sustainability of the high staking rewards and economic model remains a topic of debate.

3.1.5 Conclusion

Olympus DAO's innovative approach, and the fact that the source code for the protocol and the frontend application were publicly available on GitHub, have heavily influenced the DeFi space, leading to the emergence of many similar DAOs and reserve currency models, called "forks", a term that refers to copying the code from one GitHub repository to continue development based on the original work. It has also sparked discussions on the viability of alternative stablecoin mechanisms and the role of decentralized reserve currencies. Out of the many "OHM forks", nowadays more generically referred to as **Protocol Owned Liquidity** (POL) DAOs, only a few still operate (Fig. 3).

3.2 Case Study: Uniswap V1

Uniswap, a name synonymous with decentralized finance (DeFi), made its debut with Uniswap V1, a protocol that radically transformed the landscape of cryptocurrency trading. This case study explores Uniswap V1, its innovative mechanism, impact, and challenges it faced.

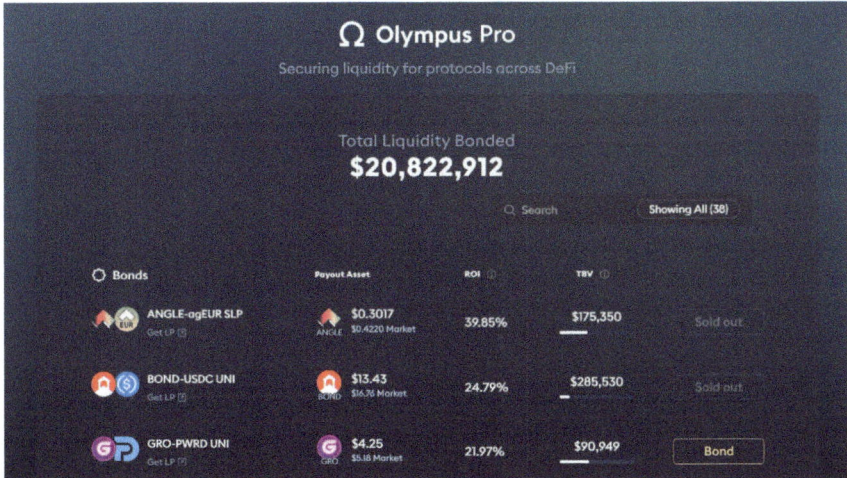

Fig. 3 Olympus DAO bonds overview page in 2021 [3]

Uniswap V1 was launched in November 2018 by Hayden Adams, influenced by an Ethereum whitepaper concept by Vitalik Buterin. The primary goal was to simplify and decentralize token exchanges on the Ethereum blockchain, removing the need for traditional order books used in centralized exchanges.

3.2.1 The Automated Market Maker (AMM) Model

Uniswap V1 introduced an AMM model, where liquidity pools replaced traditional order books. In this model, price determination is algorithmic, based on the relative balances of two assets in a liquidity pool.

3.2.2 Liquidity Pools

Users could become liquidity providers by depositing an equal value of two tokens (e.g., ETH and an ERC-20 token) into a pool. These pools facilitated trades between users, with prices set by a constant product formula: '$x * y = k$'. Anyone could create a liquidity pool for any pair of ERC-20 tokens, emphasizing the permissionless nature of the protocol. Uniswap V1 operated in a completely decentralized manner, with no central authority or intermediaries.

3.2.3 Impact and Challenges

Uniswap V1 made it easier for users to provide liquidity, democratizing access to market-making and earning fees from trades. It spurred a wave of innovation in DeFi, particularly in developing new AMM models and liquidity protocols.

However, the nature of liquidity pools introduced DeFi users to the concept of impermanent loss. Impermanent loss occurs when the price of your deposited assets changes compared to when you deposited them. The loss is 'impermanent' because it can revert if the prices return to their original state. However, if you withdraw your assets during the price disparity, the loss becomes permanent. Let's clarify this with an example:

Suppose you contribute to a liquidity pool that requires depositing an equal value of two different cryptocurrencies, say Ethereum (ETH) and a stablecoin like USDC. Imagine you put in 1 ETH and 300 USDC. At this time, 1 ETH is worth 300 USDC, so your investment is evenly balanced. Now, let's assume the price of ETH rises to 600 USDC per ETH in the open market. Arbitrage traders will see an opportunity here. They can buy ETH for cheaper in the liquidity pool and sell it at a higher price in the market. They will keep doing this until the ratio of ETH to USDC in the pool reflects the new market price. Your pool might end up with a ratio like 0.7 ETH and 420 USDC.

If you decide to withdraw your assets from the pool now, you will receive them in this new ratio, which might be less valuable than if you had just held onto your 1 ETH and 300 USDC separately. The 'impermanent' aspect comes from the fact that if ETH's price comes back down to 300 USDC, the ratios in the pool will adjust back, and the loss would disappear. However, if you withdraw while the prices are disparate, the loss becomes 'permanent'.

3.2.4 Conclusion

Uniswap V1 marked a paradigm shift in decentralized trading, introducing an innovative, user-friendly platform that democratized liquidity provision and exchange. Despite its limitations and challenges, Uniswap V1's legacy persists, laying the groundwork for the flourishing DeFi ecosystem that followed. Its impact extends beyond its technical contributions, symbolizing the potential of decentralized solutions in a traditionally centralized finance world.

3.3 Case Study: MakerDAO

MakerDAO, founded in 2015 by Rune Christensen, is a prominent player in the DeFi space and has been instrumental in introducing innovative financial solutions built on blockchain technology. It's best known for creating DAI, the first decentralized, collateral-backed cryptocurrency pegged to the US Dollar.

3.3.1 The DAI Stablecoin

The pegging mechanism of DAI, the stablecoin created by MakerDAO, is a complex and innovative system designed to maintain its value close to one US Dollar (USD). This system combines several components:

Collateralization: DAI is backed by collateral assets. Users deposit cryptocurrencies such as Ethereum (ETH) into a smart contract (known as a Vault) and generate DAI against this collateral. The collateral-to-DAI ratio must exceed a certain minimum percentage, ensuring overcollateralization. This means the value of the collateral is always higher than the amount of DAI it backs, providing a buffer against market volatility.

Stability Fee: Users who generate DAI must pay a stability fee, which can be seen as an interest rate. This fee is paid in MKR, MakerDAO's governance token, and is burned, reducing MKR's total supply. The stability fee can be adjusted by MKR token holders through governance voting, influencing the supply of DAI. If DAI's value is above $1, the fee might be decreased to make borrowing DAI cheaper, increasing its supply. Conversely, if DAI's value is below $1, the fee might be increased to make borrowing DAI more expensive, reducing its supply.

Target Rate Feedback Mechanism (TRFM): Although not always active, the TRFM is a key tool for extreme market conditions. It can adjust the target rate (the rate at which the value of DAI is expected to change over time) to bring DAI's price back to its peg. If DAI's price deviates significantly from $1, the TRFM alters the incentives for holding or creating DAI.

Pooled Ether (PETH): This was part of the single-collateral DAI system (now phased out with the introduction of multi-collateral DAI). Users would pool their ETH to generate DAI, which added an additional layer of collateralization.

Liquidation: If the collateral value in a Vault falls below a certain threshold, the collateral can be liquidated. This process is automated and ensures that the system always has sufficient assets to back the DAI in circulation.

Governance and Community Involvement: MKR token holders govern the system. They vote on critical parameters like the stability fee, collateral types, and ratios, which directly influence DAI's peg to the USD. This decentralized governance model allows for community-driven decisions to maintain the stability of DAI.

Emergency Shutdown: This is a last-resort mechanism to preserve the system's integrity. In case of a critical problem or systemic risk, an emergency shutdown can be triggered to prevent further damage. It ensures users can redeem their DAI for the underlying collateral at the correct conversion rate.

3.3.2 Conclusion

The combination of these mechanisms works to maintain DAI's value at around one USD, balancing supply and demand, and managing the risks associated with the collateral assets. This design reflects a sophisticated approach to creating a decentralized, stable, and trustable digital currency.

3.4 Case Study: Aave

Aave, a trailblazer in the DeFi sector, has emerged as one of the leading lending platforms in the cryptocurrency world. This case study explores Aave's journey, its innovative features, challenges, and impact on the DeFi landscape.

Aave was launched in January 2020, evolving from its predecessor, ETHLend, which was founded in 2017. Aave's goal was to create a more inclusive and transparent financial ecosystem by enabling users to lend and borrow cryptocurrencies without intermediaries. Aave introduced novel features like flash loans and a variety of interest rate models, setting it apart from other DeFi lending protocols:

Decentralized Lending and Borrowing. Aave operates as a decentralized, non-custodial liquidity protocol where users can participate as depositors or borrowers. Depositors provide liquidity to the market to earn interest, while borrowers can take out loans by providing collateral.

Flash Loans. One of Aave's standout innovations is the introduction of flash loans, which are uncollateralized loans taken out and repaid within a single blockchain transaction. Flash loans are used for various purposes, including arbitrage, collateral swapping, and self-liquidation.

Interest Rate Models. Aave offers both stable and variable interest rates. Borrowers can switch between these rates depending on their assessment of market conditions.

Governance Token. AAVE, the platform's native token, is used for governance, allowing token holders to vote on key protocol decisions.

3.4.1 Flash Loans

The concept of flash loans introduced by Aave represents one of the most innovative and unique features in the decentralized finance (DeFi) space. These are a type of uncollateralized loan that have become popular for their flexibility and the opportunities they offer within the Ethereum blockchain. Here's an in-depth look at how Aave's flash loans work:

A flash loan is a type of loan that allows borrowers to access substantial amounts of cryptocurrency for a very short period—within the timeframe of a single blockchain transaction. The borrower takes out the loan and is required to repay it, along with a small fee, within the same transaction block. If the loan is not repaid within this transaction, the entire operation is reversed to effectively undo all actions executed until that point. This means the funds never actually leave the protocol if the loan is not repaid.

Unlike traditional loans, flash loans do not require collateral. This is feasible because of the short duration and the atomic nature of the transaction. Flash loans are primarily used for arbitrage opportunities, self-liquidation, collateral swapping, and other DeFi strategies that can be executed quickly.

In Aave, a user initiates a flash loan transaction by requesting a certain amount of cryptocurrency from Aave's liquidity pool. Within the same transaction, the user must execute their intended operations, such as arbitrage, leveraging positions, or debt refinancing. Before the transaction ends, the user repays the loan amount plus a fee. If the user fails to repay the loan within the single transaction, the entire transaction is nullified. The process is automated and executed through smart contracts, ensuring that the conditions of the flash loan are strictly enforced.

Flash loans democratize access to substantial capital, enabling users without significant collateral to execute high-value transactions and strategies. They also foster innovation in financial strategies within the DeFi space.

The primary risk of flash loans lies in their potential use in malicious exploits. If a DeFi protocol has vulnerabilities, flash loans can be used to exploit these weaknesses due to the large amount of capital that can be accessed instantly. Such exploits have occurred, leading to significant losses in some DeFi projects.

Aave's flash loans have introduced a novel financial tool in the DeFi ecosystem, providing users with unprecedented flexibility and capabilities in executing complex financial transactions. While they offer significant benefits in terms of enabling new strategies and democratizing access to capital, they also pose risks and require a robust understanding of the DeFi space to be used effectively. As such, they represent both the innovative potential and the challenges inherent in decentralized finance.

3.4.2 Conclusion

Aave stands as a testament to the potential of DeFi to transform traditional financial services. Its innovative approach to lending and borrowing, combined with a strong emphasis on security and user engagement, has not only driven its growth but also significantly influenced the broader DeFi ecosystem. As Aave continues to evolve, it remains a key player in the ongoing development and mainstream adoption of decentralized financial services.

References

1. https://kpo-and-czm.blogspot.com/2021/05/defi-aave-quickswap-and-adamantvault-on-polygon-matic-network.html
2. https://de.wikipedia.org/w/index.php?title=Uniswap&oldid=223226470#/media/Datei:Uniswap_Screenshot.png
3. https://bankless.ghost.io/how-to-buy-discounted-tokens/

Marcelo Emmerich a seasoned technology expert with over 20 years of experience, brings profound insights into decentralized finance (DeFi) in his chapter "*A Short History of Decentralized Finance (DeFi)*" from the book "*Tokenizing the Future. A Comprehensive Guide to Web 3.0 and the Metaverse.*" Drawing from his deep expertise in blockchain, AI, and financial technologies, Marcelo unpacks the evolution and core principles of DeFi.

The chapter explores DeFi's emergence, deeply rooted in Bitcoin's foundational values—decentralization, transparency, and inclusivity—and its transformation through Ethereum's smart contract capabilities. Marcelo details DeFi's progression from its foundational phase (DeFi 1.0) to its maturation during DeFi 2.0, characterized by innovative liquidity solutions, governance enhancements, and scalability improvements. He also highlights pivotal milestones like the explosive growth during "DeFi Summer" of 2020.

Using illustrative case studies, Marcelo provides in-depth analyses of key DeFi platforms such as MakerDAO, Uniswap, and Aave, showcasing their groundbreaking mechanisms, challenges, and lasting impacts. His chapter not only chronicles DeFi's history but also projects its potential to reshape global financial systems, making it an essential read for enthusiasts and professionals aiming to understand or contribute to the decentralized revolution.

Real-World Assets (RWAS) Tokenization in Web3: Transforming Finance and Ownership

Phulchand Saraswati

1 Introduction and Background

From the early human ages to the current times, people have always focused on their assets and strived to increase them for personal and societal benefits. For most of the time, tangible objects were considered extremely valuable. Since the invention of Web 1.0, people have started to realize the potential that it holds in its lap. As the technology progressed further, people began to foresee the infinite possibility of digital assets. In the current era of a fast-paced society, creating digital assets is like a boon of technology. Real- world assets (RWAs) form the cornerstone of traditional economies, encompassing a diverse range of assets that have played a critical role in shaping ownership structures and financial systems throughout history. These assets, whether tangible or intangible, represent the tangible wealth and value inherent in various sectors of the economy.

Tangible RWAs, such as infrastructure, real estate, commodities, precious metals, and industrial outputs, have long been intrinsic to economic development. Land and property ownership have been fundamental to societal structures, dictating the power dynamics and socio-economic hierarchies that influence them. Intangible RWAs, including intellectual property, patents, and copyrights, have become increasingly significant in the modern knowledge-based economy.

The ownership of RWAs has been intertwined with economic and political power. For example, someone with ownership of lots of properties has been a symbol of influence, shaping the trajectory of societies. The emergence of corporations and the stock market further expanded the concept of ownership, allowing individuals to own company shares and participate in the profits generated by these entities. The ownership of both tangible and intangible assets has thus been a driving force behind economic progress and societal structures.

P. Saraswati (✉)
Department of Mechanical Engineering, RWTH Aachen University, Aachen, Germany
e-mail: phulchand.saraswati@rwth-aachen.de; phulchandsaraswati@gmail.com

However, traditional ownership models have faced challenges such as illiquidity, transparency, and limited accessibility. These challenges have led to exploring innovative solutions using blockchain technology to tokenize assets. In the digital era, the fusion of RWAs with blockchain technology and smart contracts opens a new paradigm, promising greater liquidity, accessibility, enhanced transparency, and inclusivity in ownership and finance. The evolution of RWAs and their transformation through tokenization is not merely a technological shift but a fundamental reimagining of ownership structures, ensuring that the benefits of economic assets are distributed more equitably in the digital age.

2 Exploring Real-World Assets (RWAs)

Real-World Assets (RWAs) encompass tangible or intangible assets with inherent value that exists in the physical realm. This category includes a diverse range of assets, spanning from real estate properties, machinery, and commodities to intellectual properties, copyrights, and specific contractual agreements. In contrast to digital or virtual assets, RWAs possess a tangible presence or real-world utility. They have played a longstanding role in traditional finance, serving various purposes such as collateral, investment vehicles, or assets held for appreciation [15].

RWAs is crucial in the global economy. They are used in the form of collateral during the lending process, thus supporting multiple businesses and transactions. Many of the assets according to the report by Bank of America confirms their illiquidity and due to that it is extremely difficult to trade. To resolve this problem, Tokenization plays a great role [5].

Tokenized Real-World Assets (RWAs) are digital tokens built on blockchain technology, representing tangible and traditional financial assets like cash, commodities, equities, bonds, credit, artwork, and intellectual property. The tokenization of RWAs signifies a significant transformation on the asset's accessibility, and trading. This shift unlocks lots of opportunities, for blockchain-driven financial services but also for a diverse range of non-financial applications supported by cryptography and decentralized consensus.

Among the most promising applications of blockchain technology, asset tokenization holds an infinite potential, covering nearly every aspect of human economic activity. The forthcoming financial landscape appears to be decentralized, featuring numerous blockchains supporting trillions of dollars in tokenized RWAs. This decentralized infrastructure is likely to be underpinned by a universal interoperability standard, connecting various blockchain and distributed ledger technology-based networks [3].

The RWA ecosystem is exceptionally diverse and expanding exponentially as more projects enter the market. Many "asset providers" focus on originating and creating demand for RWAs across various asset classes such as real estate, equities, bonds, collectibles, etc. Figure 1 depicts the vertical overview of the RWA and provides a clear picture to the users.

Fig. 1 Some of the projects list under RWAs as a vertical overview [12]

2.1 Types of Real-World Assets

RWAs can be classified under Tangible and Intangible assets. Assets such as properties (private, commercial), precious metals such as Silver and Gold, Commodities such as Oil, Gas and minerals etc. all comes under the umbrella of Tangible assets. An intangible asset has no physical form, and it cannot be handled. These assets value increase in long-term. It is extremely valuable for the owner. Brand, patent, intellectual property, trademark, or copyright can be categorized under this [7] (Fig. 2).

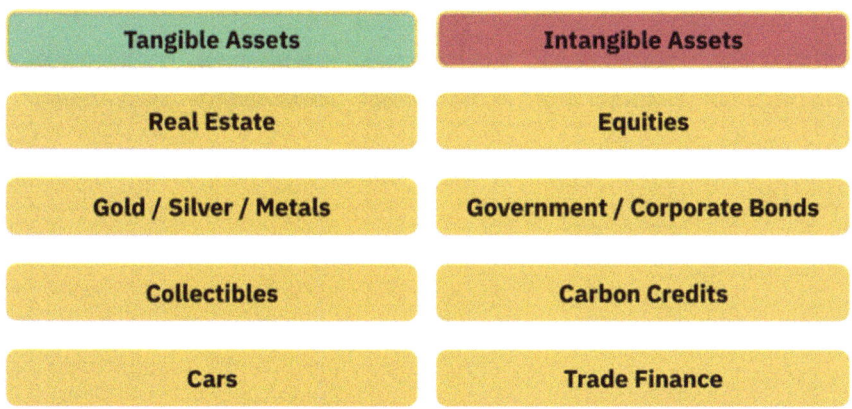

Fig. 2 Categorization of RWAs [10]

3 Web 3.0 and Tokenization of Assets

The evolution of web technologies has seen exponential and unprecedent growth. Web 3.0 represents the next stage in the evolution of the World Wide Web, where data is interconnected and comprehensible to computers, leading to the creation of more intelligent and intuitive applications. Several factors contributed to the emergence of Web 3.0, including the escalating volumes of user and device-generated data, the imperative for more efficient data processing and management, and the growing demand for enhanced privacy and security. To achieve these objectives, Web 3.0 incorporates cutting-edge technologies such as artificial intelligence, machine learning, blockchain, and decentralized protocols.

In the space of Web 3.0, data is no longer fragmented and dispersed across the internet; instead, it is interconnected and structured in a manner that machines can readily comprehend and utilize. This structural shift enables the development of more personalized and intelligent applications capable of learning from user behaviour and preferences, enhancing the web experience by making it more seamless and efficient [11] (Fig. 3).

From Web 1.0 to Web 3.0, a lot has been changed in terms of various aspects such as content, user role, interaction, and data (Fig. 4). Web 3.0 is also known as Semantic Web.

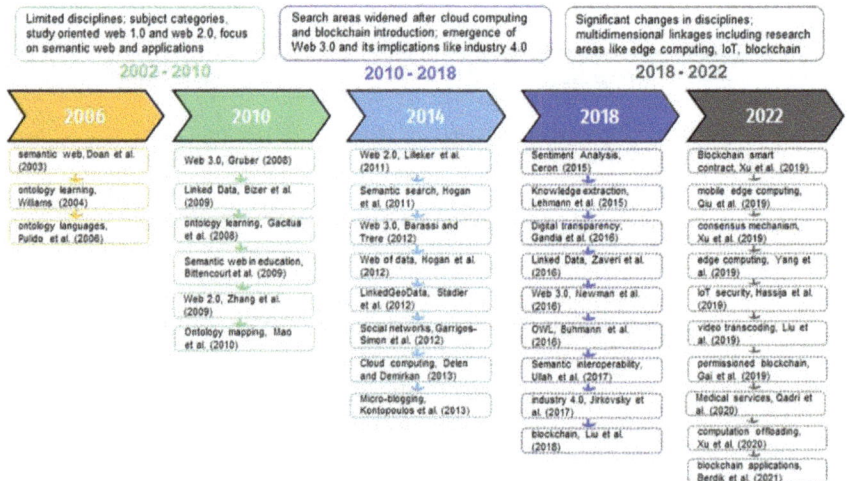

Fig. 3 Evolution of the web technologies [8]

Aspect	Web 1.0	Web 2.0	Web 3.0
Content	Static	Dynamic	Intelligent
User Role	Passive	Active	Collaborative
Interaction	One-way	Two-way	Multi-way
Data	HTML	XML, JSON, AJAX	Semantic Web
Focus	Company	Community	Individual

Fig. 4 Comparison of different aspects of Web 1.0, Web 2.0 and Web3.0 [11]

3.1 Tokenization

Tokenization, is a recent development in the realm of digital finance, involves the creation of a blockchain representation of underlying assets or securities. This process is designed to augment the inherent features and attributes of financial instruments. The tokenization of assets and securities yields innumerable advantages, encompassing a reduction in issuance and trading costs, a diminished reliance on intermediaries, increased market liquidity, and higher transparency throughout the lifecycle of an asset for all stakeholders (Fig. 5).

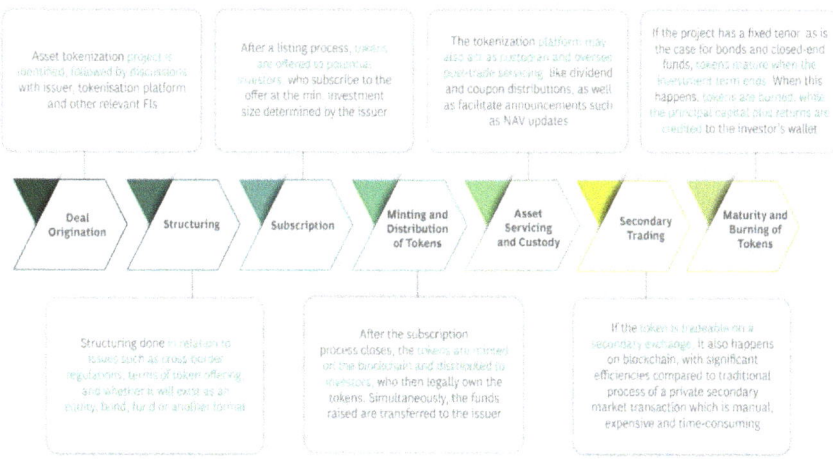

Fig. 5 Mechanism of on-chain tokenization [9]

3.1.1 Types of Tokens

a. *Security tokens*: These tokens function as digital proofs of ownership for specific assets or rights, offering a digital representation of the underlying asset. They retain all the benefits associated with traditional securities while also providing a level of programmability, which allows for customization through the expertise of a cryptocurrency developer. This programmability makes it possible to embed unique characteristics and features that meet particular requirements.
b. *Platform tokens*: They play a vital role in supporting decentralized applications (DApps) within blockchain ecosystems. For instance, the Dai token is commonly used to interact with DApps on the Ethereum network. Moreover, such tokens are widely employed across the Ethereum network, serving as a core component of its platform's functionality.
c. *Utility tokens*: This is the most fundamental type of token on a blockchain network. These tokens are used to access services, power the consensus mechanisms, pay transaction fees, and participate in voting for new blockchain developments. Additionally, they function as governance tokens, playing a key role in the decision-making processes of Decentralized Autonomous Organizations (DAOs).
d. *Fungible Tokens*: Fungible tokens are interchangeable and not unique, meaning they can be easily replicated or replaced. Converting fungible assets into tokens is straightforward since they can be divided into fractional units. A common example of a fungible token is gold. Fungible token converters include an abstraction layer that enhances interoperability and ensures platform independence.
e. *Non-Fungible*: Assets such as a diamond, a baseball, or the Mona Lisa cannot be divided into fractions. However, when converted into non-fungible tokens (NFTs), they can represent full or partial ownership. NFTs are unique and immutable, with ownership history recorded on the blockchain, ensuring they cannot be replicated. Additionally, when a non-fungible asset is tokenized, the process begins with creating an immutable digital signature, which helps establish the uniqueness of the underlying asset [1].

Asset tokenization involves issuing a blockchain token, known as a security token, that digitally represents a tangible, tradable asset. This process closely resembles traditional securitization methods. Security tokens are usually created through a type of initial coin offering (ICO) known as a security token offering (STO), which differentiates itself from other ICOs that produce tokens such as equity, utility, or payment tokens. An STO is designed to create a digital version, specifically a security token, which corresponds to a real asset. This means that a security token can represent a wide range of assets, including company shares, real estate ownership, or participation in an investment fund. These security tokens can then be traded on secondary markets [6] (Fig. 6).

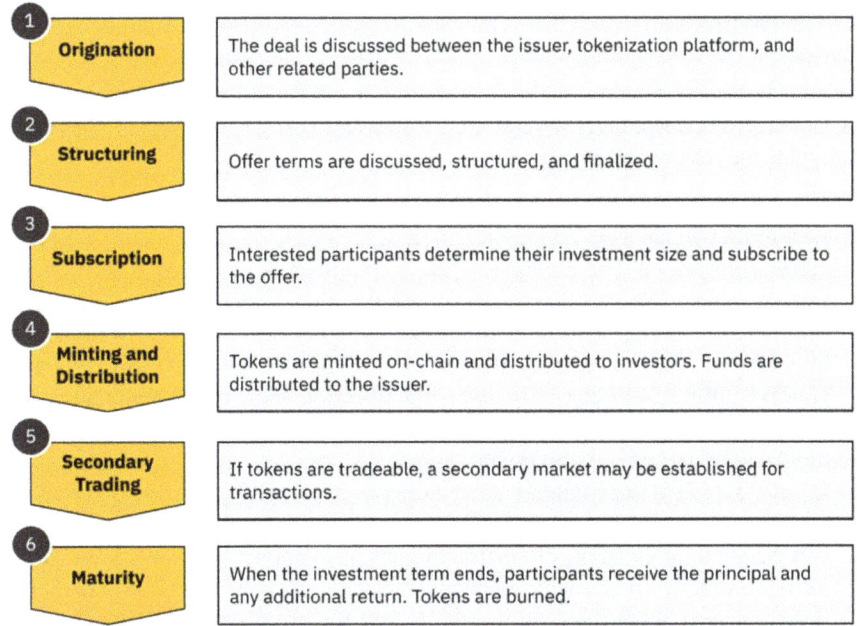

Fig. 6 Example for on-chain tokenization process [4]

The primary advantages of asset tokenization include:

1. Liquidity Enhancement: Tokenization converts traditional assets into digital tokens, allowing for fractional ownership and facilitating easier trading on secondary markets. This increases accessibility and provides investors with more opportunities to buy or sell assets.
2. Accessibility: Tokenization broadens access to investments that were previously limited to wealthy individuals or institutional investors. It allows anyone with basic internet access to own fractional shares of valuable assets.
3. Global Accessibility and Diversification: Digital tokens eliminate geographical barriers, enabling a global audience to invest in assets they might not have otherwise had access to. This facilitates portfolio diversification across various asset types, which helps to spread and reduce investment risk.
4. Cost Effectiveness: Automated, self-executing protocols streamline the trading process, minimizing reliance on intermediaries and reducing the risk of human error or disruption. This automation accelerates transactions, lowers costs, and decreases potential risks.
5. Transparency and Trust: The transparent and immutable nature of blockchain technology ensures that all transactions and ownership records are verifiable and secure from tampering. This transparency fosters trust among participants and mitigates the risk of fraud or disputes [13–15].

4 Challenges

Despite the many benefits, widespread adoption of Blockchain Technology and tokenized assets encounters several challenges and risks. BofA Global Research highlights the following key risks:

1. *Regulatory Risk*: The lack of unified global regulations poses a major barrier to widespread digital asset adoption. Different regions have varying rules, complicating compliance. For example, a token compliant with the EU's Markets in Crypto Assets (MiCA) framework might not meet US regulatory standards. Additionally, the classification of tokens as either securities or commodities introduces further uncertainty due to varying global regulatory requirements. However, this lack of clarity in token classification is expected to be less problematic for traditional asset tokenization, as these assets already have established regulatory classifications and requirements.
2. *Legal Risk*: Legal uncertainties surrounding asset ownership and smart contracts contribute to ambiguity in consumer protections. Issues about asset ownership and the legal implications of smart contracts have not yet been tested in courts. Smart contracts, which enable asset transfers without explicit legal authorization, raise concerns about their enforceability and the protection of consumers and investors. Additionally, the use of decentralized oracle networks to obtain real-world data adds further legal complexities, such as potential claims for damages in cases of data malfunctions or manipulation.
3. *Security*: Incidents of hacks, theft, and illicit activities in digital asset transactions are relatively rare, a major security breach due to exploited software bugs in smart contracts could present a significant risk. Since smart contracts are essentially software, they are prone to bugs and vulnerabilities. Despite precautions such as bug bounties, audits, and insurance to address theft risks, an exploited flaw in a private permissioned distributed ledger designed for institutional use could seriously damage the industry's credibility. Continuous efforts to strengthen security, including regular audits and bug detection, are essential to preserving the integrity of Distributed Ledger Technology (DLT) and Blockchain Technology (BCT).
4. *Global Coordination*: The lack of global coordination on standardization can lead to interoperability issues among tokenized assets and platforms. This can result in inefficiencies and higher costs related to settlement, liquidity, credit risk, financing, and cross-border transactions. When distributed ledgers are only compatible with certain platforms, it may obstruct new market entrants, reinforce existing banking structures, strengthen trading partnerships between countries, and marginalize emerging economies from the global financial system.
5. *Liquidity*: Tokenizing traditional assets can potentially establish secondary markets for assets that were previously illiquid, turning investments in private equity funds, commercial real estate, royalty streams, carbon credits, and valuable art into more liquid assets. However, according to the BofA Global Research, tokenization efforts need to ensure sufficient liquidity; otherwise, they may be inefficient in resource use. Without adequate liquidity, tokenized corporate

bonds might not offer substantial advantages over traditional bonds. Despite this, liquidity concerns are anticipated to decrease as adoption grows [2].

6. *Educational Barrier*: The adoption of tokenization introduces new technology and concepts, requiring participants to understand complex aspects such as blockchain functionality, digital wallet management, and other technical details. For those without a background in computer science or finance, learning these technical terms and concepts can be challenging and requires significant effort.

7. *Valuation Complexities and Exit Strategy*: Determining the value of tokenized assets, especially those with unique characteristics, poses significant challenges due to its inherently subjective nature. Factors contributing to these challenges include the lack of historical data and comparable benchmarks, dependence on market sentiment, and the need to assess intrinsic value. Additionally, selling tokenized assets and finding suitable buyers can be complicated, further exacerbated by the dynamic and volatile nature of the market [14].

5 Current and Future Outlook

Tokenization is a specific application within the broader field of Blockchain Technology, with the potential to create new and more efficient primary and secondary markets for both financial and non-financial assets. In the near future, investment portfolios could become more diverse, including assets such as shares in a blue-chip art fund, carbon credits, or tokens that grant the ability to influence cash flows and support smart contract-enabled blockchain platforms. Although currently in the early phases of significant infrastructure and application changes, the future of tokenization has the potential to transform the processes involved in transferring, settling, and storing value in the financial sector [2] (Fig. 7).

The Real-World Asset (RWA) market is still in its early stages but is showing signs of growing adoption and an increasing total value locked (TVL). As of the end of June, RWAs have become the 10th largest sector within decentralized finance (DeFi), according to data from DeFi Llama, rising from the 13th position just a few weeks earlier. A major factor in this growth has been the launch of the stUSDT protocol in July, which enables USDT stakers to earn yield based on real-world assets [14] (Fig. 8).

The tokenization of real-world assets (RWAs) is gaining momentum, with steady user adoption and the participation of major institutional players. Low yields in decentralized finance (DeFi), combined with rising interest rates, have driven the growing interest in RWAs, particularly in tokenized treasuries. Currently, investors are lending over $600 million to the U.S. government through the tokenized treasury market, earning an annualized yield of approximately 4.2%. The market for tokenized assets is projected to reach $16 trillion by 2030, a significant increase from $310 billion in 2022, indicating substantial growth potential. With the current total value of assets estimated at $900 trillion, this highlights the vast future potential of RWAs [14] (Fig. 9).

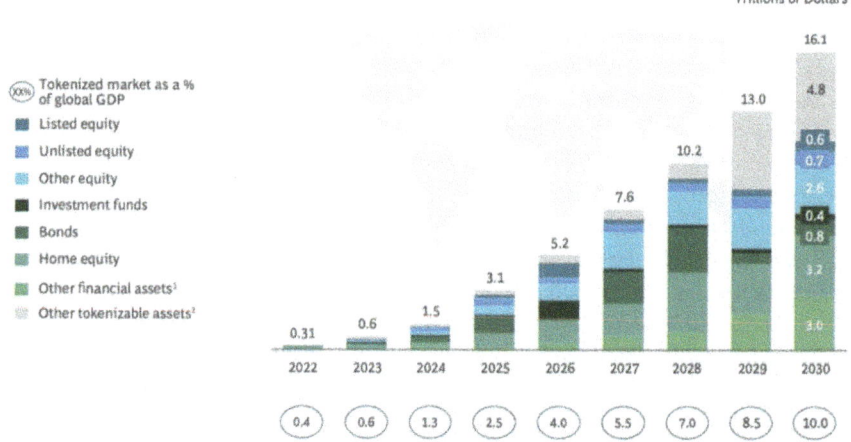

Fig. 7 Estimation for tokenization of global illiquid assets [9]

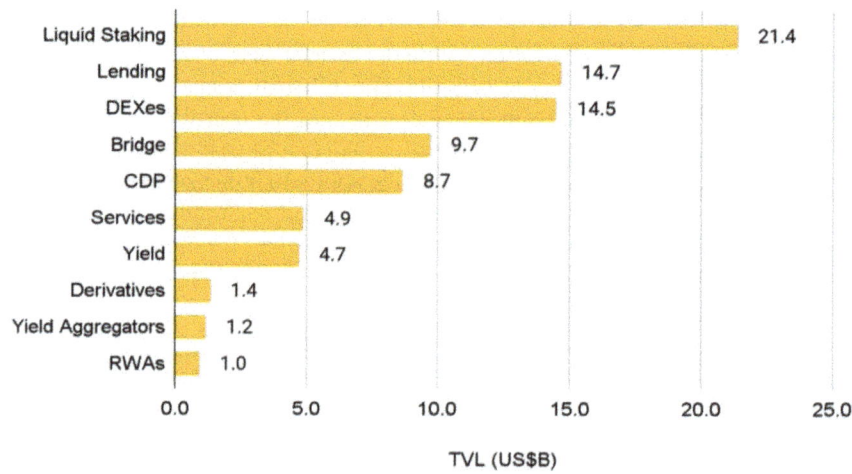

Fig. 8 RWAs are the 10th largest DeFi sector on DeFi Llama [4]

6 Conclusion

The introduction and applications of RWAs along with their integration into the Web 3.0 space has provided a profound shift in the financial paradigm. Tokenization and the principles of DeFi can democratize access to wealth, amplify financial inclusion, and reshape the traditional notions of ownership and investment. It is a great step towards the benefit of society as a whole. However, the underlying technology is already making significant contributions to various real-world industries, particularly in finance where regulations are well-defined, and assets are more readily tokenizable.

Fig. 9 RWAs potential for growth [4]

A notable example is USDC, a tokenized dollar (commonly known as a stablecoin) backed 1:1 by USD. This innovation allows the US dollar to leverage the accessibility, transparency, and low-cost transaction advantages inherent in blockchain technology. The asset tokenization industry continues to mature over the coming years, new interoperability solutions are introduced, and proper legal frameworks are put in place, digital tokens will be set to radically transform the nature of ownership, trading and liquidity. But don't be surprised if it takes a few years.

As the journey continues, the impact on global finance is transformative and helping in the dawn of an era where the benefits of tangible assets are universally accessible, irrespective of geographical or socioeconomic constraints.

References

1. A. Abrol, *What Is Tokenization? A Complete Guide* (2023). From Blockchain Council: https://www.blockchain-council.org/blockchain/what-is-tokenization/
2. T. Bowley, V. Cook, *Beyond Crypto: Tokenization* (Bank of America, 2023). From https://institute.bankofamerica.com/content/dam/bank-of-america-institute/transformation/beyond-crypto-tokenization.pdf
3. Chainlink, *Real-World Assets (RWAs) Explained* (2023). From Chainlink: https://blog.chain.link/real-world-assets-rwas-explained/#what_are_real_world_assets_(rwas)_

4. X.J. Chua, *Real-World Assets: State of the Market* (Binance Research, 2023). From https://public.bnbstatic.com/static/files/research/real-world-assets-state-of-the-market.pdf
5. H. Dauven, *Understanding Real-World Assets (RWAs)* (2023). From Dusk: https://dusk.network/news/understanding-real-world-assets-part-1/
6. A. Gupta, J. Rathod, D. Patel, J. Bothra, S. Shanbhag, T. Bhalerao, Tokenization of real estate using blockchain technology. in *International Conference on Applied Cryptography and Network Security* (Springer, Cham, 2020), pp. 77–90. https://doi.org/10.1007/978-3-030-61638-0_5
7. W. Kenton, *What Are Intangible Assets? Examples and How to Value* (2023). From Investopedia: https://www.investopedia.com/terms/i/intangibleasset.asp
8. D. Kukreja, S. Gupta, D. Patel, J. Rai, Scientometric review of Web 3.0. Sage J **2** (2023). https://doi.org/10.1177/01655515231182073
9. S. Kumar, R. Suresh, D. Liu, A. Kaul, B. Kronfellner, *Relevance of on-chain asset tokenization in Crypto Winter* (Boston Consulting Group, 2022). From https://web-assets.bcg.com/1e/a2/5b5f2b7e42dfad2cb3113a291222/on-chain-asset-tokenization.pdf
10. M. Naggar, *Real World Assets: The Bridge Between TradFi and DeFi* (Binance Research, 2023). From https://research.binance.com/static/pdf/real-world-asset-report.pdf
11. M. Nasar, Web 3.0: a review and its future. Int. J. Comput. Appl. **185**, 41–46 (2023). https://doi.org/10.5120/ijca2023922776
12. J.M. Neo, *Navigating Crypto: Industry Map* (Binance Research, 2023). From https://public.bnbstatic.com/static/files/research/industry-map-sep23.pdf
13. Polymesh, *Real-World Assets (RWAs)* (n.d.), From Polymesh: https://polymesh.network/real-world-assets
14. P. Saraswati, *Real World Asset (RWA)—Key and Critical Bridge between Decentralized Finance (DeFi) and Traditional Finance (TradFi)* (2023). From Medium: https://medium.com/@phulchandsaraswati_92184/real-world-assest-rwa-key-and-critical-bridge-between-decentralized-finance-defi-and-b38203e8439b
15. The INX Digital Company INC, *Understanding RWAs: An Introduction to Real-World Assets* (2023). From Chainlink: https://www.inx.co/learn/beginners/understanding-rwas-an-introduction-to-real-world-assets/

Phulchand Saraswati recently completed a Master's degree in Production Systems at RWTH Aachen University, where he developed a strong foundation in manufacturing and management. He holds a Bachelor's degree in Automobile Engineering. Building on these backgrounds, he has made notable contributions to the field of Web 3.0, bringing a unique interdisciplinary perspective to his work.

His research interests span battery production, project management, and the applications of Web 3.0 technology. He has been instrumental in developing the Web 3.0 ecosystem at RWTH Aachen, co-founding the Aachen Blockchain Club (ABC), and fostering collaboration with industrial partners while encouraging students to explore opportunities in this emerging field.

Beyond his academic pursuits, Phulchand's entrepreneurial spirit drives him to explore new ideas and engage with diverse topics beyond his formal curriculum. He is focused on integrating insights from multiple disciplines to generate innovative ideas and contribute to the broader academic community.

Digital Economy

From Sensors to Solutions: The Role of Helium Blockchain and LoRaWAN in AI Innovations in Environmental Monitoring, Smart Parking, Crowd Management, and Urban Gardening

Daniel Trauth, Wolfgang Prinz, and André Heryschek

1 Introduction

The digital transformation of urban spaces is an evolving frontier where technological advancements meet real-world applications. In this context, the integration of Helium Blockchain and LoRaWAN (Long Range Wide Area Network) emerges as a pivotal innovation, driving various smart city initiatives. This book chapter explores the foundational principles of Helium and LoRaWAN, followed by detailed case studies demonstrating their application in urban settings. Each use case exemplifies how these technologies enhance urban management, sustainability, and citizen engagement.

D. Trauth (✉)
dataMatters GmbH, Cologne, Germany
e-mail: d.trauth@datamatters.io; daniel.trauth@fit.fraunhofer.de

D. Trauth · W. Prinz
Fraunhofer FIT, Sankt Augustin, Germany
e-mail: wolfgang.prinz@fit.fraunhofer.de; hello@iditech.org

A. Heryschek
SWD Dormagen, Dormagen, Germany
e-mail: a.heryschek@swd-dormagen.de

W. Prinz
Institut Für Digitale Zukunftstechnologien, Hürth, Germany

2 Helium Blockchain Fundamentals

Helium Blockchain is a decentralized wireless network, designed to provide a secure and cost-effective infrastructure for Internet of Things (IoT) devices. By incentivizing users to deploy and maintain Hotspots, which serve as nodes in the network, Helium creates a widespread and reliable coverage for IoT communications. The network uses the Proof-of-Coverage consensus mechanism to ensure that Hotspots are accurately representing their service area, rewarding participants with Helium tokens (HNT) based on their contributions.

Technical Specifications

Helium operates on the LoRaWAN protocol, which is tailored for low-power, long-range communication. The network is designed to support a massive number of IoT devices with minimal power consumption, leveraging the unlicensed ISM bands (typically 868 MHz in Europe and 915 MHz in North America). Hotspots act as both miners and network routers, providing the necessary infrastructure for IoT devices to communicate.

- **Frequency Bands**: Utilizes the 868 MHz (EU) and 915 MHz (US) ISM bands.
- **Range**: Each Hotspot can cover a radius of up to 10 miles in rural areas and about 1 mile in dense urban environments.
- **Data Rate**: Supports data rates from 0.3 to 50 kbps.
- **Power Consumption**: Designed for low power consumption, suitable for battery-operated IoT devices with long lifespans (Fig. 1).

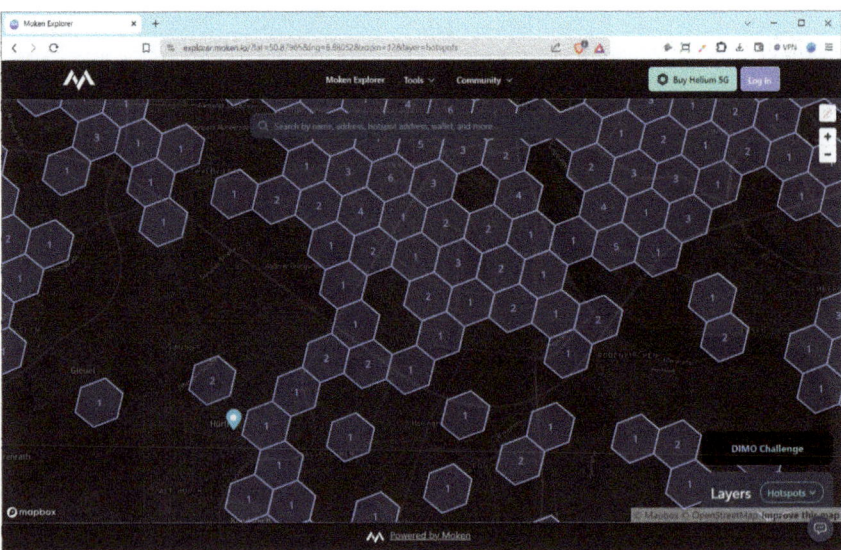

Fig. 1 Coverage of helium in the area of Hürth

3 LoRaWAN Fundamentals

LoRaWAN is a protocol for low-power, wide-area networks, enabling long-range communication between IoT devices and the internet. It operates on unlicensed radio spectrum and is optimized for low power consumption, making it ideal for battery-powered sensors and devices. LoRaWAN networks can span large geographical areas and penetrate dense urban environments, offering robust connectivity for various applications, from environmental monitoring to smart metering.

Technical Specifications

LoRaWAN's architecture consists of several key components: end devices (sensors), gateways, network servers, and application servers. The protocol supports multiple device classes (A, B, and C) to accommodate different use cases.

Device Classes

- **Class A**: Bi-directional communication where each end-device's uplink transmission is followed by two short downlink receive windows.
- **Class B**: Adds scheduled receive slots to Class A, enabling more predictable downlink communication.
- **Class C**: Continuous receive windows, only closed when transmitting, for applications requiring low latency.

LoRaWAN utilizes AES-128 encryption at the network and application layers to ensure data integrity and privacy. It uses star-of-stars topology with gateways forwarding messages between end devices and a central network server. It optimizes data rates, airtime, and energy consumption based on the network conditions.

4 Use Case 1: Environmental Monitoring in Dormagen

Dormagen, a city with a vision of becoming a Smart Industrial City, has implemented an extensive environmental monitoring project. This initiative involves deploying sensor boxes at key traffic nodes to measure CO_2 and particulate matter levels, along with temperature, humidity, and noise.

Implementation

Dormagen's use of its proprietary LoRaWAN network, complemented by the Helium Blockchain, has been instrumental in enhancing data coverage and reliability. By tokenizing network coverage through Helium, private individuals are incentivized to deploy additional Hotspots, thereby improving network reach and data granularity.

Sensor Specifications

- CO_2 Sensors: Measure carbon dioxide concentration in parts per million (ppm).
- Particulate Matter Sensors: Monitor PM2.5 and PM10 levels, indicating the concentration of particles with diameters less than 2.5 μm and 10 μm, respectively.
- Temperature and Humidity Sensors: Use digital sensing elements to provide accurate environmental data.
- Noise Sensors: Capture sound levels in decibels (dB).

The data collected by these sensors are transmitted via the LoRaWAN network to a central server, where it is processed and analyzed. The Helium network's decentralized nature ensures that data transmission is resilient and secure, with encryption ensuring privacy and integrity (Fig. 2).

Outcomes

This integration allows for real-time environmental data collection and analysis, facilitating more informed urban planning and policy-making. The encrypted data transmission ensures security and privacy, making the system robust against unauthorized access and misuse. The insights gained from the data help in identifying pollution sources, optimizing traffic flows, and improving overall urban living conditions (Fig. 3).

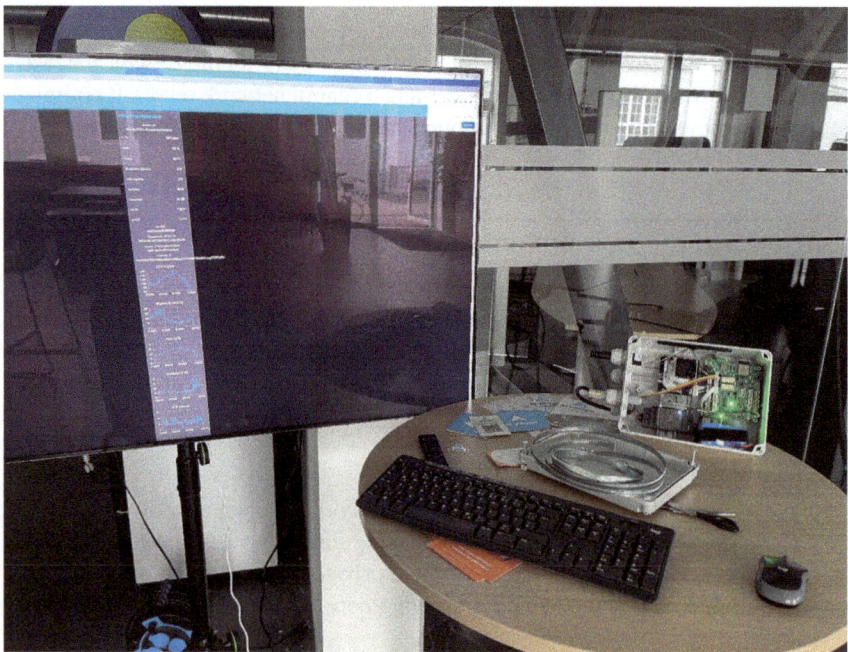

Fig. 2 The detector box of Dormagen in the Studio 6 to measure CO_2 and PM10/25

From Sensors to Solutions: The Role of Helium Blockchain ...

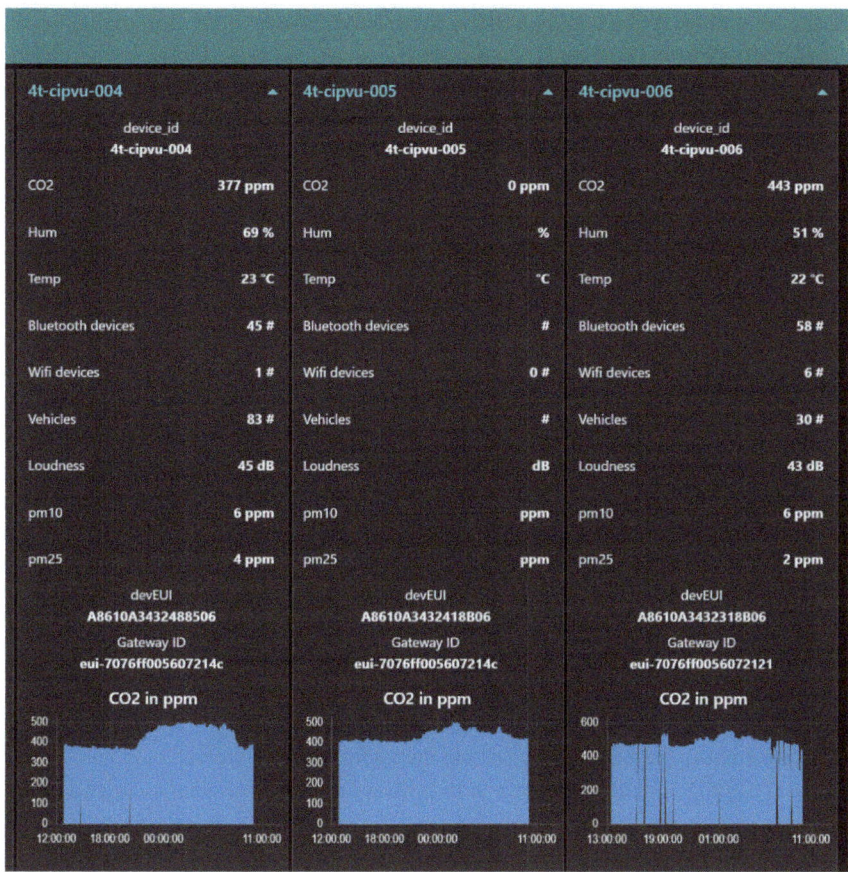

Fig. 3 Measuring results in Dormagen of three dtector boxes

5 Use Case 2: Smart Parking Solutions

The cooperation between the city of Hürth and the city of Dormagen continues. IDiTech also mediates in this use case and demonstrates how parking sensors provide important information for city planners and retailers, such as data to improve traffic flow management. Parking sensors provide real-time data on the availability of parking spaces, which helps city planners to optimise traffic flow and reduce congestion problems through targeted traffic management and dynamic parking space allocation. One example of this is Amsterdam, where parking sensors are used to control traffic in the city centre and guide drivers to available parking spaces. In addition, parking sensors contribute to environmental friendliness by reducing the time drivers spend looking for a parking space and thus reducing CO_2 emissions. San Francisco uses this technology to minimise traffic searching for parking spaces and significantly reduce emissions. Parking sensors also provide valuable services in urban planning and infrastructure. By analysing parking patterns and needs, long-term planning can make better-informed decisions, such as in Barcelona, where parking sensor data is used to determine the need and location of new car parks. Parking sensors also offer numerous benefits for retailers. They increase footfall by improving the availability of parking spaces, which increases convenience for customers. Shopping centres in Los Angeles use parking sensors to inform customers of available parking spaces in real time, which increases footfall. An improved shopping experience leads to happier customers who are more willing to stay longer and spend more. In Munich, for example, retailers in a shopping district have introduced parking sensors to alert customers to free parking spaces, which has led to positive feedback and increased sales.

Implementation

Smart parking sensors, integrated with Helium's LoRaWAN network, provide real-time data on parking space availability. This data helps city planners to manage traffic congestion and dynamically allocate parking spaces. The sensors transmit data through Helium's decentralized network, which ensures coverage even in areas with sparse traditional network infrastructure.

Sensor Specifications

- Parking Occupancy Sensors: Detect the presence of a vehicle using magnetic, ultrasonic, or infrared technologies.
- Communication: Sensors communicate occupancy status to gateways, which relay the information to a central server via the LoRaWAN network.
- Power Supply: Battery-powered with a lifespan of up to 5 years, depending on usage.

The data is displayed on digital signage and mobile applications, guiding drivers to available parking spaces and reducing the time spent searching for parking (Fig. 4).

Outcomes

The system has proven effective in reducing the time drivers spend searching for parking, thereby cutting CO_2 emissions and improving air quality. Moreover, it has enhanced the shopping experience by making parking more convenient, leading to increased foot traffic and higher sales for local businesses. The use of decentralized, blockchain-secured data transmission ensures the reliability and security of the system, fostering greater trust among users and stakeholders (Fig. 5).

6 Use Case 3: Crowd Management and Safety

Another loan from Dormagen is helping to make Hürth's public events and public buildings safer: an AI-based and GDPR-compliant people counter makes it possible. This is currently being tested in Studio 6.

Fig. 4 The parking sensor, on loan from Dormagen, in front of Studio 6 to measure the occupancy times and frequency patterns of the Fraunhofer FIT test car park

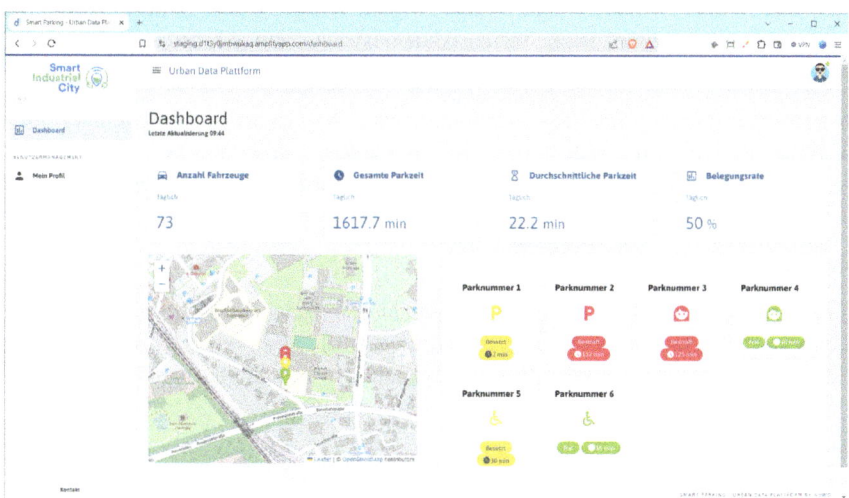

Fig. 5 Smart parking dashboard to monitor vacancies

Safety and efficiency play a key role in the management of crowds and public buildings. The use of AI-based people counters has proven to be an innovative solution that offers numerous advantages to event organisers, landlords and insurance companies:

Precise monitoring of the number of guests: with an AI-based people counter, event organisers and landlords know exactly how many people are in the building at all times. This is particularly advantageous for large events such as concerts, trade fairs or sporting events. For example, at a concert arena with a capacity of 10,000 people, an organiser can ensure that this number is not exceeded and that the safety of all guests is guaranteed.

Efficient evacuation in an emergency: In the event of an emergency, be it a fire or other dangerous situation, the people counter enables a quick and efficient evacuation. Security staff can monitor in real time how many people are still in the building and which areas have already been evacuated. In the event of a fire in a shopping centre, for example, this could be crucial in saving lives.

Compliance with **legal regulations**: Many public buildings are subject to strict capacity limits. An AI-based people counter helps to comply with these regulations and thus avoid legal problems. An example would be a museum that needs to ensure that there are never more than a certain number of visitors in the exhibition rooms at any one time to ensure safety and comfort.

Optimisation of resources: By knowing the exact number of visitors, operators can plan better and use resources more efficiently. In an amusement park, for example, the number of employees can be increased on busy days and reduced on quieter days, resulting in cost savings.

Insurance relevance: Insurance companies view the use of people counters positively, as they minimise risks. An event that demonstrably complies with capacity limits and has efficient emergency plans may be able to obtain more favourable insurance premiums (Fig. 6).

Implementation

The people counter uses Lidar technology and is connected via Helium's LoRaWAN network. This setup allows for precise monitoring of crowd density and movement, enabling real-time data analysis for event organizers and facility managers.

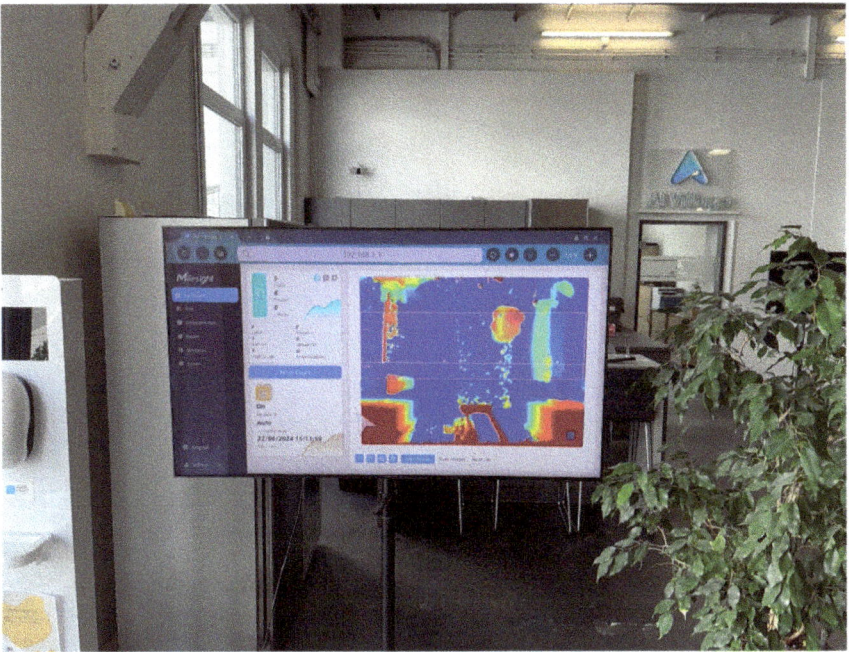

Fig. 6 View from above: how the people counter sees the entrance area of Studio 6. A person with their head and hands can be seen at the top right

Technical Specifications

- Lidar Technology: Uses pulsed laser light to measure distances and create 3D maps of the environment.
- AI Integration: Analyzes Lidar data to count and track individuals, distinguishing between adults and children.
- Communication: Data is transmitted via Helium's LoRaWAN network to a central server for analysis and action.

The system's ability to provide real-time data on the number of people in a space enhances safety protocols, facilitates efficient evacuation in emergencies, and ensures compliance with capacity regulations.

Outcomes

The deployment of these sensors has enhanced safety protocols by providing accurate data on the number of people in a given space, facilitating efficient evacuation in emergencies, and ensuring compliance with capacity regulations. Additionally, the technology helps optimize resource allocation by predicting peak times and adjusting staff levels accordingly (Fig. 7).

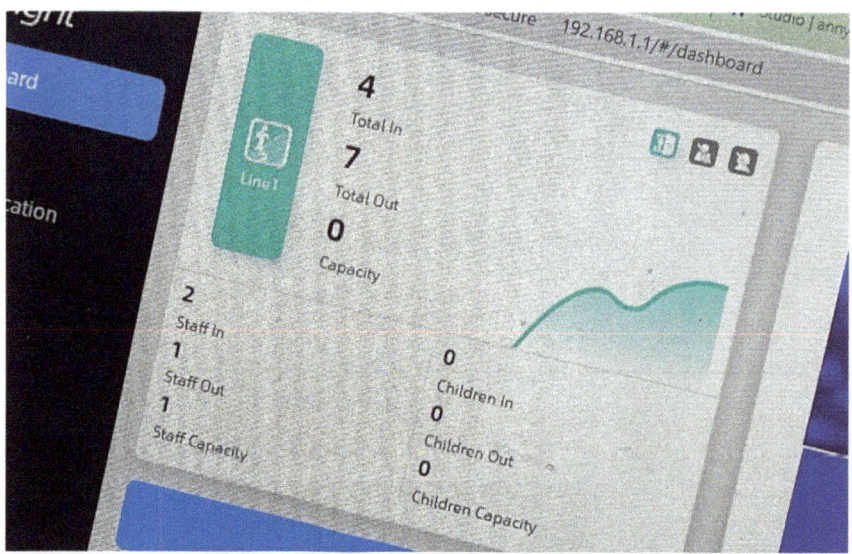

Fig. 7 The people counter recognises incoming and outgoing people and distinguishes between employees and children. This prevents employees from being counted and children from being overlooked in an emergen*cy*

7 Use Case 4: Smart Gardening

Allotment gardens are the perfect retreat for city dwellers, but also for every other hobby gardener. Far away from street noise, concrete and smog, the hobby gardener can quickly find peace and fulfilment in the tasks of the allotment garden site. But the work should not be underestimated. Especially in spring, the garden explodes and requires a lot of work. In summer, the garden is thirsty and requires a lot of water. Despite the many demands, the garden forgives nothing. Anyone who fertilises too much or too little, sows too early or too late or generally plants the wrong things will experience more nightmare than alpine idyll.

"*If only there was an AI that told me what to do and when so that I could make the best use of my time*," thought Dr Daniel Trauth when he suddenly became a gardener in the winter of 2023 and was faced with his first mountain of work in the spring. Daniel lives with his family in Cologne, and a quick watering, lawn mowing or fertilising job requires more planning with a 30-min journey and a 4-year-old. The planning should be AI-supported.

But how is AI supposed to get into the allotment garden where there is no permanent power supply and therefore no DSL or Wi-Fi? LTE or 5G routers could provide a solution, but depending on the model, they can quickly cost between EUR 100 and EUR 500, and then you need a mobile phone contract for as little as EUR 40 per month. But the power supply is still unresolved. Fortunately, Daniel spoke to IDiTech CEO Prof Wolfgang Prinz, who has been a user of the helium blockchain and the associated LoRaWAN network for years. LoRaWAN is perfect for anyone who lives away from connectivity, i.e. in the fields like farmers, in allotments away from cities, on motorways, in the mountains, anywhere where 5G and Wi-Fi are not to be expected (Fig. 8).

Why allotment gardeners need to know soil moisture, soil temperature, soil pH and soil conductivity?

It is crucial for allotment gardeners to understand the nature and conditions of the soil in order to cultivate healthy and productive plants. Some of the most important soil parameters that should be monitored include soil moisture, soil temperature, soil pH and soil conductivity. Here are the reasons why this information is essential for allotment gardeners:

Soil moisture:

- **Optimum irrigation**: knowing the soil moisture helps to control irrigation precisely. Over- or under-watering can be avoided, which is particularly important for plants that are sensitive to water stress.
- **Healthy plant development**: A constant and sufficient supply of moisture is crucial for the plants' nutrient uptake. With a good moisture level, plants can develop their roots better and grow more vigorously.

Soil temperature:

- **Seed germination**: The soil temperature influences the germination of seeds. Certain plant species require a certain minimum temperature in order to germinate

Fig. 8 LoRaWAN sensors are easy to use, robust in operation and have a service life of up to 5 years without solar support and up to 10 years with solar support. This means that LoRaWAN sensors only cost a few cent a day

successfully. The optimum time for sowing can be determined by monitoring the soil temperature.
- **Root growth**: The temperature of the soil has a major influence on root growth. Soil that is too cold can inhibit growth, while soil that is too hot can damage the roots.

Soil pH:

- **Nutrient availability**: The pH value of the soil influences the availability of nutrients. A pH value that is too high or too low can mean that important nutrients are not accessible to plants. For example, iron is poorly available in highly alkaline soils.
- **Plant health**: Different plants have different pH requirements. Some prefer acidic soils, while others require alkaline conditions. Knowing the soil pH allows you to select plants that will thrive well in the soil environment available.

Soil conductivity:

- Soil **salinity**: Soil conductivity provides information about the salt content of the soil. Too high a salt content can damage plants and impair their growth. By monitoring the soil conductivity, allotment gardeners can take measures to regulate the salt content.
- **Nutrient concentration**: A high soil conductivity can also indicate a high concentration of nutrients. This is particularly relevant when fertilising to avoid over-fertilisation, which can also be harmful.

Why should allotment gardeners operate a weather station?

It is a great advantage for allotment gardeners to operate their own weather station. By monitoring weather parameters such as air temperature, humidity, light intensity, wind direction, wind speed, rainfall and UV index, they can organise their gardening more efficiently and successfully. In addition, the weather also plays a decisive role in the well-being of people working in the garden. The following explains why these measurements are important.

Air temperature:

- **Plant growth**: Air temperature has a direct impact on the growth and health of plants. Temperatures that are too high or too low can inhibit growth or damage the plants. With accurate temperature data, allotment gardeners can better plan when to plant or harvest certain plants.
- **Frost protection**: By monitoring the air temperature, allotment gardeners can recognise the risk of frost at an early stage and take measures to protect their plants, for example by covering them or moving them to frost-free areas.

Humidity:

- **Disease prevention**: High humidity can favour the spread of fungal diseases. By monitoring humidity levels, allotment gardeners can take timely disease prevention measures such as ventilating greenhouses or reducing watering.

- **Growing conditions**: Some plants require a certain humidity level to thrive optimally. By monitoring, allotment growers can ensure that these conditions are met.

Light intensity:

- **Photosynthesis**: light is essential for photosynthesis, the process by which plants produce nutrients. With light intensity data, allotment gardeners can ensure that their plants receive sufficient sunlight.
- **Shade management**: In strong sunlight, measures such as installing shade screens may be necessary to protect plants from burns.

Wind direction and wind speed:

- **Plant protection**: strong winds can damage or uproot plants. By monitoring wind direction and speed, allotment gardeners can plan protective measures such as windbreaks or propping up plants.
- **Plant orientation**: Knowing the prevailing wind direction can help with the optimal placement of plants and structures in the garden.

Rainfall:

- **Irrigation scheduling**: by measuring the amount of rainfall, allotment gardeners can optimise their watering and avoid wasting water. In dry periods, they can schedule additional irrigation.
- **Soil erosion**: Excessive rainfall can lead to soil erosion. With accurate rainfall data, measures can be taken to protect the soil.

UV index:

- **Plant protection**: a high UV index can damage plants, especially those that are sensitive to strong sunlight. By monitoring the UV index, protective measures can be taken.
- **Gardeners' health**: A high UV index also poses a health risk to people working in the garden. Protective measures such as wearing sun cream, hats and protective clothing can be planned.

Effects on people:

- **Working conditions**: Monitoring weather conditions helps allotment gardeners to create safer and more comfortable working conditions. In extreme weather conditions, they can adjust their working hours to avoid heat stroke, hypothermia or other health risks.
- **Planning gardening**: Accurate weather data allows allotment gardeners to better plan and organise their gardening more efficiently. They know exactly when it makes sense to carry out certain tasks and can optimise their work to suit the weather conditions.

Fig. 9 Overview of the measured values of Daniel's digital garden

Fortunately, there is ready-made sensor technology for precisely these issues, which can be purchased and installed in the usual LoRaWAN shop. The data is then recorded immediately and only needs to be visualised at a suitable location (Fig. 9).

Implementation

Various sensors were deployed to monitor soil moisture, temperature, pH levels, and other critical parameters. These sensors, powered by long-lasting batteries, transmit data over the Helium network to a central system where it is analyzed using AI algorithms.

Sensor Specifications

- Soil Moisture Sensors: Measure the volumetric water content in soil, providing data on irrigation needs.
- Soil Temperature Sensors: Monitor the temperature of the soil to optimize planting and growth conditions.
- pH Sensors: Measure the acidity or alkalinity of the soil, crucial for nutrient availability.
- Conductivity Sensors: Assess the soil's ability to conduct electricity, indicating the presence of soluble salts.

The data collected by these sensors is analyzed to provide actionable insights for garden management. AI algorithms, integrated with ChatGPT, offer personalized gardening tips and alerts based on the sensor data.

Outcomes

This setup has enabled garden managers to make data-driven decisions, optimizing watering schedules, fertilization, and planting times. The integration with AI provides personalized gardening tips and alerts, enhancing productivity and sustainability in

> **Vegetation und Tiere, die profitieren:**
>
> - **Gemüse und Kräuter**: Die milden Temperaturen und die hohe Luftfeuchtigkeit sind ideal für den Anbau von Gemüse wie Salat, Spinat und Kohlrabi. Auch Kräuter wie Petersilie und Schnittlauch gedeihen gut unter diesen Bedingungen.
> - **Blumen und Stauden**: Pflanzen wie Primeln, Stiefmütterchen und Astern profitieren von der aktuellen Witterung und zeigen kräftiges Wachstum.
> - **Bodenlebewesen**: Die hohe Bodenfeuchtigkeit und die moderate Bodentemperatur fördern die Aktivität von Regenwürmern und anderen Bodenorganismen, die den Boden auflockern und fruchtbar machen.
>
> **Empfehlungen für die Gartenarbeit:**
>
> 1. **Gießen**: Trotz der hohen Luftfeuchtigkeit solltet ihr eure Pflanzen regelmäßig gießen, besonders solche, die flach wurzeln. Achtet darauf, morgens oder abends zu gießen, um Verdunstung zu minimieren.
> 2. **Düngen**: Nutzt organischen Dünger wie Kompost oder gut verrotteten Mist, um die Bodenfruchtbarkeit zu erhöhen und das Pflanzenwachstum zu unterstützen.

Fig. 10 Extract from ChatGPT with the recommendations and advice to Daniel

urban agriculture. The decentralized nature of the Helium network ensures that data transmission is reliable and secure, even in remote garden locations (Fig. 10).

8 Conclusion

The synergy between Helium Blockchain and LoRaWAN is transforming urban management by providing scalable, secure, and cost-effective solutions for a variety of applications. From environmental monitoring to smart parking, crowd management, and urban gardening, these technologies are paving the way for smarter, more sustainable cities.

Acknowledgements We would like to extend our heartfelt thanks to the City of Dormagen for their generous loans, which significantly contributed to the successful implementation of the projects described in this chapter. Your commitment and support are invaluable in advancing smart urban solutions.

A special thanks also goes to IDiTech for the free provision of the Helium account and the associated data credits. Without this generous support, the realization of the described use cases would not have been possible. Your innovation and dedication to digital transformation have been instrumental in improving the quality of life in our cities.

Daniel Trauth is a trained mechatronics engineer who went on to study both mechanical engineering and business administration at RWTH Aachen University. He also completed his doctorate at RWTH Aachen University and has been leading digital transformation projects since 2016. An accomplished entrepreneur, he has founded multiple companies focused on IoT, blockchain, and digital business models. He collaborates with Wolfgang Prinz and is actively involved in the Blockchain Reallabor at the Fraunhofer FIT. Additionally, he serves on the board of the Institute for Digital Emerging Technologies (IDiTech), contributing to the advancement of sustainable and cutting-edge digital solutions.

Wolfgang Prinz studied informatics at the University of Bonn and earned his PhD in computer science from the University of Nottingham. He is vice chair of Fraunhofer FIT in Bonn, head of the Collaboration Systems research department, and a Professor at RWTH Aachen. His research focuses on digitization, new cooperation platforms, mixed reality, and flexible communication infrastructure. In the Fraunhofer Blockchain Lab, he explores technical foundations and Blockchain-based applications. He has led various national and international research projects, including a significant European project on collaborative work environments, and serves as an editor for several journals and conferences.

André Heryschek André Heryschek is Head of Structural Change and Smart City at Stadtmarketing- und Wirtschaftsförderungsgesellschaft Dormagen, the economic development agency of the city of Dormagen. He studied public administration in Cologne and Kassel and subsequently worked in various positions as a civil servant in local authorities and at federal level.

Since 2017, he has been designing and implementing data infrastructures in Dormagen. With the conviction that data has added value and should be correlated based on use cases, a city data platform and a digital twin have been implemented. Therefore, Dormagen currently serves as a real-world laboratory in various federal research and funding projects.

Token-Based Economies in Decentralized Societies

Felix Hildebrandt

As blockchain heralds a digital renaissance, decentralized accounts become the centerpieces of our online interactions. This shift is making the development of token-based economies and social spheres increasingly important. This paper will explain the foundational principles of decentralized societies, their integration dynamics, and the embedment of identity-related tokens. Within the analysis, user data guidelines will be defined to foster consensual interactions.

1 Decentralized Societies

Social blockchain-based communities publically surfaced through initiatives like CryptoPunks [1] or Bored Ape Yacht Club [2], which indirectly cultivated the idea of decentralized societies. Within communities, identity and reputation are fostered based on assets on a global ledger. In a broad view, such decentralized societies are represented through memberships, commitments, and credentials anchored within a blockchain account, forming a concept of dynamic souls that offer a rich, multifaceted representation of an individual's affiliations [3].

1.1 Social Integration Dynamics

In 2017, blockchain initiatives began to explore the concept of a blockchain-based identity for engaging in communities. However, these efforts were limited to token ownership and needed more integration with a decentralized social framework.

F. Hildebrandt (✉)
Chemnitz, Germany
e-mail: felix.hildebrandt@gmx.net

Where Decentralized Autonomous Organizations (DAOs) facilitated governance, proposals, and token transactions on-chain, community engagement largely remained on Web2 platforms like Discord, X, Reddit, and Telegram. The shift led to a disjointed experience with scattered profiles and inconsistent reputation management.

The asset-based blockchain identity, often based on Web 2.0 associations, leads to problems confirming the authenticity and value of digital assets, as it is not natively anchored in the chain. For example, on-chain assets, like those based on the ERC-721 standardization [4], do not specify adequately linking their creators, with many NFTs originating from a zero address. Therefore, the topic of decentralized societies is seeking a more resilient connection between the relationships of accounts and their interactions. The goal is to create a blockchain environment where identities and authenticities are inherently connected, allowing the blockchain to directly confirm individual identities and the originality of digital assets using abstracted and multi-functional accounts.

1.2 Decentralizing Social Media

Decentralized societies are the solutions to patch present Web 2.0 issues of data security, privacy, and fragmented user experiences with a more open, user-centered, and secure interaction landscape. Due to self-sovereign account management, users can choose platforms and biases individually, fostering environments not governed by stringent algorithms.

This freedom of choice encourages the development of portable and open social graphs, thereby promoting a culture of diversity and inclusivity. Interoperability ensures that users have the autonomy to transfer or back up their chats seamlessly across diverse platforms, effectively preserving their digital histories and memories. Moreover, it diminishes the network effects that have kept users tethered to flawed platforms, empowering them to migrate to newer, safer platforms without losing their identities and the essence of their digital interactions.

A cardinal benefit of Web3 is the strengthened security protocols, prioritizing end-to-end encryption, significantly reducing the chances of unauthorized access and data breaches, which have become commonplace concerns today. Establishing stringent data protection norms restores faith in digital conversations, ensuring that personal spaces in the digital realm are sacrosanct and protected from unwarranted intrusions, providing a sense of reassurance and confidence.

Where abstraction separates the signing functionality from the account, decentralized societies foster the separation between service platforms and their accounts. Users gain complete control over their profiles and communication channels. This separation restores the integrity of democratic processes, mitigating risks associated with data misuse and biased content amplification, emphasizing the value and respect given to user autonomy.

1.3 Current Social Architectures

Where centralization often sidelines user needs and sacrifices individual experience for corporate gain, the Lens and Farcaster protocols, as the first pioneering open media infrastructures, are designed with a user-centric approach. They prioritize individual experiences and empower users to customize their identities within the decentralized ecosystem.

Lens manages modular and decentralized social graphs based on an NFT collection to map user profiles. Everyone can acquire a profile asset to customize an identity. While the profile's metadata and interactions are strictly defined, external apps can access the data but interpret it differently to create unique feeds or views. The protocol and its interactions are open-source smart contract standardizations. A central hub is responsible for minting profiles and keeping track of all user information, whereas custom modules are responsible for mechanisms regarding followerships or postings. Such dynamics can include time- and reputation-based restrictions or paywalls. The personal interactions are written into the profile's NFT storage and can be minted as a token. Due to the static NFT address, identity references can be maintained even if the related account gets updated. All user interactions are managed through a more scalable but separately governed subordinate network [5].

Farcaster, on the other hand, is a more minimal open protocol and network. Users acquire unrestricted and unique Farcaster IDs from a smart contract registry on-chain. However, the ID can also be referenced to a specific governed name registry for better accessibility. All data, like user information, posts, messages, or comments, is stored on nodes running a decentralized and open subnet of nodes. The system is designed so that nodes within it build and hold delta graphs created from signed messages, retaining this data for up to a year to ensure network stability and data integrity [6]. Like Lens, social media apps can aggregate deltas, build feeds, and create archives to store older messages.

Both solutions integrated proxy signing for their EOA user base, meaning they can direct the protocol's signing rights to a different wallet or device. For Lens, the profile NFT can be stored safely within a hardware wallet, while interactions can be initiated throughout other extensions. Both protocols then walk through the chain of smart contracts for verification.

Regarding recovery and permissions, Farcaster comes with a handy recovery address pointing to the Farcaster ID. This EOA can trigger the ownership transfer of the handle after a week without intervention. On the other hand, Lens infrastructure can restrict protocol-based wallets' operations with signing rights, limiting the risk of false interactions [7], but does not increase the security of the wallets overall.

Both applications are built on the EVM but are limited to the current account model. Here, identity data only exists on a separate network or within an asset that might get sold from an account. Here, a user becomes a frame of what he holds or which registry he signed up in. In the future, such media data could be directly embedded in abstracted ecosystems, further simplifying the protocols and giving the account more social presence.

2 Bound Token Economy

While account features and public identities have evolved, more user coordination and interactions are needed. All sorts of entity-related properties should be able to be shifted into the smart contract landscape and then act as the foundation of decentralized social relationships.

Concerning binding information to a persona, non-transferable Soulbound [8] Tokens (SBTs) arose to build more robust decentralized social structures. Once issued, they belong to a specific account and cannot typically be transferred or sold to a new account, making them non-financial rewards with personal value. They could imitate special, inalienable certificates, achievements, proofs of presence, social currencies and bonds, or interactions and reputations of a fictitious personality [3]. In other words, precisely what constitutes identity in the first place. Beyond the transfer limitation, SBTs are imagined to come with novel attributes, including mechanisms for social or communal recovery or exclusive issuing. One example is the study group's completion or validation of authors who worked on a specific paper- each confirming others to prove certain skills [9].

SBTs hold immense potential to elevate non-financial rewards, achievements, and reputation to a new level. While social applications offer a glimpse into the future, SBTs could catalyze a paradigm shift in the adoption of DAOs. As we've discussed, communities currently manage votes or assets within smart contracts but rely on various Web 2.0 platforms for interaction. Furthermore, NFTs with monetary value often spur community participation driven by financial gain rather than genuine interest and involvement. In this context, SBTs could emerge as a potent tool to directly manifest interactions within an organization on a chain, thereby fostering genuine engagement and decentralization.

2.1 Requirements for Bound Assets

As on-chain assets, Soul-Bound Tokens (SBTs) offer a cost-effective solution. Unlike bound goods that would require reissuing every time a key is lost or updated, SBTs eliminate this need, thereby avoiding immense additional expenses. The complexity of recovery or carryover for locked assets is also a key factor. This is why current proof of attendance issuers, with POAP [10] being the most prominent, opt for transferable certificates, even though the sale is optional. SBTs, on the other hand, create a system where attestations are accurate and verified, fostering a culture where trust is not just expected but guaranteed.

The implications of SBTs extend to various fields, with academia being a prime example. Here, attestations grounded on SBTs could potentially revolutionize the credential verification processes, making them more secure and reliable. SBTs can offer a framework where voting, open-source contributions, and attestations are grounded on verifiable and non-transferable identities.

For governance, they can ensure that tokens are linked to indisputable identities, promoting transparency and accountability. Governance tokens are often available for purchase, a practice that predominantly enables wealthy individuals to accumulate voting power or buy into communities. Their value and related asset speculation causes holders to be less willing to vote. The market connection fostered a prevailing trend of plutocratic governance systems, where the affluent disproportionately influence decision-making processes. Furthermore, these structures seldom incorporate safeguards for the protection of minority interests, thereby contradicting the principles of equality professed by many protocols. Using SBTs, participation or achievements could be needed to vote, get airdropped goods, or join certain groups. By holding individuals accountable through a system that is incorruptible and transparent, SBTs facilitate a governance model that is more fair, democratic, and participation-friendly. However, such structures also provide the opportunity to reflect negative aspects—for example, unfair behavior or the accumulation of debts.

In the open-source community, contributors' identities are integral to fostering a collaborative and reliable ecosystem. SBTs facilitate the recognition and accreditation of contributors by tethering contributions to immutable identities. The affiliation ensures that contributors are recognized and rewarded for their input and promotes a culture of trust and mutual respect, which are the bedrock of open-source environments.

There should also be a consensus between the issuing and receiving parties. Otherwise, unintentionally credible identities would be at risk of spam without a way to remove it from their accounts. Selling entire accounts is also dangerous: objects would linger on the exact identity address even though the owner changes. SBTs only unfold their power in trustworthy social networks and should be linked to specific requirements. If SBTs were transferred, people would likely recognize it and outlaw the seller's reputation if he "sold its soul." In the long term, ways must be found to restrict bots from forming social circles and fake relationships.

For decentralized societies, however, there is a chicken-and-egg problem when using regular wallets: Externally Owned Accounts (EOAs) need SBTs to become valid identities. However, these cannot be issued if one quickly loses access to them. On the other hand, EOAs need authentic social group networks for identity recovery, which can be implemented exclusively with SBTs. An EOA is just the sum of its particles: Since they cannot carry any information, the social space is only about what the address has acquired. Such accounts become hollow placeholders without the anchoring personality with depth and content.

The intertwining of SBTs and identities steers toward Name Bound Tokens or Account-bound Tokens with similar goals but derivates from the soul concept. However, a substance of personality needs to be available and integrated into the network, creating a holistic identity ecosystem.

Abstracted smart accounts solve the problem by providing a static address as an identity, delivering natively upgradable security, and linking the account to public user data that can generate reputation: It enables data to be stored and shared more securely from the outset. With configurable permissions, variable recovery methods would be conceivable. Here, the emphasis is no longer on what a user accumulates but on who they are and how they present themselves to the outside world. Universal Profiles solve the identity problem and make up the ideal soul framework.

Concepts for community-based backups are possible results. In this way, "lost souls" could be helped back on their feet without requiring action from the user. Vaults or rights could be cleverly combined to embed backups in social structures. SBTs can strengthen the dynamics of a DAO and blur the boundaries between Web 2.0 and Web3.

2.2 Advising Social Dangers

While promising, bound tokens introduce significant concerns and potential social dangers. Blockchains are inherently immutable and permanently available. With their data persistence, blockchains violate the present GDPR's right of deletion [54] on the custodial medium. This problem is deeply rooted in blockchain technology and has once again brought the issue of privacy to the forefront. In a realm where each crypto user's financial history is already immutably stored on platforms, social goods add another layer of complexity by tightly linking tokens to individual identities.

Unlike traditional Internet environments, where posts can be stored as screenshots and retained by third parties, the blockchain system allows users to prove the hash and details of the message indefinitely, creating immutable accountability that can be dangerous. Even if data is kept off the blockchain, transactions or connectors can be used for movement analysis. Due to the finality of assets, subjective social statements could become objective truths, potentially misleading individuals into accepting claims at face value and bypassing the evaluation process [11]. Later, the links could result in an unwanted negative reputation [12]. In the future, individuals may overshare information that could be viewed negatively, such as a low credit score, religious connections, or political parties.

Data restrictions could further result in exclusion with social or political consequences, potentially becoming a systematic problem for minorities. Here, the surrounding dynamics of personal data markets could become dangerous. While intended to offer enhanced choice and control over personal data, some users may feel compelled to give up their privacy, a compromise not demanded for financially secure groups of people. Such inequality could pose individual risks and challenge the integrity and equity of technologies dependent on personal data [13]. Thus, the analytics issues that arise in centralized content moderation could become even more complex in decentralized contexts, as monitoring and auditing information flows in a global space becomes much more difficult, if not impossible [14]. Especially in an open network, handling bound assets should be reconsidered carefully.

Regarding protection and obliviousness, efforts should be made to incorporate the right to dissociate and block non-consensual minting [15], as stated by Tim Daub (TD), creator of multiple bound token standards. It would be possible for the connection to the user account to be capped and for an association to be made only after mutual consent. However, this step presents several difficulties, especially in scenarios involving more than two people, which can lead to consensus blocking and power imbalances.

Another common argument against social tokens is their susceptibility to censorship, mainly because the issuer retains the power to control reissuing. Procedures should safeguard against unilateral control and manipulation. Shared control could be solved through protocols or previous permission managers in proxy accounts, extending equally to tokens.

Standards should include a broader view of potential abuse to provide a roadmap for constructing a self-managed identity that balances control, access, and protection, among other things. Such a nuanced approach is essential to balancing anonymity and privacy that respects individual preferences while ensuring community safety.

There have been some outstanding standardization attempts in the past. Among others, the Web of Trust Initiative (WoT) established ten principles for self-sovereign identities [16]. A comprehensive table of key regulations was derived with identity rules such as the Laws of Identity [17] by Kim Cameron (KC), Identity and Access Architect at Microsoft, and the Principles of a Digital Being [18] by the Privacy Standardization Architect Natsuhiko Sakimura (NS), to form a global guideline. Table 1 shows the user-centered principles applied to token development.

These principles advocate the notion of a user having independent existence and control over their identity, ensuring access to personal data, and advocating for transparency in systems and algorithms. They emphasize the importance of long-lived, transferable, interoperable identities and stress the need for user consent, data minimization, and protection of user rights. By focusing on decentralization in social tokens, a robust mechanism can be implemented to protect against undue influence and ensure a fair playing field for all actors.

2.3 Differentiating Tokens and Claims

Despite the inherent risks in the dynamically evolving digital landscape, tokens promote interactions and trust within decentralized environments. However, the security concerns lead to the question of why interactions should become tokenized assets instead of data claims [55] in the first place.

SBTs and claims [56] rely on a shared, potentially decentralized data model but differ in their motivations and privacy considerations. SSIs inherently protect users' private and sensitive information and aim to decouple entity authentication from centralized registries, identity providers, and certificate authorities, thus enabling a decentralized approach to identity verification. In contrast, SBTs seek a blended approach to private data to encode social trust networks on the blockchain, creating provenance and reputation within decentralized entities such as DAOs. Therefore, the overarching goal of SBTs is to foster a personality-based token ecosystem that mirrors trust relationships and affiliations in real-world networks, which indispensably need transparency. As with the current internet, the data economy is rarely understood as private property [3], and it is up to the user to decide how much to disclose. Some

Table 1 Guiding table for identity-based token development

Aspect	Principles	Derived guidance
Sovereignty	Independent existence and control (WoT), complete control and consent (CL), accountable expressions (NS), and dissociation (TD)	Users must have control over their identity-bound asset, creating, managing, and expressing their digital being autonomously as long as the association is wanted
Management	Access, transparency, and minimization regarding data (WoT), minimal disclosure for constrained use (CL), and fair data handling (NS)	Users must be able to control and limit access to their tokens, focusing on transparency while minimizing unnecessary data collection
Agreements	Consent on all actions (WoT), control and consent for justifiable parties (KC), and consent during issuing (TD)	Users must express consent with a clear description of parties justified in using the identity-bound information
Adaptation	Portability and interoperability (WoT), diversity of operators with consistent experience across contexts (KC), and adoption friendliness (NS)	The identity-bound asset must promote portability and interoperability across various platforms, ensuring a consistent experience
Safety	Data protection (WoT) and universal benefit (NS)	Users' rights and data must be protected, fostering a system beneficial for all entities, including individuals, companies, and governments
Lifecycle	Data persistence (WoT), directed identity data (KC), upholding the right not to be forgotten (NS), and dissociation as the end of data lifecycle (TD)	Long-lived asset data must be assured while respecting the right to be dissociated and facilitating the directed mechanisms that are secure and private
Experience	Human integration (KC) and human-friendly design (NS)	Systems must be developed with a human-centric approach, catering to individual differences and focusing on inclusive, integrative solutions

data has to be on-chain intentionally as a common dataset. Otherwise, centralized companies will create services for this, constraining gained freedoms. Universal Profiles are a prime example of keeping everything open without restrictions or governance but making it possible to operate globally under services.

However, it is essential to make a distinction regarding the storage of identity-related data. While data claims are the data whose signed reference is written to a decentralized ledger, SBTs only represent metadata to describe the asset. All concrete information that may be attached as a separate link is still stored off-chain, as with SSI, and allows for additional protection. Still, SBTs could unintentionally empower harmful intermediaries by exposing too much control and data on chains. Nevertheless, having an always-available token with metadata may help prevent the provider from removing the associated record.

Regarding custody, SBTs are a crypto-native approach, as the specific token form requires the blockchain to unify value and ownership in a decentralized context. Claims do not necessarily need to be anchored in the blockchain layer. Storing credentials on the blockchain brings several benefits by enabling cost-effective and reliable data retrieval while coordinating every participant in a neutral environment. It uses blockchains to avoid the tradeoff between inconvenience and centralization [12], even if the initial cost exceeds those of signed claims.

SBTs envision a pluralistic network of values where personal information is programmable. It provides a flexible approach to granting access to underlying data, social credentials, affiliations, or government-issued documents. This method opens avenues for a bottom-up, decentralized coordination mechanism that can redefine the coordination of social groups and communities and overcome the limitations imposed by government-issued identity documents [14].

In summary, while SSI and SBTs share a common goal of decentralizing identity and trust mechanisms, they follow different paths. SSI tends to take a more privacy-oriented approach that promotes individual control over personal data. In contrast, SBTs are heading toward a decentralized reputation system anchored in the blockchain. As the digital landscape evolves, harmonizing these approaches and mitigating their inherent challenges remains vital in fostering a trusted and user-centric digital identity ecosystem. However, as the requirements for identity-bound tokens have shown, the SBT landscape needs a robust account system with good asset integration to securely and automatically query participant connections and facilitate approvals.

2.4 Standardization Landscape

In 2022, the topic of restrictive tokens experienced a considerable upswing, mainly as the idea was spread by Vitalik Buterin, referencing soul-binding from the gaming industry [8], where items are bound to characters after completing in-game challenges. Based on the initial idea, many projects have been proposed to solve barriers for several use cases. The biggest NFT marketplace, OpenSea, introduced such a locking as a possible feature for trading [19]. Another group addressed the private data issue by combining ZK proofs with an SBT [20]. The following sections present and evaluate a historically ordered list of SBT and restriction-related ERCs to weigh opportunity and use cases. Until the 10th of October, 2023, 26 individual restriction-based standardizations were analyzed.

The two main topics of discussion are how to solve the transfer permissions and give consent. A common practice experienced early on was using regular asset standards with an empty or reverting transfer function. While complying with the general setup, digital assets would throw an error once they were called, breaking interfaces that could not detect the lock before calling the function. Therefore, many interfaces and life cycles were established with the goal that marketplaces, recovery services, and social apps can represent tags or interactive buttons accordingly.

2.4.1 Fungible Proposals

In 2018, ERC-1132 introduced the first locking mechanism for fungible assets, creating an extension for ERC-20 tokens to self-lock token amounts for a specific period. Even if it did not come with its interface, users can check locked and transferable token balances during a future timestamp or increase time or amounts on the fly [21]. Continued management becomes particularly useful for any governance or participation-specific field where identities earn a reputation.

Years later, ERC-3643 introduced a multi-layered standardization to facilitate regulation for fungible securities. While SBTs were not introduced, the goal was built to ensure that assets remain compliant with various restrictions across jurisdictions while interacting with the DeFi world. However, this also suits the topic of socials, ensuring that certain types of tokens are traded with adherence to guidelines. At its core, the standardization functions as a token permission system, ensuring that only approved and registered identities can perform specific actions. At the same time, allowlisted agents can impose token restrictions, updates, and backups or even freeze and retain tokens. In the context of SBTs, the authorization party could specify which addresses can mint, restore, or disassociate a token. The standard can also manifest licensing authorities in a DAO during restructures, imposing re-issuing or removals [22]. Because the identities exist independently of the wallet addresses, community leaders could implement token freezing and instance backups for members. The same applies to the tokens, meaning all token regulations can be updated and bound separately to allow updated restrictions or government rules.

Another attempt regarding bound assets was made with ERC-6808 and ERC-6809, representing the equal backward compatible standardization for FTs and NFTs. The concept splits the responsibility between the holder and owner, authorizing how a second wallet can spend owned assets. This separation as an enhanced security approach works using time-bound restrictions and transfer approvals [23]. Here, both standardizations feature multi-layered concepts that allow for authorizing, adding, or removing multiple designated key owners and allow for several transfers for a specific period.

2.4.2 Non-fungible Proposals

Where FTs always focus on time-locking or regulation, the first NFT-exclusive concept of restriction was a static, non-tradable token. The original idea stemmed from ERC-1238 in 2018, later converted into a multi-token standardization. However, ERC-4671 continued the original name to represent personal possessions handed out by institutions. At its core, the proposal includes a minting and revoking mechanism by which the receiver has to approve the minting before it can be issued. Once minted, the asset is bound to a specific user account but can permanently be revoked by the owner. If the owner revokes the statement, all information and ownership remain unchanged, but the token will return an invalid status that can be publicly queried.

Besides its transaction behavior, the standard also features an interface so marketplaces can differentiate and adjust their front ends based on the binding. It also comes with a whole set of modular extensions regarding token-specific metadata, a global storage contract to keep track of multiple assets and their states or address renewals, where users can pull over their tokens to a new address [24].

After Vitalik Buterin outlined the original idea of SBTs from the gaming industry in 2022, ERC-4973 implemented an account-bound token without any transfer functionality. The token or item can be consensually given out or taken to an account acting as a soul. If the user does not want to show the asset, it can be unequipped by the owner upon receipt. However, users can always re-equip it [25]. Here, the handout functionality, in particular, acts like an airdrop to the signed EOA address, equal to in-game behavior.

ERC-5058 also presents an extension for NFTs, where owners can grant approval to issuers and lock it up to a future block time. While locked, the transfer is prohibited until the issuers unlock the asset or its locking period is over. Without user actions, assets stay tradable as regular. The standardized interface also outlines the idea of bound NFTs, replica tokens of the original assets, and their metadata created during the locking process. The twin could then be handed out to rent the original asset until the expiration time is reached and the twin destroyed [26]. With this functionality, NFTs could be locked securely in hardware wallets while still being used on dApps, supporting token standardization. The security approach is similar to ERC-6809 [27], where the holder and owner are separated, with the only difference being that it allows for multiple instead of singular owners at a time.

Due to various specifications, social dangers, and criticism of SBTs, the community tried to define multiple minimal proposals, removing the context discussion of how to bind and focusing on locking. Developers can then choose how to implement it strictly, only knowing that the transfer event has to revert whenever the asset is locked. ERC-5192 became one of the first SBT-related standards that was finalized. With this minimal approach, the standardization is an extension and specifies two simple events for locking and unlocking an asset [28]. Marketplaces can then check the interface and lock status within a single view.

The second final minimized standardization ERC-5484 focused explicitly on the consensus part of an SBT for regular NFTs. Before a soulbound token is issued, the issuer and the receiver must agree on who can burn the token. This authorization is permanent and cannot be altered post-issuance. The operator must then present the token metadata to the receiver and obtain the receiver's signature to bind the asset. This method would force the data and removal settings to be available off-chain or by having an inheriting contract where the recipient can retrieve and sign the metadata and burn authorization before creating the individual SBT. Any changes to the metadata post-issuance are prohibited. To support various use cases, both parties can choose to enable the disassociation from the owner, issuer, or both [29].

More niche, ERC-5753 outlined a minimal locking interface similar to ERC-5192. Within the proposal, the lock and unlock functions were split and defined as a single address to free the asset once locked [30].

ERC-6147 proposed another step to enable new NFT use cases by providing expiration dates for specific guards of an asset. The specification allows the owner to retain holding rights while temporarily giving transfer functionality to a time-limited guard until full control returns [31]. What was initially meant to decrease the risks of losing the EOA key could give institutions or protocols the right to remove or stake personal accomplishments in an SBT context.

In addition to the minimal extensions already discussed, ERC-6454 brings another approach to transferability determination. Here, the standard introduces an interface with a single function to check transferability from a specific issuer to a recipient. By including both addresses, the standard provides a more agile way of allowing transfers for backup reasons. It does not come with individual events to restrict implementation or combination with other mechanisms. Where other standards utilize separate burn or revoke functions, ERC-6454 embeds them directly into the regular transfer function, as an initial assignment or final removal can be utilized by leaving the issuer or recipient as the zero address [32].

ERC-6982 proposed an interface for minimal and efficient locking by minimizing on-chain events to optimize operation costs further. The standard features a default event stating the status for all future minted tokens whenever a token contract is initialized. In correlation, there is also a separate function to set the default locking value for all existing or future tokens later. Individual token IDs can also be modified [33].

Last but not least, another minimal mechanism was added by ERC-7066, assigning a single, infinite unlocker to an individual token ID. The standard describes two main functions that allow retroactive locking for regular assets or restrictions during transfer [34]. Here, SBT use cases are seen as a glimpse, mainly reaching for wallets to lock their valuable assets to a hardware device or multi-signature contract. Other than ERC-5753, however, a single function is used to transfer and lock simultaneously.

2.4.3 Hybrid and Multi-fungible Proposals

Non-transferable tokens were initially proposed with ERC-1238 and turned into a multi-type standard. Within the proposal, so-called badges or experience points were lifted off as fixed statements about someone's EOA. Those were later seen as primitive for SBTs, so the standard was further fine-tuned [35]. Although the tokens cannot be transferred, the idea was that they could be staked and potentially lost post-staking or even set to expire, with manifold use cases across reputation, achievements, or DAO integrity. The initial draft focused on NFTs only [36], while the later architecture was heavily inspired by ERC-1155 [37] and its capacity to manage multiple token types under a single contract. In addition, the standard makes it possible for each token ID to attach a custom data link for the badge or experience currency they gained. Besides letting the owner remove the badge at all times, the core part of this standardization is the authorization during the mint time [38], with a receiver interface [39] that returns an EOA-specific approval value for an EIP-712 signature [40]. Overall, the standardization became a broad mixture of features, aiming to enable extensions like storage links, expirable properties, or holder separation for any fungibility type [41].

In 2020, ERC-3525 introduced a semi-fungible, restricted token. While it also comes with regular token IDs like NFTs, the standardization adds slots and values. While the ID ensures that every token can be distinguished separately, a slot represents the property or characteristic differentiating tokens within the same ID category. On top, the value field distinguishes the number of tokens within a specific slot. A classic use-case example would be having voting tickets with different weights for a specific election. DAO members could acquire tickets of multiple weights, fungible within the same weight ranking. The included locking mechanism introduces approvals, so owners can allow operations for specific slots and entire IDs or attach custom metadata to them [42]. Related to SBTs, such approvals help restrict minting and manage social member trees.

As ERC-5058 did for NFTs, ERC-5516 presented an interface to bind assets from multiple types to several owners. On top of that, the standardization uses a pending state for the binding directly on-chain. Before the minting, the single token is transferred to one or multiple recipients and enters a pending state where the received asset can be signed or rejected. Parties can agree individually, resulting in an on-chain event and the fixed binding to the signed address. If not, the token is shown as rejected but unaltered metadata. As it integrates the previously mentioned ERC-1155, the standard also allows the blend of fungible and non-fungible characteristics by supporting the transfer of multiple tokens in a single transaction. However, the proposal does not provide content-disassociation [43].

ERC-5633 initiated another extension for tokens of different fungibility. By having a non-obligatory soulbound property, bound and non-bound assets could coexist within the same contract, as each token ID can be addressed individually. Once bound, all transfers except creation and removal will be denied [44]. One great use case would be games that generate rare or unique items of the same asset type that can never be sold while leaving the majority unaffected.

Quickly after, ERC-5727 took the approach of slots from ERC-5325 and mixed it with previous SBT concepts from ERC-5192 and ERC-5484 to create a multi-tool close to real-world use cases. It is a semi-fungible, non-transferable token standard with removal permission, including separated issuer and verifier parties. Each token ID can be individually verified through an external party and even gain or lose credit, stating the importance or rarity of the accomplishment. Moreover, it also comes with a set of add-ons to approve statuses for shared governance, token expiration, and backups for SBTs [45].

Another minimal for multi-token add-on was proposed on ERC-6268, outlining a minimal untransferable indicator [46]. As within ERC-5633, a locked parameter can be assigned for unique IDs. However, this time, the standardization also comes with individual events to lock or unlock one or multiple tokens, making queries for dApps more efficient.

2.4.4 Atypical Proposals

During the last two years of SBT development, even unconventional binding approaches have come to light. Because of the previous hurdles of having a strict and static private key for every EOA address and the chicken-and-egg problem, another idea was that soulbound items should relate to a name or property with rotatable keys, just as smart accounts have solved. The original idea of ERC-5107 was to bind an NFT to an ENS name that acts as a universal anchor point on-chain but has yet to become a concrete specification [47]. Here, splitting the entity from the account to be maintained externally was the right amount of modularity until upcoming account changes face adoption. By doing so, the proposal aimed to inherit all security features from the name registry [48].

After years of the badge idea originally proposed with ERC-1238, ERC-5114 made another attempt at irrevocable soulbound badges and picked up the question of how an on-chain identity may look. As profile pictures were the main driver of the digital asset narrative, the proposal standardizes how non-transferable badges can be bound to NFTs. Here, unique pictures of on-chain characters could bind particular traits, clothing, or accessories to their souls, while the attached content cannot be censored or altered later on [49]. This idea can become an even broader concept of creating marketplaces for hybrid SBTs that only get bound once redeemed to another asset. More importantly, the parent NFT can stay transferable while its properties are individually bound.

As DeFi remains the most common use case of blockchain, SBTs were also seen as a significant opportunity to port over banks' real-world use cases. ERC-5252 outlines the architecture to connect an account-bound token with the DeFi realm to enable reputation or even hand out uncollateralized loans. The standard describes a design pattern for account-bound finance. An investor's deposit is associated with a bound NFT and directed to a personalized finance contract, maintaining an investor entity across multiple wallets. Depending on specific transfer and operator approvals, this NFT can be minted and burnt [50].

Speaking of connecting more real-world use cases, ERC-6239 tries to embed the Resource Description Framework (RDF) from the World Wide Web Consortium into the metadata of tokens. Most tokens use regular fallback metadata for NFTs like ERC-712. The World Wide Web Consortium, also responsible for SSI development, developed a scheme to capture, store, and manage social metadata in a structured and interconnected manner. Included JSON and XML structures enable better creation of relationships between attributes and facilitate the integration and sharing of social data across various applications. Most significantly, the proposal aims to close the gap between the regular and decentralized areas of the Internet. The related contract implementation, therefore, houses different events and calls to create, update, remove, or access token data similar to regular database or API fetching in Web 2.0 [51].

ERC-5114 already mentions a radical shift for bound items attached to an NFT, but ERC-6551 goes further by mixing SBTs and abstracted accounts. While an asset usually gets bound to accounts, the standard describes the opposite concept of how accounts can be bound to a single NFT, meaning a digital asset can have a wallet.

The idea behind this mechanism is to allow an asset to own data and interact with other smart contracts. The Token-Bound Account (TBA) then has its signature using ERC-1271 to verify the transactions. From an architectural standpoint, each TBA is a minimal proxy account, ensuring a deterministic address within a registry. This account then delegates execution to the external business logic whenever a transaction comes in [52]. With the concept, the owner of the NFT can individually grow the asset's social interaction graph while the creators designed its functional capabilities beforehand.

ERC-6956 proposes a more abstract standard for digital twin NFTs as the interest in connecting blockchain and real-world assets grows. The proposal aims to bind physical and digital assets with NFTs while each asset is connected through a unique anchor. An oracle must then verify and attest this anchor, ensuring that control over the asset equals control over the NFT. Afterward, the anchor can be transferred or destroyed in relation to its real world's lifecycle, creating an asset-to-asset binding [53].

While most atypical proposals do not directly fit the SBT construct, they outline similar locking mechanisms. In a broader sense, they also help develop restriction— and identity-based on-chain economies.

2.5 Evaluation of Token Standards

As seen through the standard analysis of SBT-related models, many proposals include features regarding the current lack of account abstraction. However, as the broad adoption of smart accounts has yet to be present, security-related topics often get integrated into tokens, hindering proposals to focus on the core feature. On top of that, while locking mechanisms increase security, they can also lead to more overhead by not losing the secondary seed phrase.

Making SBTs an extension to regular tokens benefits the already-built adoption. However, almost all gathered proposals focus on NFTs. As restriction can be seen as the foundation for many different use cases, once personas can act through secure and static smart accounts, the field of SBTs can be expected to grow much more comprehensively than non-fungible assets. Here, regular standards often lack the farsightedness to open them up for increased possibilities regarding issuer, owner, and combinability. Fig. 1 shows the complete survey of analyzed asset standards.

Another primitive often seen during the standards analysis is the variety of add-ons or extensions that seem minimal in the first place. However, on closer inspection, these only offer superficial capacities. These should be independent standardizations referencing their base concept to foster organized and individual development. As seen in Fig. 1, many standards are in a stagnant or year-long review state. The difficulty of progress could correlate to the opacity of features, as 3 out of 8 minimal standardizations could be finalized within a few months. Compared to all other approaches, only 16.7% of ERCs have the status "finalized."

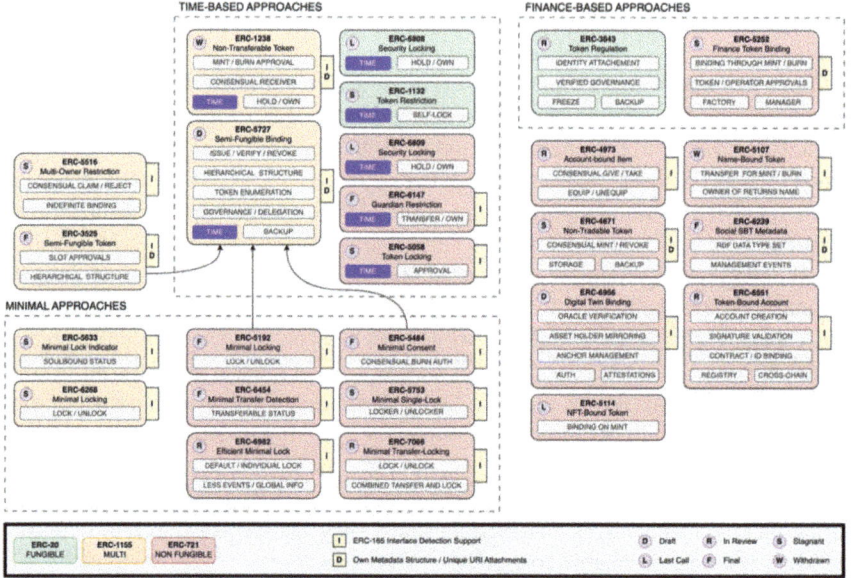

Fig. 1 Classification of restriction standards

The tendency to tokenize assets with features has sparked an essential dialogue about utility and necessity. While tokens, particularly soulbound tokens, are perceived as integrating rights and permissions directly into user accounts, their widespread use raises the fundamental question about their need. Many rights, obligations, voting, and login permissions for DAOs and other blockchain-based entities are already encapsulated in a single contract that anyone can check individually. Tokens should not be used to map direct rights, only indirect permissions or achievements that might be used for benefits. Compared to real life, users might show friends their goals or pictures of their work but not the account login or key to their office, which they manage on the backend. If backend circumstances change while they have permissions, they need to be changed, ending in meaningless parallel maintenance of two instances. With the right frameworks, only some pieces of data have to become an expensive token. Attendance or acknowledgments could be a simple claim or non-divisible currency with additional metadata. On top of that, developers should continually optimize for systematic and cost-effective approaches regarding decentralized data management.

Addressing the adaptability of permissions and rules, a structural connection between an account manager and the DAO could facilitate similar dynamic adjustments based on users' achievements and time allocations. If the account address remains static, the application of ranks and permissions could be mapped directly to accounts, enhancing the adaptability and responsiveness to member achievements. Similarly, consent functionality could become a direct part of communities if their abstracted accounts provide suitable interfaces. Users could then look up if there is a notification regarding their global account instead of relying on individual token implementations.

2.6 Guiding Principles for Development

To establish a suitable standard, listing guidelines to comply with is essential. These principles are based on the conducted guiding table for identity-based tokens and the finalized proposals from the analyzed token standards, giving a direction for standard best practices. The universal restriction rules for social and identity-bound content should, therfore, look like the following:

1. Binding the asset to an account address must be possible.
2. Only metadata should be anchored directly.
3. The link to personal data must be completely removable by the owner.
4. The recipient must confirm transfer rules before the handout is executed.
5. The recipient must confirm removal rules before the handout is executed.
6. The recipient must confirm the token contents before executing the handout.
7. Bound assets can no longer be transferred without further intermediate steps.
8. Bound assets and rules can no longer be modified without reauthorization.
9. Restriction types must allow for final, hybrid, or temporary binding.
10. A variety of allowed operators must be supported to perform transfers.
11. Various people must be able to remove the asset from the account.
12. The owner must always be able to remove the asset.
13. The standard must exist as an extension to regular tokens.
14. The standard must not commit to a fungibility type.
15. The bound asset must have a detectable interface.

Based on this rule set, future development should occur, combining current identity technologies, law, and smart contract technology. User data should only exist off-chain and be linked to the blockchain to maintain the right to be forgotten.

3 Future of the Social Data Economy

With pioneers like Lens Protocol, Farcaster, and Common Ground, social media applications are arousing interest and demonstrating how public blockchain-based identity integration can look like. While they deliver exceptional benefits such as self-sovereignty, individual feeds, open-source code, interchangeable frontends, and censorship-resistant base layers, they are generally limited and hindered by the widespread EOAs and storage networks still in their initial stage. Security and convenience are outsourced to tokens, external registries, or centrally governed environments to counteract the restraints. This phenomenon leads to a mixed nest of different identity systems for each service, similar to Web 2.0, as every ecosystem generates and manages its profile setup.

The social development front runs in parallel to the almost decade-long evolution of abstract smart accounts, trying to remove the limitations and unleash the full potential of blockchain for the next wave of adoption. Smart contracts allow on-chain accounts to be filled with rich social context and convenience features. Contract-based computation can be simplified by seamlessly merging the asset interactions into the core center of the account. Both help to move beyond previous plutocratic and anonymous governance systems.

With the rapid account changes, new gates are opening to embed user data. In the process, various token and claim-based solutions are emerging to allow for identity-restricted assets. The extended analyses of the current token restriction standards have shown the importance of reconsidering data sharing through blockchains to navigate risks.

If managed improperly, multiple previously anonymous or pseudonymous blockchain accounts could quickly be linked in undesirable ways to marginalize or misuse social groups for attacks, as decentralized social circles do not rule out manipulative communities. Conversely, too many private SBTs and connections could lead to hidden communication channels off-chain. This dichotomy can hardly be contained due to the lack of national regulation within global blockchain realms, so developers should strictly follow presented protection rules.

Society-wise, a balance of transparency and privacy is needed to build trust and integrity. Activities of powerful institutions should remain transparent and accountable while the privacy of individuals is strictly protected. While SBTs offer programmable privacy features and great potential to navigate in-group dynamics, they are no panacea. Flexibility should not lead to an environment devoid of legal and ethical boundaries. As the paper discussed different data-sharing approaches, it is often enough to use signed claims like VCs instead of SBTs and allocate rights directly within DOAs, alternatively to the assets themselves.

In conclusion, the paramount challenge for decentralized societies remains to empower individuals with complete control over their content while maintaining the user-friendliness they have come to expect from conventional web services. The emerging nexus of abstract profiles with intertwined digital goods is fundamental to achieving these goals. However, projects must first counteract the additional operation costs of heavy smart contract usage to build up network effects. Therefore, economies will likely spread within separate or subordinate networks. Nevertheless, more significant engagement opportunities across identities will lead to a renewed upswing of DAO and NFT technology and increased user interaction. As interaction amplifies, it will attract more customers outside the regular blockchain field. By gaining considerable adoption, the prevailing decentralization of accounts and new data-sharing schemes lead to more democracy on the Internet and in society. The rise of decentralized organizations, profiles, and associated SBTs represents a positive shift, transforming the financially driven focus of the blockchain industry into a domain that prioritizes social engagement and community building. In this context, decentralized networks and societies empower equality and pluralism by retaining control over their data exchange and shifting focus to what is truly important: individuals and the authentic, unique relationships they foster.

References

1. CryptoPunks. Yuga Labs. Accessed: 30 May 2024 (Online). Available: https://cryptopunks.app/
2. Bored Ape Yacht Club. Yuga Labs. Accessed: 30 May 2024 (Online). Available: https://boredapeyachtclub.com/
3. G.E. Weyl, P. Ohlhaver, V. Buterin, Decentralized Society: Finding Web3's Soul (2022, pp. 15–19). Accessed: 30 May 2024 (Online). Available: https://papers.ssrn.com/sol3/papers.cfm?abstract_id=4105763
4. W. Entriken, D. Shirley, J. Evans, N. Sachs, ERC-721: Non-fungible Token Standard (2018). Accessed: 30 May 2024 (Online). Available: https://eips.ethereum.org/EIPS/eip-721
5. Lens Protocol. Aave. Accessed: 27 Oct 2023 (Social Network Protocol). Available: https://docs.lens.xyz
6. Farcaster: A Decentralized Social Network. Farcaster. Accessed: 30 May 2024 (Peer to Peer Network). Available: https://github.com/farcasterxyz/protocol/blob/main/docs/OVERVIEW.md
7. Lens Dispatcher. Aave. Accessed: 27 Oct 2023 (Social Network Protocol). Available: https://docs.lens.xyz/docs/dispatcher
8. V. Buterin, Soulbound. Vitalik Buterin's website. Accessed: 30 May 2024 (Online). Available: https://vitalik.ca/general/2022/01/26/soulbound.html
9. L. Shin, G.E. Weyl, P. Ohlhaver, How Soulbound Tokens Could Reduce Speculation and Improve DAO Voting. Accessed: 30 May 2024 (Online). Available: https://youtu.be/lKKgP2wS39U
10. Proof of Attendance Protocol. POAP Inc. (Ethereum). Available: https://poap.xyz/
11. K. Sills, Soulbound Tokens (SBTs) Should Be Signed Claims. Kate Sills' website. Accessed: 30 May 2024 (Online). Available: https://katelynsills.com/blockchain/soulbound-tokens/
12. V. Buterin, Where to Use a Blockchain in Non-financial Applications? Vitalik Buterin's website. Accessed: 30 May 2024 (Online). Available: https://vitalik.ca/general/2022/06/12/nonfin.html
13. D. Allen, E. Frankel, W. Lim, D. Siddarth, J. Simons, G.E. Weyl, Ethics of decentralized social technologies: lessons from Web3, the fediverse, and beyond. The Justice, Health, and Democracy Impact Initiative (2023). Accessed: 30 May 2024 (Online). Available: https://verimedia.io/wp-content/uploads/2023/03/ethics-decentralized-social-tech.pdf
14. Soulbound Tokens (SBTs) Study Report Part 1: Building and Embracing a New Social Identity Layer? Blockchain Governance Initiative Network (2023). Accessed: 30 May 2024 (Online). Available: https://bgin-global.org/documents/20230201_SBT.pdf
15. T. Daubenschütz, What are Account-bound tokens? Proof In Progress. Accessed: 23 Oct 2027 (Online). Available: https://proofinprogress.com/posts/2022-05-30/what-are-account-bound-tokens.html
16. C. Allen, S. Appelcine, A Primer on Self-Sovereign Identity. Web of Trust Info. Accessed: 30 May 2024 (Online). Available: https://github.com/WebOfTrustInfo/rwot5-boston/blob/master/topics-and-advance-readings/self-sovereign-identity-primer.md
17. K. Cameron, The Laws of Identity. Kim Cameron's Identity Weblog. Accessed: 30 May 2024 (Online). Available: https://www.identityblog.com/stories/2005/05/13/TheLawsOfIdentity.pdf
18. S. Natsuhiko, Seven Principles of Digital Being. Nat Zone Digital Identity and Privacy. Accessed: 30 May 2024 (Online). Available: https://nat.sakimura.org/2023/01/19/seven-principles-of-digital-being/
19. Metadata standards: disable trading for staked or locked tokens. in *Developer Tutorials*. OpenSea. Accessed: 30 May 2024 (Online). Available: https://docs.opensea.io/docs/metadata-standards#disable-trading-for-staked-or-locked-tokens

20. E. Bottazzi, S. Jain, Zero-Knowledge Soul-Bound-Token (ZK SBT). Accessed: 30 May 2024 (Online). Available: https://github.com/enricobottazzi/ZK-SBT
21. N. Goel, ERC-1132: Extending ERC20 with Token Locking Capability (2018). Accessed: 30 May 2024 (Online). Available: https://eips.ethereum.org/EIPS/eip-1132
22. J. Lebrun, T. Malghem, K. Thizy, L. Falempin, A. Boudjemaa, ERC-3643: T-REX—Token for Regulated EXchanges (2021). Accessed: 30 May 2024 (Online). Available: https://eips.ethereum.org/EIPS/eip-3643
23. M. Onila, N. Zeman, N. Cotaie, ERC-6808: Fungible Key Bound Token (2023). Accessed: 30 May 2024 (Online). Available: https://eips.ethereum.org/EIPS/eip-6808
24. O. Aflak, P.-M. Le Bris, M. Martin, ERC-4671: Non-tradable Tokens Standard (2022). Accessed: 30 May 2024 (Online). Available: https://eips.ethereum.org/EIPS/eip-4671
25. T. Daubenschütz, ERC-4973: Account-Bound Tokens (2022). Accessed: 30 May 2024 (Online). Available: https://eips.ethereum.org/EIPS/eip-4973
26. Tyler, Alex, John, ERC-5058: Lockable Non-fungible Tokens (2022). Accessed: 30 May 2024 (Online). Available: https://eips.ethereum.org/EIPS/eip-5058
27. M. Onila, N. Zeman, N. Cotaie, ERC-6809: Non-fungible Key Bound Token (2022). Accessed: 30 May 2024 (Online). Available: https://eips.ethereum.org/EIPS/eip-6809
28. T. Daubenschütz, ERC-5192: Minimal Soulbound NFTs (2022). Accessed: 30 May 2024 (Online). Available: https://eips.ethereum.org/EIPS/eip-5192
29. B. Cai, ERC-5484: Consensual Soulbound Tokens (2022). Accessed: 30 May 2024 (Online). Available: https://eips.ethereum.org/EIPS/eip-5484
30. F. Makarov, ERC-5753: Lockable Extension for EIP-721 (2022). Accessed: 30 May 2024 (Online). Available: https://eips.ethereum.org/EIPS/eip-5753
31. 5660-eth and Wizard Wang, ERC-6147: Guard of NFT/SBT, an Extension of ERC-721 (2022). Accessed: 30 May 2024 (Online). Available: https://eips.ethereum.org/EIPS/eip-6147
32. B. Škvorc, F. Sullo, S. Pineda, S. Bogosavljevic, J. Turk, ERC-6454: Minimal Transferable NFT detection interface (2023). Accessed: 30 May 2024 (Online). Available: https://eips.ethereum.org/EIPS/eip-6454
33. F. Sullo, A. Spataru, ERC-6982: Efficient Default Lockable Tokens (2023). Accessed: 30 May 2024 (Online). Available: https://eips.ethereum.org/EIPS/eip-6982
34. P. Chittara, StreamNFT, S. Joshi, ERC-7066: Lockable Extension for ERC-721 (2023). Accessed: 30 May 2024 (Online). Available: https://eips.ethereum.org/EIPS/eip-7066
35. R. Roullet, PR for ERC-1238: Non-transferable Token Standard (2022). Accessed: 30 May 2024 (Online). Available: https://github.com/ethereum/EIPs/pull/5617/files
36. N. Greco, ERC-1238: Non-transferrable Non-Fungible Tokens (badges) (2018). Accessed: 30 May 2024 (Online). Available: https://github.com/ethereum/EIPs/issues/1238
37. W. Radomski, A. Cooke, P. Castonguay, J. Therien, E. Binet, R. Sandford, ERC-1155: Multi Token Standard (2018). Accessed: 30 May 2024 (Online). Available: https://eips.ethereum.org/EIPS/eip-1155
38. R. Roullet, ERC1238 Implementation. Violet Protocol (2022). Accessed: 30 May 2024 (Online). Available: https://github.com/violetprotocol/ERC1238-token/blob/main/contracts/ERC1238/ERC1238.sol
39. R. Roullet, ERC1238 Receiver Implementation. Violet Protocol (2022). Accessed: 30 May 2024 (Online). Available: https://github.com/violetprotocol/ERC1238-token/blob/main/contracts/ERC1238/IERC1238Receiver.sol
40. R. Bloemen, L. Logvinov, J. Evans, EIP-712: Typed Structured Data Hashing and Signing (2017). Accessed: 30 May 2024 (Online). Available: https://eips.ethereum.org/EIPS/eip-712
41. R. Roullet, N. Greco, C. Chung, ERC-1238: Non-transferable Token (NTT) Standard (2022). Accessed: 30 May 2024 (Online). Available: https://erc1238.notion.site/
42. W. Wang, M. Meng, Y. Cai, R. Chow, Z. Wu, AlvisDu, ERC-3525: Semi-fungible Token (2020). Accessed: 30 May 2024 (Online). Available: https://eips.ethereum.org/EIPS/eip-3525
43. L.M.G. Ramos, M. Arazi, ERC-5516: Soulbound Multi-owner Tokens (2022). Accessed: 30 May 2024 (Online). Available: https://eips.ethereum.org/EIPS/eip-5516

44. HonorLabs, ERC-5633: Composable Soulbound NFT, EIP-1155 Extension (2022). Accessed: 30 May 2024 (Online). Available: https://eips.ethereum.org/EIPS/eip-5633
45. A. Zhu, T. Chen, ERC-5727: Semi-fungible Soulbound Token (2022). Accessed: 30 May 2024 (Online). Available: https://eips.ethereum.org/EIPS/eip-5727
46. Y. Aoki, ERC-6268: Untransferability Indicator for EIP-1155 (2022). Accessed: 30 May 2024 (Online). Available: https://eips.ethereum.org/EIPS/eip-6268
47. T. Daubenschütz, ERC-5107: Initial Draft for Name-Bound Tokens (2022). Accessed: 30 May 2024 (Online). Available: https://github.com/ethereum/EIPs/pull/5107
48. T. Daubenschütz, T. Cohen, E. Bottazzi, ERC-5107: Name-Bound Tokens (2022). Accessed: 30 May 2024 (Online). Available: https://github.com/ethereum/EIPs/blob/ad528431af47054626b468edabecc0dcec91bd54/EIPS/eip-xxxx.md
49. M. Zoltu, ERC-5114: Soulbound Badge (2022). Accessed: 30 May 2024 (Online). Available: https://eips.ethereum.org/EIPS/eip-5114
50. H. Kang, V. Pernjek, ERC-5252: Account-Bound Finance (2022). Accessed: 30 May 2024 (Online). Available: https://eips.ethereum.org/EIPS/eip-5252
51. J. Chang, ERC-6239: Semantic Soulbound Tokens (2022). Accessed: 30 May 2024 (Online). Available: https://eips.ethereum.org/EIPS/eip-6239
52. J. Windle et al., ERC-6551: Non-fungible Token Bound Accounts (2023). Accessed: 30 May 2024 (Online). Available: https://eips.ethereum.org/EIPS/eip-6551
53. T. Bergmueller, L. Meyer, ERC-6956: Asset-Bound Non-fungible Tokens (2023). Accessed: 30 May 2024 (Online). Available: https://eips.ethereum.org/EIPS/eip-6956
54. General Data Protection Regulation. GDPR-EU (2023), p. Article 17. Accessed: 30 May 2024 (Online). Available: https://gdpr-info.eu/art-17-gdpr/
55. Decentralized Identifiers (DIDs) v1.0 (19 July 2022). Accessed: 30 May 2024 (Online). Available: https://www.w3.org/TR/did-core/
56. Verifiable Credentials Data Model v1.1 (2022). Accessed: 30 May 2024 (Online). Available: https://www.w3.org/TR/vc-data-model/

Increasing Economic Performance Through Digital Application

Arpad Djuraki

1 Introduction

In the following paper, the topic of economics within the metaverse is explained, exploring its evolution and transformative impact. The metaverse emerges from the progression of the internet, traversing Web 1.0 to Web 3.0, with distinct characteristics at each phase. From the one-way information flow of Web 1.0 to the interactive user-generated content of Web 2.0, culminating in the decentralized and blockchain-driven Web 3.0, the metaverse represents a fusion of virtual and real worlds. Its development unfolds through four stages: budding, growth, acceleration, and maturity, signifying advancements in technology and societal impact.

Foundational technologies such as blockchain and smart contracts could facilitate the convergence of virtual and real worlds within the metaverse. Digital assets, particularly non-fungible tokens (NFTs), play a pivotal role in enabling value exchange and reshaping traditional business models. Metanomics, focusing on the economics of the metaverse, presents diverse opportunities, from virtual goods and digital fashion to virtual real estate. The metaverse's technological structure, comprising infrastructure, hardware, software, content, application, and the economic system, underpins its multifaceted nature. Business prospects span gaming, education, research, and health, fostering varying impacts across industries. Virtual goods derive value from functional, hedonic, social, and programmable aspects, with scarcity influencing their value through techniques like NFTs.

Poised as a transformative force in commerce and communication, the metaverse unlocks substantial economic opportunities whose impact is challenging to quantify due to emerging technologies. Estimates from various researchers present a wide range of potential market sizes, while projections suggest profound economic

A. Djuraki (✉)
RWTH Aachen University, Aachen, Germany
e-mail: a.djuraki@googlemail.com

impacts. Analysis Group anticipates a potential US$560 billion annual contribution to the US GDP by 2031, forming part of a projected global impact of around US$3 trillion. McKinsey suggests the metaverse sector could generate a substantial USD 5 trillion impact by 2030, ranking it as the world's third-largest economy. The Metaverse Value Chain shows robust growth potential, fueled by significant investments, emerging as a major growth opportunity across various industries. Projections indicate the metaverse could generate between $4 trillion and $5 trillion by 2030, underlining its transformative role in the global economy.

2 The Idea of the Metaverse

The second chapter provides an overview of the evolution of the internet from Web 1.0 to Web 3.0, emphasizing the emergence of the metaverse as a significant concept. It outlines the characteristics of each web phase, from the one-way information flow of Web 1.0 to the interactive and user-generated content of Web 2.0, leading to the decentralized and blockchain-driven Web 3.0. The metaverse, described as a fusion of virtual and real worlds, is explored with diverse definitions and potential innovation. The development of the metaverse is divided into four stages: budding, growth, acceleration, and maturity. Each stage signifies advancements in technology, applications, and societal impact, leading to a more immersive and interdependent digital-physical reality in the future. The following section also notes the challenges and uncertainties associated with the metaverse's development, highlighting its potential to reshape various industries and aspects of human life.

2.1 Evolution of the Internet

The Internet, a network of interconnected computer systems that spans the globe, first emerged in 1969 [1]. This era was called the Web 1.0. It was mainly characterized by the user's access to information provided in a single direction. Users may click on the links on the web page to browse the text, images, and other content set by the developer [2]. It provided many people an easy way to share read-only data around the world quickly and inexpensively. Therefore Web 1.0 is also commonly referred to as the "read-only web" [3]. It lasted from the 1980s until around 2005, when the transition to Web 2.0 took place [4].

The read-only Web 1.0 did not solve business integration problems nor provide end-user web applications other than simple information-sharing on Websites. New standards evolved that allowed for a "read-write" internet, also called Web 2.0 [3]. To be more specific, the introduction of the smartphone brought Web 2.0 into existence. In 1999, millions of people had access to the Internet a handheld device on which they could access the internet, watch videos and make photos [5]. It is characterized by user-created personalized recommendations for interaction, where users are not limited to browsing the Web, but can also create their own content and upload it to the Web. The original purpose of Web 2.0 was to bring the Internet closer to democracy and to make users more interactive [2].

In the near future, Web 3.0 will enable even more interaction and more than just a place where people can share information [4]. According to Giaglis et al. [3] Web 3.0 will be a return to the fundamentals of a decentralized and open Web 1.0, but with all the modern capabilities of Web 2.0. Further Web 3.0 means a decentralized Internet that is no longer determined by a few platforms. Instead, it uses decentralized technologies. A token-based economy, including Non-Fungible-Tokens (NFTs), Decentralized Autonomous Organizations (DAOs) and decentralized finance (DeFi) and self-determined identities are a central part of the concept. This goes back to Gavin Wood, co-founder of the Ethereum blockchain [4]. Some definitions of the metaverse posit it as the next evolutionary step of the Internet or Web 3.0. The metaverse refers broadly towards virtual worlds in which users can interact with each other and deal with applications and services [6].

Moreover, some companies claim they have started to work on the next generation of an immersive Internet called the "Internet of Senses", in which the user could smell, feel and taste things. In such a digital world, the user will be able to experience everything in their digital environment even more realistically and perceive a blurring of worlds [7] (Tables 1 and 2).

The evolution of the Internet has led to the emergence of Web 3.0, a decentralized solution that empowers users and returns control over the Internet to them. With billions of internet users worldwide, the need for a better internet that protects privacy and data ownership has become increasingly pressing. [1]. For a clear overview of the different Web stages, please look at Fig. 1 and Table 3 in the attachment.

Table 1 Global estimates of the metaverse: global market sizing estimates [23]

Researcher	Estimated value (global market)
Grant View Research	US$ 678.8 billion per year in 2030
Fortune Business Insights	US$ 1.5 trillion per year by 2029
Goldman Sachs	US$ 2.6 trillion to 12.5 trillion annually in roughly 20 years or more
Citi GPS	US$ 8 trillion to 13 trillion opportunity per year by 2030
Grayscale	US$ 1 trillion in annual revenue

Table 2 Global estimates of the metaverse: global GDP contribution estimates [23]

Researcher	Estimated value (GDP)
Analysis Group	US$ 3 trillion per year
PwC	US$ 1.5 trillion per year by 2030
Deloitte	US$ 1.9 trillion to 3.6 trillion per year

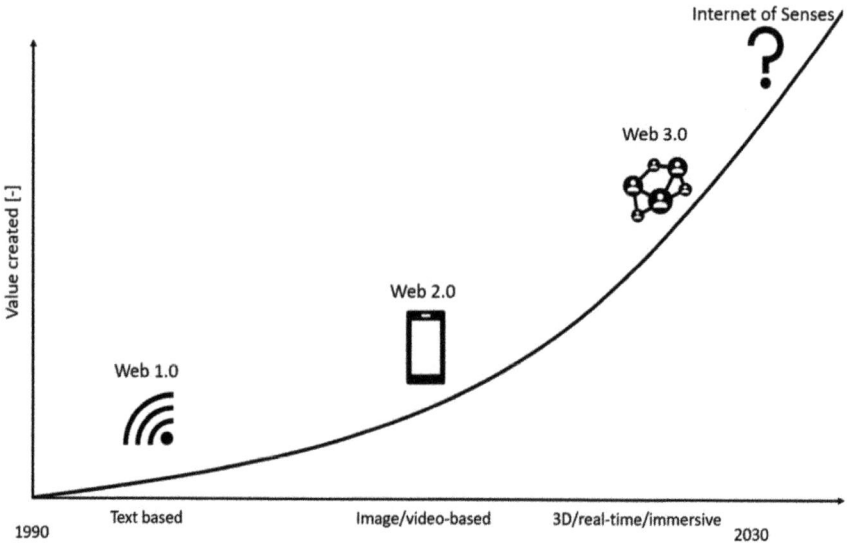

Fig. 1 Evolution of the web [1] and [5]

Table 3 Evolution of the web [2–4]

Web	1.0	2.0	3.0
Interface	Website with limited interaction	Website and mobile centric user interaction	Extended reality (VR/AR/MR)
Logic	Hypertext, standard HTML	JavaScript in collaboration with CCS3 and HTML5, cookies, use of statistics and algorithms	Semantic logic, also through the use of artificial intelligence
Data	Low data storage, mostly text files	Data provision via frontend and mostly central backend	Secure and transparent transaction from Smart Contracts and decentralized data via the blockchain
Control	Centralized	Centralized	Decentralized
Social	Primary information on static websites, that only could be consumed Linking knowledge and Exchange of mostly scientific content Beginning of open communities and forums	Expansion to almost everyone Areas of work and life Users you can create your own content such as texts, Post pictures, videos, comment and share on social media	Interaction in virtual rooms using avatars Users can create your own 3D content and create assets Connected ecosystems/DAOs with different roles, identities and digital assets
Interact	Read only	Read–Write	Read–Write–Own

2.2 Definition of Metaverse

At first is the term "metaverse", a combination of "meta" (beyond) and "verse" (abbreviation of "universe"), which describes a world consisting of both virtuality and reality beyond the physical world built by human beings using digital technologies [8]. The concept was initially introduced by Neal Stephenson in his 1992 published book "Snow Crash". In this context, the metaverse is described as a three-dimensional space where humans interact by means of digital avatars [9]. So, the metaverse is not a new concept, it actually predates the Web itself. The idea of a virtual world that people could visit in order to interact with others dates back over 40 years in science fiction, becoming more mainstream thanks to films such as Tron and Total Recall in the 1980s and 1990s [10].

A major problem surrounding the term metaverse is that a generally accepted definition is still missing. The missing definition should not be seen as a weakness. Rather, this shows the different facets that make up a metaverse could have and thus shows what potential for innovation exists [4].

In that case, Elmasry et al. [11] cited that the metaverse should not be defined rigidly, because that limits the imagination of creators, who work to create it.

The main points through Giaglis et al. [3] of all studied definitions are as follows:

- General and significant change across norms, disciplines, cultures or other barriers creates new opportunities.
- Sustained, coherent, shared experiences provide the feeling of a new "world".
- Can be immersive and interactive, but users can also interact with it to a limited extent. In other words, the metaverse is flexibly immersive.

2.3 Stages of the Evolution

According to Li and Jia [12] the development of the metaverse can be roughly divided into four stages, the budding period, the growth period, the acceleration period and the maturity period.

Phase I: Budding Period (from 2003 to 2021)

The metaverse concept has evolved since the launch of "Second Life" in 2003, followed by the introduction of Roblox (2006) and Minecraft (2012). Various virtual games and platforms like Avakin Life, Rec Room, VRChat, and Fortnite Creative emerged in subsequent years. Oculus Rift's 2016 launch marked a pivotal moment in virtual reality. In 2021, Roblox went public, Facebook rebranded as Meta, and global VR device shipments surpassed 10 million units. Capital investments fueled metaverse growth in entertainment, education, medicine, tourism, and military sectors, turning it from a concept into a burgeoning industry [12].

Phase II: Growth period (from 2022 to 2029)

As of 2022, the metaverse is experiencing a slowdown in the capital market, moving into a growth phase of the hype cycle. The industry's understanding becomes more conservative, and technological directions clearer. Underlying technologies like 5G, edge computing, blockchain, artificial intelligence, and chip manufacturing provide foundational support. Although not mature, these technologies enable the development of the metaverse's basic form and business iterations. In terms of hardware, 2022 sees mass production of new optical and display technologies, optimizing parameters for extended reality (XR) hardware and integrating virtual reality (VR) and augmented reality (AR). The launch of Apple and Meta's new generation XR terminals in 2023–2024 is expected to drive progress in the hardware market [12]. A report by the Next G Alliance predicts the use of one billion XR glasses and sensing devices by 2030, potentially replacing mobile terminals [13]. Entertainment and B2C fields like tourism, education, medicine, and office applications are leading in adoption. Digital personas gain popularity, becoming a significant aspect of work and life. Artificial intelligence and parametric models play a profound role, with AI's capabilities approaching those of humans. Major Internet giants launch metaverse platforms, but full cross-platform integration and interaction protocols are still pending [12].

Phase III: Acceleration Period (from 2030 to 2035)

Around 2030, the metaverse is expected to undergo significant advancements in supporting technologies. This phase, characterized by digital twins complementing the real with the virtual, will witness virtual experiences becoming closer to reality. The time gap between technology generations is narrowing, with the 6G network anticipated to start commercialization around 2030. 6G is poised to create an intelligent, green, and low-carbon network, integrating space, air, and ground with sub-millisecond air interface latency and advanced capabilities. Carbon-based chips, especially IoT carbon nanotube chips, are expected to see commercial use in the next 3–5 years, with broader applications around 2030. Infrastructure replacement and upgrade, following a pattern similar to 5G development, are seen as the foundation for accelerated metaverse development, taking approximately 5 years to reach maturity. In hardware, breakthroughs in holographic display and brain-computer interface are anticipated. While 2D holography is mature, 3D holography may lag behind by 2–3 years, depending on 6G network capacity and chip technology. China's progress in brain-computer interface, adopting a non-invasive approach, shows promise with test projects in medical and military fields. However, large-scale commercialization in consumer terminals is expected to take 8–10 years. Application scenarios in the metaverse include the integration of all non-material production links in industrial production, pushing industrialization forward. Business-to-business (B2B) and business-to-consumer (B2C) interactions are integrated into metaverse scenarios, connecting data and generating new industries, markets, and spaces. AI-generated content becomes a primary source, contributing to the improvement of the metaverse economic system. The comprehensive influence of the metaverse on work and life leads to an era where the Metaverse permeates all aspects. Native metaverse platforms, adapted to metaverse information technology, are evolving to enable better integration, potentially forming a monopoly pattern [12].

Phase IV: Maturity Period (from 2035 to ?)

The current phase of the metaverse is marked by virtual nativity and a struggle between the real and the virtual. The potential for merging with revolutionary technologies like quantum computing, strong AI, brain-computer interface, and robotics exists, creating a more controllable and imaginative form. However, this evolution is anticipated to be lengthy and uncertain. Driven by these technologies, the metaverse is gradually establishing autonomous operation rules, will, and ecology. Its impact on real society is shifting from passive to active, posing ethical and legal challenges and hinting at the formation of a new civilization ecology. There's a risk of losing control over human ontological civilization, leading to potential termination or premature retreat in this period [12].

In contrast to earlier internet development stages, the metaverse demonstrates a higher level of interdependence between the physical and digital worlds. The future metaverse is expected to see a greater fusion of the digital and the physical, with a significant qualitative difference from today's mobile Internet. The most extensive development envisions an almost complete duality of the physical and virtual world, resembling a digital twin of the physical world in digital space [14]. The development phases according to Peters et al. 2022 are shown in Fig. 2, similar to Li and Jia [12].

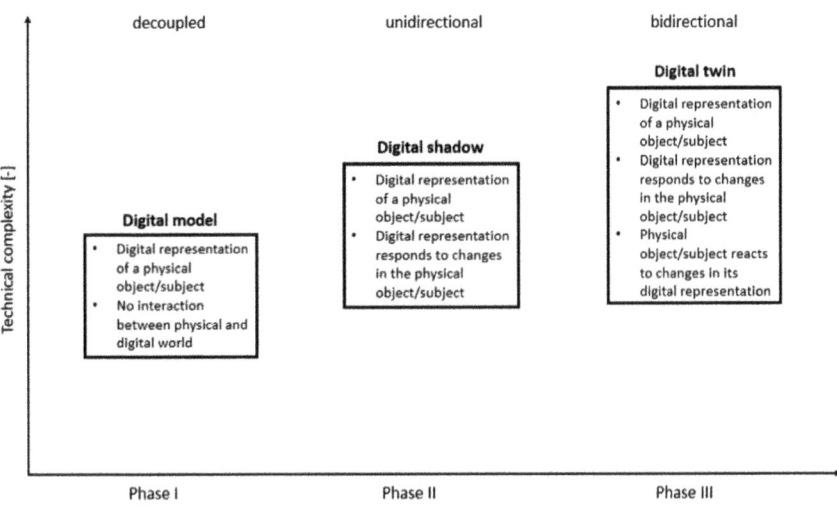

Fig. 2 Evolution of the metaverse [14]

3 Digital Economy

Chapter three examines the metaverse's economic system, focusing on pivotal decentralized technologies like blockchain, smart contracts, and crypto assets, chosen to empower the economics of the metaverse, providing digital sovereignty to users. Smart contracts, introduced in the mid-nineties, are highlighted for their role in automatically executing agreements in code logic, transforming traditional contractual processes. Digital assets, particularly tokens and non-fungible tokens (NFTs), serve as the medium for value exchange within the metaverse, offering transparency and verifiability. The metaverse's economic system, comprising digital creation, currency, assets, and markets, presents a dynamic landscape shaping innovative business models across various sectors.

3.1 Underlying Technology

To better understand the economic system of the metaverse, in this section, we first introduce the technologies that implement the underlying logic of the economic system (what is shown in Fig. 3), i.e., blockchain and smart contracts. In addition, the fuel for user activity in the economic system, i.e., crypto assets, also needs to be focused on [15].

Blockchain: In Web 3.0, Banerjee et al. [16] propose a shift from storing application data in private databases to utilizing blockchains that enable universal reading and writing. This move to decentralized storage through blockchain restores digital

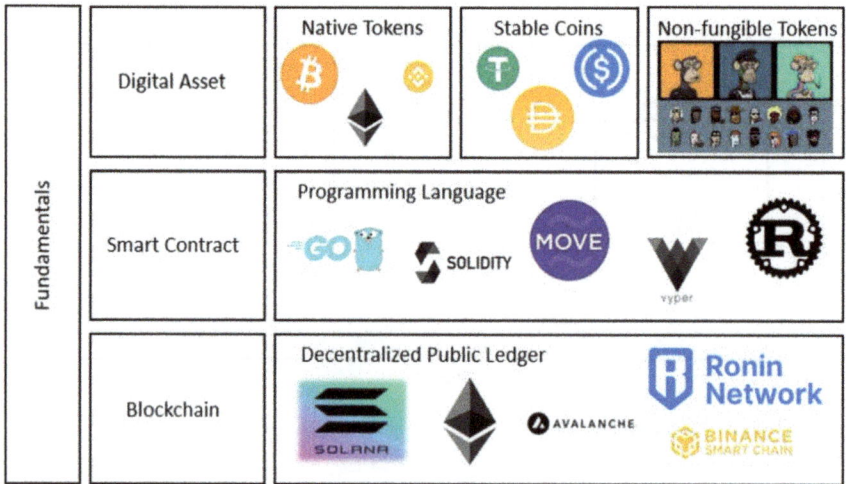

Fig. 3 Several fundamentals of the metaverse according to Wu et al. [15]

sovereignty to users. Blockchains come in three main types: public, private, and consortium. Public blockchains, exemplified by Bitcoin, are highlighted for their typical applications. The text emphasizes the decentralized nature of blockchain technology and its role in ensuring digital autonomy for users [15].

Smart Contract: In the mid-nineties, Szabo [17] introduced the concept of smart contracts, proposing the incorporation of contract logic into code. Unlike traditional contracts documented on paper and enforceable by law, smart contracts are automatically executed agreements between two different parties, with their terms embedded in code logic. Various programming languages, such as Solidity, Go, Java, among others, currently support the development of smart contracts, as discussed by Wu et al. [15].

Digital Assets and Tokens: Digital assets, characterized as intangible digital objects with verifiable and ownable values, include tokens, a prominent representative. Tokens, functioning as digital assets within smart contracts, serve as the medium for storing and exchanging value in the metaverse. The advantages of digital assets encompass a ubiquitous ledger, transparent updates, and verifiable payments without the need for centralized settlement. In the metaverse, blockchain automatically records human interactions in a tamper-proof public ledger, rewarding block miners or validators with tokens. Tokens come in two types: fungible, interchangeable tokens, and non-fungible tokens (NFTs), which are unique and not interchangeable [15].

3.2 Metanomics

In 1994, American economist Tapscott introduced the concept of the digital economy in his book "The Digital Economy." During the early stages of the Internet, the key feature of the digital economy was the utilization of information for digital flow and transmission within the network. Tapscott anticipated a transformative impact on the global economy and society. The term "digital economy" gained official recognition in 1998 when the U.S. Department of Commerce released a report on the new digital economy. Since then, the concept has become widely acknowledged by governments and scholars, as highlighted by Chen and Zhang in 2022 [2].

When you think about the economics of the metaverse or also called "metanomics" there are opportunities in almost every market area. Imagine you have an online avatar and you want to change what it/you are wearing, you will be able to buy limited-edition, digitally branded clothing that you pick after browsing a virtual showroom. Or you may start your own small business, such as an art gallery where you display your latest and greatest collections, or a virtual private club [18].

The concept of Metanomics traces back to 2007 when Cornell professor Rob Bloomfield hosted a course on the subject in Second Life. Themes from that time, such as parallels between physical and digital real estate markets, continue to resonate today. Notably, the ownership economy, driven by the emergence of Web 3.0, has introduced new dynamics. In the current landscape, individuals can personalize virtual homes by purchasing tokenized digital assets, including original art and even the land on which the virtual house is built. Platforms like Decentraland are selling virtual plots for development, contributing to the growth of the virtual real estate market. The average price of a land parcel across major Web 3.0 metaverses doubled from $6,000 in June to $12,000 by December 2021. Brands are also contributing to this growth by purchasing space to create virtual stores and experiences. Notably, in June 2021, a land package in Decentraland was sold for $913,000, transformed into the Metajuku shopping district by the developer Everyrealm (inspired by Japan's Harajuku). Moy et al. [18] highlighted these developments in the evolving landscape of Metanomics.

The rapid growth of the digital economy in recent years has been fueled by a fresh round of technical revolution and is playing an increasingly significant role in the global economy [2].

3.3 Hierarchy of the Ecosystem

The development of the metaverse relies on a multifaceted technological foundation, according to studies by CBInsights [19], Gartner [20], Lee et al. [21] and Elmasry et al. [11]. There is no central enabler for the metaverse's success; instead, it results from the interaction of various factors, as emphasized by Büchel and Klös in 2022 [22].

The metaverse's technological structure comprises several layers: infrastructure, hardware, software, content, application, and economic system (Fig. 4).

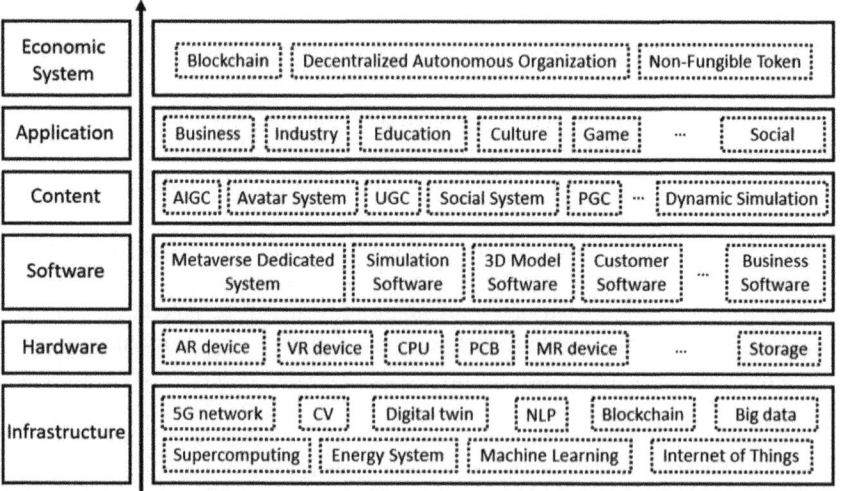

Fig. 4 Six-layers structure of metaverse ecosystem [8]

- The foundation layer can provide basic infrastructures for metaverses, such as network infrastructure, 5G/6G communication, Cloud, data centers and edge computing, computing power, blockchain technology, connectivity infrastructure [23].
- The hardware layer mainly involves human–computer interaction devices in the metaverse, such as AR/VR/XR devices, smartphones and other user devices, operating systems and sensors [23].
- The software layer can provide system software for the metaverse, such as the operating systems dedicated to the metaverse. In addition, application software such as 3D modeling software, synchronous simulation software, business software, and customer-oriented application software are also included [8].
- The content layer can generate core content for the metaverse. For example, AI technologies are enabling user-generated content (UGC), which helps create avatar systems, social systems, and personalized content. The core technologies involved in the content layer include 3D modeling, real-time rendering, dynamic simulation, spatial computing, holographic imaging, and decentralized technology [8].
- The application layer can support various industries at the social level, such as industry, commerce, and agriculture. Some typical scenarios include gaming, education, culture, sports, socializing, and so on [8].
- The top layer of the metaverse ecosystem is the economic system, which is mainly composed of four basic elements, i.e., digital creation, digital currency, digital assets (e.g., Non-Fungible Token), and digital markets [8].

Building a metaverse requires different essential levels, each serving a crucial role. Technical infrastructure, hardware, software, and practical means of payment are all necessary components for a functional metaverse. Developers should possess expertise in related areas such as gaming, virtual concerts, or social media, going beyond technical knowledge to understand which products and services are valuable and in demand in society. The levels not only contribute to the current functionality but also shape the future of the metaverse, with access (hardware) and virtualization (software) identified as key areas. The success of the metaverse relies on the interaction and integration of these diverse levels [22].

3.3.1 Business Driver

The metaverse is anticipated to create significant business opportunities, extending beyond the hardware sector to areas like events. It envisions virtual spaces where people can gather, showcase virtual property, enjoy music, videos, and engage in activities such as dancing and playing together. The potential for diverse and immersive experiences within the metaverse suggests promising prospects for those who capitalize on these opportunities [3].

There are several potential business drivers of economic opportunity.

- First X-reality experiences, encompassing virtual reality (VR), augmented reality (AR), and artificial intelligence (AI), have the potential to revolutionize both business-to-consumer (B2C) and business-to-business (B2B) interactions. Notably, companies like Nike are leveraging VR to sell virtual goods and create additional revenue streams. In the entertainment sector, experiments with metaverse applications, such as the GRAMMY Week on Roblox, showcase the offering of virtual experiences to fans. The integration of metaverse technologies is transforming various industries and opening new possibilities for innovative experiences and revenue generation [23].
- The second aspect involves using the metaverse for enterprise simulations, potentially leading to efficiency improvements. Firms are exploring ways to optimize physical processes through metaverse technologies. For instance, NVIDIA is developing technologies to streamline the training of self-driving cars, while Ericsson collaborates with NVIDIA to optimize the layout of 5G radio networks. Companies like Amazon and PepsiCo are utilizing metaverse technologies, such as 'digital twins,' to simulate and optimize the design of distribution centers. Beyond businesses, cities are also employing the metaverse to aid in planning and visualizing new developments, showcasing its versatility in various contexts [23].
- Lastly, augmented workforce experiences offer opportunities for training, remote work enhancement, and improved learning. Surgeons, for instance, can simulate 3D surgical procedures and receive real-time feedback. The metaverse has the potential to enhance remote working by introducing more lifelike communication, and in the realm of education, it can improve quality by providing more immersive and interactive learning experiences [23].

The initial potential applications of the metaverse highlight its business case. In a professional setting, smart glasses can simplify object recognition and authorization. The metaverse also promises improved remote work and socializing experiences compared to existing software like Zoom and Microsoft Teams. Considering ongoing factors like the pandemic and the push for Net Zero, with a potential decrease in long-distance trips, remote interactions are likely to become more commonplace. This shift could significantly benefit the metaverse. Similar to the advantages brought by the rapid development of the Internet, the metaverse may offer additional digital sales channels, making its presence valuable for companies. This aligns with the current standard of having an online presence for many businesses [22].

3.3.2 Business Models

The metaverse is in its early stages, and contributors worldwide are actively involved in planning its inclusive design and construction. The ongoing development includes the implementation of infrastructure and technologies that are expected to support the metaverse in the future [24].

The business models in the metaverse can be categorized into advertising financing, subscription models, and transactional models, essentially evolving from those in the platform economy. The metaverse operates within the framework of a platform economy, and while it won't introduce a new dimension, additional monetization methods, inspired by the gaming industry, may be incorporated into the business context. Viewing the metaverse as a continuous development of the current World Wide Web, it offers the potential for various business models to transition into this new environment [4].

It's important to note that business models in the metaverse are dynamic, and how businesses generate revenue will evolve over time with the continuous development of the [4].

This section discusses the experience continuum, emphasizing new usages and business models across virtuality and reality, categorizing them into three areas:

- Consumer metaverse: Referring to all applications and experiences designed for and accessed by accessed by individual consumers (e.g. gaming, entertainment, events, social interaction) [10]. Sahgal et al. [6] describe this as a "real-time multisensory social interaction". Given its experience creating virtual platforms, the gaming industry has been at the forefront in this field. Roblox launched in 2006, had on average 66.2 million daily active users in March 2023 [6].
- Enterprise metaverse: Referring to non-industry-specific applications used across businesses for interaction. These are driven mainly by the need for corporate collaboration among employees (e.g. training and education, meetings collaboration working) [3, 10, 25]. For example, the Stanford Virtual Heart is a 3D model that allows medical students to better understand how the organ works [6].
- Industrial metaverse: Focused on technical collaboration among employees and machines. These applications are often industry- or business-specific (e.g. design and development, simulation and optimization, operational improvement) [10]. In the context of simulation, digital twins can be used to run simulations, understand performance issues and propose potential improvements [6, 8].

It can be assumed that the metaverse will gradually expand into more and more industries recorded—for example according to the following scheme (Fig. 5).

For a more detailed overview refer to Table 4 in the attachment.

As the metaverse evolves, certain industries are likely to emerge as winners, particularly in areas like work, education, research, and health. In the field of medicine, the digital medical record allows for the tracking of user movements, presenting significant opportunities for preventive measures and diagnostics. However, some industries, such as traditional banking, travel, and tourism, may face disruption and need to be cautious to avoid obsolescence. On the other hand, there are industries at risk of becoming obsolete, such as traditional mobility and the construction industry. The impact of the metaverse will vary across sectors, with potential winners and losers based on adaptability and innovation (e.g. classic mobility or construction industry) [4].

The metaverse offers a vast range of applications across various industry sectors, extending beyond monetary value. Four main areas of emerging impact have been

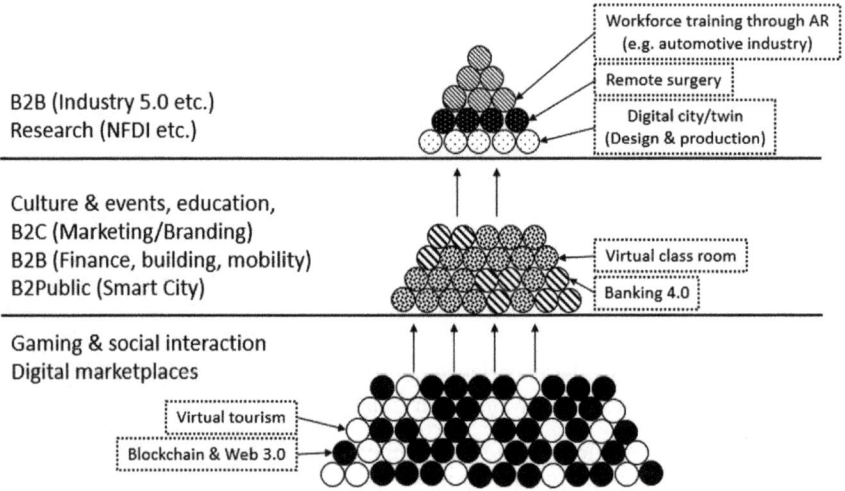

Fig. 5 Usage-diffusion-pyramid [4]

identified: learning and education, work and collaboration, digital twins, and social interaction and entertainment. While some use cases are already emerging, others may take years to become mainstream, showcasing the diverse and evolving potential of the metaverse across different domains [6].

3.4 Virtual Goods and Its Value

In the real world, the value of a commodity is determined by a combination of two factors. On the one hand, commodities must have a use value and be able to be used to satisfy some aspect of people's needs. On the other hand, commodities must be scarce. If there is no scarcity of a certain type of good, then it will not have value, even if it is very useful [26].

In the realm of virtual economics, as proposed by Lehdonvirta and Castronova [27], the use value of virtual goods is categorized into three aspects: functional, hedonic, and social values. In the metaverse, these three types of values coexist and contribute to the overall value of virtual goods.

3.4.1 Functional Values

Virtual goods have functional values that include aiding users in gaining attention, a valuable and scarce resource. Attention is crucial in the digital economy, and companies strive to attract and retain users' attention through advertising and promotions.

Table 4 Current proto-metaverse applications within the consumer, enterprise, and industrial metaverses [10]

	Travel an transportation	Energy and utilities	Healthcare	Retail and consumer	Aerospace and defense	Financial services	Manufacturing and automotive	Entertainment	Education
Early adopters/ key player	ANA, Matterport	vedanta, groupe	DASSAULT SYSTEMES, Johnson & Johnson, Pfizer	GUCCI, Nike, Coca-Cola	BOEING, RR	BNP PARIBAS, J.P.Morgan, BANK OF AMERICA	NVIDIA, HYUNDAI, BMW	Meta, unity, DISNEY, ROBLOX	LIFE, STRIVR
Typical focus of applications consumer	In-journey entertainment Customer interface		Virtual care	Digital assets Virtual try-ons/shops/ actions Virtual experiences		Virtual lounges Virtual support for clients VR apps for account opening		Virtual events experiences/ simulators Virtual assets Virtual worlds E-sports/music	

(continued)

Table 4 (continued)

	Travel and transportation	Energy and utilities	Healthcare	Retail and consumer	Aerospace and defense	Financial services	Manufacturing and automotive	Entertainment	Education
Enterprise	VR/AR training		Virtual meetings and events			Remote collaboration and workshop tools			
Industrial	Digital twins for asset design, manufacture, operation, maintenance	VR to design, inspect, test and validate equip. Live operation data. Operator/engineer training	Remote monitoring. Digital human simulations to test therapies. Digital twins of facilities	Product customization and comparison	Digital twins to optimize operations/maintenance. Future combat air system	Decentralization finance	Digital twins of factories. Human behavior simulation. Simulations of supply chains. Asset mgt. and maintenance		Campus digital twins
Remarks	Sector was early adopter of VR for training. Future potential for virtual tourism is a threat/opportunity	Digital twins already present in some facilities. Levels of energy consumption of the metaverse could be an issue	Better diagnostics and solutions. Personalized medicine. Show patients the surgery before performing it	Boost brand awareness. Collaborations with platforms. New payments. NFT creation. User behavior tracking	Reduce costs of training. Increase pilot safety. Enable predictive maintenance. Reduce emissions	Enhance personalization, "human approach" for banking. New financial products arising from metaverse	Enable performance improvement across supply chain. Digital twinning already established Industry 4.0 will help metaverse adoption	Convergence of social experiences and entertainment. Gaming companies among most active in shaping metaverse	Generate more realistic simulations for students. Enable new forms of collaboration among students/teachers

In the metaverse, elements similar to the real world, such as location and zones, influence attention. For instance, central locations like the capital city center and popular themed areas in the metaverse attract significant attention. In projects like Second Life, Decentraland or other metaverse games staging areas act as attention hubs, analogous to real-world city centers, and are considered valuable for advertising. Overall, the value of attention is essential in the metaverse, with strategic locations commanding higher sale and lease prices for advertising space [26].

3.4.2 Hedonic Values

Unlike functional value, which focuses on "usefulness," hedonic value emphasizes "fun" and the pleasure a good can bring. This concept holds true in the metaverse, where user-created buildings in platforms like Minecraft may lack practical use but evoke admiration and affection. The hedonic value of these works, rooted in enjoyment and emotional response, determines their potential as commodities and their ability to command high prices, even if they lack inherent functional value [26].

3.4.3 Social Values

Social values in consumption involve buying things for others, a concept known as "Conspicuous Consumption." This behavior is analyzed by the economist Van Buren, who notes that it serves to display social status, satisfy vanity, and boost self-esteem. In contrast to ordinary consumption, conspicuous consumption often involves wasteful spending, emphasizing luxury over practicality. The metaverse, as a virtual space with open access, provides an ideal platform for people to showcase possessions. Here, extravagant and showy items may be even more effective in fulfilling their display function than in the real world. This understanding helps explain why virtual items, such as a Gucci bag, can command higher prices in the metaverse than in reality [26].

3.4.4 Programmable Values

Virtual objects possess programmable value, a significant use value that stems from their programmability. Unlike real-world assets, virtual assets, often built on blockchain technology, offer greater programmability, allowing easy adaptation to users' needs. This contrasts with real-world scenarios, such as obtaining a loan using a house as collateral, which involves complex processes and fees. With blockchain-based virtual assets, programmability simplifies usage and allocation, eliminating the need for extensive appraisals and facilitating secure transactions through smart contracts. This distinctive feature enhances the utility of virtual assets in the metaverse [26].

3.4.5 Supply of Virtual Good

The value of virtual goods is linked to their scarcity, akin to the real world. Scarcity in virtual goods is derived from both material input requirements and intentional construction. Material input, despite being digital, is essential for producing virtual goods, and their scarcity can result from constraints on server capacity. Operators often intentionally create artificial scarcity, limiting the availability of plots or branding virtual items as unique through techniques like digital watermarks, digital rights management (DRM), and NFTs. The deliberate creation of scarcity serves two main purposes: enhancing user experience by providing differentiation and personalization, and generating revenue to support ongoing metaverse development. This intentional scarcity aligns with real-world economic laws, breaking down the dimensional wall between the metaverse and reality. Virtual goods, possessing both use value and scarcity, establish a value basis and subject themselves to economic principles such as the law of demand in the metaverse [26].

4 Global Estimates

The metaverse, with its transformative impact on commerce and communication, presents a substantial economic opportunity, albeit challenging to precisely quantify due to nascent technologies. Various estimates for potential market size and economic contribution differ, but collectively indicate the metaverse's significant global economic potential. Deloitte's [23] assessment recognizes the uncertainties surrounding size and value but emphasizes the overall significance of the metaverse in the global economy (Tables 1 and 2).

According to Analysis Group's projections, the metaverse has the potential to make an annual contribution of around US$560 billion to the US GDP by 2031. This estimate is part of a broader global impact on GDP, which is anticipated to reach approximately US$3 trillion. The basis for this forecast rests on the idea that the metaverse's expansion worldwide could mirror the historical growth pattern observed in mobile telecommunications technology. Deloitte [23] Estimates regarding the economic impact of the metaverse vary widely, much like the diverse interpretations of its definition. These estimates range from exceptionally high figures, such as being '10 times the total value of the entire current global economy,' to more conservative projections. For instance, Bloomberg Intelligence suggests that metaverse revenues 'could approach USD 800 billion in 2024.' This disparity is evident in industry reports as well. On one side, a comprehensive 177-page report from Citibank in March 2022 predicts an USD 8 trillion metaverse by 2030, accounting for 10% of today's global economy. In contrast, a similar report from JP Morgan's Onyx lowers this estimate to USD 1 trillion [3].

Certainly, the dynamics of the metaverse landscape are still in flux, and its economic prospects exhibit variability that might experience fluctuations. However, as per insights from a McKinsey report, the metaverse sector is anticipated to generate a substantial USD 5 trillion impact by 2030, equivalent to the magnitude of Japan's economy, ranking as the world's third-largest. While the metaverse's potential varies across industries, there are universal implications for all sectors. For instance, there is a projected market impact ranging between USD 2 trillion and USD 2.6 trillion on e-commerce by 2030, with the advertising market expecting an impact within the range of USD 144 billion to USD 206 billion [3].

The segments encompassing the components and physical aspects within the metaverse Value Chain are anticipated to exhibit a robust growth trajectory, with a projected Compound Annual Growth Rate (CAGR) of 43% in the coming decade. By the year 2025, this growth is expected to culminate in approximately US$300 billion, and by 2028, it is forecasted to escalate to US$830 billion. Certain predictions even posit the possibility of this market evolving into a trillion-dollar industry within the next decade. An alternative perspective reveals that the combined value of the worldwide hardware, broadband, and gaming industries was approximately US$2.6 trillion in 2021. Projections indicate an anticipated Compound Annual Growth Rate (CAGR) of 7.4% in these sectors up to the year 2025. Analysts at Citi have estimated that the total addressable market (TAM) of the metaverse could reach US$8–13 trillion by 2030. This is based on a device agnostic definition, assuming the metaverse accounts for 30–40% of the digital economy, which itself could account for 20–25% of global GDP by 2030. Their narrower definition based on users with VR/AR devices, gives a TAM of US$1–2 trillion [5].

By 2030, simulated systems for addressing real-world industrial challenges are projected to achieve a market value of US$100 billion. Immersive business collaboration technologies, encompassing productivity tools and virtual workspaces, are anticipated to reach a market value of US$30 billion. Additionally, digital worlds and immersive experiences for activities such as shopping, gaming, socializing, and entertainment are expected to attain a market value of US$50 billion, as per Bienert et al. [28].

The metaverse is poised to become a major growth opportunity for various industries in the next decade due to its vast range of potential applications and substantial investments from tech giants, venture capital, and corporations. While current initiatives largely focus on marketing, industries are increasingly exploring diverse uses. Projections suggest that by 2030, the metaverse could generate between $4 trillion and $5 trillion across consumer and enterprise applications, as outlined by Elmasry et al. [11].

5 Conclusion

In conclusion, the metaverse stands at the forefront of a digital revolution, marked by its evolution from Web 1.0 to the decentralized and blockchain-driven Web 3.0. Its transformative impact extends beyond technological advancements, shaping a new economic landscape with profound implications for various industries. Foundational technologies like blockchain, smart contracts, and digital assets, particularly NFTs, play pivotal roles in creating a convergence of virtual and real worlds, introducing innovative business models.

Metanomics, exploring the economic dimensions of the metaverse, reveals a plethora of opportunities, from virtual goods and digital fashion to virtual real estate, redefining traditional concepts within the platform economy. The metaverse's multifaceted technological structure, spanning infrastructure, hardware, software, content, application, and the economic system, signifies its potential to reshape human experiences in gaming, education, research, and health.

As the metaverse unfolds through its developmental stages, it presents a trajectory marked by growth, acceleration, and maturity, offering a glimpse into a more immersive and interdependent digital-physical reality. The economic potential of the metaverse is vast, with estimates ranging from trillions to potentially rivaling the entire global economy. Despite uncertainties and diverse projections, the metaverse is anticipated to become the world's third-largest economy by 2030, generating trillions in impact.

Looking ahead, the metaverse's value for the world lies not only in economic terms but also in its capacity to redefine how individuals interact, businesses operate, and societies function. With significant investments and major growth opportunities projected across industries, the metaverse is poised to become a catalyst for innovation, collaboration, and economic prosperity on a global scale. As it continues to mature, the metaverse is likely to influence not just commerce but also communication, education, and various aspects of human life, making it a cornerstone of the digital future. The journey of the metaverse holds promise for a world where virtual and real seamlessly coexist, shaping a future that transcends current technological boundaries.

"You take the blue pill—the story ends, you wake up in your bed and believe whatever you want to believe. You take the red pill—you stay in Wonderland, and I show you how deep the rabbit hole goes. Remember: all I'm offering is the truth. Nothing more—Morpheus (The Matrix 1999).

Attachment

See Tables 3 and 4.

References

1. J. Zheng, D.K.C. Lee, in *Understanding the Evolution of the Internet: Web1.0 to Web3.0, Web3 and Web 3 plus* (Singapore, 2023)
2. C. Chen, L. Zhang, Y. Li, T. Liao, S. Zhao, Z. Zheng, H. Huang, J. Wu, When digital economy meets Web3.0: applications and challenges, in *IEEE Open Journal of the Computer Society*, Bd. 3, pp. 233–245 (2022)

3. G. Giaglis, L. Dionysopoulos, M. Charalambous, A. Ntouzgou, T. Damvakeraki, Metaverse report, in *EU Blockchain Observatory and Forum* (2022)
4. Bitkom, Wegweiser in das Metaverse—Technologische und rechtliche Grundlagen, geschäftliche Potenziale, gesellschaftliche Bedeutung (Bitkom, Berlin, 2022)
5. T. Roberts, J. McDermottroe, P. Garvin, J. Morris, in *Metaverse—New Horizons for the Digital economy* (2022)
6. V. Sahgal, W. Pasquarelli, S. Pandey, E. Saliba, J. Gantz, A. Simms, M.A. Gonzalez, Toward a successful metaverse—the case for measuring enabling factors, in *Economist Impact* (2023)
7. ART, Metaverse—virtual world, real challenges, in *Council of the Eurooean Union—General Secretariat* (2022)
8. J. Li, S. Cai, Q. Yang, H. Huang, How to enrich metaverse? Blockchains, AI, and digital twin, in *From Blockchain to Web3 & Metaverse* (Springer, Berlin, 2023), pp. 27–61
9. O. Küpeli, T. Gürpinar, in *Towards a Definition of the Industrial Metaverse Applied in Context of the Blockchain and Web3 Ecosystem* (Corfu, 2023)
10. D.A. Meige, J. Abascal, R. Eagar, M. Papadopoulos, in *The Metaverse beyond fantasy*, ed. by A.D. Little (2022)
11. T. Elmasry, H. Khan, L. Yee, E. Hazan, G. Kelly, R.W. Zemmel, S. Srivastava, in *Value Creation in the Metaverse—The Real Business of the Virtual World* (2022)
12. L. Li, H. Jia, in *Analysis of Metaverse Development in the Context of Digital Economy* (Beijing, 2023)
13. atis, atis.org (2024) (Online). Available: https://www.atis.org/press-releases/new-report-by-atis-next-g-alliance-helps-the-industry-prepare-the-future-6g-network-for-the-impact-of-multi-sensory-extended-reality-applications/ (Zugriff am 19 Apr 2024)
14. R. Peters, B. Schmietow, B. Krieger, *Zwischen Hype und Zukunftsthema: Auf dem Weg ins Metaverse? Bestandsaufnahme und Handlungsperspektiven für die Gestaltung des Metaverse* (Institut für Innovation und Technik (iit), Berlin, 2022)
15. J. Wu, K. Lin, D. Lin, Z. Zheng, H. Huang, Z. Zheng, Financial crimes in Web3-empowered metaverse: taxonomy, countermeasures, and opportunities. IEEE Open J. Comput. Soc. **02**(03), 37–49 (2023)
16. A. Banerjee, R. Byrne, I. D. Bode, M. Higginson, in Web3 beyond the hype (2022) (Online). Available: https://www.mckinsey.com/industries/financial-services/our-ins ights/web3-beyond-the-hype (Zugriff am 31 Dec 2023)
17. N. Szabo, Formalizing and Securing Relationships on Public Networks. in *First Monday*, Bd. 2, Nr. 9 (1997)
18. C. Moy, A. Gadgil, N. Parina, D. Bettinger, J. Kent, E. Bowyer, H. Nasseri, M. Bedawala, in *Opportunities in the Metaverse—How Businesses can Explore the Metaverse and Navigate the Hype vs. Reality*, ed. by J.P. Morgan (2022)
19. CBInsights, *The Metaverse could be Tech's Next Trillion-Dollar Opportunity: These are the Companies Making it a Reality* (2022)
20. Gartner, gartner.com (2022) (Online). Available: https://www.gartner.com/en/articles/what-is-a-metaverse (Zugriff am 31 Dec 2023)
21. L.-H. Lee, T. Braud, P. Zhou, A.W. Lin, D. Xu, Z. Lin, A. Kumar, C. Bermejo, P. Hui, in *All One Needs to Know about Metaverse: A Complete Survey on Technological Singularity, Virtual Ecosystem, and Research Agenda* (Hong Kong, 2021)
22. J. Büchel, H.-P. Klös, in *Metaverse: Hype oder "next big thing"? Potenziale und Erfolgsbedingungen* (Institut der deutschen Wirtschaft Köln e. V, Köln, 2022)
23. Deloitte, in *The Metaverse and its Potential for the United States* (2023)
24. L. Christensen, A. Robinson, in *The Potential Global Economic Impact of the Metaverse* (Analysis Group, 2022)
25. M. Nigam, A. Metuku, D. Mitchelson, G. Saito, H. Chou, H. Maekawa, H. Eguchi, J. Su, J. Pitzer, K. Fong, R. Abrams, K. Wong, M. Walker, M. Binetti, P. Chen, P. Winslow, S. Badri, S. Shin, in *Metaverse: A Guide to the Next-Gen internet* (Credt Suisse, 2022)
26. Y. Chen, H. Cheng, The economics of the metaverse: a comparison with the real economy (2022)

27. V. Lehdonvirta, E. Castronova, *Virtual Economies—Design and Analysis* (The MIT Press, Cambridge, Massachusetts, 2014)
28. V. Bienert, I. Fisher, M. Grieves, A. Hauptvogel, S. Kögl, S. Köklü, D. Lange, K. O'Donovan, H. Sagi, L. Shannon, L. Signe, M. Ziegler, The emergent industrial metaverse, in *MIT Technology Review Insights* (2023)

The Potential of Web3 in the Data Economy and AI Opportunities

Kai Schmitz-Hofbauer

1 Introduction

The importance of data for almost every digitization endeavor is undisputed—in the business as well as in private and social contexts. However, a significant challenge arises from the fact that a few large companies, due to their market power and capital, have access to vast amounts of data.[1] These companies possess a substantial volume of data, and many of them offer free services through which they gain access to user data and the right to monetize that data—of course on a voluntary basis. The use of these free services is essentially paid for with the user data. Although government regulations and policies of course define important framework conditions, this is a challenge to the competitive landscape, particularly for individuals and, notably, for small and medium-sized enterprises (SMEs). The imbalance in data access, control and monetization can become an obstacle for these entities, impacting their competitiveness in the digital age.

Addressing this challenge requires a rebalance in focus towards **data ownership** and **sovereignty**. A potential solution lies in placing greater emphasis on data ownership through data sovereignty and give the original data owners the power to participate more from data utilization and commercialization. Artificial Intelligence (AI), which is in a greater spotlight again since the beginning of the year 2023, relies heavily on vast amounts of data to learn from, making predictions or generating content. The quality and quantity of data needed directly influence the quality of the corresponding AI systems. Data sovereignty is therefore not only a question of control, but also a decisive factor for the transparent and fair development and application of modern AI systems.

---

[1] For a further discussion, see [2, 20].

K. Schmitz-Hofbauer (✉)
Munich, Germany
e-mail: research@schmitz-hofbauer.de

Within the paradigm of emerging **decentralized technologies**, often labeled as Web3, a sweeping potential emerges. Web3 envisions a decentralized, secure, and user-centric digital ecosystem. It involves sharing power and benefits through decentralization. Decentralized approaches can complement and add value to centralized approaches. The potential spans a wide spectrum. Regarding data and AI, it has the potential of shifting sovereignty over data, artificial intelligence and their advantages to the owners of the data—regardless of whether they are individuals, companies, organizations or communities. Simultaneously, it allows companies and organizations to continue harnessing this data for business purposes.

This article analyzes the intersection of decentralization with data and AI from two perspectives:

1. Approaches that can be used to achieve a new potential for **data-driven business**—especially in terms of sovereignty, control, transparency and economical benefit for the data owners.
2. How recent advances in generative AI have the potential to improve Web3 solutions.

2 Building Blocks for Decentralized Data Economy

A significant challenge lies in rebalancing the data ecosystem to empower data owners. Web3 is combining powerful concepts that have the potential to change how we manage data by giving individuals and communities more control over their personal information. This shift to data sovereignty ("from users to owners") brings many advantages for both users and businesses. It supports creating a fair, transparent, and user-centric data eco-system that can be used in different scenarios. Before diving into specific examples, some key concepts and technologies will be introduced.

2.1 Self-sovereign Identity

Self-Sovereign Identity (SSI) is a decentralized model for managing and using digital identities. It puts the individual at the center of identity management, allowing people to have greater control over their identity information.

The concept includes that individuals have full control and ownership over their identity and personal data. It allows individuals to manage and share their data with others, granting or revoking access as there are legitimate reasons for an access. SSI relies on decentralized technologies, often based on blockchain or distributed ledger technology, to enable individuals to create, own, and control their digital identities without the need for a central authority. However, it is essential to note that while blockchain is a common technology associated with SSI, not all SSI implementations rely on blockchain.

As this article is primarily concerned with data economy and commercialization of data, please refer to [13, 15, 17] for more detailed information on SSI.

2.2 Data Sovereignty

Data Sovereignty focuses specifically on the data owners' control over their data. The concept applies to individuals as well as organizations or companies. An overview about different definitions can be found in [4]. Data Sovereignty primarily concerns data ownership and control, with implementation often involving consent management and secure data storage. It also considers the laws and regulations of the countries where the data is collected, stored and processed. Data sovereignty has an intersection with Self Sovereign Identity (SSI). In general, data sovereignty is a broader concept that applies to all data, not just personal or identity data.

One important enabler of SSI and data sovereignty is **selective disclosure** [15]. This is the ability of a person to determine which specific personal identity data they choose to disclose to others. This means that individuals have the autonomy to decide which facets of their identity they wish to share with others, while still being able to engage in online interactions and transactions that necessitate a form of data sharing. Especially, selective disclosure allows individuals to share only the minimum amount of information necessary for a specific purpose, such as proving their age, residence, health information or qualification.

While data sovereignty is focused on empowering data owners with control over their data, self-sovereign identity extends this concept to encompass a **decentralized and user-centric approach to managing a person's entire digital identity**. The two concepts are interconnected, as controlling personal data is a fundamental aspect of managing a self-sovereign digital identity.

2.3 Decentralized Storage

Decentralized storage is a method of storing data across multiple nodes in a network. It eliminates the need for traditional centralized storage services by distributing information across a network of participants. Two examples of decentralized storage are Filecoin[2] and Storj.[3] Based on blockchain technology, Filecoin aims to create a decentralized marketplace for storage services. Users can store their data securely on the Filecoin network and pay for the storage. Storage providers earn Filecoin tokens in exchange for contributing their storage capacity. One of the biggest advantages is the transparency, especially that it can be used to create a tamper-proof log of all data transactions. Furthermore, this decentralized model not only enhances data security

---

[2] https://filecoin.io/.
[3] https://www.storj.io/.

and privacy but also promotes efficiency and scalability by leveraging unused storage resources from around the world while reducing reliance on centralized entities. It should be noted that with central managed cloud storage providers, storage is also usually distributed across a large infrastructure in multiple physical locations and not necessarily stored in a single location, but in this case a fixed provider sets the rules as part of terms and conditions.

Even though decentralized managed storage has advantages for data exchange use cases, current implementations also face some limitations compared to cloud storage services. These include latency, speed, convenience [12] regulatory and legal issues, and interoperability [1].

Decentralized storage is still an emerging field. The future will show whether decentralized storage will reach the level of performance and acceptance of centralized (cloud) storage providers.

2.4 Micropayments

Payment procedures also play an important role in connection with data exchange and data economy—in the sense of a decentralized marketplace. Depending on the use case, the payments can be small—especially when it comes to "direct trade" between an organization and an end consumer. These small payments are also referred to as micropayments.

Cryptocurrencies and blockchain technology are suitable building blocks to automate **micropayments** because of:

- Decentralized, open infrastructure, meaning they are not dependent on any central authority (preventing vendor buy-in).
- Low transaction fees—extremely important for micropayments.[4]
- Fast, highly automated and secure transactions.
- Security and transparency: more resistant to fraud and hacking attempts.

When choosing a concrete technology, aspects like scalability (transaction volume per time), regulatory compliance (complying with financial regulations and laws), user adaption (probability of use) and environmental impact (energy consumption and carbon footprint) should also be considered.

Since blockchain-based solutions currently still have limitations in the aspects mentioned, it is important not to rely on blockchain-based solutions for "ideological reasons", but to always focus on the benefits for users when choosing a payment channel and, in case of doubt, choose the best solution backwards from the user's perspective.

---

[4] Depending on the blockchain used and current gas price level.

3 Putting Theory into Action: Practical Applications

3.1 Sovereign Data Exchange and Data Monetization

Controlling data access is not the only benefit of being in control of your data. It also makes **commercialization** of data for the original owners easier. This applies to private users, organizations, and companies alike. The following use cases can be implemented using sovereign data exchange, so that all parties involved benefit from the data access and/or monetization of data:

- **Purposeful provision of data for research and marketing**: data is provided to a company or an organization in a controlled manner, e.g. for brand research purposes or in connection with consent to receive advertising. Users also could provide health data to a pharmaceutical company that is developing new drugs. In return, the data owner can receive a micropayment, which can be easily transferred automatically as a cryptocurrency.
- **Loyalty programs** based on self-sovereign data would also fall into the aforementioned category. One example could be that a person's purchase (e.g. in a café or at the hairdresser) is recorded and loyalty points or tokens are credited to the user's wallet—without the need to collect data from a central loyalty program.
- Users can agree to **generate data on request**, e.g. filling out questionnaires at the request of market research companies. Again, in return, the data owner can receive a **micropayment**.
- **Customized and commercially optimized personal offers** can be created based on your own data, which are also tailored to you. Discounted car insurance rates, for example, are conceivable in return for access to driving behavior data.

In this context, it is also important to consider the ethical implications of using personal data for commercial purposes as well. For example, health insurance companies should not offer discounts based on an individual's health data without explicit consent of the users and taking overall fairness considerations into account. This could potentially lead to discrimination and unequal access to healthcare. It is crucial to maintain **ethical guidelines** and regulations when using personal data to ensure fairness and protect individuals' rights.

- Advances in digitization powered by sovereign data approaches could also lead to **preferential access to services and offers**—such as age verification or the provision of data for automatic credit checks when applying for a loan or when renting an apartment.

Figure 1 depicts a possible data and micropayment flow (Table 1).

This requires the use of appropriate software—a data **wallet**—which manages the secure storage and access to the data. The wallet is the place where the user stores and manages their own data. Of course, both the data owner and the data consumer typically use data wallets.

In addition to data wallets, end users also use crypto wallets. General requirements for crypto wallets as described in [14] must be considered.

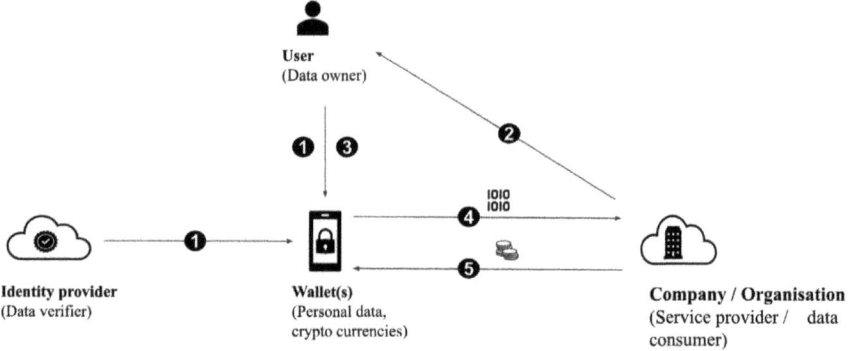

Fig. 1 Typical data and payment flow

Table 1 Description of data and payment flow

Step	Execution
1	Provision of verified and verifiable identity features or self-maintained data with subsequent storage in a wallet
2	Request for data usage
3	Explicit approval sharing of requested data with requested service
4	Data transfer (peer-to-peer)
5	*Optional*: Micropayment

Blockchain technology also makes it easy to implement **data pay-per-use models** in which payment is made per use of a data point. For this purpose, suitable **smart contracts** can be provided to orchestrate the processing of these transactions.

3.2 Data Cooperatives

In general, a sovereign data exchange can be implemented with two different modes:

1. Exclusive storage of data in the user's area of responsibility combined with peer-to-peer sharing.
2. Use of **data cooperatives/escrow services** for storing data and managing data access.

In a vision of a purely decentralized digital ecosystem, data is stored exclusively in an individual's or organization's own environment with full control over their own digital identity and personal data environment and exchanged peer-to-peer when needed. This corresponds to the approach described in the previous section. Data is stored securely in a wallet and exchanged with a trusted third party on request and explicit approval.

The Potential of Web3 in the Data Economy and AI Opportunities 295

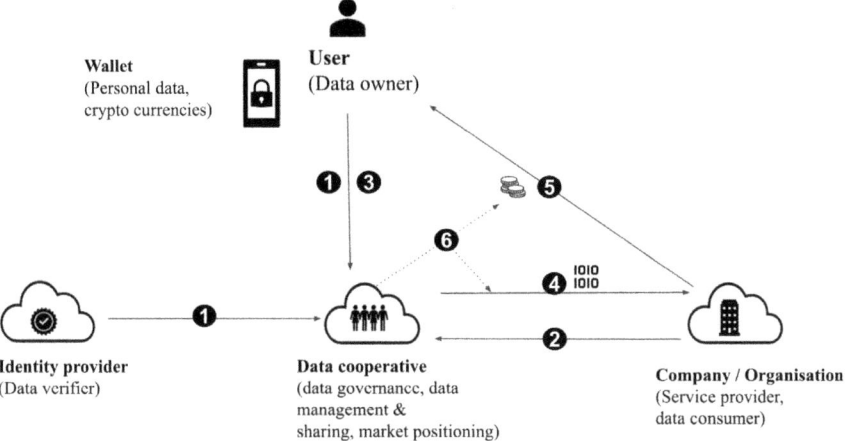

Fig. 2 Data cooperative as a governance entity

This approach comes with the following challenges:

- The widespread adoption of self-sovereign identity requires cooperation and collaboration among various stakeholders, including individuals, organizations, governments, and technology providers. Overcoming resistance to change and achieving consensus can be difficult. Standardization and interoperability of SSI-specific protocols are the most relevant challenges in the area of governance [17, p. 42]. Although approaches such as ID Union[5] are attempting to close this bottleneck, they have not yet achieved widespread market adoption.
- The secure storage of data and cryptographic key material in wallets is still a major challenge for the mass majority of the users with potential risk for the non-technological layperson.

To overcome such challenges, it would be straightforward to have a 3rd party in place to securely store and manage the data and take care of **data governance**. This 3rd party could be a community that acts on behalf of its members and supports them to hold copies of the data, safeguard data, manage access to the data, represent the members in negotiating and performing audits. In reference to [10, p. 28], we refer to this type of community as a **data cooperative**. This data cooperative can also orchestrate the exchange of data with micropayments. The following illustration depicts a possible data and micropayment flow using a data cooperative (Fig. 2; Table 2).

At first glance, such a data cooperative appears to be a classic centralized entity. However, a key difference is that the community is equipped with a decentralized governance structure in which the members always have the opportunity to shape the community and preserve their sovereignty. A **DAO (Distributed Autonomous**

---

[5] https://idunion.org.

Table 2 Description of data and payment flow (data cooperative)

Step	Execution
1	Provision of verified and verifiable identity features or self-maintained data with subsequent storage managed by a data cooperative
2	Request for data usage
3	Explicit approval sharing of requested data with requested service (if necessary, only once, so that it can be used for future queries)
4	Data transfer (data cooperative—data consumer)
5	*Optional*: Micropayment
6	*Optional*: Monitoring of data and payment flow

Organization) [11], for example, could be used as a decentralized form of organization and governance structure, where the rules of the organization are written in code and members can vote for specific decisions based on their share.

However, there is still the challenge of how to bring a DAO to life. One viable way is to start with a classic company structure that has the task of bringing the offer to the market. This includes finding first mover users and participating companies and organizations. Governance tokens can be issued to reward users with a high level of engagement, depending on the intensity of use. In this way, control could be gradually transferred to the community.

A company transitioning to a DAO model can still earn money from their initial investment and beyond through several methods, even as they transfer control and governance to the community. A few strategies are listed below:

- **Token sale**: The company can create and sell governance tokens to early adopters and investors. The funds raised from this initial token sale can provide capital for development, marketing, operations and other expenses.
- **Transaction fees**: If the platform enables transactions, the company may take a small fee from each transaction. These fees could continue even after the DAO takes over, providing ongoing revenue.
- **Service and subscription fees**: The platform could charge for premium features, services, or subscriptions that provide users with additional benefits.
- **Development and maintenance contracts**: Even as a DAO, there will be a need for ongoing development, maintenance and support. The initial company could serve as a service provider to the DAO.

The structures discussed in this section can also be used for business-to-business data exchange. While large companies often exchange data with each other bidirectionally based on individual agreements, such approaches offer immense potential for small and medium-sized businesses.

3.3 Data Rooms

In the context of a decentralized data economy, **data rooms** (or **virtual data rooms**) are virtual spaces where organizations or individuals can securely store and share their data with authorized parties. They play a crucial role in facilitating data exchange and collaboration while ensuring data privacy and security.

Data rooms usually allow setting granular access controls and permissions, ensuring that only specific users or groups can view, edit or download the data (**data sovereignty**). This is done via policies which are evaluated automatically. This helps to maintain confidentiality and privacy while enabling collaboration among authorized parties.

In the data economy, data rooms play a crucial role in facilitating secure and controlled data sharing. They are commonly used in various business transactions such as mergers and acquisitions, financial due diligence, legal proceedings, and intellectual property management. Data rooms can vary in terms of decentralization or centralization based on their design, purpose, and the technologies they employ. Decentralized data rooms are usually built based on decentralized storage (see Sect. 2.3).

3.4 Decentralized Data Curation for (Generative) AI

AI systems rely heavily on high-quality data for learning and decision-making. This represents another application of decentralized data sharing and economy, beneficial to both individual and organizational holders of Intellectual Property (IP).

AI models are only as good as the data on which they are trained on. If the data is of high quality, e.g. it is accurate, relevant, comprehensive and unbiased, the AI system can learn effectively and produce reliable results. On the other hand, if the data is of poor quality, e.g. incomplete, outdated or biased, the AI system may produce inaccurate or biased results. High-quality data is therefore essential to ensure the effectiveness, reliability and fairness of AI systems. The outputs of AI systems worsen as the quality of data diminishes. In a **data-centric AI approach**, the focus is on improving the quality of the data itself rather than tweaking the model or algorithm.

Modern AI approaches like **Generative AI**, **Large Language Models (LLMs)** or image generation algorithms like **Stable Diffusion** are dependent on a huge amount of high-quality training data. As explained in the paragraph before, the AI's performance is tied to the quality of the data it relies on.

As of today, some of the data used for training these models come from diverse sources, which can lead to inconsistencies and inaccuracies. Besides that, **biases present in the data**, whether they are due to the data source or the data collection process, can also affect the quality of the training input—and subsequently also to model outputs. This is particularly the case if—as practiced today—the training corpus contains documents scraped from the internet.

In addition, there are also uncertainties regarding the copyright of data from the internet, which can potentially have legal implications for the results of Generative AI models and thereby financial consequences for the users and providers of Generative AI systems. Even though this is in case of ChatGPT users now covered by GenAI's copyright shield [9], the root of the problem still exist—the **missing (legal) quality of the training data.** The importance of clarifying the rights of use for data in the AI environment is also demonstrated by the legal action taken by The New York Times against Microsoft and the ChatGPT provider OpenAI for copyright infringement at the end of 2023. This arose from the use of millions of articles published by The Times to train the AI models [3].

Another aspect is that upcoming AI regulations, like the proposed EU AI Act, will also have obligations for data quality and transparency [6]. For example, the Act would require stringent **transparency obligations** for providers of generative foundation AI models. Providers of general-purpose AI systems must also ensure a robust protection of fundamental rights, health, safety, the environment, democracy and the rule of law. High-quality training data and transparency about the origin of the data can make an important contribution in this regard.

Lastly, the dynamic nature of language, with constantly evolving slang, idioms, and cultural references, can make it difficult to maintain the relevance and accuracy of the training data over time. Therefore, it is crucial to employ robust data management strategies to ensure the quality of training input for AI models like LLMs.

One measure to ensure the **quality and transparency of AI training data** is also via decentralized communities as in principle described in Sect. 3.2. However, the participants would in most cases not be private users but organizations that participate in the creation and curation of data. Such communities could be, for example, public projects such as [19], specific communities that want to preserve a language [5] or private companies such as publishing companies that have researched and created high quality content. Exclusive content—such as that from a legal publishing house—could provide high quality topics for certain specialized areas—e.g., for AI with a legal application area.

Vogel [22] is presenting a decentralized Web3-based approach called Corpus Civitates to connect owners of curated high-quality content and Generative AI providers. Corpus Civitates is taking care of governance, the mediation and documentation of usage rights and the orchestration of payment flows. In this context, micropayments via cryptocurrencies are also a very viable option. Here, too, pay-per-use models can be considered alongside traditional commercial license models (Fig. 3).

In summary, decentralized approaches can also provide valuable services in the development of powerful AI by providing high quality curated data, and in a way that respects the rights and (commercial) interests of content, data and rights owners.

The Potential of Web3 in the Data Economy and AI Opportunities 299

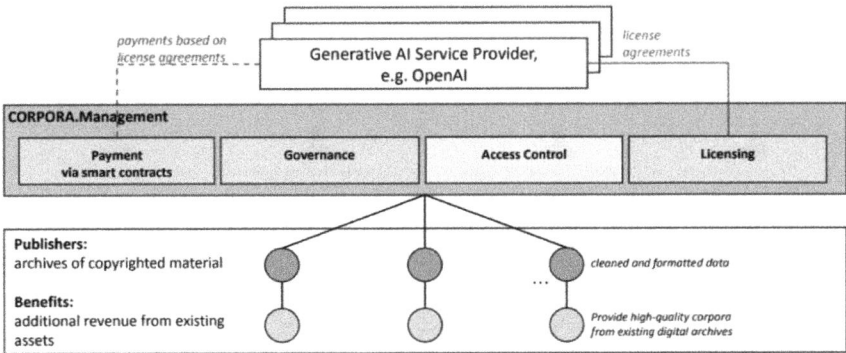

Fig. 3 Corpus civitates: decentralized curation of high-quality content for AI purposes [22]

4 Tokenization and Monetization Models for Data in Web3

In order to monetize and trade data, a tokenization in a Web3 sense is required. This means a mapping of the data assets to tradable crypto tokens. One option to tokenize data and the usage rights is via **NFTs (Non-Fungible Tokens)**. The usage rights—especially for a bigger data set like a document corpus—can be described as metadata as part of an NFT. With the purchase of the NFT, the buyer simultaneously acquires the rights of use, which in turn are documented in a traceable manner via blockchain mechanics [21]. This also contributes to greater transparency of the AI systems that are trained with this kind of exchanged data.

Even though the usage rights are traded on a public blockchain, the **data itself remains private**. Besides documenting usage rights from a legal perspective, the NFT can also be used as an access mechanism to the data itself, while at the same time ensuring data privacy and data protection and preventing the public from accessing the data.

The handling of data rights in connection with NFTs requires that the scope of the intellectual property rights granted is known. Creators and sellers of NFTs may need to consider compliance with consumer protection laws. Different jurisdictions have different legal requirements, so platforms trading NFTs need to ensure that they have addressed this before operating in a particular jurisdiction [7]. It is recommended to seek advice from a legal professional to understand all the implications before implementing such an approach.

A practical example of how this can be implemented is the Ocean Protocol,[6] which aims to enable data sharing and monetization in a decentralized and trustless way. It uses smart contracts to create data tokens on the blockchain, which are tokens that represent access to data sets. Those who own data can create data tokens and trade them with data consumers, who, in turn, can use these tokens to gain access to the corresponding data.

---

[6] https://oceanprotocol.com/.

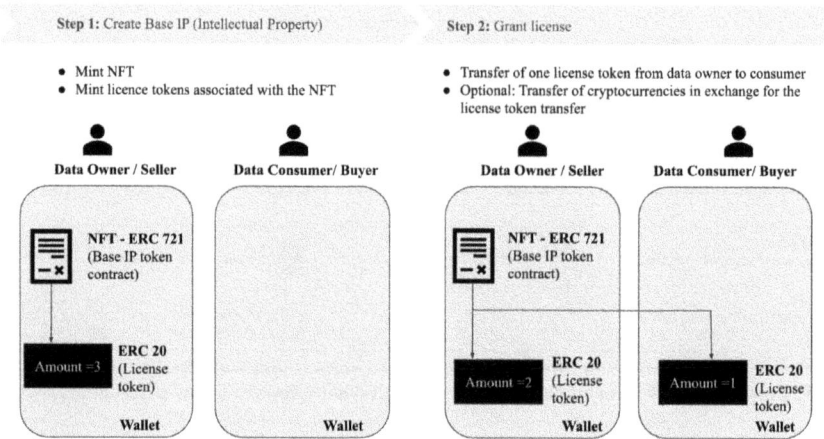

Fig. 4 Data NFTs and data tokens, based on [8]

The Ocean Protocol [8] uses two types of tokens: data NFTs and data tokens. Data NFTs are ERC721[7] tokens that represent the unique ownership of a data asset. Data tokens are ERC20[8] tokens that represent the access rights to a data service, e. g. for querying, analyzing, or visualizing the data asset. Each data service has its own data NFT and one or more types of data tokens. To access a data service, you need to hold at least one data token of the same type as the service. For example, if you want to access a dataset that is represented by an ERC721 token called my_dataset, you need to hold at least one my_dataset data token. You can buy data tokens from data providers.

Data tokenization with the Ocean Protocol enables decentralized access control, provenance tracking, and interoperability between different data assets and services. The following graphics depicts how data tokenization with the Ocean Protocol works (Fig. 4).

The Ocean Protocol is also bringing its own data marketplace. A further interesting feature is called compute-to-data, which allows data owners to monetize their data without revealing it to the consumers, by allowing them to run computations on the data and receive the results as data tokens.

---

[7] ERC721 is a standard for creating unique, non-fungible tokens (NFTs) that represent ownership of a specific, indivisible asset.
[8] ERC20 is a standard for creating fungible tokens, meaning each token is identical and interchangeable with others.

5 B2C and B2B Perspectives: How SSI and Sovereign Data Impacts Consumers and Businesses?

5.1 Impact for End Users and Data Owners

In the previous chapters, numerous examples illustrate how data exchange and a data economy can be supported by decentralized Web3 approaches. In the following, the advantages that arise for all parties will be summarized.

For end users and data owners this brings the following benefits:

- **Empowered control**: Consumers have greater control over their personal data, enabling them to make informed decisions about sharing, access, and privacy settings.
- **Monetization**: Consumers can choose to sell their data to interested businesses and organizations, essentially becoming data providers in the digital economy. Companies may pay consumers for access to their data and individuals can earn money by sharing it selectively.
- **Informed decision-making**: Self-sovereign data allows consumers to have more control over their personal information. By making informed choices about data sharing and access, they can avoid unnecessary costs or subscriptions and ensure that they receive fair compensation for their data when shared.

These advantages can lead to greater acceptance of data sharing, which potentially can accelerate many digitization endeavors.

5.2 Impact on Business and Organizations

Business and organizations in the role of a data consumer benefit first and foremost from simplified data access, while at the same time ensuring user acceptance and ensuring that the intended use is transparent. In addition to avoiding associated risks and negative press, this offers the following benefits:

- **Enhanced consumer and end user trust**: Implementing self-sovereign data practices can build trust with customers, as they perceive the company as respecting their data ownership and privacy rights. This can lead to stronger customer relationships and brand loyalty.
- **Innovation and personalization opportunities**: Self-sovereign data can foster new business models and services that leverage consumer data while respecting their autonomy, potentially leading to innovative, user-centric solutions.

6 AI Opportunities

6.1 Generative AI and Blockchain

The previous chapter explored how Web3 strategies can enhance modern (generative) AI models by improving data quality and transparency. The potential also exists in the reverse direction: Generative AI can also improve Web3 solutions—especially in terms of quality and suitability for mass use.

6.2 Explain Smart Contracts in Natural Language

One of the fundamental problems with Web3 and smart contract-based applications is that they are unfortunately still the domain of experts. The transparency of a blockchain and the smart contract running on it is hidden from those who do not understand the underlying technology and are able to read for example Solidity program code. Measured against the overall addressable potential user base for decentralized applications, this is probably only a small proportion.

The vast majority of users must rely on the experts to use their knowledge of the details for the benefit of all. It would be better if everyone could understand how a smart contract works. Generative AI in particular can help to **demystify** this and translate Smart Contract code into natural language. The explanation generated by the AI model can even take into account the user's level of knowledge, so that the description varies depending on the level of understanding. This is a form of AI-driven personalization.

This is explained in more detail in the following using a practical example and easily accessible tooling. We use Open Interpreter,[9] which is a versatile tool leveraging a large language model. Open Interpreter breaks down a natural language prompt into small tasks, converts them into executable code, and runs them locally on the system. As an example, the following prompt is executed using Open Interpreter, describing the task in pure natural language without providing details about the execution or technical smart contract development.

```
Act as a Solidity expert explaining a smart contract
for a non-technical target group.
Read the following smart contract ~/codeanalyse/
smart-contract-examples/lottery/contracts/
CharityRaffle.sol.
Explain the smart contract in natural language
for a non-technical target group. Start with a
```

---

[9] https://openinterpreter.com/.

Charity Raffle Smart Contract

This page provides a non-technical explanation of the Charity Raffle Smart Contract.

Summary

This smart contract is a digital version of a charity raffle game. In this game, people can participate by paying an entrance fee. The game keeps track of all the participants and the charities involved. The game has different stages. It starts with the 'Open' stage where people can join. After a certain time, the game moves to the 'Closed' stage where no more people can join. Then, the winners are chosen and the game moves to the 'Drawn' stage. Finally, the winners are paid and the game moves to the 'Paid' stage. There are two types of winners in this game: a participant and a charity. The participant winner is chosen randomly from all the participants. The charity winner is the charity that received the most donations. The game also allows anyone to view the current stage of the game, the list of participants, the list of charities, the donations made to each charity, and other game details.

Detailed Explanation

The smart contract CharityRaffle.sol is a digital implementation of a charity raffle game. Here's a detailed explanation of how it works:

- **Entrance Fee:** To participate in the game, players need to pay an entrance fee. This fee is stored in the contract.
- **Game States:** The game has four states - Open, Closed, Drawn, and Paid. The game starts in the Open state where players can join by paying the entrance fee. After a certain duration, the game moves to the Closed state where no more players can join. Then, the winners are drawn and the game moves to the Drawn state. Finally, the winners are paid and the game moves to the Paid state.
- **Winners:** There are two types of winners in this game - a player and a charity. The player winner is chosen randomly from all the players who joined the game. The charity winner is the charity that received the highest donations.
- **Donations:** The contract keeps track of the donations made to each charity. The charity that receives the highest donations becomes the charity winner.
- **Getter Functions:** The contract includes several getter functions that allow anyone to view the current state of the game, the list of players, the list of charities, the donations made to each charity, and other game parameters.

Fig. 5 AI generated natural language explanation of a smart contract

```
summarization of what the contract is doing and then
describe the details.
Create a visually appealing HTML and deploy it to the
local Apache server.
```

As source code we use `CharityRaffle` Smart Contract from [16]. Executing this prompt leads to the following, human readable explanation of the analyzed smart contract (Fig. 5).

It is important to recognize that machine learning and AI algorithms work on the basis of probabilities, which means there is a possibility that the results may not be entirely accurate. Keeping this in mind, it improves the ability for Web3 novices to engage with and understand Web3 topics.

6.3 Create Smart Contracts by Natural Language Specification

Generative AI can of course not only be used to explain smart contracts, but also to create smart contracts from natural language. For example, a user story or any other type of functional specification could serve as a basis. The use of Generative AI in software development has become increasingly popular since the beginning of 2023.

The best-known tool in this area is probably GitHub Copilot,[10] which integrates with leading development environments.

It should be noted that this approach is not suitable for non-experts, as logical errors can also occur in the generation process, which can have profound consequences, particularly due to the immutability of smart contracts.

However, for professionals, generative AI provides valuable services in this area for increasing productivity.

6.4 Smart Contract Audits

Smart contracts are typically **immutable** once deployed. For this reason, **careful audits** before deployment are incredibly important. In a professional environment, this is normally carried out by an independent team of experts—not by the actual developers. Again, generative AI can provide valuable services to support such a team of experts in their work, automate parts of the work and thus significantly reduce the costs of professional Web3 projects.

The Generative AI opportunity for smart contract audits lies in harnessing the power of AI models to automatically generate and assess smart contract code. This can enhance the audit process by identifying vulnerabilities, loopholes or inefficiencies that human auditors might miss due to the complex and tedious nature of the task. This can potentially add value in the following ways:

- **Automated vulnerability detection**: Generative AI can analyze existing smart contracts and learn from their structure, function, and past vulnerabilities. It can then use this knowledge to identify potential security issues in new smart contract code more rapidly than human auditors.
- **Code suggestions and optimizations**: After having analyzed the smart contract code, Generative AI can suggest enhancements or optimizations to improve efficiency, gas usage (in the case of Ethereum-based contracts) and adherence to best practices.
- **Generating test cases**: An AI system can automatically generate test cases that cover a wide range of scenarios for thorough contract testing. This ensures that contracts react appropriately to both typical and edge-case inputs.
- **Compliance checks:** Generative AI could be trained to understand various regulatory requirements and automatically check smart contract code for compliance issues, reducing legal risks.

As an example, the following prompt is once more executed using Open Interpreter:

---

[10] https://github.com/features/copilot.

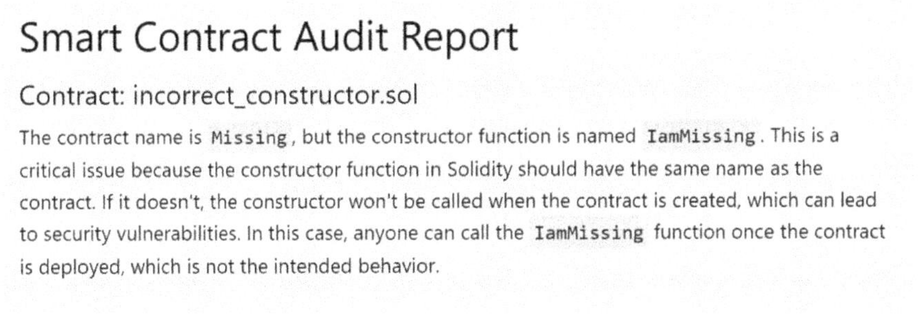

Smart Contract Audit Report

Contract: incorrect_constructor.sol

The contract name is `Missing`, but the constructor function is named `IamMissing`. This is a critical issue because the constructor function in Solidity should have the same name as the contract. If it doesn't, the constructor won't be called when the contract is created, which can lead to security vulnerabilities. In this case, anyone can call the `IamMissing` function once the contract is deployed, which is not the intended behavior.

Fig. 6 Example of an AI generated smart contract audit

```
Act as an auditor for smart contracts.
Go to the directory ~/code/not-so-smart-contracts/
wrong_constructor_name.
Read all contracts.
Find all critical issues and explain what the problem
is.
Create a visually appealing HTML and deploy it to the
local Apache server.
```

The explicitly faulty smart contracts from the "(Not So) Smart Contracts repository" [23] are used for demonstration purposes.

Executing the above prompt produces the following example result, which indicates a serious and commercially disastrous programming error. The following picture shows the result generated by Generative AI. In this case, the tool takes over the orchestration of the required calls and the creation of a final HTML report (Fig. 6).

Such kinds of automated quality checks can easily be integrated in CI/CD pipelines (continuous integration/continuous deployment).

Finally, it should be explicitly noted that these functionalities can support experts in their respective domains, making work more efficient and cost-effective. Final decisions—whether technical or non-technical in nature—should, however, be made exclusively by qualified experts. Human verification of the results is still and remains important.

7 Summary, Conclusion and Outlook

Decentralized approaches can make significant contributions to the data economy and the development of high-quality AI models. In this context, there is particularly great potential for a decentralized and sovereign data economy as well as for delivering high-quality data that builds the basis for modern artificial intelligence services.

Conversely, AI aids in the creation, quality, understanding, and execution of efficient, decentralized applications.

The examples discussed in this article represent just a small excerpt from the intersection of decentralization, data, and AI. There are further areas of application like:

- **Market research based on publicly available transaction data**: Publicly available blockchain transaction data can be a powerful tool for market research. By analyzing patterns, volumes, and timings of transactions, insights into market participant behavior can be gleaned. This data can indicate market sentiment, identify active participants for competitor analysis, and reveal trends through the study of money flows. However, the pseudonymous nature of blockchain data is a challenge in associating actions with specific individuals or entities. Despite this, careful analysis can yield valuable information for market research.
- **Smart contract monitoring**: AI can play a crucial role in monitoring the execution of smart contracts on a blockchain. AI can be used to analyze the patterns and behaviors of smart contracts, identifying any anomalies or potential security threats. This is particularly useful in the context of decentralized finance (DeFi) applications, where smart contracts potentially can handle transactions worth millions of euros. By monitoring these contracts in real-time, AI can help prevent fraudulent activities and ensure the integrity of the blockchain.

Moreover, AI can also assist in optimizing the execution of smart contracts. By analyzing past transaction data, AI can predict the gas fees for contract execution, helping users to save costs.

- **AI-driven smart contracts**: AI could in the future make smart contracts smarter. Artificial Intelligence (AI) has the potential to make smart contracts much better by helping them to adapt automatically to a changing environment. With real-time data, AI lets smart contracts change terms and actions to stay useful in different situations. In the future, these AI-supported contracts could be able to make complex decisions taking many factors into account. They can also use external data through oracles to respond well to real-world events.

Bringing AI and smart contracts together might make processes smoother and improve how automated agreements work. But as of today, existing blockchains, such as Ethereum, have constraints on their computing capabilities, storage capacity, and data transfer capacity. This makes it still challenging to incorporate sophisticated AI models into smart contracts. Takyar [18] provides an in-depth exploration of the convergence of AI and Web3 technologies, discussing the opportunities and challenges that lie ahead.

The current landscape at the intersection of decentralization, data and AI is rich with diverse applications, and a rapid pace of innovation. The application area of data and AI in particular will open up numerous additional opportunities. Decentralized approaches are not to be understood as a replacement for existing centralized approaches, which have their strengths in terms of performance, a very high degree of maturity and acceptance, but rather as a very useful supplement to the existing practices. In summary, the convergence of data, AI and Web3, can have a significant impact on future digitalization endeavors, driven by continuous innovation, improved data sovereignty and transparency.

References

1. A Deep Dive into Decentralized Storage Systems (2023). https://www.nervos.org/knowledge-base/decentralized_storage_systems_(explainCKBot). Accessed 12 Dec 2023
2. H. Gu, Data, big tech, and the new concept of sovereignty. J. Chin. Polit. Sci. (Available via Springer Political, 2023). https://doi.org/10.1007/s11366-023-09855-1. Accessed 11 Apr 2024
3. M. Grynbaum, R. Mac, The Times Sues OpenAI and Microsoft Over A.I. Use of Copyrighted Work (New York Times, 2023). https://www.nytimes.com/2023/12/27/business/media/new-york-times-open-ai-microsoft-lawsuit.html. Accessed 29 Dec 2023
4. M. Hellmeier, J. Pampus, H. Qarawlus, F. Howar, Implementing data sovereignty: requirements & challenges from practice, in *ARES'23: Proceedings of the 18th International Conference on Availability* (ACM, 2023). https://doi.org/10.1145/3600160.3604995. Accessed 28 Dec 2023
5. How Iceland is using GPT-4 to preserve its language. In: Customer stories. OpenAI (2023). https://openai.com/customer-stories/government-of-iceland. Accessed 20 Nov 2023
6. T. Madiega, Artificial Intelligence Act, in *A Europe Fit for the Digital Age*. European Parliament (2023). https://www.europarl.europa.eu/legislative-train/carriage/regulation-on-artificial-intelligence/report?sid=7501. Accessed 16 Dec 2023
7. F. Mukaddam, NFTs and Intellectual Property Rights. Norton Rose Fulbright (2021). https://www.nortonrosefulbright.com/en/knowledge/publications/1a1abb9f/nfts-and-intellectual-property-rights. Accessed 23 Oct 2023
8. Ocean Protocol, Data NFTs and Datatokens. The Ocean Protocol (2023). https://docs.oceanprotocol.com/developers/contracts/datanft-and-datatoken. Accessed 17 Dec 2023
9. OpenAI, New models and developer products announced at DevDay. OpenAI (2023). https://openai.com/blog/new-models-and-developer-products-announced-at-devday Accessed 05 Dec 2023
10. A. Pentland, Building a new economy: data, AI, and Web3, in *ACM|Communications of the ACM*, pp 27–29 (Online). Available via MIT Open Access Articles (2022). https://hdl.handle.net/1721.1/147636. Accessed 02 Nov 2023
11. O. Rikken, M. Janssen, K. Zenlin, The ins and outs of decentralized autonomous organizations (DAOs) unraveling the definitions, characteristics, and emerging developments of DAOs, in *Blockchain: Research and Applications. Royal Society of Chemistry*. Available via ScienceDirect (2023). https://www.sciencedirect.com/science/article/pii/S2096720923000180. Accessed 16 Dec 2023
12. P. Santapur, J. Li, Centralized clouds or distributed ledgers: evaluating tradeoffs in storage architectures. J. Stud. Sci. Res. **5**. George Mason University (2023). https://journals.gmu.edu/index.php/jssr/article/view/3909. Accessed 16 Dec 2023
13. F. Schardong, R. Custódio, Self-sovereign identity: a systematic review, mapping and taxonomy, in *Advances in Blockchain Technologies towards Identity Management and Its Applications in ICT and IoT*. Available via MDPI (2022). https://www.mdpi.com/1424-8220/22/15/5641. Accessed 30 Oct 2023
14. K. Schmitz-Hofbauer, A. Vogel, Best Practices for Enterprise Crypto Account Management—Part 1 and Part 2. Available via Medium (2022). https://medium.com/@andreas.vogel/best-practices-for-enterprise-crypto-account-management-b6bc2d1e6544, https://medium.com/@schmitz-hofbauer/best-practices-for-enterprise-crypto-account-management-part-ii-4c7418acd744. Accessed 16 Dec 2023

15. H. Siddiqui, Self Sovereign KYC in Blockchain. Available via Medium (2022). https://medium.com/@hira.siddiqui/self-sovereign-kyc-in-blockchain-474c2c2ba9f3. Accessed 23 Oct 2023
16. Smart Contract Examples and Samples. Available via GitHub (2023). https://github.com/smartcontractkit/smart-contract-examples. Accessed 15 Dec 2023
17. J. Strueker, N. Urbach, T. Guggenberger, J. Lautenschlager, N. Ruhland, V. Schlatt, J. Sedlmeir, J. Stoetzer, F. Voelter, Self-Sovereign Identity—Grundlagen, Anwendungen und Potenziale portabler digitaler Identitäten. Projektgruppe Wirtschaftsinformatik des Fraunhofer. Instituts für Angewandte Informationstechnik FIT, Bayreuth (2021). Available via https://www.fit.fraunhofer.de/content/dam/fit/de/documents/Fraunhofer%20FIT_SSI_Whitepaper.pdf. Accessed 12 Nov 2023
18. A. Takyar, AI In web3: how AI manifests in the world of web3. Available via LeewayHertz (2023). https://www.leewayhertz.com/ai-in-web3/. Accessed 28 Dec 2023
19. The Common European Language Data Space. The European Commission (2023). https://language-data-space.ec.europa.eu/about_en. Accessed 16 Dec 2023
20. M. van Rijmenam, Privacy in the age of AI—risks, challenges and solutions (2023). https://www.thedigitalspeaker.com/privacy-age-ai-risks-challenges-solutions. Accessed 11 Apr 2024
21. A. Vogel, Empower the Creator of NFTs with Respect to Royalty Payments. Available via Medium (2022). https://medium.com/@andreas.vogel/empower-the-creator-of-nfts-with-respect-to-royalty-payments-160c90b6e2e4. Accessed 15 Dec 2023
22. A. Vogel, Self-determination of cultural and linguistic communities in artificial intelligence. Institute Corpus Civitatis (2023). https://corpuscivitatis.org/approach/. Accessed 15 Dec 2023
23. (Not So) Smart Contracts. Available via GitHub (2023). https://github.com/crytic/not-so-smart-contracts. Accessed 16 Dec 2023

Kai Schmitz-Hofbauer holds a PhD in engineering with a focus on software engineering. Over the past 20 years, he has excelled in various positions as a researcher, entrepreneur and product manager, focusing on innovative domains. His field of expertise spans artificial intelligence, data management, cloud computing, and Web3 technologies. Throughout his career, he has been developing and managing innovative software solutions, focusing on integrating cutting-edge technology to drive business growth and efficiency.

In today's IT landscape, centralized approaches dominate. The motivation behind the contribution to this book is to provide the reader with an impetus on how decentralized approaches can complement existing centralized approaches in the area data and AI—balancing the interests of end users and the legitimate commercial interests of enterprises seeking to monetize data.

Web3 Gaming and NFT-Based In-Game Items

Marcus Rump and Oliver Nolden

In this chapter, we would like to give an overview of in-game items, their creation, their importance for the long-term motivation of players and the business models of the manufacturers. We will then look at the possibilities of NFT-based in-game items and provide an assessment of how we believe Web3 technologies can be used to improve the gaming experience, increase security and at the same time enable completely new applications.

As the name Web3 suggests, it is a further development of Web1 and subsequently Web2. Web1 represents the first and earliest form of the Internet from around 1989, which could be used by normal users. Web1 is generally understood to mean only the reading of information. Users were not yet able to create information themselves, as the infrastructure was complex and generally not available privately. Web1 was the emergence of the Internet, which was characterized by search engines, reading emails and viewing websites.[1]

The term Web2, also known as the current web, was first used in 2004 by Dale Dougherty and Craig Cline.[2] Web2 added writing to reading. The beginning of the social era was heralded. Nowadays, in 2024, Web2 is still the most widely used.

This is where Web3 comes into play, with the addition of owning for reading and writing. The term Web3 was coined in 2014 by Gavin Wood, a co-founder of Ethereum. However, it should also be mentioned that Web3 is still in its infancy and is the subject of controversial debate.[3] It is not accepted by all gamers. Nevertheless, it is to be expected that it will continue to develop and become established in many areas.

---

[1] Web3 versus Web 1.0 and Web 2.0.

[2] Web 2.0—What was that again?

[3] Web3 Controversies.

M. Rump (✉) · O. Nolden
TicketHash Technology UG (Haftungsbeschränkt), Lünen, Deutschland
e-mail: marcus@tickethash.com

In gaming in particular, the web3 seems to be able to solve the ownership, privacy, control and security problems of the web2 with the help of decentralized ledgers based on blockchain technology.[4] Digital wallets, in-game items, payments and NFT marketplace APIs are key developments in so-called Web3 gaming.[5]

1 From the First Game to the Digital In-Game Item

Since the early days of mankind, playing games has been one of the most popular pastimes and is practiced for pleasure, relaxation and also as a profession, subject to the rules of the game. Games are usually competitive in nature and end with a winner who receives a prize or honor.

In the course of human history, games have developed steadily from the first traditional games with the hands, such as Morra, which can be compared to today's game of rock, paper, scissors or even and odd, to the Olympic Games and games such as chess, a checkers and the first game of the year, Hare and Hedgehog, from 1979.[6]

In 1961, the first digital computer game Spacewar was created, programmed by American student Steve Russell. This game heralded the golden age of digital games from 1972 to 1983. Many of the classics, some of which are still revered today, were created during this period.

From 1983 to 1991, Japanese consoles and home computers came onto the market, followed by technological progress with more powerful processors from 1992 to 2000. The years 2001–2010 were characterized by online gaming, high-resolution graphics and motion control. From 2011 to today, trends such as esports, mobile gaming, virtual reality and indies (abbr. for independent games) computer games followed, which were developed with a significantly lower budget than games with a high publisher budget. From 2017, games based on blockchain technology followed.[7]

Collecting in-game items has a long history in games. Even in games such as Pac Man or Super Mario in the 80s and 90s, digital fruit and coins were used to fuel players' urge to collect items, even though they only had limited use in the game and were mainly used to achieve a higher score.

It became more sophisticated from 1996 onwards in the Diablo series, developed by Blizzard Entertainment, or Blizzard for short. This is a hack-and-slay game with a fantasy theme. The user chooses a character and explores different worlds while fighting monsters. If a monster is defeated, it can drop weapons, armor and other items.

Even in the first edition in 1996, collecting better and better equipment was a driving force, as it helped progress against ever stronger opponents. The in-game

---

[4] Web3 Games are for everyone: A complete explanation.
[5] A Beginner's Guide to Web3 Gaming.
[6] Game—Wikipedia.
[7] Game history of digital games.

item system became more and more sophisticated from edition to edition. In 2000, there were already almost endless possibilities for variation. The items had randomly generated quality levels (normal, magic, rare, unique and set). Depending on the quality level, a certain number of properties and associated values were assigned, such as damage bonus, defense, dexterity, attack speed and much more. Various types of weapons were available, such as swords, daggers and bows, as well as shields, armor, helmets, belts, shoes, rings and amulets. Some special ones could be combined into sets to unlock additional bonuses. There was also a crafting system for further improvement options. This is a game mechanic that allows the player to create new items from other items or raw materials. For example, a weapon or armor can be crafted from four required and different items according to certain crafting recipes.

The choice of equipment had a significant influence on the game and in-game items were cheerfully traded from player to player in online multiplayer mode. Outside the game, the trade in in-game items also flourished on platforms such as Ebay, where rare items changed hands for several hundred euros in some cases. As a result, publisher Blizzard increasingly had to contend with hackers who were out to make a quick buck with duplicated in-game items.[8]

World of Warcraft, WoW for short, was also published by Blizzard in 2004. It is a role-playing game of the MMORPG (massively multiplayer online role-playing game) class, in which the player chooses a character and continues to level it up by completing quests and individually expanding its abilities. In WoW, obtaining better and better equipment also plays a key role in motivating players. The game quickly became the most successful game of all time[9] (over 10 billion US dollars in sales by 2012). This game also developed a market for in-game items and WoW gold, the in-game currency—both inside and outside the game. In China, veritable gold farms[10] sprang up in which workers played for a living in the virtual world in order to collect WoW gold and use it to earn a living in the real world.

Epic Games took a completely new approach to marketing in 2017 with the survival shooter Fortnite. The game is offered free of charge and is also basically free to play. Instead of the usual one-off purchase price, the publisher has opted for a business model[11] based entirely on the sale of in-game items. The basis for this is the in-game currency V-Bucks, which is purchased in advance for real money and can then be exchanged in the game for skins, equipment and dances (emotes) in addition to passes for additional content. With this concept, the game has broken sales and user records within just a few years. Over 20 billion US dollars in revenue were generated within the first 6 years.[12] The number of players reached the threshold of 500 million worldwide in 2023.[13] The trading of in-game items is not officially

---

[8] Diablo II's Economy: Lessons in Trust and the Promise of Bitcoin.

[9] Here Are The 10 Highest Grossing Video Games Ever.

[10] The Life of the Chinese Gold Farmer.

[11] Fortnite: The Business Model Pattern Behind the Scene.

[12] Fortnite Revenue and Growth Statistics (2024).

[13] Registered users of Fortnite worldwide from August 2017 to March 2023.

supported, but due to the popularity and rarity of some skins in particular, full player accounts are sometimes traded. However, the risks of fraud are high when using external platforms for trading.

These examples make it clear that digital in-game items have been and continue to be an essential aspect of some of the biggest games in recent decades. Furthermore, they contribute significantly to their popularity and economic success. The transfer of in-game items between players is currently only supported within the respective platform. Outside of this, they are neither usable nor transferable in any meaningful way, apart from the usually unwelcome trade via detours on platforms such as eBay. If a game is shut down at some point, the associated in-game items will also lose their significance.

1.1 How Web3 Concepts Are Changing Gaming

On the one hand, Web3 technologies have the potential to solve some of the difficulties mentioned, but on the other hand they also offer a whole host of completely new possibilities. However, their use also creates new challenges that need to be solved.

1.2 Non-fungible Tokens

The most obvious solution is probably the tokenization of in-game items as non-fungible tokens (NFTs). Various standards exist for this on different blockchain platforms, such as ERC-721[14] or ERC-1155[15] on Ethereum.

As a rule, fulfilling an NFT standard means implementing a smart contract with certain standardized interfaces and providing metadata (images, description texts) in a standardized form. In-game items that fulfill one of these standards automatically receive the following features:

- They are clearly identifiable by means of an ID and can be traced back to an owner. Duplication is impossible.
- Due to the publicly provided metadata, NFTs are visible outside the game. This means that they can theoretically be integrated into third-party applications, such as trading platforms, at will.
- Sharing between users is possible outside the game. All that is required is interaction with the standard interfaces of the smart contract.
- The item can be traded on a whole range of established NFT platforms. For example OpenSea, Blur or Rarible, etc.

---

[14] ERC-721 Standard.
[15] ERC-1155 Standard.

By supporting the NFT standard, it may be possible to "outsource" some problems from the game at low cost. Provided that the manufacturer considers these functions to be useful for its players, solid trading functions can be offered without the need to implement separate in-game marketplaces.

In addition, new trends and interesting approaches have emerged in the gaming sector as a result of the further development of Web3 technologies.

1.3 GameFi

One notable paragraph is GameFI. The term is a combination of game and finance. At its heart is the idea that players can earn real money in the game (play-to-earn). To this end, the game is combined with cryptocurrencies and financial mechanisms. Money is then usually earned by trading the NFTs or cryptocurrencies earned.

The oldest play-to-earn game and thus a pioneer is the game Cryptokitties, in which users first have to buy kittens, which can be exchanged and sold, as well as crossed with each other. The descendants can then be offered to new users in exchange for ether. In the 2017 bull market, enormous prices were achieved for cryptokitties. The most expensive Cryptokitty was auctioned for $140,000 in May 2018.[16]

One of the best-known play-to-earn games is certainly Axie Infinity. This game was the first game to achieve a trading volume of more than USD 1 billion and at times two million active players during the coronavirus pandemic. It provided gamers in the Philippines, for example, with a new source of digital income after many small businesses came to a standstill due to lockdowns.[17] The traded NFTs and the internal SLP token increased enormously in value due to the strong growth in user numbers. Things went downhill just as quickly afterwards. This was partly due to a US$ 625 million hack of the underlying Ronin network,[18] an Ethereum Virtual Machine (EVM)[19] blockchain built specifically for Axie, but also due to the inflationary nature of the SLP token (the game's cryptocurrency), as the enormous growth in the number of new gamers caused a lot of capital to flow out of the game onto exchanges. These outflows are a major problem for many games with their own internal tokens. In response to the market dynamics, Axie Infinity is transforming away from the play-to-earn model to Axie Infinity: Originals, without the use of cryptocurrencies.[20]

---

[16] Cryptokitties.
[17] Play to Earn: You should know these blockchain games.
[18] Axie Infinity's blockchain was reportedly hacked via a fake LinkedIn job offer.
[19] ETHEREUM VIRTUAL MACHINE (EVM).
[20] Play To Earn: A Real Alternative to a Work?

1.4 DAO

Some games form a Decentralized Autonomous Organization (DAO). This is an organization run by the community without a central authority. The community is in control. Smart contracts define the essential rules and when the conditions are met, the decisions made by the community are executed. The central aspect is usually a voting mechanism implemented in the smart contract that enables participants to submit proposals and vote on them, with voting rights linked to the possession of governance tokens that were initially issued to raise capital and can later be freely traded on crypto exchanges. The mechanism is therefore similar to that of a public limited company, where shares also carry voting rights. However, a much more direct influence is possible, as decision-making by voting takes place almost in real time. In principle, it is possible to operate only parts of an organization as a DAO, for example to involve the community in the conceptual development of a game.[21]

1.5 Decentralized Metadata

The life cycle of classic in-game items ends when the associated game is no longer playable. In the cloud age, in which players are connected to each other via servers, the time for this is determined by the manufacturer. Assuming that an in-game item not only has value for the player in the game, but possibly also outside of it, this can be a big disappointment. How nice would it be if you could keep your in-game items forever?

Unfortunately, NFT-based in-game items do not automatically solve this problem. Although the smart contract remains permanently available on a blockchain platform for as long as it exists, the NFT's metadata is usually still stored on the manufacturer's centralized servers. If this server is switched off, all that remains of the NFT is essentially the identification number. The associated image and descriptions are lost and the entire NFT is no longer usable. In order to solve this problem, various solutions have been developed with which the metadata can be secured in a decentralized manner. The best known is currently the InterPlanetary File System (IFPS).

The IPFS is a content-addressable peer-to-peer method of storing and sharing hypermedia in a distributed file system.[22] A common system and part of NFT technology, but still a relatively centralized system, as the NFTs can also be stored on just one node, which makes them more vulnerable to cyberattacks compared to storage in a decentralized blockchain, which is distributed across many nodes and is therefore more secure.

The Filecoin project is a decentralized storage system in which the files are broken down into small pieces and stored on several nodes. In this way, the system provides

---

[21] The Rise of Gaming DAOs.

[22] The InterPlanetary File System (IPFS).

a backup plan in case one of the nodes used goes offline. Filecoin is an open-source, public cryptocurrency and digital payment system and an incentive system for the IPFS peer-to-peer storage network, which is intended as a blockchain-based cooperative digital storage and data retrieval method.[23]

There are other approaches to store NFTs fully decentralized on blockchains, such as Aleph.im's[24] blockchain-based storage solution, which is a DApp for NFT backups to secure NFTs on specific blockchains. This technology is already integrated into Ethereum, Solana and Polygon and currently supports smart contracts from OpenSea, Rarible and Superrare.[25]

2 Are Web3 Games Ready for the Mass Market?

Web3 games are currently still struggling with many challenges that stand in the way of mass market adaptation. These range from poor user experience and security concerns to technical and economic inefficiencies. Nevertheless, we believe that Web3 games have the potential to conquer the mass market. However, well thought-out concepts are needed to unleash the full innovative power of the new technology. In the following, we would like to outline the biggest challenges and possible solutions.

2.1 Crypto Knowledge as a Prerequisite

NFTs have their origins in blockchain technology, which requires a high level of technological understanding and personal responsibility. Since these are decentralized technologies, there is no way to turn to a central authority in the event of difficulties, for example to restore access to your digital wallet.

In order to achieve successful adaptation in the mass market, the technological barriers to entry when using Web3-based gaming must be kept as low as possible. Although most projects, not only in the gaming sector, are working to improve and simplify the user experience of this technology, it is still necessary to have a certain level of crypto know-how in order to be able to use the game at all. This means that there must be at least one wallet address that is loaded with cryptocurrency. In addition, there must be knowledge of how to connect a crypto wallet to the game. However, such know-how is far from standard among the broad mass of gamers, which is why corresponding games still lead a niche existence despite some impressive successes. Low-threshold access is crucial to reaching a large number of users, which is what makes a game economically successful.

---

[23] Coinbase—Info Filecoin.
[24] Aleph.im.
[25] Are NFTs really secure? Why we need decentralized backup solutions.

We at TicketHash therefore favor a hybrid approach in which linking a wallet is only necessary if the player wants to use external functions such as trading on external NFT marketplaces.

2.2 Fragmented Wallet Market

The range of NFT standards, blockchain platforms and wallet implementations is currently difficult to keep track of. Market leaders have emerged, but new products are constantly being added due to ongoing innovations. Consolidation would be desirable here from a user and integration perspective. However, there are products such as Wallet Connect that allow several blockchain platforms and wallets to be integrated into one with a single integration.

We make use of this at TicketHash so that we don't have to integrate each wallet individually. Currently, over 500 different wallets are supported for around 30 blockchains, including Ethereum, Solana, Polygon and the Binance Smart Chain.

2.3 High Fees

Another possible problem is excessive gas fees that arise when minting or trading NFTs on the blockchain. In principle, every writing interaction with a smart contract triggers a transaction that is subject to fees. The amount depends primarily on the underlying blockchain, but also on the amount of data and current utilization of the network. Especially during periods of high activity, these gas fees skyrocket due to inefficiency and congestion on the blockchain. Even simple in-game actions can become economically unattractive for players.

Our recommended approach here is not to solve basic game functions with smart contracts. Only the core functions of NFTs, such as exchange and trading, should be mapped over them.

2.4 Unstable Infrastructure

Another problem is the stability of blockchain networks. Blockchains should provide a reliable and secure infrastructure for games, but many existing networks suffer from stability, which can lead to buggy gameplay and frustrating user experiences. Players cannot be expected to play games that do not work a significant percentage of the time because the logic is provided via smart contracts on an unstable blockchain.

For this reason, we would never develop a game that is based purely on smart contracts. The issues of scaling and uninterrupted operation are sufficiently solved with cloud platforms such as AWS, Azure and GCP. If trading is viewed as an external additional function, it is easier for the player to accept if this function is temporarily restricted or not available at all.

2.5 Monetary Focus

A common misconception of many current blockchain games is the overemphasis on monetary rewards in game design. Many blockchain games prioritize aggressive tokenomics over creative and fun gameplay experiences, which can turn off players looking for real entertainment value rather than just financial gain. Instead, they attract a crypto clientele that is primarily interested in speculation. The 2021 bull market and subsequent bear market have shown that an over-reliance on in-game token rewards to drive user engagement is short-lived at best. Game developers should look at the trading that takes place in traditional, fun games like Diablo, WoW or Fortnite and try to use blockchains to get the most out of these gaming experiences.

We believe it is important that games are intuitively understood, the user is not overwhelmed, and the speculative nature is not the main purpose of the game.

2.6 With Hybrid Blockchain Games into the Mass Market

Game developers of blockchain games should familiarize themselves with the processes of traditional and successful games, in particular understand the internal trading processes and creatively and sensibly use blockchain technologies where added value, e.g. with regard to the security of transactions when trading NFTs, brings an improvement. Not everything has to be done with a blockchain. Developers of classic games can see Web3 as an additional building block that can add useful additional functions to existing game concepts.

If a game has an excellent game concept, appeals to a broad gaming target group, integrates Web3 functions in a way that is helpful for the gamer and only requires a minimal number of wallet interactions, then in our view NFT games can step out of their niche existence and achieve greater acceptance.

2.7 TicketHash NFT League

In the previous chapters we have already teased that we ourselves are actively developing games with Web3 functions for the mass market. NFT League is a digital trading card game with which we want to digitize the thrill of analog trading card games with the help of NFTs.

Successful trading card games from the analog world, such as Pokemon or Magic the Gathering, have appealed to fans' passion for gaming and collecting for decades. The paper cards are exchanged and traded among friends and on special platforms. In the digital world, such long-term trading card games or trading card games have not yet really caught on. One of the reasons is certainly that, due to technical progress and a lack of interoperability, building up an elaborate collection of digital trading cards makes no sense. This is where NFT technology comes in. Trading cards can be traded and exchanged as NFTs outside of the game.

For adaptation in the mass market, it is essential, in our view, that it is first and foremost a good game with sustainable gameplay. The gamer should identify with the content of the game, which in this case we provide in cooperation with real football clubs.

The aim is to offer and develop an NFT-based trading card game that can be fully used without any NFT know-how. Starting with the purchase of the card packs, also called booster packs, through the game, to exchanging and trading. Basically, the trading cards used in the game are located internally in a closed system of the app. The gamer can carry out and experience all actions, including mini-games, in the trading card game app. Even trading and exchanging is possible via internal functions.

However, if the user wants to offer his in-game items globally, i.e. externally on NFT marketplaces, he can optionally do this at any time. Only from this point on does the user need a basic understanding of blockchain, how to set up a digital wallet and link it to the app in order to then mint his trading cards as NFTs. If he does this, the trading cards can be used, exchanged and traded externally from that moment on.

3 Web3 Gaming with NFT-Based Digital Collectibles—An Outlook

Projects that use a relatively new technology usually go through an evolutionary process. New concepts for digital business models emerge. As a result, there are a wealth of advantages and disadvantages. The concepts that emerge from the perspective of NFT-based trading card games often still have to prove themselves and be accepted by the player base.

NFTs are particularly interesting for smaller independent indie gaming development studios, as they represent another source of income and specifically offer in-game items that the community is demanding.

NFTs have also been announced by major companies such as Ubisoft, Electronic Arts (EA), Sega, Konami, KRAFTON and Square Enix. Ubisoft entered into two partnerships with major players in the blockchain gaming sector in November 2023.

Firstly, with the Ethereum-based Layer 2 Scaling Solution from Immutable X and with the blockchain gaming studio Animoca Brands. KRAFTON from South Korea already entered into a partnership with Solana Labs in 2022.

EA is certainly imagining trading cards for the FIFA series as a supplement in the future, and Square Enix announced the game Symbiogenesis in 2023, in which players collect art NFTs along a story as characters in their own universe.

In response to criticism from gamers, Square Enix President Matsuda said that it is the group of play-to-have-fun gamers who are critical of NFT technology and its possibilities. This group is more interested in pure entertainment and has little interest in the new decentralized possibilities of this technology. Ubisoft also agrees with this opinion and says that ignorance about NFTs is a main reason for the negative attitude of many long-time fans.[26] However, lessons learned in the past are that new technologies should be introduced with caution to avoid immediate rejection on the grounds of further commercialization at the expense of users.

What can we expect from the blockchain gaming industry in the future? Market corrections are always perceived negatively, but they can also have positive sides. One positive of these is that these phases rid the market of unsuccessful projects, which often form the majority. It can be observed that many tokens of most projects have plummeted in price and only the tokens of the strongest gaming projects that have value and offer a lasting mature game concept have remained above their initial offering price.

Projects that can stay afloat during the volatile market invest a lot of time in revision and design. Some projects launch products on other blockchains in the hope of using their existing branding to capitalize on it and build on the trust of their users.

In terms of future developments, it is likely that the gaming sector will see many new and improved mechanisms. This is underlined by the commitment of large gaming companies that are driving innovation.[27]

The diverse challenges must be overcome in order to drive the development and acceptance of blockchain games. In particular, gas fees that are economically too high in times of high transactions and the associated losses that often come with them at the expense of blockchain stability. A simple, intuitive integration of in-game purchases with FIAT currencies via credit card, based on trading processes in successful traditional games. The focus should be on the best possible gaming experience, along with simplifying the complexity and only using blockchain technology where it provides noticeably positive added value.

In our view, a new technology can only establish itself permanently if it solves a real problem for a broad group of users. Users don''t want to buy fantasy monkey NTFs without a utility, they want to do something active with them, i.e. play. The sustainable application and benefit must be recognizable. Web3 technology will not solve this on its own, but if used correctly, it will provide added value.

---

[26] NFTs in games: What is it and why is it criticized?
[27] Are Play-to-Earn Crypto Games Dying? It's Not All Doom and Gloom.

This is not necessarily the case with every implementation of digital collectibles, but we think that our TicketHash NFT-League digital trading card game will be a good example of this.

Marcus Rump studied geography at the universities of Bochum and Freiburg. He focused on topics related to physical geography and geoinformatics during his studies. After graduating, he gained international experience working as a sales and business development manager for companies in the defense, intelligence, and new space sectors. These companies develop C4ISR applications and utilize geoinformatics as a key technology.

Rump has been involved with blockchain technology since around 2016 and, at the end of 2018, he decided to found TicketHash Technology as its CEO and co-founder. Since then, TicketHash has developed real-world use cases that utilize blockchain and Web3 technologies without focusing on them. Blockchain technologies should provide support where it makes sense, rather than replacing proven technologies.

Currently, TicketHash is developing a trading card app based on gamified NFT collectibles. The app uses content from soccer clubs and can be used as a fan engagement and digital marketing tool. Player photos are turned into trading cards. This opens up access to new digital business models for the club, including a fair share of revenue. Gamers can identify with their favorite club and trade their most valuable cards. These collectibles can be collected, exchanged, traded, and played with in the app. TicketHash prioritizes an intuitive and user-friendly experience to create an application suitable for the mass market.

Oliver Nolden holds a degree in Computer Science from the Carl von Ossietzky University of Oldenburg. Between 2006 and 2021 he supported large software development projects in Germany and across industries in the areas of development, software architecture, dev-ops, technical analysis and requirements management, first as an employed consultant and later as a partner at Neusta.

At the end of 2021 he founded his own company and has been working as a freelancer ever since.

As co-founder and CTO of TicketHash Technology, Oliver Nolden has been developing mass-market solutions in the Web3 environment since 2018. In his view, it is often necessary to reconcile two worlds: on the one hand, the technology based on the right mix of relatively new, dynamic Web3 technologies and hyperscalable cloud platforms. On the other hand, there are the users, with their desire for exciting applications that are understandable and functional yet offer completely new possibilities.

He sees the NFT League digital trading card game as an ideal example of this.

Decentralized Science (DeSci): How Web3 is Revolutionizing Science

Lukas Weidener

1 Introduction

Web3 is considered a transformative innovation that reshapes the landscape of numerous sectors and industries. The profound technological advancements inherent in Web3 are also expected to have a significant impact on the scientific system. Given the complexities associated with Web3 and the scientific system, a detailed analysis is warranted to understand the concept of Decentralized Science (DeSci) and its potential implications for modern science. This chapter aims to provide a comprehensive analysis of how Web3 is revolutionizing science as an essential part of DeSci while providing clear terminology and explanations.

1.1 Web 3.0 and Web3

The conceptualization and subsequent development of the World Wide Web (Web 1.0) in the early 1990s instigated profound transformations in global communication and information sharing [1]. Web 1.0 was primarily a 'read-only' platform characterized by static webpages that allowed users to retrieve information but offered limited, if any, capability for user-generated content or interaction. This iteration of the web is marked by a unidirectional flow of information [2].

The emergence of Web 2.0 marked a significant paradigm shift from this static model and 'read-only' concept. The second iteration of the World Wide Web is not just a technological upgrade but has also led to profound social evolution, fostering a user-centric environment that emphasizes interactivity and participatory culture. It

---

L. Weidener (✉)
Research Unit for Quality and Ethics in Health Care, UMIT TIROL—Private University for Health Sciences and Health Technology, Hall in Tirol, Austria
e-mail: lukas@weidener.eu

enabled a 'read-write' web where users could easily create, share, and collaborate on content, epitomized by the rise of social media platforms, blogs, and wikis [2]. Web 2.0, therefore, represents the dynamic and interactive versions of the web familiar to contemporary users.

The initial concept of the third iteration of the web, Web 3.0, also known as the 'semantic web,' can be traced back to the founder of the World Wide Web, Tim Berners-Lee [3]. Web 3.0 is envisioned to be a highly interconnected, decentralized, and intelligent version of the Web, utilizing technologies such as artificial intelligence (AI) to facilitate a more personalized user experience by understanding and interpreting the meaning and context of data [3]. This vision of a semantic web, which focuses on data meaning and connectivity, should be differentiated from a more recent conceptualization known as Web3. Web3, enabled by the development of Distributed Ledger Technology (DLT) and more specifically, blockchain technology, encapsulates the 'read-write-own' concept [2]. This highlights the importance of user sovereignty over data and digital assets, enabling an unprecedented degree of control in content creation, consumption, and ownership. This paradigm shift toward decentralization in Web3 underscores the substantial differences compared with traditional centralized models of online interaction and governance.

Although 'Web3' and 'Web 3.0' are frequently used interchangeably in discourse, it is essential to clearly distinguish between them. Web 3.0, as initially conceived, is concerned with the semantic understanding and automated processing of information, whereas Web3 is focused on the economic and governance layers of the web, facilitated by DLT. Web3 further emphasizes self-sovereignty, ownership, transparency, and interoperability, aiming to create a user-centric Internet, where data privacy and financial transactions are controlled by individuals rather than centralized entities [2]. Given the technological advancements in Web3, it offers profound transformative opportunities for various sectors and industries, including science.

1.2 Science

To contextualize the prospective impact of Web3 on scientific methodologies and the broader domain of science, it is imperative to delineate the concept of science itself. The Cambridge Dictionary defines 'science' as: *"(knowledge from) the careful study of the structure and behaviour of the physical world, especially by watching, measuring, and doing experiments, and the development of theories to describe the results of these activities"* [4]. Therefore, science is a systematic and organized pursuit of knowledge that aims to uncover the mechanisms behind natural and artificial events. It encompasses an iterative process of proposing hypotheses, conducting controlled experiments, and arriving at verifiable explanations and predictions.

Science is commonly subdivided into different scientific disciplines, such as natural and social sciences, which are based on and utilize different methodological approaches [5]. For example, the natural sciences, such as medicine and biology, typically employ empirical methods to study physical phenomena involving observation, experimentation, and quantitative analysis. Social sciences, such as psychology and sociology, frequently employ qualitative methods such as case studies or surveys to understand human behavior and societal structures [5].

While the methodological approach of different scientific disciplines may differ, all scientific disciplines share a common commitment to the pursuit of knowledge. This pursuit of knowledge is based on different values and principles such as objectivity, transparency, replicability, and honesty, enabling continuous advancement through science.

Science has been fundamental to the progression of modern civilization, driving human advancement through a series of groundbreaking discoveries and technological innovations. Examples of such transformative influences are the Internet, which has revolutionized the way we communicate and access information [6], and antibiotics, which have drastically improved health outcomes by combating infectious diseases [7]. These achievements not only represent important milestones of technological and medical progress but also reflect the profound effect of science on societal development, economic growth, and the collective understanding of the world. Despite the immense impact of science on civilization, modern science faces numerous challenges [8, 9].

1.2.1 Challenges of Modern Science

The challenges of modern science are a commonly cited reason for the employment of technological advancements such as DLT and Web3, and should therefore be introduced in more detail. These challenges include, but are not limited to:

- **Funding**: The mechanisms governing the allocation of funds within the scientific community frequently provoke debate over their transparency and efficacy [10]. Constrained funding resources intensify competition among researchers, which may inadvertently incentivize data manipulation or compromise research integrity [8]. Moreover, the prevailing funding structures can disproportionately favor research with immediate commercial potential, often at the expense of exploratory or high-risk studies that, while less immediately lucrative, may hold transformative potential [11]. Additionally, the current funding paradigm offers limited opportunities for external stakeholders to contribute to or comprehend the intricacies of financial allocations in science, thereby reducing the democratization of science and the potential for diverse investments [12].
- **Replicability**: The ability to replicate research results is fundamental to scientific integrity, ensuring that the findings are robust, reliable, and can be independently verified. Despite the importance of replicability, a growing number of studies have highlighted the 'replication crisis,' in which researchers are unable to reproduce a substantial proportion of the results from previous studies, raising concerns about the robustness and reliability of scientific knowledge [9]. Factors contributing to this crisis include selective reporting, pressure to publish novel findings, and a lack of incentives for conducting replication studies, which are often considered less prestigious [13].

- **Scientific Publishing**: The current model of scientific publishing is frequently criticized for its inaccessibility and inefficiency [14]. As a substantial part of aca-demic research is subject to paywalls, limiting access to those without subscriptions or financial means, the current scientific publishing system promotes disparity in knowledge dissemination [15]. This inaccessibility of knowledge impedes researchers from lower-income institutions and the general public from participating in scientific processes [16]. Furthermore, the peer review process, while fundamental to maintaining quality and credibility, often lacks transparency and can be subject to bias, delay, and inconsistency [14, 17]. The long publication times can lead to a diminished pace of research dissemination and innovation, negatively affecting the potential impact for the general public.
- **Ownership**: Governance of ownership and protection of intellectual property rights present significant challenges in modern science [18]. Although the assignment of ownership is crucial for acknowledging scientific contributions and upholding quality, it can impose limitations on the dissemination of knowledge and scientific collaboration. For example, in the field of biomedical research, where the development of new treatments and the associated ownership of intellectual property rights are closely linked to major financial interests, transparency can be compromised in favor of proprietary advantage [19].
- **Collaboration**: The demand for frequent publication and the complexities surrounding ownership rights within the current scientific system discourage interdisciplinary collaboration [20]. Moreover, traditional scientific processes tend to be insular, offering only limited opportunities for engagement from external stakeholders, including patients and the broader public. This exclusivity not only decreases the diversity of research perspectives but also limits the potential for inclusive, community-driven scientific research [21].

1.2.2 Open Science

In light of the challenges of modern science, the so-called 'open science movement' aims to democratize the accessibility of scientific knowledge and research processes. While there are various definitions of open science, one of the most predominant and comprehensive definitions was developed by the United Nations Educational, Scientific, and Cultural Organization (UNESCO). UNESCO defines open science as follows:

> an inclusive construct that combines various movements and practices aiming to make multilingual scientific knowledge openly available, accessible and reusable for everyone, to increase scientific collaborations and sharing of information for the benefits of science and society, and to open the processes of scientific knowledge creation, evaluation and communication to societal actors beyond the traditional scientific community. It comprises all scientific disciplines and aspects of scholarly practices, including basic and applied sciences, natural and social sciences and the humanities, and it builds on the following key pillars: open scientific knowledge, open science infrastructures, science communication, open engagement of societal actors and open dialogue with other knowledge systems. [22]

This definition underscores the significance of open science in not only making scientific knowledge universally accessible but also in enhancing collaborative efforts across various disciplines and promoting inclusivity within the scientific community. It embodies the removal of barriers to information sharing and the active engagement of various stakeholders, both scientific and nonscientific. Based on the principles of openness and transparency, open science aspires to create a more equitable platform for discovery and dialogue, where diverse contributions and participation in the scientific process are possible, and the benefits of research are shared more broadly [22].

While aiming to address some of the challenges of modern science, such as enhancing collaboration and revolutionizing scientific publishing, open science does not directly provide potential solutions for other significant issues. For instance, the complexities of research funding or scientific ownership remain largely unaddressed. These gaps highlight the need for continued evolution of the management and dissemination of scientific knowledge.

In response to these enduring challenges, there is growing advocacy for leveraging technology to enable the goals of open science. Innovations such as DLT and Web3 offer new pathways for managing scientific data and ensuring their integrity, transparency, and accessibility. These technologies can facilitate a more equitable distribution of scientific knowledge and resources, thereby fostering a more inclusive and collaborative research environment. This intersection of open science and technological advancements is considered to be the key drivers for the emergence of 'Decentralized Science (DeSci).'

2 Decentralized Science (DeSci)

Since the conceptualization and emergence of Distributed Ledger Technology (DLT), particularly blockchain technology, as introduced by the pseudonymous entity Satoshi Nakamoto with the publication of the Bitcoin whitepaper in 2008 [23], there has been ongoing interest in applying DLT in science [24]. An example of this application is Gridcoin, a project initiated in 2013 that rewarded users to contribute their computational power to various scientific research projects [25, 26]. This includes, but is not limited to, simulations of protein folding aimed at understanding diseases and developing potential treatments [26].

Following the initial efforts of applying DLT in the scientific field, the term 'DeSci' is reported to have been first coined in 2021 on the online social networking platform 'X' (formerly known as 'Twitter') as an analogy to the more established term of 'DeFi' (Decentralized Science) [27]. DeFi aims to create a more open, transparent, and accessible financial system by leveraging DLT, a movement largely led by the inception of Bitcoin [28]. Preceding the claimed coinage of the term in 2021, 'Decentralized Science' was used as the short title of an academic publication published in 2018 [29]. As the terms 'DeSci' or 'Decentralized Science' are not explicitly used in the publication, their origin remains unclear.

Given the novelty of the concept and movement of DeSci, there have been inconsistencies in the definition of the term 'Decentralized Science'. In 2021, the concept and movement of DeSci was introduced to a wider audience by a correspondence article in the journal 'nature.' In the 'Call to join the decentralized science movement,' DeSci is described as a movement that:

> aims to harness new technologies such as blockchain and 'Web3' to address some important research pain points, silos and bottlenecks [30]

This description emphasizes the use of technology, particularly blockchain and Web3. Another well-known definition emphasizing the importance of Web3 in the context of Decentralized Science was proposed by the Ethereum Foundation:

> Decentralized science (DeSci) is a movement that aims to build public infrastructure for funding, creating, reviewing, crediting, storing, and disseminating scientific knowledge fairly and equitably using the Web3 stack [31]

Although both definitions explicitly emphasize Web3 and the 'Web3 stack,' they fail to provide additional specifications for Web3. As Web3 is stated to be foundational to DeSci as one of the key differences to the concept of open science, more nuanced considerations are required. While both definitions refer to DeSci as a movement, a recent publication aimed to provide a more comprehensive and holistic definition, referring to DeSci as an approach to science [32].

> Decentralized Science (DeSci) represents a collaborative and decentralized approach to science, leveraging technological and infrastructural advancements such as Distributed Ledger Technology (DLT), Web3, cryptocurrencies, and Decentralized Autonomous Organizations (DAOs) to enable permissionless, open, and inclusive participation, facilitating collective governance, equitable incentivization, unrestricted access, shared ownership, and transparent funding of the scientific process [32]

This definition not only aims to provide a comprehensive understanding of the term 'Decentralized Science (DeSci)' but also specifies important values and principles, such as decentralization, transparency, openness, and accessibility. This definition further addresses the challenges of modern science and reinforces its purpose. Moreover, Web3 is referenced explicitly to underscore its importance in DeSci. The definition also references aspects such as DLT, cryptocurrencies (which include tokens), and Decentralized Autonomous Organizations (DAOs), which are commonly referred to in the context of Web3.

While cryptocurrencies and tokens are pivotal in facilitating governance, incentivization, and funding within Decentralized Science (DeSci), Decentralized Autonomous Organizations (DAOs) form a foundational component of the current DeSci ecosystem. Given their significant role, the forthcoming section is dedicated to the Decentralized Autonomous Organizations active within the DeSci ecosystem, hereinafter referred to as 'DeSci-DAOs'.

2.1 DeSci-DAOs

Decentralized autonomous organizations (DAOs) are a novel organizational structure enabled by DLT and its inherent functionalities. The concept of decentralized operations is not entirely unprecedented, and various entities have historically embraced decentralized management approaches. For instance, the Organization for Economic Cooperation and Development (OECD), with its distributed network of member countries, operates through a collaborative and decentralized structure in which policies and standards are developed through consensus among its members [33]. Similarly, the United Nations Educational, Scientific, and Cultural Organization (UNESCO) functions through a network of national commissions in its member states, allowing decentralized participation and implementation of its programmes and educational strategies [34]. These organizations showcase how decentralized structures can facilitate international cooperation and policymaking, although they still rely on centralized governance bodies for final decisions.

In the context of Decentralized Autonomous Organizations (DAOs), decentralization extends traditional methods by leveraging Distributed Ledger Technology (DLT). DLT enables a transparent, verifiable, and immutable record of (trans-)actions, with the potential to facilitate more inclusive and permissionless participation in the decision-making and governance of a DAO [35]. Decentralization in the context of DAOs, therefore, not only refers to the potential decentralization of its members but also to a more inclusive possibility for governance. For example, DAOs commonly feature a so-called 'treasury', which represents financial assets of a DAO [35, 36]. Unlike traditional entities, this treasury is not controlled by a centralized entity, but rather governed by the collective decisions of DAO members. Using DLT, the voting process and all associated (trans-)actions are transparent, thus improving fairness and traceability.

Another innovative feature of the DAOs is their autonomous functionality. Taking the shared treasury as an example, once a decision-making process is successfully completed and the required quorum is met, the distribution of financial assets can be executed autonomously through DLT capabilities [36]. This ensures that once the collective will of the DAO members is determined, the subsequent distribution of resources is carried out independently of any central authority or manual intervention. This can be ensured by smart contracts, which are self-executing contracts with predefined terms of agreement implemented in the code. Therefore, smart contracts automatically enforce and execute agreed-upon terms when certain conditions are met (e.g., decentralized decision-making within a DAO) [36].

Autonomy is an important aspect of the current definition of a DAO. For example:

> A decentralized autonomous organization (DAO) is a limited liability company with special provisions allowing the company to be algorithmically run or managed (in whole or in part) through smart contracts executed by computers [37]

This definition is especially important because DAOs have been increasingly recognized by governmental institutions and regulations. Based on this definition, DAOs can operate and be officially registered as LLCs in Wyoming. By allowing

DAOs to be officially registered as Limited Liability Companies (LLCs), a bridge is formed between the innovative decentralized operations of DAOs and the established legal structures that govern corporate entities.

Cryptocurrencies, particularly tokens, are fundamental to DAO governance mechanisms. These governance tokens often represent voting rights within the organization, enabling token holders to participate in decision-making processes [36]. The allocation of governance tokens within a DAO generally reflects the varied contributions of its members, which go beyond mere financial input to include all forms of participation and support. As such, token ownership in a DAO in the form of governance tokens is often earned rather than purchased, aligning the influence within the DAO with active and meaningful contributions [36]. This system ensures that those who are invested in the DAO by any means have a correspondingly greater influence on governance and future decision-making.

DAOs are usually organized around a central concept or objective. In the context of DeSci, DAOs are commonly centered around scientific interests, such as the funding and advancement of research in the field of longevity. A DeSci-DAO can be defined as:

> A DeSci-DAO is a decentralized, blockchain technology-based organization that aims to advance scientific knowledge and innovation, by facilitating (interdisciplinary) collaboration, participation, diversity, and communication. DeSci-DAOs leverage smart contracts and tokens to facilitate transparent and tamper-proof blockchain transactions, interactions, and decision-making, ensuring fair and democratic distribution of resources through built-in treasuries and governance mechanisms. [32]

This definition references the scientific objective of DeSci-DAOs, while emphasizing the importance of governance. A definition of DeSci-DAOs is essential for understanding and evaluating the current DeSci ecosystem with the aim of addressing the challenges of modern science.

2.2 The DeSci Ecosystem

Given the novelty of DeSci, the current ecosystem is subject to ongoing and rapid change. In the current DeSci ecosystem, many projects and initiatives have been self-identified as DAOs. However, according to the definition provided by DeSci-DAOs, many fail to address important aspects, such as decentralized governance or shared treasuries [32]. This is attributable to the concept of 'progressive decentralization,' which refers to the process of founding a project with a high degree of centralization, but with decentralization in mind [38]. Therefore, having 'DAO' in their project name does not necessarily represent the current status but rather the overarching goal.

To illustrate the current DeSci ecosystem, selected projects with their corresponding number of Discord members are presented in Table 1. The messaging and social interaction platform was chosen because the majority of projects and DAOs in DeSci are centered around Discord servers.

Table 1 DAOs and projects in DeSci with focus area and the number of members on the Discord server (as of November 8, 2023)

Project name	Focus area	Number of discord members
VitaDAO	Longevity research and funding	10.535
AntidoteDAO	Funding cancer research initiatives	3.368
LabDAO	Tools and infrastructure for computational biology	2.009
TalentDAO	Decentralized publication protocol for social sciences	1.711
DeSciWorld	Connecting decentralized science communities	1.233
ValleyDAO	Synthetic biology technology funding and access	1.214
CerebrumDAO	Brain health and preventing neurodegeneration	995
AthenaDAO	Research and funding dedicated to women's health	995
HairDAO	Research and funding dedicated to hair loss	851
ResearchHub	Accelerating scientific research and publishing	811

Table 1 underscores the emerging nature of DeSci, with VitaDAO, despite being the project with the largest Discord community, hosting only a relatively modest member count of 10.535 on their server. This reflects the developmental stage of community building within the DeSci ecosystem. The number of active contributors in the DeSci ecosystem is expected to be substantially smaller, as not everyone who joins a Discord server is an active contributor. Furthermore, the number of unique members may also be significantly lower, as the current Discord members are likely to be part of more than one server.

3 Addressing Scientific Challenges Through DeSci

Despite its early development phase, the DeSci ecosystem is expected to address some of the challenges in modern science through the adoption of DLT and Web3. To illustrate the potential impact of DeSci in addressing the challenges of modern science, potential solutions and improvements to each of the challenges introduced in Sect. 1.2.1 is provided subsequently.

- **Funding**: DeSci addresses funding challenges by leveraging DLT for transparent, immutable record-keeping and ensuring clarity in fund allocation. It democratizes funding through token-based governance, allowing diverse stakeholders including patients, to participate in funding decisions. Projects such as AthenaDAO [39]

and VitaDAO [40] exemplify this shift, directing resources towards traditionally underfunded domains such as women's health and longevity research.
- **Replicability**: DeSci addresses replicability issues by utilizing decentralized storage solutions to ensure long-term accessibility and verifiability of scientific publications and data. By utilizing DLT and its associated capabilities, DeSci ensures long-term tamper-proof accessibility to research data, thereby promoting transparency and verification [18]. This not only enhances the reliability of the findings, but also encourages the replication of studies by providing an immutable reference point. Additionally, through token-based and reputation incentives, contributions that validate research can be rewarded, facilitating the replication of studies.
- **Scientific Publishing**: DeSci aims to transform the existing scientific publishing system by enabling a more collaborative environment through the integration of immutable record-keeping and transparent peer review processes. By leveraging decentralized storage, DeSci enables unrestricted, permissionless access to research irrespective of one's affiliation or financial resources [18]. Platforms such as ResearchHub exemplify this new paradigm by encouraging the early and open sharing of research results without fear of misappropriation [41]. This not only enhances inclusive engagement in scientific discourse, but also reinforces the collective integrity of the scientific process.
- **Ownership**: In the context of DeSci, the concept of ownership has been reimagined to embrace collective governance and to enhance scientific collaboration. For example, by utilizing governance tokens, stakeholders within the scientific community can exert collective control over a treasury, which may include intellectual property (IP) rights related to funded research [18]. Furthermore, the introduction of Intellectual Property Non-Fungible Tokens (IP-NFTs) offers an innovative approach for managing and protecting IP rights [42]. These digital tokens represent ownership and usage rights for scientific findings, creating a transparent and secure method for attributing contributions and sharing the benefits of scientific advancements in an online environment [42].
- **Collaboration**: To address the challenge of collaboration in science, DeSci aims to facilitate interdisciplinary cooperation and reduce scientific exclusivity. By leveraging the principles of Web3 and DLT, facilitates transparent and efficient collaborations, where contributions and discoveries are immutably recorded, ensuring clear attribution and incentivizing collective advancement over individual benefit [18]. This openness extends to engaging a wider range of stakeholders, including patients and nonscientific community members, in the research process [43]. Consequently, DeSci's approach to incentivization, governance participation, and equitable compensation are key to lowering barriers to scientific engagement, thereby facilitating more inclusive and participatory research.

While DeSci aims to address these challenges of modern science, potential improvements extend beyond those previously delineated. Using DLT, Web3, and associated capabilities, substantial improvements in data sovereignty are anticipated. This advancement will enable researchers to maintain control and agency over their scientific data, while being able to collaboratively sharing data on clearly defined terms of usage. Moreover, improvements in data sovereignty are not limited to researchers alone, but could also contribute to a more inclusive model where all stakeholders, including patients, can actively participate with the assurance of privacy and control over their personal data. A democratized and secure approach to data could unlock new possibilities in collaborative science, ensuring that participation and contribution are broadly accessible.

4 Discussion

By leveraging the inherent capabilities of Distributed Ledger Technology and Web3, Decentralized Science presents a transformative opportunity in the scientific field. It targets the resolution of critical issues such as funding disparities, lack of data accessibility, and transparency in research processes. DeSci redefines funding mechanisms to be more inclusive and transparent, utilizing governance tokens and decentralized decision-making to democratize resource allocation. One successful example is the case of Vita-DAO, which pioneered the funding of longevity research, showing how these models can effectively direct resources to underfunded domains [44]. Furthermore, DeSci has the potential to revolutionize scientific publishing by employing decentralized storage and immutable record keeping. These technologies ensure the integrity of data and facilitate a transparent peer review process. By granting permissionless access to research outputs, DeSci encourages broader participation and mitigates the risk of data theft. In terms of ownership, DeSci introduces collaborative governance models, with Intellectual Property Non-Fungible Tokens (IP-NFTs) as a novel approach to managing and protecting intellectual property. These digital tokens represent ownership and usage rights and offer a secure method for attributing contributions and sharing the benefits of scientific advancement. Improved data sovereignty is expected to facilitate (interdisciplinary) collaboration while ensuring that contributors maintain control and ownership over their data.

Although DeSci and the increased use of technology are claimed to be new revolutionary approaches to science, there are limitations, challenges, and risks that need to be acknowledged. Given the novelty of the movement and the currently limited size of the DeSci ecosystem, there is increased potential for issues such as security vulnerabilities, data privacy concerns, and the risk of collusion. Furthermore, while DeSci aims to improve inclusivity, the increased use of technology raises the question of a potential digital divide. Therefore, the use of and reliance on technology may exacerbate the gap between well-funded entities, such as institutions or individuals, and those with fewer resources.

Moreover, while the decentralized nature of DeSci aims to improve research funding through token-based governance and participation in decision-making processes, it could make it susceptible to manipulation by individuals with significant token holdings. Furthermore, the concept of IP-NFTs also poses legal and ethical considerations regarding the true ownership and commercialization of scientific knowledge. The implications of assigning ownership to discoveries that are often the result of collective intellectual efforts may challenge the foundational principles of open science.

Furthermore, the potential reliance on token-based incentives could introduce biases in the scientific process, where research directions are influenced by market trends or the financial value of a token rather than scientific merit. The need for robust governance structures and checks and balances within DeSci is therefore imperative to mitigate the risks of malpractice. The decentralization of scientific research could pose substantial ethical risks due to a lack of oversight by ethics committees or regulatory bodies.

5 Recommendations

Given the transformative potential offered by DeSci to improve the current scientific system, it is important to establish a set of recommendations to facilitate its growth while mitigating potential risks and limitations. The following recommendations are proposed:

- **Awareness**: Increased awareness of DeSci through targeted outreach programs and advocacy. This involves campaigns to educate the broader scientific community and the public about the potential benefits and improvements of DeSci.
- **Education**: Implementing comprehensive educational programs at various levels to enhance understanding of DeSci, Web3, and DLT. This includes creating accessible learning materials and training workshops tailored to different literacy levels.
- **Onboarding**: Develop streamlined onboarding processes for new users to lower the entry barrier into DeSci. This includes simplifying user experience and providing guided support for newcomers.
- **DeSci, Web3, and DLT Literacy**: Establish methods for assessing and validating literacy in DeSci, Web3, and DLT, ensuring that participants are adequately equipped to engage with these technologies.
- **Research on Governance**: Encourage and fund research on effective governance models within DeSci ecosystems. This can help in understanding the best practices for decision-making and conflict resolution in decentralized systems.
- **Ethical Standards**: Formulate and implement a set of ethical standards for all DeSci activities. These standards should address the ethical use of data, fair practices in token-based incentives, and responsible research conduct.
- **Regulatory Engagement**: Engage proactively with regulatory bodies to shape policies that support the growth of DeSci while ensuring compliance with existing legal frameworks. This engagement should also focus on the potential implications of DeSci for intellectual property laws and data privacy regulations.

6 Conclusion

In conclusion, the emergence of Decentralized Science (DeSci) marks a potential revolutionary turning point in the history of science. By leveraging Distributed Ledger Technology and Web3, DeSci aims to address the highly relevant challenges in modern science, such as funding, replicability, publishing, ownership, and collaboration. Despite its novelty and associated risks, DeSci offers a promising vision of a more inclusive, trans-parent, and equitable scientific system. By embracing recommendations on awareness, education, governance, ethical standards, and regulatory engagement, the scientific community can navigate the complexities of this paradigm shift, ensuring that DeSci's potential is fully realized for the advancement of science and society.

References

1. M. Leiner et al., A brief history of the internet. ACM SIGCOMM Comput. Commun. Rev. **39**(5), 22–31 (2009). https://doi.org/10.1145/1629607.1629613
2. S. Voshmgir, *Token Economy: How the Web3 Reinvents the Internet*. 2020. Accessed 08 May 08 2023. [Online]
3. T. Berners-Lee, J. Hendler, The semantic web. *JSTOR*, 2001. Accessed 08 May 2023. [Online]. Available: https://www.jstor.org/stable/26059207
4. T.E. Cambridge Dictionary, Science|English meaning—Cambridge Dictionary. Accessed 08 May 2023. [Online]. Available: https://dictionary.cambridge.org/dictionary/english/science
5. A.F. Chalmers, What is this thing called science? (2013)
6. M. Graham, W. Dutton, Society and the internet: How networks of information and communication are changing our lives. Accessed 09 Nov 2023. [Online] (2019)
7. M.A. Cook, G.D. Wright, The past, present, and future of antibiotics. Sci. Transl. Med. **14**(657). American Association for the Advancement of Science, Aug. 10, 2022. https://doi.org/10.1126/scitranslmed.abo7793.
8. D. Howard, The new normal in funding university science. JSTOR, 2013. Accessed 08 May 2023. [Online]. Available: https://www.jstor.org/stable/43315823
9. J.P.A. Ioannidis, Why most published research findings are false, in *Getting to Good: Research Integrity in the Biomedical Sciences*, vol. 2, no. 8 (Springer International Publishing, 2018), pp. 2–8. https://doi.org/10.1371/journal.pmed.0020124
10. P. Stephan, *How Economics Shapes Science* (2015)
11. J.-M. Fortin, D.J. Currie, big science versus little science: how scientific impact scales with funding. PLoS One, **8**(6), e65263 (2013). https://doi.org/10.1371/journal.pone.0065263
12. T.M. Bubela, T. Caulfield, Role and reality: technology transfer at Canadian universities. Trends Biotechnol **28**(9), 447–451 (2010). https://doi.org/10.1016/j.tibtech.2010.06.002
13. M.R. Munafò et al., A manifesto for reproducible science. Nat. Hum. Behav. **1**(1), 1–9 (2017). Nature Publishing Group. https://doi.org/10.1038/s41562-016-0021
14. D. Murray et al., Author-reviewer homophily in peer review. *bioRxiv*. (Cold Spring Harbor Laboratory, 2018), p. 400515. https://doi.org/10.1101/400515.
15. D.J. Solomon, B.-C. Björk, A study of open access journals using article processing charges. J. Am. Soc. Inform. Sci. Technol. **63**(8), 1485–1495 (2012). https://doi.org/10.1002/asi.22673
16. V. Larivière, S. Haustein, P. Mongeon, The oligopoly of academic publishers in the digital era. PLoS ONE **10**(6), e0127502 (2015). https://doi.org/10.1371/journal.pone.0127502
17. R. Smith, Peer review: a flawed process at the heart of science and journals. J. R. Soc. Med. **99**(4), 178–182. SAGE Publications Ltd. https://doi.org/10.1258/jrsm.99.4.178

18. S. Shilina, Decentralized science (DeSci): Web3-mediated future of science. Accessed 04 Sep 2023. [Online]. Available: https://www.researchgate.net/publication/367044419_Decentralized_science_DeSci_Web3-mediated_future_of_science
19. F. Murray, S. Stern, Do formal intellectual property rights hinder the free flow of scientific knowledge? An empirical test of the anti-commons hypothesis. J. Econ. Behav. Organ. **63**(4), 648–687 (2007). https://doi.org/10.1016/j.jebo.2006.05.017
20. D.H. Sonnenwald, Scientific collaboration. Ann. Rev. Inf. Sci. Technol. **41**(1), 643–681 (2007). https://doi.org/10.1002/aris.2007.1440410121
21. S. Lee, B. Bozeman, The impact of research collaboration on scientific productivity. Soc. Stud. Sci. **35**(5), 673–702. Sage Publications Sage CA, Thousand Oaks (2005). https://doi.org/10.1177/0306312705052359
22. UNESCO, UNESCO Recommendation on Open Science (2021). https://doi.org/10.54677/MNMH8546
23. S. Nakamoto, Bitcoin: a peer-to-peer electronic cash system. Cryptography Mailing list at https://metzdowd.com (2009)
24. S. Leible, S. Schlager, M. Schubotz, B. Gipp, A review on blockchain technology and blockchain projects fostering open science. Front. Blockchain **2**, 486595 (2019). https://doi.org/10.3389/fbloc.2019.00016
25. Gridcoin, Gridcoin (GRC)—first coin utilizing BOINC—official thread. Accessed 08 May 2023. [Online]. Available: https://bitcointalk.org/index.php?topic=324118.0
26. Gridcoin, Gridcoin whitepaper—the computational power of a blockchain driving science and data analysis (2018). Accessed 23 Nov 2023. [Online]. Available: https://gridcoin.world/assets/docs/whitepaper.pdf
27. D. Koepsell, A DeSci origin story. Accessed 08 May 2023. [Online]. Available: https://medium.com/coinmonks/a-desci-origin-story-b6b234f7b1a3
28. E.A. Meyer, I.M. Welpe, P. Sandner, Decentralized finance—a systematic literature review and research directions (2022). Accessed 09 Nov 2023. [Online]. Available: https://aisel.aisnet.org/ecis2022_rp/25
29. M. Etzrodt, Advancing science through incentivizing collaboration, not competition (2018). https://doi.org/10.5281/ZENODO.1156360
30. S. Hamburg, Call to join the decentralized science movement. Nature **600**(7888), 221–221 (2021). https://doi.org/10.1038/d41586-021-03642-9
31. Ethereum Foundation, Decentralized science (DeSci). Accessed 04 Sept 2023. [Online]. Available: https://ethereum.org/en/desci/
32. L. Weidener, C. Spreckelsen, Decentralized science (DeSci): definition, shared values, and guiding principles. Frontiers in Blockchain, 7, 1375763 (2024). https://doi.org/10.3389/fbloc.2024.1375763
33. P. Carroll, A. Kellow, The OECD: a study of organisational adaptation (2011). Accessed 09 Nov 2023. [Online]
34. A. Seeger, Understanding UNESCO: a complex organization with many parts and many actors. J. Folk. Res. **52**(2–3), 269–280 (2015). https://doi.org/10.2979/jfolkrese.52.2-3.269
35. C. Ziegler, I. Welpe, A taxonomy of decentralized autonomous organizations, in *Forty-Third International Conference on Information Systems, Copenhagen 2022* (2022)
36. J. Han, J. Lee, T. Li, DAO governance. SSRN Electronic J. (2023). https://doi.org/10.2139/ssrn.4346581
37. State of Wyoming, Decentralized autonomous organizations—SF0038. Accessed 08 May 2023. [Online]. Available: https://www.wyoleg.gov/Legislation/2021/SF0038
38. J. Walden, Progressive decentralization: a playbook for building crypto applications—a16z crypto. Accessed 13 June 2023. [Online]. Available: https://a16zcrypto.com/posts/article/progressive-decentralization-crypto-product-management
39. L. Minquini, Introducing AthenaDAO. Accessed 09 Nov 09 2023. [Online]. Available: https://www.athenadao.co/blog/athenadao
40. T. Golato, P. Kohlhaas, VitaDAO Whitepaper. Accessed 09 Nov 2023. [Online]. Available: https://www.researchhub.com/paper/1266843/vitadao-whitepaper

41. P. Joyce, B. Armstrong, P. Lu, The researchcoin whitepaper. Accessed 09 Nov 2023. [Online]. Available: https://www.researchhub.com/paper/819400/the-researchcoin-whitepaper
42. T. Golato, IP-NFTs for Researchers: a new biomedical funding paradigm. Accessed 08 May 2023. [Online]. Available: https://medium.com/molecule-blog/ip-nfts-for-researchers-a-new-biomedical-funding-paradigm-91312d8d92e6
43. N. Dehouche, L. Della Ventura, R. Blythman, J. Koury, S. Hamburg, removing the barriers for participation in decentralized science from traditional academia (2023). https://doi.org/10.55277/ResearchHub.8c7ndflk
44. C. Ortlepp, Announcing the first biopharma IP-NFT Transaction. Accessed 09 Nov 2023. [Online]. Available: https://www.molecule.xyz/blog/announcing-the-first-biopharma-ip-nft-transaction

Lukas Weidener (Dr. med. Dr. phil.) is a German physician and researcher specializing in emerging technologies, including artificial intelligence and distributed ledger systems. After earning his undergraduate degree in Biology, he completed master's programs in eHealth & Communication and Blockchain Technology, followed by two doctoral degrees. Dr. Weidener now serves full-time at BIO (bio.xyz), a company dedicated to advancing scientific innovation through blockchain technology-based solutions.

Quantifying MEV NFT Arbitrage

Matthias Franz Krekeler

1 Introduction

Blockchain technology emerged in the age of digital trading. A decentralized network provides a digital ledger secured by cryptography to store transactions. The first publicly mined blockchain transaction was executed by Satoshi Nakamoto [22]. In 2014 smart contracts were introduced in Ethereum's white paper by Buterin [7], facilitating the way for new decentralized blockchain applications, including decentralized finance (DeFi) and non-fungible tokens (NFT). DeFi refers to currencies and exchanges functioning without a central intermediary. Its importance is demonstrated by the total value locked in on Ethereum, currently being at $26 billion [11]. Following the inception of DeFi, some market participants started dealing directly with the block producers rather than making transactions publicly available to be written into the ledger [9]. This practice is common in combination with currency arbitrage. Notably, 60% of DeFi activity involves arbitrage [29].

In contrast to tokens traded in DeFi, NFTs are uniquely identifiable tokens, comparable to collectible items such as baseball cards with individual information and value. This work aims to conduct an empirical analysis of NFT arbitrage bots on Ethereum, providing data on the profitability and competitiveness of NFT arbitrage, questioning the relationship between NFT arbitrage activity, MEV, hype and network factors.

- **RQ1**: How profitable are NFT arbitrage opportunities? If profit exists, how competitive is the field?
- **RQ2**: What is the impact of NFT arbitrage opportunities on the Ethereum network? Are these opportunities correlated with the price of Ethereum and a component of MEV?
- **RQ3**: How can bots be identified and labeled within this market?

M. F. Krekeler (✉)
Hamburg, Germany
e-mail: f@think-complete.com

2 Background Concepts

This section explains the basic concepts and fundamentals. It explores NFT arbitrage, motivated by the intersection of high-frequency trading (HFT), maximal extractable value (MEV), NFTs, and blockchain technology. Definitions of Ethereum, tokens, and NFTs are provided in previous chapters of the book.

Mempool, PoW and PoS The mempool (short for memory pool) is a collection of transactions waiting to be confirmed and included in a block by miners [1]. In NFT arbitrage, the mempool can provide the bot information to programmatically execute the next trade. For block validation, Proof of Work (PoW) and Proof of Stake (PoS) are two distinct consensus algorithms used in blockchain networks. PoW, employed in networks like Bitcoin, involves miners competing to solve complex mathematical problems. The first miner to solve the problem gains the right to validate transactions and add a new block to the blockchain. The original concept of PoW was introcuded by Dwork et al. [14] to combat email spam, it requires energy and compute intensive. In contrast, PoS selects validators deterministically based on their staked holdings, reducing the high energy costs associated with PoW. However, this approach raises concerns about cryptocurrencies being classified as securities since validators can earn interest on staked tokens in most protocols [26].

Maximal Extractable Value (MEV) Originally termed "Maximal Extractable Value" by Daian et al. [9], MEV is a concept that quantifies the total amount of value miners can derive from reordering, adding or censoring transactions before writing them into the blockchain. When the user submits a decentralized trade, the transaction first goes into the mempool. Miners and traders reading the mempool can theoretically insert their own transactions before and after. This can be used to extract value from DeFi trades. Flashbots introduced a modified validator software that allows direct transaction submission to the validator, avoiding exposure to the mempool [10]. Private transactions and no gas costs for failed transactions are key benefits. The impact of MEV can be seen in 87% of miners/validators adopting techniques to extract further value per block [10]. Although the term was initially specific to miners, it has since evolved to a general concept. MEV is often executed through smart contracts facilitating multiple transfers, ensuring that trades are executed and recorded in the block only if they are profitable. According to the data of Eigenphi.org around half of all MEV was arbitrage (eigenphi.io, 2022). Many of these trades executed atomically. This is described in detail by Wang et al. [35].

Off-Chain, On-Chain auctions, Arbitrage An on-chain auction takes place entirely on the blockchain. Off-chain auctions happen outside the mempool, similar to online platforms such as eBay. However, a hybrid approach can also be used, where bidding happens off-chain, and settlement occurs on-chain.

"Arbitrage is the simultaneous purchase and sale of the same or similar asset in different markets in order to profit from tiny differences in the assets listed price. It exploits short-lived variations in the price of identical or similar financial instruments in different markets or in different forms." [17]. The more efficient the market the lower the arbitrage opportunities in theory. While described as risk free, in practice order execution failure or fees can diminish profits.

NFT arbitrage deals with less liquid, unique NFTs, during the on-chain settlement of the off-chain auctions. Many NFT markets enable collection offers, which allow buyers to make a single offer that applies to all NFTs in a given collection. These collection bids are often exploited by arbitrage bots, if they can find a cheaper NFT on a different market.

3 Related Work

This section reviews academic literature on NFT market dynamics, blockchain analysis, MEV and arbitrage. The chapter concludes by summarizing the literature gap this study aims to address.

NFT markets Recent literature on NFTs predominantly explores market dynamics and pricing factors. For instance, [4] Ante demonstrates that while Bitcoin and Ethereum prices impact the NFT market, the reverse is not true. This research is supported by the work of Apostu et al. [5], which states no direct influence of NFTs on the pricing of Ethereum. Similar to Dowling [13], who noted little spillover measurement between NFT markets and cryptocurrencies. Another key finding by Yousaf and Yarovaya [39] highlights that trading volume has the strongest connection to NFT returns and volatility during extremely positive market conditions, when prices are rising rapidly. The relationship between volume, returns, and risk is asymmetric and behaves differently in bull versus bear markets. Overall, these findings suggest that NFT markets are evolving, with trading patterns. Moreover, comparisons with traditional assets and market measures like gold price [6] or interest rates [32] indicate no correlation. In contrast [3] concludes, that different NFT collections can influence each other's pricing, indicating market inefficiency and immaturity. Dowling's [13] case studies on specific collections also point to pricing inefficiencies.

On the other hand, indicators for pricing factors are found in "TweetBoost: Influence of social media on NFT valuation" by Kapoor et al. [19]. Their results reveal that social media data improves the ordinal classification accuracy by 6% over baseline models that use only NFT platform features. This is in line with the findings of Pinto-Gutierrez et al. [25] towards a hype driven market approach.

Scam and market manipulation research, for example by Victor and Weintraud [33] and von Wachter et al. [34], highlight the issue of wash trading in the NFT market. These manipulations may lead to inflation of authentic trading volumes. It provides a lower bound estimate for NFT wash trading on Ethereum (3.93% of addresses, processing a total of 2.04% of sale transactions, trigger suspicions of market abuse).

MEV Previous research focuses on miners / validators manipulating the order flow in DeFi for profit. One of the first papers "Flash Boys 2.0: Frontrunning in Decentralized Exchanges, Miner Extractable Value, and Consensus Instability" by Daian et al. [9]

introduces the term MEV and studies the breadth of decentralized exchange (DEX) arbitrage bots and their profit-making strategies. Aditionally, it highlights blockchain-specific elements such as priority gas auctions (PGAs) and the potential security risks posed by MEV. The paper argues that these issues threaten the fair and transparent trading ecosystems promised by blockchains and smart contracts.

In the field of private mempools, A Flash(bot) in the Pan by [36] measures the popularity of Flashbots and provides insight on how private pools and mining activity can affect the overall market conditions. Piet [24] provides a heuristic for detecting private transaction, based on monitoring mempool activity. The author further proposes a graph algorithm for profit measurement of MEV activity.

3.1 Research Gap and Contribution

In conclusion, while existing literature focuses on DeFi arbitrage and sources of MEV, no academic research has quantified the prevalence of NFT arbitrage strategies specifically or examined their role within the broader MEV ecosystem. This study aims to address this gap through an empirical analysis of on-chain NFT transactions. The results aim to provide greater transparency into the scope of NFT arbitrage and its impact on market dynamics.

4 Methodology

In the methodology, we present our data acquisition, pattern detection and analysis concept. It provides design choices and the considerations taken prior to the actual implementation. The theoretical and systematic approach for this work is underlined by the steps for an exploratory data analysis (EDA) proposed by Schutt et al. [27]. With the focus on an iterative. Following a data exploration we conduct confirmatory data analysis to verify our initial research questions. However, further hypotheses might emerge during EDA. To test the scientific significance of the target metric is correlated to market data. "Technically, independence implies zero correlation, but the reverse is not necessarily true." [28] Pearson correlation coefficient is chosen for its widely accepted application in measuring the linear relationship between two datasets, as proposed by Umar et al. [30]. For further investigation in the time-series analysis, we choose the Granger Causality Test. This determines whether one time series is useful in forecasting another Granger [18]. Significance Testing: To reject the null-hypothesis based on randomness, the t-test is used. It tests the null hypothesis, that there is no correlation. By rejecting the null hypothesis (with a p-value less than 0.05), we can determine if there is relationship between the two datasets Umar et al. [31]. Our methodology integrates extensive data collection with data analysis. The core data is enriched with our target metric—profit, followed by a comparison to our secondary metrics, including Flashbots, Ethereum network data and trend data. The

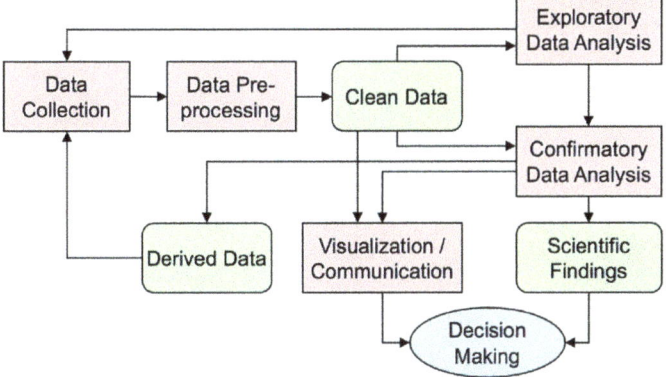

Fig. 1 The exploratory data analysis (EDA) process applied in our methodology-Schutt and ONeil [27]

analysis encompasses both visual and statistical techniques, aiming to reveal insights into NFT arbitrage patterns and testing our hypothesis (Fig. 1).

5 Implementation

In this section, we outline the technical implementation and its known limitations. We begin with data collection and preprocessing, proceed to profit calculation, and conclude with data cleaning and final remarks.

5.1 Data Collection

The process begins with identifying a few NFT arbitrage trades, initially hinted at by a tweet [21] describing a profitable trade, which leads us to find similar transactions and start pattern detection. We identify that two ERC-721 transfer event logs occur in a single transaction. The first emitted "Transfer" event has a smart contract address as the destination, and the second emitted "Transfer" event has the smart contract as the sender. After trying out various SQL solutions, a table was found on Covalent that enables querying and joining event logs, with NFT market data. Figure 2 depicts the entity relationship diagram with NFT and Blockchain tables, highlighting the relationships between the entities and their attributes.

The dataset exclusively contains on-chain data from August 2021 to May 2023. Offchain market data, which could highlight missed opportunities, is not included. Additionally, the dataset does not provide details around token-specific traits such as rarity or overall project popularity.

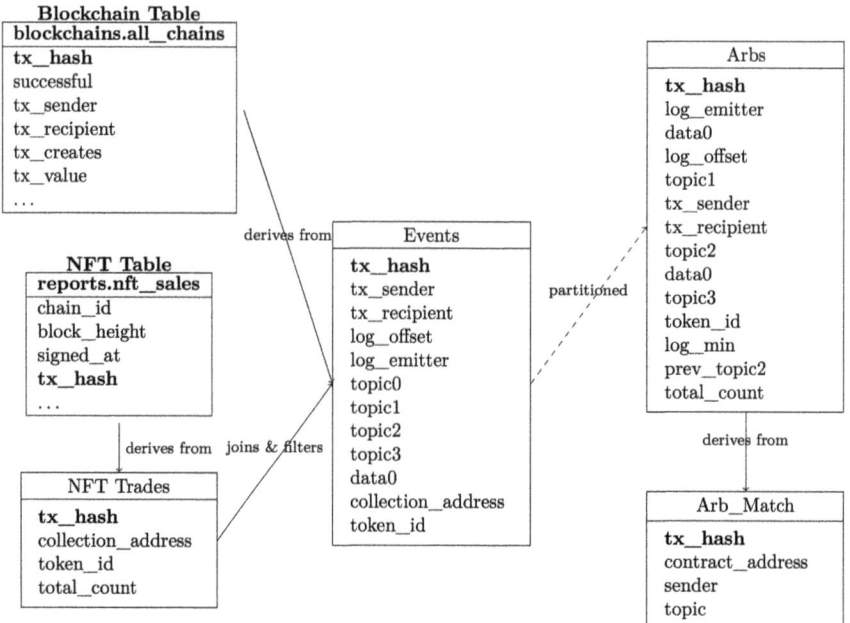

Fig. 2 Entity relationship diagram with NFT and blockchain tables use in the SQL query

5.2 Profit Calculation

In the Ethereum network, the cost of executing a transaction requires paying a "gas" fee. This fee is determined by execution costs of the transaction and current "gas" price. It is paid in Ether (ETH). It's necessary to not only calculate the profit of the trade but to account for all occurring costs. After collecting data from a custom SQL query and merging it with transaction tracer data (phalcon.xyz), it is possible to calculate the realized profits. Invalid or failed transactions are discarded. For trades with no direct USD profit estimate, the Ethereum amount is multiplied by the historic ETH/USD price from an external API to calculate equivalent USD profit. In total over 26,000 transactions resulted in valid NFT arbitrage trades.

5.3 External Data

The supplemental data (1) allows analysis of the relationship between NFT arbitrage profits and wider market factors.

Specifically, the Flashbots dashboard [16] was queried to acquire historical statistics on total miner extractable value (MEV) and the percentage of network gas consumed by MEV transactions. Defillama [12] provided the data for market size.

Table 1 Summary of data sources and their contributions

Data source	Data provided	Purpose/Use
Flashbots	MEV percentage of total gas, Daily MEV profit	Compare MEV profits with NFT arbitrage
Wikimedia	Daily Wikipedia page views for "NFT" and "OpenSea"	Relate to hype cycles and consumer demand (2019–2023)
Defillama	Daily trading volume and count of NFT trades	Reflect market size, emphasize number of transactions
Coingecko	Real-time and historical Ethereum prices	Provide USD/ETH value and see if cost has an effect
Etherscan	Average gas price for Ethereum transactions	Insight into block demand

Coingecko (coingecko.com [8]) provided historic pricing information. Etherscan, provides data on the average gas price for transactions in the Ethereum network (etherscan.io [15]). Using the average gas price can provide insights into block demand (Table 1).

6 Results

In this section, we conduct an in-depth exploratory and statistical correlation analysis on the previously cleaned dataset.

6.1 Exploratory Data Analysis

We begin by comparing the daily NFT arbitrage profits to our key metrics, as presented in Fig. 3. The chart covers the period from August 2021 to May 2023. The plot reveals frequent spikes in daily NFT arbitrage profits, surpassing more than ten thousand dollars. Particularly, there is a significant rise in profits observed at the beginning of August 2022. The Ethereum price below indicates a downturn during May 2022, influenced by the cryptocurrency Terra Luna's crash, leading to a surge in average daily gas prices. MEV profits also spiked during these highly volatile market periods. The daily NFT volume in USD data from Defilama has multiple change points, with a downturn after the Terra Luna crash and a revival in the spring of 2023. Regarding data completeness, it's worth noting that the Flashbots MEV profits data is cut off after the transition of Ethereum from Proof of Work (POW) to Proof of Stake (POS) in the middle of August 2022. Furthermore the NFT marketplace OpenSea was founded in December 2017, its Wikipedia page started only at the end of 2021. The median profit from NFT arbitrage is approximately $200. The distribution of wealth is skewed, with a small number of highly profitable trades generating the majority of profits. The total profit derived from NFT arbitrage appears to be relatively small compared MEV arbitrage. In comparison, total MEV arbitrage profits reached approximately $500 million in 2022 [10], whereas our measurements indicate NFT arbitrage profits were $1–2 million that year. In order to further investigate whether NFT arbitrage can be considered a subset of MEV, we calculate the ratio of NFT arbitrage profits to Flashbots MEV profits, expressed as a percentage. For consistency of both datasets, the timeframe between August 1, 2021, and August 1, 2022, is selected. According to the results, the mean ratio of NFT arbitrage profits to Flashbots MEV profits is approximately 4.65%. Note that the distribution of this ratio is skewed (Figs. 4 and 5).

$$\overline{Ratio} = \frac{1}{n} \sum_{i=1}^{n} \frac{NFT\ Arbitrage\ Profits_i}{Flashbots\ MEV_i}$$

NFT Collections The assessment of different collections uncovers that particular NFT brands, notably Moonbirds, Otherdeed, and BoredApeYachtClub are arbitraged frequently, reflecting their popularity. However, an outlier trade of a Moonbird NFT generated $550,000 and distorts collection averages. Some NFT collections saw persistent arbitrage after launch, while others spiked around launch then declined. We estimate the profit margin by dividing the daily NFT arbitrage profit per collection

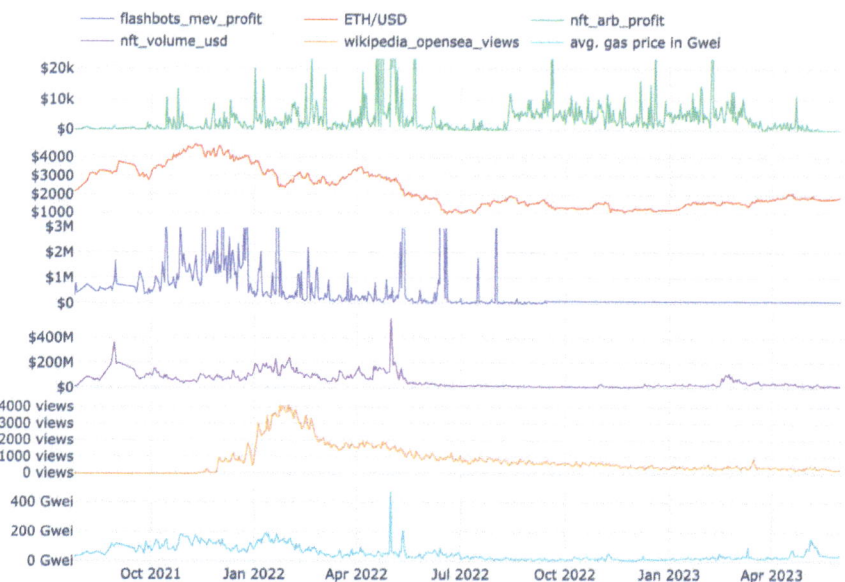

Fig. 3 Overview of daily NFT arbitrage profits (green) compared to our metrics. *Note* Flashbots MEV arbitrage profits data is cut off at August 2022

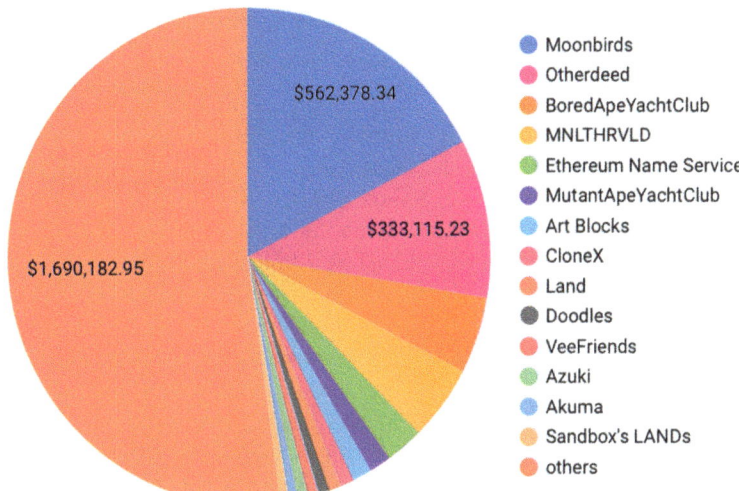

Fig. 4 Total arbitrage profit per NFT collection

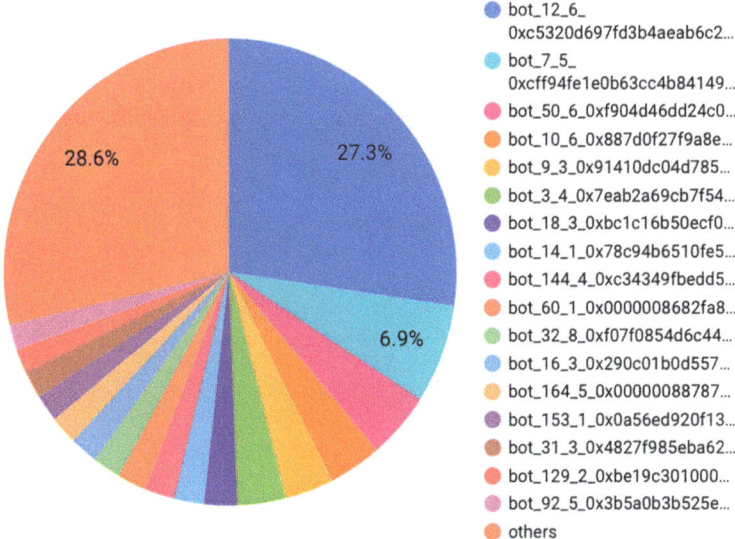

Fig. 5 Market share of profits per bot

through daily NFT floor price per collection. The mean profit margin is around 20%. Next, we will look at the bots itself.

Bot Analysis The analysis reveals that the top 10 bots accumulated most profits, earning around 50% of all measured profits. In total, approximately 180 unique bots were identified. Bot profits exhibit positive correlation to number of trades executed, with more trades likely leading to higher total profit. In terms of lifetime activity most bots remained active for under 3 months, generating around $20–25,000 in gains on average. The total profit seems to be higher the longer the bot is active.

Profit Analysis As depicted in Fig. 6, trades that generate profits ranging from $10 to $25 are totaling 9000. On the other hand, a small quantity of trades (450) yields exceptionally high profits exceeding $10,000. The small quantity of trades contributes to the majority of profits, amounting to $1.5 million. This suggests a skewed distribution of profits, characterized by a 'long-tailed' nature. Our box plot analysis Fig. 7 illustrates temporally, median profit per trade peaked in 2021, as did total profit per bot. 2021 appeared the most lucrative year for NFT arbitrage amidst rising mainstream interest. To evaluate competitiveness, we measure the wealth or profit distribution. The Gini coefficient is a measure of inequality, where 0 signifies perfect equality and 1 perfect inequality. We plot the inequality using the Lorenz curve in Fig. 8. The x-axis depicts cumulative percent of the bots from least accumulated profit to most accumulated profit, the y-axis shows their cumulative profit owned. The 45-degree Equality Line represents perfect equality, where wealth percentage equals population percentage. Total equality plots precisely along this line. If it bows significantly away, it indicates high inequality.

Our results indicate a Gini coefficient of 0.818, this suggests a high degree of inequality. In the context of income distribution, it means that there is significant income disparity, with a small proportion of the population controlling a large proportion of the total profit. On the other hand, interestingly, the profit margins are on

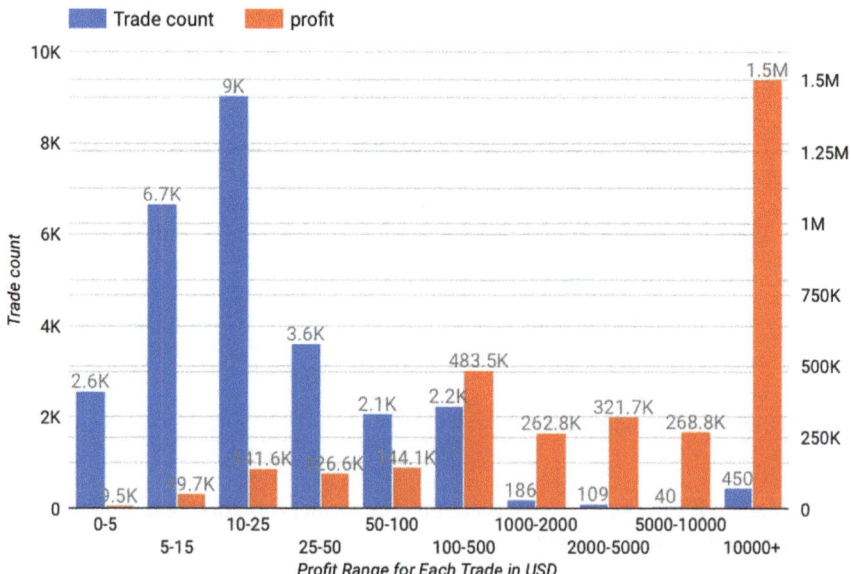

Fig. 6 Comparing the trade count (blue) and the accumulated profit (red) for different profit ranges in USD

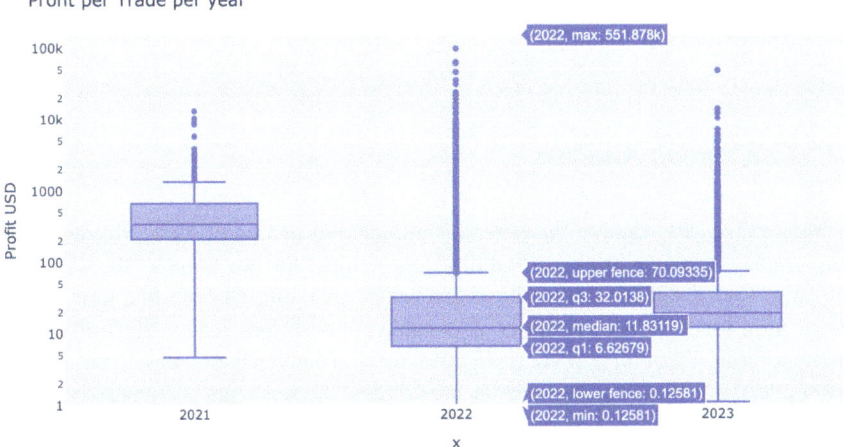

Fig. 7 Box plot displaying NFT arbitrage profit per trade, on a logarithmic scale ranging from $1 to $150,000. The average profit per trade declined from 2021 to 2023

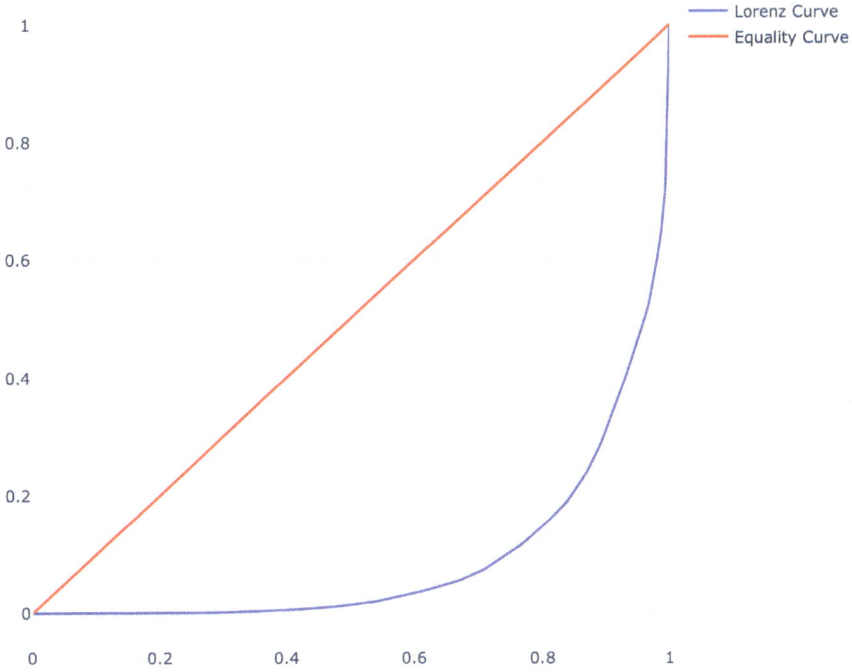

Fig. 8 Lorenz curve for wealth distribution of bot profits

average 28.34%, which is quite high compared to 1–2% in DeFi. Further, suggesting inefficiencies and low liquidity.

6.2 Correlation Analysis

In this section, we examine the potential influencing factors of NFT arbitrage and their predictive power over profitability. For evaluation, we employ the Granger Causality Test, a statistical hypothesis test for determining whether one time series can be used for forecasting another. A p-value below 5% indicates rejection of the null hypothesis. The null hypothesis in this case is, if a given metric has no prediction power over daily NFT arbitrage profits. We conduct the test using lag lengths of 1 and 2 4. The Granger test reveals predictive relationships, therefore we further conduct a Pearson correlation test to quantify the linear relationships between metrics. The study period was from January 1st to September 1st, 2022 (Table 2).

Flashbots Profits The Granger test found no causality between Flashbots MEV profits and NFT arbitrage profits for both lag lengths (1 and 2). The p-values were significantly above the 5% significance level, indicating no predictive power of Flashbots profits on NFT arbitrage profitability. Pearson correlation also indicated a slight, statistically insignificant negative correlation.

Ethereum USD Price and Average Gas Price No significant causality was found between the Ethereum price and NFT arbitrage profits, with high p-values at both lag lengths. Similarly, average gas prices demonstrated no predictive power over NFT arbitrage profits. However, the Ethereum price had the strongest positive correlation with NFT arbitrage profits, though not statistically significant.

Wikipedia Views The Granger causality test presented no significant causality for Wikipedia views on NFTs and OpenSea at both lag lengths, indicating no predictive relationship between Wikipedia views and NFT arbitrage profits.

Table 2 Correlation between daily profit and our metrics

Metric	Profit
Flashbots MEV Profit in USD	− 0.033537
Wiki Views (NFT)	− 0.016274
Wiki Views (OpenSea)	0.016633
Avg. Gas Price in ETH	0.050155
ETH/USD	0.118021
NFT Volume ETH	0.174304
NFT Volume USD	0.184333

During the timeframe 2022-01-01 and 2022-09-01

Table 3 T-Test results for Pearson correlation. P = 0.05

Metric (daily)	Profit
NFT volume USD	0.00332
NFT volume ETH	0.00553
ETH/USD	0.06138
Avg. gas cost	0.42794
Flashbots MEV profit in USD	0.59619
Wiki views (OpenSea)	0.79275
Wiki views (NFT)	0.79712

During the timeframe 2022-01-01 and 2022-09-01

Table 4 Granger causality test results for various metrics and NFT arbitrage profit

Metric	Lag-1 P-value	Lag-2 P-value
Flashbots MEV profit in USD	0.4268	0.7427
Ethereum (ETH/USD) price	0.5122	0.8123
Average gas price in ETH	0.956	0.8865
Wiki views (NFT)	0.7976	0.9779
Wiki views (OpenSea)	0.1228	0.4977
NFT volume in ETH	0.028	0.232
NFT volume in USD	0.036	0.289
Total number of NFT trades	0.9873	0.9893
NFT arbitrage trades and NFT trades	0.064	0.940

NFT Market Metrics Strong evidence of causality is found between NFT arbitrage profit and NFT market trading volume. The Pearson correlation test results in Table 3 depicted the strongest correlation and significance for NFT trading volume in USD. Table 4 applies the Granger causality test and confirms that NFT trading volume drives arbitrage profits at one lag, while evidence weakens at two lags. In contrast, the total number of NFT trades does not show strong causality.

6.3 Case Studies

In this section, we examine historical trades discovered through our exploratory data analysis, providing context on NFT arbitrage opportunities.

In the first case from May 2022, a Bored Ape Yacht Club NFT was mistakenly listed for 10 ETH instead of 105 ETH on OpenSea. An arbitrage bot quickly purchased and resold the underpriced NFT for 110 ETH, profiting from the seller's mispricing.

This case demonstrates how pricing mistakes can happen due to human errors. The user later admitted to the mistake on Twitter: *"Yep, was a fat finger. Was trying to list at 105. Never thought it'd happen to me. Devastating."* @nft_metaman [23].

The second case involves a suspicious transaction of an Ethereum Name Service (ENS) NFT, which functions like a domain name. The specific blockchain domain, *buymydomainplz.eth*, was bought and sold by the same person in July 2022, hinting at potential market manipulation. According to social media posts (@address_eth [2]), the individual listed the ENS for 5 ETH, self-bid 95 ETH as a joke, but an arbitrage bot purchased it for 5 ETH and immediately sold it back to the original owner profiting the difference.

The third case reported by White [38] involves miscommunication during NFT launches. In January 2023, Porsche released 7500 Porsche 911 NFTs, each priced at 0.911 ETH. Poor sales resulted in Porsche announcing an early halt of the minting process, before actually doing so. This caused a temporary price surge on the secondary market. Traders took advantage of this, minting new NFTs at the original price and selling them at the inflated market rate, making a profit. In another instance also reported by White [37], during January 2022, a bug allowed OpenSea users to buy NFTs at old prices from supposedly canceled listings, causing $1.8 million in losses. One Bored Ape sold for 0.77 ETH then resold for 84.2 ETH. The exploit revealed flaws in OpenSea's off-chain auction settlement process. These exemplary case studies, demonstrate how human error, potential manipulation, system bugs and market dynamics can create arbitrage opportunities.

7 Evaluation

This work presents an empirical analysis of high-frequency trading in the domain of NFT markets built on Ethereum. Specifically, it focused on quantifying and analyzing bots conduct arbitrage strategies to profit from pricing discrepancies across NFT platforms. Section 5 documented the implementation to collect and process relevant transaction data. Section 6 revealed that NFT arbitrage accounts for less than five percent of overall MEV profits on Ethereum. Approximately 150 trading bots generated around three million US dollars in total profit over the last three years. The analysis of profit distribution reveals a Gini coefficient of 0.818. Notably, around 17 bots were responsible for approximately 70% of all profits. When measuring correlations, the Granger causality test fails to reject the null hypothesis for MEV profits being useful for predicting NFT arbitrage profits. A stronger correlation is found by measuring public hype through Wikipedia views on the term "OpenSea". However, the leading indicator is NFT trading volume itself, with a p-value below 5%. In this concluding section, the findings of the study are reviewed in terms of the initial research questions and hypotheses.

7.1 Discussion

The findings slightly support prior research emphasizing the role of hype and social media in driving NFT markets [19, 26]. They confirm the decoupling between the NFT and DeFi markets [4, 5]. However, this study provides new context by quantifying the profitability of NFT arbitrage. High profit margins and low transaction counts confirm the described inefficiencies within these markets [3, 13]. This work contributes the observation that private transactions aren't always utilized for NFT arbitrage, underlining distinct dynamics of the NFT markets.

With the obtained results, we can answer our research questions. For the first question ("How profitable are NFT arbitrage opportunities and if it's profitable, how competitive is it?"), the detected NFT arbitrage profits are interpreted as being a small market compared to DeFi, highly profitable only for a small number of bots due to low liquidity and high competition. Most bots are only active for two months. These short-lived actions are supported by looking at NFT collections, they are mostly exploited for arbitrage during the initial hype phase. In our dataset approximately every 90 days there is a trade that is extremely profitable, as the field is full of outliers, one trade can make up over 50% of the profit of a bot. The lack of efficiency is further backed by looking at the ratio of floor prices and profit, with a margin of around 20%, which is very high compared to traditional arbitrage. That competition is increasing in the NFT market was expected.

A strong domain knowledge is often required to understand shifts in patterns. For example, consider Ethereum's transition from PoW to PoS. Some key reasons for NFT arbitrage are the introduction of a new marketplace, disparities between 'mint' and 'floor' prices for an NFT collection, liquidity fragmentation, and unintentional front-running by other buyers or sellers. For the second research question *(What is the impact of it on the Ethereum network and MEV?)*, no statistically significant influence was measured. Although NFT arbitrage can be considered a subset of MEV profits, it only represents a small fraction (less than five percent). Interestingly the NFT market is not as strongly affected by market shocks compared to DeFi. The ETH/USD rate is weakly correlated with profits, which is expected when profits are denominated in ETH. The key driver for NFT arbitrage profits, is the daily NFT volume itself. The public interest, measured through Wikipedia views for "OpenSea", holds a weak correlation, but the chart concludes this correlation fades over time. This could be due to users not looking up the article on Wikipedia anymore, as they are already familiar with the topic. Or the introduction of a new marketplace, changing dynamics.

For the last research question *(How can bots be identified and labeled within this market?)*, it was found that most arbitrages can be detected via a double transfer action, though the evolution of smart contracts adds more complexity over time. For example, often the state changes in the EVM are more reliable than calculations from log events.

7.2 Limitations

This study has limitations including the absence of mempool and off-chain order book data which could provide additional insights, as well as potential distortions from outlier trades or misclassified transactions. When evaluating potential harms of NFT arbitrage, the future activities of NFT buyers should be considered regarding whether they will profit or incur losses from pricing differences between markets. Developers and marketplace operators should aim to improve market efficiency and fairness, such as by adding conditions in smart contracts to prevent single exchange arbitrage opportunities.

8 Conclusion

This work presents an empirical analysis of high-frequency trading bots that conduct arbitrage strategies in Ethereum-based NFT markets. The research aims to quantify the prevalence and profitability of NFT arbitrage, as well as to examine its relationship to the broader ecosystem of MEV trading and to gain insight into market efficiency. For this purpose, a literature review on MEV, exploratory data analysis, and NFT market factors are conducted, highlighting a lack of research on NFT arbitrage. Due to the absence of existing datasets on NFT arbitrage, exploratory data analysis is chosen as the initial methodology. This workflow comprises four phases: dataset creation, data cleaning and enrichment, data analysis, and correlation analysis. For the data collection process, necessary sources of information, such as event logs and state changes, are identified. After the data provider is selected, the correlation tests for the target metrics are chosen. The implementation begins with the dissection of a small sample of NFT arbitrage trades. Based on this analysis, a SQL query is designed to fetch an initial dataset. Afterward the data is cleaned and enriched, additional metrics are gathered through APIs. Lastly, profit calculation is applied to the filtered data. Collecting and filtering the dataset enables analysis of more than 26,000 on-chain NFT transactions identified as arbitrage trades, which together have generated over US\$3 million in total profits over three years. This analysis reveals that, while lucrative for a handful of dominant players, NFT arbitrage remains a niche activity generating a small fraction (under 5%) of total MEV profits. The distribution of NFT arbitrage gains appears highly unequal, with a few bots earning the majority of profits. No strong statistical evidence is found to suggest that gas prices or total MEV profits directly influence NFT arbitrage profitability.

This suggests NFT trading may be partially decoupled from the wider DeFi ecosystem dynamics, especially since NFT trading volumes exhibit the highest statistically significant correlation with NFT arbitrage profits. Further case studies provide domain knowledge, indicating that structural inefficiencies, information asymmetries, human errors, and hype cycles contribute to the pricing differences between NFT markets. On the one hand, sudden spikes in profitability typically occurs around major NFT launches. On the other hand, the short activity lifespan of most bots highlights the challenges of staying profitable long-term. The visualizations are accesible via an interactive tool [20].

This research provides initial empirical insight into the intersection of digital blockchain assets and bot trading. The study is limited by its primary focus on a subset of on-chain data, yet it highlights market inefficiencies, correlations, and competition within the evolving NFT landscape.

8.1 Future Research

This research provides the first dataset for understanding NFT arbitrage activity on the Ethereum blockchain. Our suggestions for future research fall into two categories: technical and economic. To enhance economic understanding, future studies could improve the collected data by including the costs of failed bot transactions. This would make overall profitability measurements more accurate. To improve measurements, incorporating off-chain data sources like market APIs could reveal missed arbitrage opportunities. Further highlighting how markets might use aggregation to improve user experience. In terms of technical research, analyzing the public smart contract code of identified arbitrage bots could provide context. Decompiling could reveal common algorithms, optimizations, and vulnerabilities. This could enhance the technical understanding of arbitrage bots and provide context for executing detected strategies. Finally, future studies could explore other blockchains, such as NFT ordinals on Bitcoin.

References

1. (2018). Mempool explained. https://www.mycryptopedia.com/mempool-explained/. Accessed 01 June 2022
2. @address_eth (2022). Just found out the buymydomainplz.eth sale for 95 eth was similar to the fail. The guy did a selfbid of 95 eth while his domain was listed for sale for 5 eth. a bot bought it for 5 eth and accepted the offer in the same transaction. https://twitter.com/address_eth/status/1551533719915307008
3. L. Ante, Non-fungible token (nft) markets on the ethereum blockchain: Temporal development, cointegration and interrelations. IRPN: Innov. Behav. Econ. (Topic) 10–20 (2021)
4. L. Ante, The non-fungible token (NFT) market and its relationship with bitcoin and ethereum. FinTech **1**(3), 1–9 (2022)
5. S.A. Apostu, M. Panait, L. Vasa, C. Mihescu, Z. Dobrowolski, Nfts and cryptocurrenciesthe metamorphosis of the economy under the sign of blockchain: a time series approach. Mathematics 7–10 (2022)
6. O. Bani-Khalaf, N. Taspinar, The role of oil price in determining the relationship between cryptocurrencies and non-fungible assets. Invest. Anal. J. **52**, 64–66 (2023)

7. V. Buterin, A next generation smart contract decentralized application platform, pp. 13–14 (2014). https://ethereum.org/669c9e2e2027310b6b3cdce6e1c52962/Ethereum_Whitepaper_-_Buterin_2014.pdf. Accessed 22 May 2023
8. coingecko.com (2022). Coingecko: Ethereum. https://www.coingecko.com/en/coins/ethereum. Accessed 01 June 2022
9. P. Daian, S. Goldfeder, T. Kell, Y. Li, X. Zhao, I. Bentov, L. Breidenbach, A. Juels, Flash boys 2.0: frontrunning in decentralized exchanges, miner extractable value, and consensus instability, in *2020 IEEE Symposium on Security and Privacy (SP)* (2020), pp. 1–2
10. dashboard.flashbots.net (2022). Transparency dashboard flashbots. https://dashboard.flashbots.net/. Accessed 01 June 2022
11. DefiLlama (2023). Defillama. https://defillama.com/chain/Ethereum. Accessed 01 July
12. defillama.com (2022). Defillama: Nft marketplaces. https://defillama.com/nfts/marketplaces. Accessed 01 June 2022
13. M. Dowling, Fertile LAND: pricing non-fungible tokens. Financ. Res. Lett. **44**(C), 4–6 (2022)
14. C. Dwork, M. Naor, Pricing via processing, or combatting junk mail, in *CRYPTO'92: Lecture Notes in Computer Science No. 740*. Springer. eigenphi.io (2022), pp. 139–147. Eigenphi. https://eigenphi.io Accessed 01 June 2022
15. etherscan.io (2022). Etherscan gas price. https://etherscan.io/chart/gasprice/. Accessed 01 June 2022
16. explore.flashbots.net (2022). Flashbots: explore. https://explore.flashbots.net/. Accessed 01 June 2022
17. J. Fernando, Arbitrage: how arbitraging works in investing, with examples (2023). https://www.investopedia.com/terms/a/arbitrage.asp. Accessed 10 June 2023, Reviewed by Samantha Silberstein, Fact checked by Pete Rathburn
18. C.W. Granger, Investigating causal relations by econometric models and cross-spectral methods. Econometrica **37**(3), 424 (1969)
19. A. Kapoor, D. Guhathakurta, M. Mathur, R. Yadav, M. Gupta, P. Kumaraguru, Tweetboost: influence of social media on nft valuation (2022)
20. F. Krekeler, Mev nft arbitrage dashboard (2023). https://lookerstudio.google.com/reporting/a5052ec4-a2c7-4f40-8043-769f0eed2e47. Online; Accessed 23 June 2023
21. B.C. Miller, New mev just dropped: moonbirds_xyz arbitrage (2022). https://twitter.com/bertcmiller/status/1517278228918018049. Accessed 20 July 2023
22. S. Nakamoto, Bitcoin genesis, 6–8. Mined on January 03, 2009 10:15:05 (2009)
23. @nft_metaman. Yep, was a fat finger. Was trying to list at 105. Never thought it'd happen to me. Devastating (2022). https://twitter.com/nft_metaman/status/1531120182621523972
24. J. Piet, J. Fairoze, N. Weaver, Extracting godl [sic] from the salt mines: Ethereum miners extracting value, pp. 18–19 (2022). https://arxiv.org/abs/2203.15930
25. C.A. Pinto-Gutierrez, S. Gaitán, D.M. Jaramillo, S. Velasquez, The nft hype: What draws attention to non-fungible tokens? Mathematics 11 (2022)
26. F. Saleh, Blockchain without waste: proof-of-stake. Rev. Financial Stud. **34**(3), 7–9 (2021)
27. R. Schutt, C. ONeil, *Doing Data Science: Straight Talk from the Frontline* (OReilly New York, NY, USA, 2013)
28. H.J. Seltman, Experimental design and analysis. *Book is on the World Wide Web*, pp. 76–77 (2018)
29. D. Sui, Uniswap volume—mev activities (2023). https://dune.com/danning.sui/uniswap-volume-mev-activities. Version: 2014. Accessed 22 May 2023
30. Z. Umar, M. Usman, S.Y. Choi, J. Rice, Diversification benefits of nfts for conventional asset investors: evidence from covar with higher moments and optimal hedge ratios. Res. Int. Bus. Financ. **3** (2023)
31. Z. Umar, M. Usman, S.Y. Choi, J. Rice, Diversification benefits of nfts for conventional asset investors: evidence from covar with higher moments and optimal hedge ratios. Res. Int. Bus. Financ. **5** (2023b)
32. P. Vartanian, A. Alves de Moura Junior, J. Racy, R. Neto, Nonfungible token (nft) prices, cryptocurrencies, interest rate and gold: an econometric analysis (jan. 2019-aug. 2022). Int. J. Econ. Financ. **15**(1) (2022)

33. F. Victor, A.M. Weintraud, Detecting and quantifying wash trading on decentralized cryptocurrency exchanges, in *Proceedings of the Web Conference* (ACM, 2021), pp. 8–9
34. V. von Wachter, J.R. Jensen, F. Regner, O. Ross, Nft wash trading: quantifying suspicious behaviour in nft markets. SSRN Electronic J. 7–8 (2021)
35. Y. Wang, Y. Chen, S. Deng, R. Wattenhofer, Cyclic arbitrage in decentralized exchange markets. SSRN Electronic J. 7–8 (2021)
36. B. Weintraub, C.F. Torres, C. Nita-Rotaru, R. State, A flash(bot) in the pan, in *Proceedings of the 22nd ACM Internet Measurement Conference* (ACM, 2022), pp.12–13
37. M. White, Opensea users lose a collective $1.8 million to an issue allowing people to buy nfts at low prices from old opensea listings the sellers thought theyd deleted. Web3 Is Going Just Great (2022). https://web3isgoinggreat.com/single/opensea-users-lose-collective-1-8-million. Accessed 26 July 2023
38. M. White, Porsche bungles nft roll-out. Web3 Is Going Just Great (2023). https://web3isgoinggreat.com/single/porsche-bungles-nft-roll-out. Accessed 26 July 2023
39. I. Yousaf, I.L. Yarovaya, The relationship between trading volume, volatility and returns of non-fungible tokens: evidence from a quantile approach. Financ. Res. Lett. **50**(C), 7–10 (2022)

Franz Krekeler studied Human–Computer Interaction at Hamburg University and Machine Learning with Y-Data (Yandex, now Nebius) in Tel Aviv. He participates in coding competitions, teaching, and open-source. Currently, he is actively exploring market analysis, arbitrage, language models and applied data observability.

Blockchain-Based Data Security and Enhanced Transparency in the Digital Signage Industry

I. Dimitrov, D. Trauth, and W. Prinz

1 Introduction

The following research aims at showing how the unprotected digitalisation of the data in the advertising sector can have a negative impact on its business side. According to a new research, privacy is one of the central problems for the digital data-driven advertising, as it creates negative perceptions [1]. Therefore when working with data structures, it has to be carefully considered how this information and value produced out of it is stored and managed. The digital advertising industry is not different from its privacy concerns. Its complex environment makes the entire process expensive and ineffective. Issues like lack of transparency, accountability or even fraud can lead to massive financial problems for the business. However, with its capacity to provide secure and immutable data verification and confirmation, blockchain technology has the potential to address those problems and advance the future of the advertising sector.

In many industries and businesses, the lack of trust and transparency of information can play a major problem in their development. In the advertising sector, data is collected and exchanged between multiple parties like advertisers, publishers, suppliers, etc. Due to this massive flow, its integrity and veracity is often questioned. This can result in a variety of problems such as fraud, lost advertising revenue, and poor targeting [2].

These problems can be addressed with the help of blockchain technology, which offers a transparent and secure verification of data. Blockchain is fundamentally a distributed ledger technology that allows the secure and public sharing of data [3]. This is possible by utilizing smart contracts, which are self-executing programs

---

I. Dimitrov (✉) · D. Trauth · W. Prinz
Fraunhofer Institute for Applied Information Technology FIT, Sankt Augustin, Germany
e-mail: ivan.dimitrov@fit.fraunhofer.de

I. Dimitrov · W. Prinz
RWTH Aachen University, Aachen, Germany

© The Author(s), under exclusive license to Springer Nature Switzerland AG 2025
W. Prinz and D. Trauth (eds.), *Tokenizing the Future*,
https://doi.org/10.1007/978-3-031-91405-8_23

that may be used to automate the process of data verification. Smart contracts enable real-time data tracking and produce an immutable record of information that is transmitted, which can be used to build trust and transparency in the digital advertising industry by allowing open access to it.

The ability to easily track and verify data more precisely and reliably is one of the main advantages of adopting blockchain technology for data verification in many industries. In order to build trust and confidence, it is crucial to confirm that the data is accurate and has not been altered by anyone.

The ability of blockchain technology to minimize fraud is another significant advantage for using it in the current research paper. Tracking and verifying data in a form that is more transparent and safe is one of the main features that will be proposed here. The technique which is used will aim at making it more difficult for shady characters to tamper with data or engage in fraudulent actions, that can lead in the prevention of fraud.

Due to its immutable property, fraudulent behaviors including ad fraud and viewability fraud are not possible because any data that is recorded cannot be changed or modified once being stored [4].

Advertisers, publishers, and other stakeholders can have more confidence in the data they are using to make critical business choices and provide trust among each other. Better outcomes for all stakeholders as well as increased efficiency and effectiveness in the digital advertising ecosystem may result out of this.

2 Motivation

Since the popularisation of the digital technology, traditional retail businesses have begun to struggled with the reduction of client traffic and the rise in operating expenses due to the increased amount of online competitors [5]. Therefore, it has become more and more challenging for street vendors to compete with the e-commerce resellers. However, visits to local stores still do have benefits, like a smaller carbon impact, the chance to prevent returns, and faster delivery. Even though there are multiple marketing strategies for sales growth, many retailers have still underutilized their advertising space. For example, large storefront windows or extensive floor plans, are still not being used effectively to attract customers. Utilizing a variety of digital devices, such as large screen TVs, monitors, and tablets, enables retailers to create targeted, flexible, and dynamic advertising campaigns. For instance, a local bakery on a main street could quickly advertise a special offer or promotion at a nearby market.

2.1 Problem Statement

The scenario described above presents a simplified understanding of the challenges faced by retailers. In reality, the market is even more complex, with many competitors offering untransparant pricing models that include additional costs for unnecessary advertising strategies. These pricing models make it complex for retailers to evaluate the value of the proposed advertising solutions, or in some cases, retailers are paying for advertising that is not aligned with their goals, such as increasing customer traffic and revenue. This makes it challenging for retailers to make proper decisions and achieve optimal results from their advertising campaigns.

Furthermore, proper and detailed information about the advertisement impact and results is often stored on centralised servers, often controlled by a single authority. Such kind of centralised storage can present a security concern since it creates a single point of failure, making the critical business data vulnerable to modification or leak of information. Therefore, an efficient solution that addresses the security issues related to centralized data storage in the advertising industry is required to safeguard important company information from malicious actors.

2.2 Related Work

Related work in the field of Blockchain has primarily focused on utilising its benefits in variety of use cases. Studies have shown that in the field of information storage, it allows trustless users to agree on an immutable and auditable piece of data without third party integrations [6]. The paper is built upon studies from different industries like sharing financial information [7] or storing of Medical Records [8] using the blockchain technology and will address their limitations and improve upon the current state of research.

In the literature the term Blockchain is defined in many different ways, therefore we will try to simplify it and use the most accurate one which corresponds the best to our research question. *An ongoing chain of recorded information, formed in blocks with a fixed size, which is sequentially linked with hashed values and stays immutable.*

The Blueprint of the new economy is called Blockchain 3.0 [9]. It extends the business side of the technology by encompassing the political sectors, government, healthcare, etc. Blockchain 3.0 envisions a more advanced form of "smart contracts" to establish a distributed organizational unit that makes and is subject to its own laws and which operates with a high degree of autonomy [10].

3 Proposed Solution

This chapter will present a conceptual overview of the proposed solution. The goal will be the development of data storage system that utilizes a combination of centralization and decentralization to ensure the security and integrity of stored information. Our proposal suggest saving input of advertising meta data to a centralised server, which allows for a scalable retrieval and management. Additionally, saving the same information to a blockchain, a decentralized and distributed digital ledger. The use of blockchain technology will enable an extra layer of security as it allows for transparent and tamper-proof record keeping. This dual storage concept ensures that data is both easily accessible and protected, making it suitable for usage in a variety of applications.

Figure 1 illustrates the concept which was developed in order to validate the researched idea. It has been build around the idea of mobile advertisement.

The past few years have been marked by a global pandemic caused by the coronavirus. It has had a significant impact on the global health, leading to widespread of illnesses and death. As a result, there has been a significant increase in awareness of the importance of hand hygiene as a means of preventing the spread of the virus.

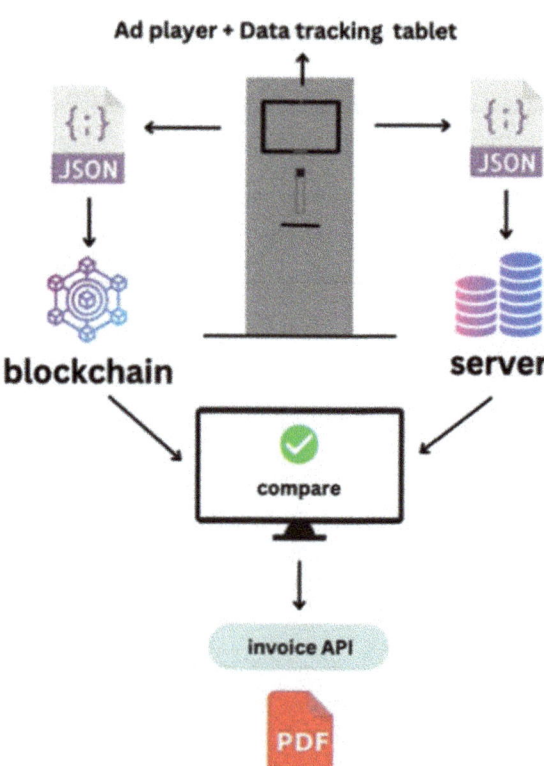

Fig. 1 Concept illustration

This increase in demand has created new business opportunities for entrepreneurs and companies that produce and sell these products. The proposed in this paper solution, has been evaluated and tested in real-world situations to ensure its effectiveness and practicality.

Figure 1, illustrates visually how the idea has been developed. In the middle, there is a hand sanitising device, that is used to release cleaning gel automatically when a hand is detected. On top of it, there is a build-in mobile tablet integrated with face detection camera, that plays video advertisement every time there is a human standing in front of it. The amount of time, every video promotion has been played is proportional to the time spent from each person in front of the sanitiser cleaning its hands. As we can see now, the recorded time is a valuable metric, because it determines how much money will the promotion provider pay. Therefore it has to be carefully stored and protected, as it will be used to ensure the integrity and accuracy of the generated payment invoices.

In the traditional way, this data is being sent to a centralised server which keeps a record of timestamps for each time an advertisement video has been played. However this approach has several drawbacks such as:

- A centralized server is a single point of failure and can be targeted by hackers, therefore it is vulnerable to attacks. If a hacker is able to access the server, they can potentially steal or alter the data stored there.
- With centralized servers, it can be difficult to determine who has access to the data and who is responsible for any unauthorized access, which leads to a lack of transparency.
- In a centralized system, the organization that controls the server has complete control over the data and can make changes or deletions without the knowledge of its users or clients.

In order to prevent those issues from occurring and increase the overall security and reliability, our research proposes the adoption of the blockchain technology as additional level of protection and transparency.

3.1 Objectives

The research's primary goal is to utilize the Blockchain infrastructure as a secure back-bone for building trust among the advertising sector. Considering the proposed concept described in the previous section, the main objectives of this academic study are the following points:

- Developing a secure smart contract for storing business meta data.
- Developing a dashboard for storing and retrieving information from both server and blockchain sides.
- Platform for analysing and utilising gathered information for business development.

4 Implementation

Considering the structure of the described proposal, the realisation of the solution is divided into two phases: Blockchain and Demonstrator development.

The Blockchain part involves the development of smart contracts using Solidity as a main programming language. They are designed to be compatible with all Ethereum Virtual Machine (EVM) blockchains and work with Solidity 0.7.0 or later. The smart contracts are also engineered to store series of transactions with predefined attributes, such as (Hash, Month, Year, ID). A visual illustration can be found on Fig. 2. Their purpose is to collect the precise information of each client using the sanitizer and hash that data using Keccak256 algorithm. When that succeeds, the data will be stored. At the end of each month, when an invoice is ready to be generated, the data collected from the smart contracts is compared with the data recorded on the server and if they match, an invoice will be generated accordingly. Important feature is that all transactions per month are permanently stored on the public distributed ledger and can be queried for verification from everyone if needed. Furthermore, each transaction can be queried individually per its ID or Hash value.

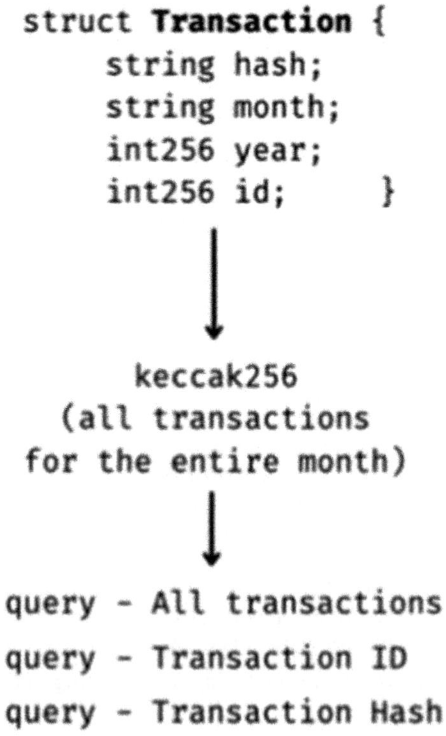

Fig. 2 Hashing feature of the smart contract

```
struct Transaction {
    string hash;
    string month;
    int256 year;
    int256 id;        }
            │
            ▼
        keccak256
    (all transactions
    for the entire month)
            │
            ▼
    query - All transactions
    query - Transaction ID
    query - Transaction Hash
```

Fig. 3 Work direction between the users and the system

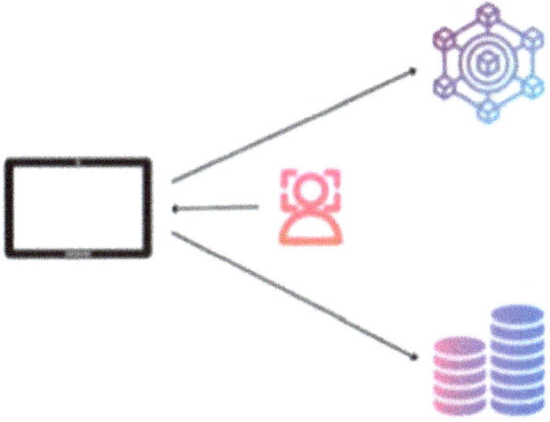

As a planned feature, is the development of an advertising marketplace. Its objective is to enable advertisers to upload their video promotions and allow the sanitizing tablets to select from the highest-paying video available on the marketplace. Its blockchain architecture enables the collective decision-making and management of the marketplace without the need for a central authority. A smart contracts will be used to automate the process of uploading and selecting video promotions, as well as the disbursement of payments. Furthermore, it can generate its own ecosystem, that utilizes a token-based economy, where advertisers can purchase tokens to submit their video promotions and sanitiser owners can receive tokens every time they access the marketplace and select video promotions they wish to display.

This approach provides several benefits such as increased transparency and security, as well as the ability to operate without a central regulator. This allows for a more efficient and decentralized decision-making process, leading to a more scalable and effective advertising marketplace. The smart contract ensures that the transactions are transparent, immutable and efficient.

The demonstrator construction is considered to be the hardware connection of the sanitiser with the tablet and the software development of its features and dashboard.

To begin with, let's look at the hardware. The sanitiser that is releasing the cleaning gel is mounted to a construction which is easily transportable and can be placed in different locations. Its primary idea is to be located on the building's main entrances. Mounted to the construction is also the tablet which is wired to a router in order to have real time connection with internet. On the tablet itself, there is a built-in artificial intelligence front camera which can recognise every time when someone is standing in front and will therefore launch an advertising video. The idea is to launch these videos directly from the blockchain powered decentralised marketplace (Fig. 3).

The prototype running on the tablet is powered by an Android application that has the primary function to provide informational data, promotions and detections of the people crossing by. The application is connected to both centralised database and blockchain with the goal to transmit the collected information.

Each sanitiser owner has an access to a dashboards where there is a direct control to the Android application settings. Every owner has a direct overview of what promotions are being played, what is the total viewed time, how many users has passed by, etc. This data is in real-time collected and transmitted to the centralised database as well as to a smart contract. At the end of the month, when the sanitiser owner requests to withdraw the collected from advertisement videos payments, a simultaneous request to both server side and blockchain is triggered in order to compare the view time data. If it matches, an invoice is generated directly to the users dashboard and a payment is triggered. If the request is not in crypto payments, a traditional API like PayPal or Stripe can be used to release fiat money payments.

The future implementation of the project will focus on automating the entire process so that the owners of the sanitisers do not need to choose the promotional videos played in their locations, but use a built-in smart contract algorithm, that automatically selects the best monetised video per location and time duration, so that they can maximise their profits and save administration time.

5 Evaluation

A real functional prototype has been built in order to measure the effectiveness and performance of the system. To simulate the test transactions, a smart contract deployed to Göerli Ethereum testnet was used.

Figure 4 shows the time needed from the point of sending the data from the Android tablet to recording it on the Ethereum testnet. As one can see, the time is fluctuating among the different attempts. After performing 20 attempts, the average time is 19.2 s. There are two reasons correlated to the latency. First is the internet connection and second is the transaction fee. Self-evident, the connection speed might differ due to the mobile behaviour of the nodes, however a bigger role plays the transaction fee. In order for the contract to be called, the amount of gas required for its execution needs to be paid. The exact gas price is determined based on the supply and demand of transactions at the time of calling the contract. If their supply is high, the miners give priority to those who pay higher gas fees, because this is their reward for validation. That is why if the smart contract is called with a relative lower gas price to the average during peak hours, the deployment may take a longer time. However, according to the test, there was no transaction that took longer than thirty-five seconds to be accomplished.

Table 1 shows the average Transaction (TXN) Fee paid per interaction with the smart contract. It is calculated by multiplying the current gas price at the time of calling the contract with the amount of gas used by the transaction (Gas price x Gas needed for TXN). The fee is averaged based on 4 iterations of 5 attempts each. For example, the first five calls of the smart contract consumed on average 0.000019 Ethers, the second 0.008856, etc.

In summary the prototype was built to simulate real situations in order to evaluate the performance of the proposed system. The observations show that using the

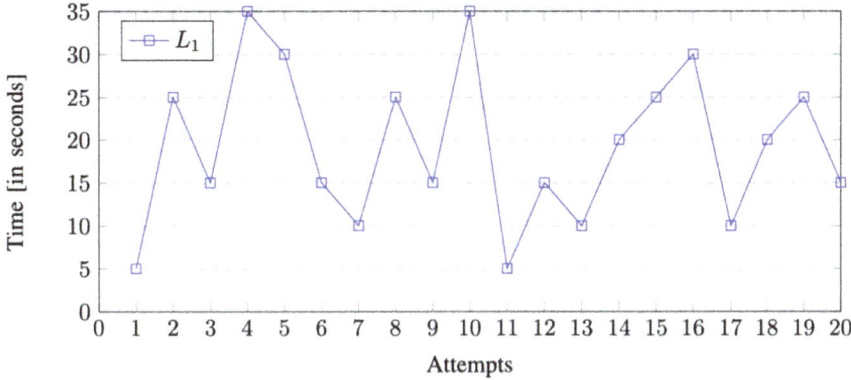

Fig. 4 Deployment to Ethereum testnet

Table 1 Average TXN fee paid per interaction

Number of attempts	1–5	6–10	11–15	16–20
Average fee (in Ether)	0.000019	0.008856	0.010459	0.005508

blockchain technology can improve the transparency and security when dealing with valuable data, because it prevents its vulnerability of manipulation. However, when dealing with large amount of real time data transmitted in a relatively close interval of time, a delay is observed which might be an issue for some industries. Taking as an example the advertising sector used in this research, we can conclude that our experiments provided very satisfactory results and using blockchain technology for data verification and automatisation can lead to substantial improvements in the industry.

6 Monetisation

In the reference to the proposed solution, let's take a look at how can it be monetized in a real business environment. The proposed prototype will profit the advertisers by enabling them to effectively measure the impact of their campaigns on targeted audience and receive payment accordingly. One possible monetization could be the pay-per-view model, where advertisers pay based on the number of views their advertisements receive.

To accurately calculate the number of impressions, the designed with this research demonstrator could utilize computer vision and facial recognition technologies to play video advertisements based on the demographics of the people standing in front of it. As reviewed previously, the system could record both the video time played and the number of people shown on a centralised server as well as on a blockchain for further immutability. The concept will ensure that the data is tamper-proof and provides a secure way for advertisers to confirm the authenticity of their metrics when generating invoices at the end of each promotion period. The implementation of blockchain will provide an extra layer of security, as all data will be saved in a decentralized manner, making it very difficult for the screen owner or anyone else to manipulate the records. In addition, the blockchain technology could be used as legal authority stamp to verify that the data stored on the server is the same as the data from the invoice. That will help to eliminate any discrepancy or any attempt for fraud in the monetisation process by assuring the advertisers that they are paying for actual view time of their promotions.

7 Conclusion

This paper aims to provide a secure data exchange model using blockchain technology that can be used to enhance the productivity and innovation in the digital advertising sector. In this proposed model, instead of a flat rate remuneration system, advertisers should be compensated based on the actual viewing times of the advertisement. This approach can also prevent the inclusion of irrelevant or low-performing promotions. It reduces the barriers to entry for retailers, as there are no variable costs for advertising in the worst-case scenario. However, as the model proves to be efficient, the variable costs will increase alongside the revenue. In a system with multiple parties involved in variable costs and profit, trust issues may arise. It can also be difficult to verify the accuracy of invoiced viewing times retrospectively. Utilizing blockchain technology can address these trust issues by providing a transparent and secure way for storing information.

8 Future Work

Various aspects can be improved upon this research, which forms a basis for future work. However, the first and foremost should be providing higher scalability. As seen inside the evaluation part, the transaction throughput time can slow down the entire system when working with real time data sets. Additionally, secured automation of the entire process should become a priority for research. The usage of blockchain technology has the potential to improve the future of digital advertisement by eliminating the need for trust between participants. Our demonstrator can be extended to cover a wider range of use cases, including the decentralization of traffic advertisements. For example, billboards located at bus/train stations or high-traffic areas can be owned by private individuals instead of companies. By utilizing blockchain technology, a Decentralized Autonomous Organization (DAO) can be established, allowing individuals to purchase shares of a billboard and a smart contract can allocate earnings generated from itself during the month or any other chosen time period.

This could significantly disrupt the advertising industry by shifting control from large companies to individuals who do not need to trust each other. However, the realization of such innovative ideas relies on intensive further research and development.

Acknowledgements The successful completion of this research would not have been possible without the help of our colleagues. We would like to express our gratitude to company Ruhrkraft for their support throughout the project. Their contribution in building the prototype, testing the software, and working in parallel with our team was highly appreciated. We would like to specially thank the Ministry of Economy, Industry, Climate Protection, and Energy in North Rhine-Westphalia, Germany for providing the program "Mittelstand Innovativ und Digital (MID)" and contributing financially to the construction of our prototype. Special thanks also to all of our colleagues at Fraunhofer and Blockchain Reallabor who were involved in the Blockchain research during this period. Your commitment to the project has made a significant impact on the success of this work.

References

1. J. Sutanto, E. Palme, C. Tan, C.W. Phang, Addressing the personalization–privacy paradox: an empirical assessment from a field experiment on smartphone users. MIS Q. **37**(4), 1141–1164 (2013)
2. K. Rahman, Applications of blockchain technology for digital marketing: a systematic review, in *Blockchain Technology and Applications for Digital Marketing* (2021), pp. 16–31
3. F. Masood, An overview of distributed ledger technology and its applications. Int. J. Comput. Sci. Eng. **6**(10), 422–427 (2018)
4. R. Kumar, Fraud detection: a review on blockchain (2022)
5. Saha, A study on the impact of online shopping upon retail trade business. IOSR J. Bus. Manag. **2**, 74–78 (2015)
6. N. Alexopoulos, J. Daubert, M. Mhlhuser, S.M. Habib, Beyond the hype: on using blockchains in trust management for authentication. InTrust-com/ BigDataSE/ICESS 2017, vol. 1 (IEEE, 2017), pp. 546–553
7. Z. Su, H. Wang, H. Wang, X. Shi, A financial data security sharing solution based on blockchain technology and proxy re-encryption technology, in *2020 IEEE 3rd International Conference of Safe Production and Informatization (IICSPI)* (Chongqing City, China, 2020), pp. 462–465. https://doi.org/10.1109/IICSPI51290.2020.9332563
8. R. Rajadevi, E.M.R. Devi, R.S. Latha, S. Harshini, K. Ajay, M. Abinash, Secured storing and sharing of medical records based on blockchain, in *2022 International Conference on Computer Communication and Informatics (ICCCI)* (Coimbatore, India, 2022), pp. 1–5. https://doi.org/10.1109/ICC-CI54379.2022.9741070
9. Swan, *Blockchain: Blueprint for a New Economy* (2015), ISBN 1491920491, 9781491920497
10. L.D. Nunzio, F. Fallucchi, M. Raso, P. Alessandra, N. Scarpato, Smarter city: smart energy grid based on blockchain technology. Int. J. Adv. Sci. Eng. In-Form. Technol. **8**(1), 298 (2018)

Ivan Dimitrov is a Ph.D. student and researcher at Fraunhofer FIT, specializing in the innovative field of smart contract wallets and account management on EVM-based blockchains. With a keen interest in the Ethereum Virtual Machine (EVM) ecosystem, Ivan is drawn to these chains due to their expansive community, diverse opportunities, and the pioneering advancements they enable in decentralized technology. His work aims to contribute to the evolution of secure, user-friendly blockchain solutions that leverage the flexibility of EVM platforms. Beyond his research, Ivan is an active member of the RWTH Blockchain Club, where he collaborates with other blockchain enthusiasts to explore, promote, and expand the reach of blockchain technology. Ivan's commitment to advancing EVM-based applications reflects his belief in the transformative potential of blockchain for finance, governance, and beyond. As a strong supporter of decentralized technology, he is dedicated to bridging the gap between cutting-edge research and real-world application, fostering a more accessible and secure blockchain environment for all.

The Impact of Blockchain on Transparency and Trust in Sustainable Agri-Food Supply Chains

Thuy Tien Nguyen Thi, Mandana Gharehdaghi, Maximilian Austerjost, and Axel T. Schulte

1 Climate Change Impact on Agriculture

The year 2023 broke a new record as the average global temperatures reached new heights ever since recordings started in 1940 [1]. Especially the agriculture sector is sensitive to climate change as yields and crops are influenced by increased temperatures, extreme weather events (drought, floods, wildfire) that could pose a threat to securing the global food supply in the future [2–4]. At the same time, agriculture is not only one of the main victims but also one of its key contributors. It emits around a quarter of global greenhouse gas emissions, requires substantial amounts of freshwater and the growing demand for food lead to large parts of forests and wildlands converted into agricultural land. This in turn results in a reduction of the world's biodiversity at an unparalleled rate, that threatens human well-being as it interferes with important ecosystem services such as crop pollination [5, 6].

Regenerative agriculture is one approach that is increasingly gaining popularity to mitigate the negative impact of the agriculture industry on land and climate [7, 8]. However, until the promises are realized, the life of farmers have already been negatively impacted by climate change through the loss of income among others [9]. End-consumers as well become more environmentally aware and demand information on the origin and quality of the products they buy. As a result, not only end-consumers but also retail customers place additional pressure on current supply chain networks by demanding more transparency and traceability that back up the sustainability claims on products, which in turn implies improved access to real information about the products [10, 11].

T. T. Nguyen Thi (✉) · M. Austerjost · A. T. Schulte
Fraunhofer Institute for Material Flow and Logistics IML, Dortmund, Germany
e-mail: thuy.tien.nguyen.thi@iml.fraunhofer.de

M. Gharehdaghi
University of Pannonia, Veszprém, Hungary

2 Product Certificates, Labels and Greenwashing in Agri-Food Supply Chains

Due to the on-going environmental challenges, sustainability and fair working conditions are becoming an increasingly important selection criteria when choosing a product. Studies show that end-consumers are willing to pay a premium for sustainable and fair products [12, 13] as well as for relevant product information, e.g., on the living conditions of the chicken when buying an egg [14]. It is therefore lucrative for manufacturers to advertise sustainability issues to promote a product for sale. However, it is often challenging for end-consumers to recognize how sustainable a product really is. Unfortunately, terms like "natural" or "from controlled cultivation" are not legally protected (unlike "organic" for example) and can be used in any context [15]. Internationally recognized labels and certifications from various organizations help to understand where and under what conditions food was produced. By using labels or certificates, the producers indicate to end-consumers as well as to wholesale and retail companies that they keep their promises. Such food certifications are often created by industry associations, NGOs or multi-stakeholder committees, which define the guidelines and specifications based on international conventions and agreements [16]. Examples of such conventions are International Organization for Standardization (ISO) guidelines, UN conventions, guidelines of the International Labour Organization (ILO) or Hazard Analysis and Critical Control Points (HACCP) [17–20]. The better-known certifications originate from testing organizations (e.g., Institut Fresenius or TÜV) as they are authorized to carry out audits and award certifications. Furthermore, there are many labels supplied by producer associations or individual retail chains [16].

By using certificates, labels and related associations, the producers try to reassure end-consumers during the buying decision process for sustainable products. The challenge that arises is, that distinguishing between truly sustainable products and companies on one side, and companies that only appear to follow a sustainable approach on the other, becomes more complicated due to the growing number of sustainability claims [13]. In general, the end-consumer considers multiple factors to base their trust and buying decision in, such as brand name, reputation of the retailer or producer, packaging, pricing, communication, the sales store and the certificates or labels with their related associations [21–27].

However, especially in the agrifood sector the effectiveness of labels is still open for discussion as end-consumers cannot directly verify whether the products fulfill the sustainability claims [28]. In fact, studies for organic food products show that end-consumers are usually not aware of the meaning of the label [23]. Some companies even follow misleading communication practices, in which they promise but cannot prove sustainable claims through concrete measures. This behavior is commonly known as *greenwashing*, which aims to form an unfounded overly positive belief among stakeholders about a company's environmental impact [7, 28, 23, 13]. Greenwashing negatively impacts consumer trust, that serves as an important prerequisite to develop a market for green and sustainable products [28, 29].

3 Blockchain: Enabling Technology for Transparency and Trust in Sustainable Food Supply Chains?

In literature, improving transparency and enhancing the security of supply chain networks are both viewed as crucial elements for effective supply chain management (SCM). However, since transparency necessitates the sharing of information among supply chain actors, increased openness results in less secretiveness. Hence, there is a trade-off between transparency and secrecy. The capacity to maintain secrecy and a refusal to divulge sensitive commercial data indicate a lack of trust. Trust is built when supply chain members continuously carry out their anticipated tasks and disclose correct information [30]. This would imply that volatile supply networks, with relatively short buyer–supplier relations, do not have the time to build up enough trust to comply with this aspect. Since transparency necessitates information sharing among supply chain members [31, 32], a lack of trust may considerably impede supply chain transparency.

This is where blockchain-technology could provide special benefits to SCM. The characteristics that link blockchain to the supply chain area are often trust, technology, traceability and transparency making blockchain a great fit with food supply chains [33]. Blockchain increases SCM transparency and traceability since time-stamped blocks may be constructed for transactions that follow the product's digital footprint and each transaction is completely auditable [34]. Two significant drivers of blockchain adoption include "creating more openness and visibility" as well as "enhancing procedures and decreasing costs" [35]. Jindong, a Chinese retailer, has partnered with Kerchin, a Mongolia-based beef producer, to use blockchain-technology to compile digital product information such as farm details, batch numbers, factory and processing data, expiration dates, storage temperatures, and shipping details that are digitally connected to trace every step of the food items' processing. Customers may use their system to trace information on frozen meat, such as a cow's breed, weight, and diet, as well as the location of farms, by scanning the QR code on the box [36]. The contribution of blockchain to sustainable SCM is greater confidence in an information supply chain movement. Customers are concerned about the environment and the safety of those involved, making blockchain an enabler of trust [37]. Trust can be defined as the interaction between expectations and the manifestation of those promises in the actual behavior of the people involved [38]. Therefore, the idea is that a blockchain should enable stakeholders to establish the quality of products in particular supply chains [39] that effectively contributes to sustainable activities including a circular economy, less waste, and lower emissions [40, 41].

The idea of blockchain-based sustainability food labels to overcome trust issues and fulfill the end-consumers demand on product information and transparency regarding the environment and social aspects has already been explored [42, 43]. Even though the creation of trust and transparency are main drivers of blockchain adoption in food supply chains [43, 44] the current applications are mainly limited to food retailers and producers that track their private-label food products. Furthermore, end-consumers are generally not aware and therefore need to be educated on the way how to use such blockchain-based solutions. The complexity of blockchain solutions represents one barrier to their understanding of the advantages and their limitations [42, 43]. This shows that besides growing interest from researchers and practitioners alike, there is still limited understanding of blockchain's concrete benefits and potential. Besides a large body of existing conceptual studies, discussion papers and literature reviews on the topic that assert the immense potential from a theoretical point of view, only few studies exist, especially in the agrifood domain that take an empirical approach [45]. Therefore, our research questions are as follows:

RQ 1: What different aspects of trust exist in supply chains and how does blockchain-technology support it?

RQ 2: What are the limits and the key challenges in achieving blockchain-based transparency and trust in sustainable agrifood supply chains?

4 Methodology

To conduct this type of research, the application of either qualitative or quantitative methods alone does not answer the research questions. For that reason, a stepwise qualitative-quantitative exploratory research method (mixed method) was adopted along with an analysis of existing blockchain use cases.

A blockchain use case database involving 317 cases, which focuses on the agriculture sector among others [46] served as the foundation for a systematic approach to search use cases that involve trust. The findings were collected into a matrix, identifying similarities and commonalities in trust understanding across different scenarios. This methodical process provided valuable insights into trust's foundational importance for blockchain adoption and its different meanings.

In the Mixed Methodology section, it was imperative to ensure alignment with the factors identified from the literature review. To achieve this, a total of 26 experts were engaged in the qualitative phase utilizing a semi-structured questionnaire, followed by the quantitative stage where data from the interviews were analyzed using Atlas.ti software to formulate a model grounded in the analyzed data. This step holds significant importance as it serves to validate the thorough coverage of essential factors within the scope of the research.

5 Results

Figure 1 presents the final result of a model for understanding blockchain-related trust in sustainable agrifood supply chains from the perspective of the end-consumer, which will be further detailed in the following sections.

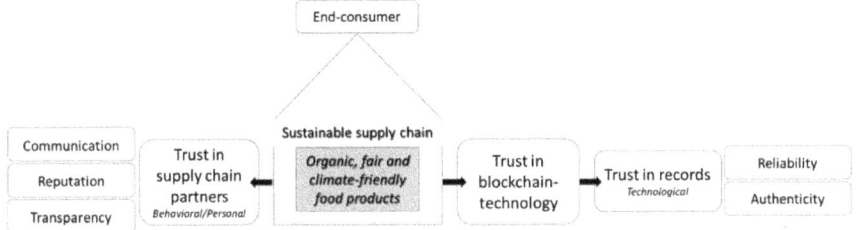

Fig. 1 Model for blockchain-related trust in sustainable agrifood supply chains [own representation based on [47]

5.1 Aspects of Trust Based on Blockchain Use Case Analysis

The use case database provided a short description for each application of which 40 were identified to include the word trust (two cases were excluded as trust was part of the company name). This however does not imply that other use cases do not consider trust as an important determinant just because it was not included in the short description. Rather, the keyword analysis shows that trust related to the use of blockchain encompasses several meanings and is therefore ambiguous (Fig. 2). First, transparency related to and as an enabler of trust is mentioned by half of the cases that shows the intricate relationship between trust and transparency to be implemented by blockchain-technology. Second, for some the motivation behind using blockchain is the creation of trust in the partners as a fundamental element of collaboration. Third, some mention the creation of trust by different means than transparency, e.g., verification, authentication, immutability or by providing a secure blockchain-based platform. Fourth, to signal trust and consequently improve their brand image towards the customers is another factor for companies to adopt the technology. Finally, contrary to the majority that see blockchain as an enabler of trust in the brand, process, or partners, two cases refer to the technology as an eliminator of trust to pave the way for trustless interactions as blockchain-technology replaces the need for trust. This however shifts the trust to the technology itself. In conclusion, there appears to be a broad application of the word trust that is actively communicated in blockchain use cases, which underlines the requirement for further clarification on the meaning of trust related to blockchain use cases.

5.2 Aspects of Trust Based on Expert Interviews

This part of the study presents the findings from the mixed method. In the qualitative section, relevant literature has further identified factors affecting trust (two aspects: *transparency* and *reputation*) in the supply chain of agri-food using blockchain-technology on both sides: advantages and disadvantages (Fig. 3). After extracting relevant factors from the viewpoint of the literature reviewed, we finalized our

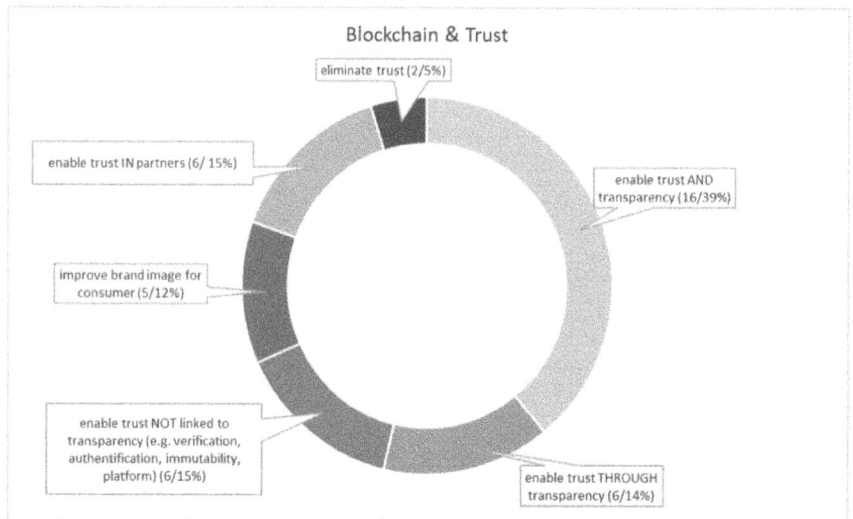

Fig. 2 Application area of blockchain-based trust

questionnaire structure and interviewed 26 experts from different European and Asian countries, including upstream and downstream actors with a background in blockchain, HRM, IT, the food supply chain, and end-consumers. In the next step, the quantitative part of the study—the result of the interview analysis using Atlas.ti software—shows (Fig. 3) how different backgrounds and interests lead to different priorities when ranking transparency, trust, blockchain role, sustainability's aspects importance, and reputation. *Transparency* in trust classification has the highest score for related interviewees with 61.5%. In the sustainability field, *social aspects* related to actors are the most important topics (49%), and the least important part is *economical aspects*, with only 9.5% focusing mainly on investment as the first step to start using blockchain-technology and how they deal with it by using natural resources or sponsorship. The results of using blockchain technology show only 23.7% of the *disadvantages* related to energy consumption and investment issues that are controllable based on the tech-interviewees advice. Since *transparency* scored 61.5% on a 100-point scale and *reputation* 38.5%, managers may see these as an instrument to obtain trust and a way in which the use of blockchain-technology in sustainable agri-food improves trust among actors. It answers the question of whether trust would be a precondition for using blockchain or if blockchain use would be a precondition for raising trust through transparency. The study shows how the total decision flow takes place in individual companies as well as in networks of companies ("actors"), where the power of reputation may play a decisive role in the adoption of blockchain.

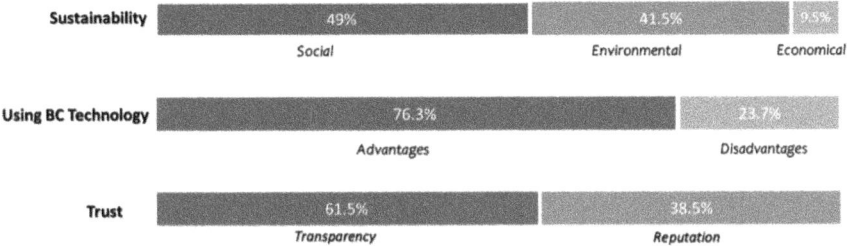

Fig. 3 Expert interview results

6 Discussion

Another model for blockchain-based trust for sustainability in the sector of metal supply chains confirms that trust needs to be viewed from different perspectives when applying blockchain [47]. Based on that understanding, which served as the foundation for the presented model in Fig. 1, trust in sustainable supply chain partners refers to aspects such as reputation, communication and transparency, whereas the trust that is provided by the blockchain-technology refers to the trust in the records stored on it. Trust in records refers on one side to the reliability of the given statement as a matter of fact, how it was created, who created it and how it is subsequently maintained, i.e., reliability represents the trust in the truth of the recorded facts. Additionally, trust in records also refers to authenticity, which in comparison to reliability, shifts the focus from the content of the record to the trustworthiness of the record itself and that it is free from tampering and corruption [48].

The use of blockchain to provide trust in sustainability information therefore can be distinguished from a technological perspective in two ways. This includes the truthfulness of the information that is recorded on the blockchain (reliability) and the blockchain itself as a medium for tamperproof data storage (authenticity). However, as depicted in Fig. 4, even though the blockchain as a means of record can be trusted, for the case of sustainable agrifood supply chains it must be considered that the reliability of the data before being put on the blockchain depends entirely on the events in the physical world, i.e., the source of the recorded data and how it was created. Blockchain-oracles are entities to provide external data to a blockchain and the blockchain-oracle problem summarizes the challenges with guaranteeing the reliability of external off-chain and physical data. However, using external data sources introduces again a single-point of failure when the data entered from the external data provider is not correct [49].

Overall, the use of blockchain-technology by food producers and retailers serves to signal trust to the end-consumer in sustainability claims by avoiding for instance greenwashing. This blockchain-based trust can be communicated by retailers to the end-consumer among others in the form of blockchain-based labels to show transparency in the supply chains and improve a company's reputation towards the end-consumer, which together serve as relevant determinants of blockchain-adoption.

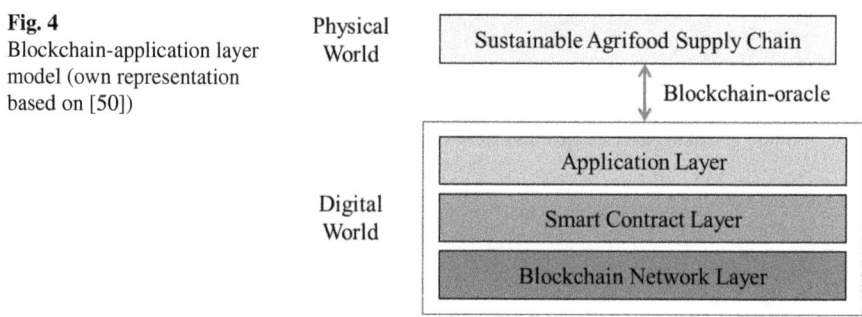

Fig. 4 Blockchain-application layer model (own representation based on [50])

However, as mentioned previously the trust lies not so much in the data itself that is stored on the blockchain but rather serves as an additional communication signal by companies to improve their reputation.

7 Conclusion

Altogether, the proposed research agenda with the associated step-by-step implementation appears to be a successful method for answering the two research questions posed. Literature shows that blockchain technology certainly has the potential to facilitate secure information transfer, meet demands for transparency and traceability, and build trust among actors. However, it also shows that the expected usefulness is determined by human behavior rather than technical aspects (as exemplified by concerns about data reliability).

In our research agenda, we present a refined perspective rooted in empirical evidence regarding the factors influencing managers' decisions to adopt or abstain from adopting blockchain-technology. We accomplish this by unveiling a model derived from our case study and interview findings. This model not only sheds light on the intricacies of managerial choices but also empowers us to formulate practical recommendations. Furthermore, it allows for the anticipation of human behavior, particularly the actions of key stakeholders, which significantly impact the successful implementation of blockchain within the agrifood supply chain. Transparency and reputation play a vital role in building trust in sustainable supply chains. Even though trust cannot be fully captured by blockchain-technology alone, it can serve to complement it.

Acknowledgements This work has received funding from the European Union's Horizon 2020 research and innovation programme under the Grant agreement No. 101060534.

References

1. C. van Campenhout, 2023 on track to become hottest year on record, says EU climate service. In *Reuters Media*, 6 Oct 2023 (2023). Available online at https://www.reuters.com/business/environment/2023-track-become-another-record-breaking-year-temperature-hits-new-high-2023-10-05/, checked on 18 Nov 2023
2. N.K. Arora, Impact of climate change on agriculture production and its sustainable solutions. Environ. Sustain. **2**(2), 95–96 (2019). https://doi.org/10.1007/s42398-019-00078-w
3. EPA: Climate Change Impacts on Agriculture and Food Supply (United States Environmental Protection Agency, 2022). Available online at https://www.epa.gov/climateimpacts/climate-change-impacts-agriculture-and-food-supply. Updated on 16 Nov 2023, checked on 18 Dec 2023
4. G. Malhi, M. Kaur, P. Kaushik, Impact of climate change on agriculture and its mitigation strategies: a review. Sustainability **13**, 1318 (2021). https://doi.org/10.3390/su13031318
5. G. Ceballos, P.R. Ehrlich, A.D. Barnosky, A. García, R.M. Pringle, T.M. Palmer, Accelerated modern human-induced species losses: entering the sixth mass extinction. Sci. Adv. **1**(5), e1400253 (2015). https://doi.org/10.1126/sciadv.1400253
6. H. Ritchie, P. Rosado, M. Roser, Environmental Impacts of Food Production. Our World in Data (2022). Available online at https://ourworldindata.org/environmental-impacts-of-food. Checked on 18 Nov 2023
7. FAIRR: The Four Labours of Regenerative Agriculture. Paving the way towards meaningful commitments (2023). Available online at https://go.fairr.org/FAIRR_Report_The_Four_Labours_of_Regenerative_Agriculture_2023. Checked on 6 Nov 2023
8. T. Kurth, B. Subei, P. Plötner, F. Bünger, M. Havermeier, S. Krämer, *The case for regenerative agriculture in Germany -and beyond*, (Boston Consulting Group; Nature And Biodiversity Conservation Union, 2023). Available online at https://www.nabu.de/imperia/md/content/nabude/landwirtschaft/230323-the_case_for_regenerative_agriculture_longversion-engl.pdf. Checked on 16 Nov 2023
9. Bayer, Farmer Voice (2024). Available online at: https://www.bayer.com/sites/default/files/farmervoice2024-report-digital-final.pdf
10. W. Nikolakis, L. John, H. Krishnan, How blockchain can shape sustainable global value chains: an evidence, verifiability, and enforceability (EVE) framework. Sustainability **10**(11) (2018). https://doi.org/10.3390/su10113926
11. A. Scuderi, B. Pecorino, Protected designation of origin (PDO) and protected geographical indication (PGI) Italian citrus productions. Acta Hortic. **1065**, 1911–1918 (2015). https://doi.org/10.17660/ActaHortic.2015.1065.245
12. S. Dekhili, M.A. Achabou, Price fairness in the case of green products: enterprises' policies and consumers' perceptions. Bus. Strat. Environ. **22**(8), 547–560 (2013). https://doi.org/10.1002/bse.1763
13. R. Torelli, F. Balluchi, A. Lazzini, Greenwashing and envi-ronmental communication: effects on stakeholders' perceptions. Bus. Strat. Environ. **29**(2), 407–421 (2020). https://doi.org/10.1002/bse.2373
14. M. Gharehdaghi, D.-J. Kamann, Behavioural and organisational factors determining blockchain adoption. CJAST **42**(7), 24–41 (2023). https://doi.org/10.9734/cjast/2023/v42i74077
15. Ehlert: At what point is a food "organic"? Requirements and legal info (2020). Available online at https://www.ehlert-gmbh.de/en/blog/news/organic-food-when-is-a-food-considered-organic. Checked on 6 May 2024
16. U. Eberle, A. Spiller, T. Becker, A. Heissenhuber, I.U. Leonhäuser, A. Sundrum, Political Strategy for Food Labelling. Joint Statement of the Scientific Advisory Boards on Consumer and Food Policy and on Agricultural Policy at the Federal Ministry of Food, Agriculture and Consumer Protection (Federal Ministry of Food and Agriculture, 2012). Available online at https://www.bmel.de/SharedDocs/Downloads/EN/_Ministry/Scientific_Advisory_Board-Food_Labelling.pdf?__blob=publicationFile&v=2. Checked on 6 May 2024

17. FDA: HACCP Principles & Application Guidelines (Food and Drug Administration, 2024). Available online at https://www.fda.gov/food/hazard-analysis-critical-control-point-haccp/haccp-principles-application-guidelines. Checked on 6 May 2024
18. ILO: Decent Work for Food Security and Resilient Rural Livelihoods. Decent work in the rural economy policy guidance notes (2019). Available online at https://www.ilo.org/media/405646/download.
19. ISO: Food for the future (International Organization for Standardization, 2021). Available online at https://www.iso.org/news/ref2732.html. Checked on 6 May 2024
20. UNFSS, Food and Agriculture Organization (FAO) (2019). Available online at https://unfss.org/home/projects-of-partner-agencies/food-and-agriculture-organization-of-the-united-nations-fao. Checked on 6 May 2024
21. G. Carrà, I. Peri, G. Vindigni, Diversification strategies for sustaining small-scale fisheries activity: a multidimensional integrated approach. Rivista Di Studi Sulla Sostenibilita 79–99 (2014). https://doi.org/10.3280/RISS2014-001006
22. Y.-P. Lin, J.R. Petway, J. Anthony, H. Mukhtar, S.-W. Liao, C.-F. Chou, Y.-F. Ho, Blockchain: the evolutionary next step for ICT E-agriculture. Environments **4**(3) (2017). https://doi.org/10.3390/environments4030050
23. F. Perrini, S. Castaldo, N. Misani, A. Tencati, The impact of corporate social responsibility associations on trust in organic products marketed by mainstream retailers: a study of Italian consumers. Bus. Strat. Environ. **19**(8), 512–526 (2010). https://doi.org/10.1002/bse.660
24. H. Subramanian, Decentralized blockchain-based electronic marketplaces. Commun. ACM **61**(1), 78–84 (2017). https://doi.org/10.1145/3158333
25. A. Vastola, P. Zdruli, M. D'Amico, G. Pappalardo, M. Viccaro, F. Di Napoli et al., A comparative multidimensional evaluation of conservation agriculture systems: a case study from a Mediterranean area of Southern Italy. Land Use Policy **68**, 326–333 (2017). https://doi.org/10.1016/j.landusepol.2017.07.034
26. G. Vindigni, I. Peri, A. Urso, Biodiversity and ecosystem approach as political discourse. Qual. Access Success **14**, 11–17 (2013)
27. C. Zarbà, S. Bracco, A. Zarba, The progress and competitive-ness of SMEs from SBA up to Horizon 2020. A new approach of innovation and technological advancement. Issue Suppl. **1**(15), 202–206 (2014)
28. F. Henglein, Modern blockchain technology to become our weapon in the green transition (2023). Available online at https://science.ku.dk/english/press/news/2023/modern-blockchain-technology-to-become-our-weapon-in-the-green-transition/. Updated on 13 Oct 2023, checked on 18 Dec 2023
29. K. Nuttavuthisit, J. Thøgersen, The importance of consumer trust for the emergence of a market for green products: the case of organic food. J. Bus. Ethics. **140**, 323–337 (2017). https://doi.org/10.1007/s10551-015-2690-5
30. M. Zhang, B. Huo, The impact of dependence and trust on supply chain integration. Int. J. Phys. Distrib. Logist. Manage. **43**(7), 544–563 (2013). https://doi.org/10.1108/IJPDLM-10-2011-0171
31. R.C. Lamming, N.D. Caldwell, D.A. Harrison, W. Phillips, Transparency in supply relationships: concept and practice. J. Supply Chain Manage. **37**(3), 4–10 (2001). https://doi.org/10.1111/j.1745-493X.2001.tb00107.x
32. T.R. Morgan, R.G. Jr Richey, A.E. Ellinger, Supplier transparency: scale development and validation. IJLM **29**(3), 959–984 (2018). https://doi.org/10.1108/IJLM-01-2017-0018
33. M. Pournader, Y. Shi, S. Seuring, S.L. Koh, Block-chain applications in supply chains, transport and logistics: a systematic review of the literature. Int. J. Prod. Res. **58**(7), 2063–2081 (2020). https://doi.org/10.1080/00207543.2019.1650976
34. Y. Wang, J.H. Han, P. Beynon-Davies, Understanding block-chain technology for future supply chains: a systematic literature review and research agenda. SCM **24**(1), 62–84 (2019). https://doi.org/10.1108/SCM-03-2018-0148
35. R. van Hoek, Exploring blockchain implementation in the supply chain. IJOPM **39**(6/7/8), 829–859 (2019). https://doi.org/10.1108/IJOPM-01-2019-0022

36. N. Kshetri, 1 Blockchain's roles in meeting key supply chain management objectives. Int. J. Inf. Manage. **39**, 80–89 (2018). https://doi.org/10.1016/j.ijinfomgt.2017.12.005
37. S. Saberi, M. Kouhizadeh, J. Sarkis, L. Shen, Blockchain technology and its relationships to sustainable supply chain management. Int. J. Prod. Res. **57**(7), 2117–2135 (2019). https://doi.org/10.1080/00207543.2018.1533261
38. K. Blomqvist, P. Stahle (2004): Trust in technology partnerships. in *Trust in Knowledge Management and Systems in Organizations* eds. by M.L. Huotari, M. Iivonen, M.-L. Huotari: (Idea Group Publ, Hershey, Pa, 2004), pp. 173–199
39. R. Adams, B. Kewell, G. Parry, Blockchain for good? digital ledger technology and sustainable development goals. in *Handbook of Sustainability and Social Science Research* eds. by W. Leal Filho, R. Marans, J. Callewaert: (World Sustainability Series. Springer, Cham, 2018). https://doi.org/10.1007/978-3-319-67122-2_7
40. V.K. Manupati, T. Schoenherr, M. Ramkumar, S.M. Wagner, S.K. Pabba, R. Inder Raj Singh, A blockchain-based approach for a multi-echelon sustainable supply chain. Int. J. Prod. Res. **58**(7), 2222–2241 (2020). https://doi.org/10.1080/00207543.2019.1683248
41. D. Zhang, Application of blockchain technology in incentivizing efficient use of rural wastes. A case study on Yitong system. Energy Procedia **158**, 6707–6714 (2019). https://doi.org/10.1016/j.egypro.2019.01.018
42. C. Contini, F. Boncinelli, G. Piracci, G. Scozzafava, L. Casini, Can blockchain technology strengthen consumer preferences for credence attributes? Agric. Food Econ. **11**(1), 27 (2023). https://doi.org/10.1186/s40100-023-00270-x
43. H. Treiblmaier, M. Garaus, Using blockchain to signal quality in the food supply chain: the impact on consumer purchase intentions and the moderating effect of brand familiarity. Int. J. Inf. Manage. **68**, 102514 (2023). https://doi.org/10.1016/j.ijinfomgt.2022.102514
44. H. Xiong, T. Dalhaus, P. Wang, J. Huang, Blockchain technology for agriculture: applications and rationale. Front. Blockchain **3** (2020). Article 7. https://doi.org/10.3389/fbloc.2020.00007
45. S.L. Bager, C. Singh, U.M. Persson, Blockchain is not a silver bullet for agro-food supply chain sustainability: insights from a coffee case study. Curr. Res. Environ. Sustain. **4**, 100163 (2022). https://doi.org/10.1016/j.crsust.2022.100163
46. TRUSTyFOOD, Database Preliminary Analysis (2023). Available online at https://www.trustyfood.eu/framework-of-services/dashboard/. Updated on 7 Mar 2023, checked on 18 Dec 2023
47. A. Batwa, A. Norrman, A. Arvidsson, How blockchain interrelates with trust in the supply chain context: insights from tracing sustainability in the metal industry. in *Hamburg International Conference of Logistics (HICL) 2021* (epubli, 2021), pp. 329–351. Available online at https://tore.tuhh.de/entities/publication/00338085-7fe0-482e-9769-9d8677454913
48. V.L. Lemieux, Trusting records: is Blockchain technology the answer? RMJ **26**(2), 110–139 (2016). https://doi.org/10.1108/RMJ-12-2015-0042
49. G. Caldarelli, C. Rossignoli, A. Zardimi, Oracle trust models for blockchain-based applications. An early standardization. in *2023 IEEE International Conference on Technology Management, Operations and Decisions (ICTMOD)*, pp. 1–6, IEEE (2023). https://doi.org/10.1109/ICTMOD59086.2023.10438142
50. E.P. Kechagias, S.P. Gayialis, G.A. Papadopoulos, G. Papoutsis, An ethereum-based distributed application for enhancing food supply chain traceability. Foods **12**(6), 1220 (2023). https://doi.org/10.3390/foods12061220

Thuy Tien Nguyen Thi works as a research associate at the Fraunhofer Institute for Material Flow and Logistics IML to explore and implement blockchain-technology for companies and their supply chains. In national and international research projects, her current focus is on blockchain for the creation of sustainable food supply chains. Other areas of interest are the digitalization of international trade processes, experimentation spaces and the coordination of complex supply chains using key technologies such as Blockchain & AI. She is actively involved in Web3, corporate and political groups to promote collaboration between them and thereby advance the adoption of blockchain for businesses. She likes to keep her knowledge on the rapid evolving technology up to date by participating in hackathons and programming bootcamps. Her academic background includes a master's degree in International Business Administration focusing on Supply Chain Management & Logistics from European University Viadrina and one in Blockchain & Distributed Ledger Technologies from the University of Applied Sciences Mittweida.

Mandana Gharehdaghi , originally from Iran, earned her PhD in *Blockchain Technology Adoption in Supply Networks* from the University of Pannonia in Hungary. Since beginning her doctoral journey in 2021 under the supervision of Professor Dirk-Jan F. Kamann, her research has explored the transformative impact of blockchain on supply networks across various industries including food, fashion, pharmacy and automotive focusing on both micro- and meso-levels dynamics that introduce the 3 Arenas Model. She has published seven scholarly works in leading academic journals and she is an active member of IPSERA and the DLT Talents communities. She approaches technology adoption through a behavioral lens, emphasizing the critical role of individuals as decision-makers whose actions drive organizational change and broader societal innovation. Recently, she has begun working on the **AI-Blockchain Nexus (AIB Nexus)** at the organizational level, aiming to uncover how the integration of these technologies can enhance decision-making, transparency, and sustainable innovation in complex supply ecosystems.

Maximilian Austerjost is a specialist in industrial engineering and digital transformation in value-added networks. From 2014 to 2020, he was a research associate at the Chair of Corporate Logistics at TU Dortmund University. His research and teaching focused on design principles for resource-efficient intralogistics, IT solutions for the sharing economy in production and warehousing, and realising IT ecosystems to facilitate knowledge exchange between companies. He completed his PhD in this area with distinction. Since 2021, he has worked as Head of Blockchain at the Fraunhofer Institute for Material Flow and Logistics IML in Dortmund. In this role, he and his team develop blockchain-based application software for sustainable, decentralised logistics networks. Since 2022, he has also been the research coordinator for Enterprise Logistics and a member of the Fraunhofer IML Scientific Board. Most recently, in 2024, he became deputy head of department.

Axel T. Schulte is a Department Head in the area of Enterprise Logistics at the Fraunhofer Institute for Material Flow and Logistics IML. This applied research performed by Axel and his team of 20 + full-time researchers represents the increasing demand for the stronger incorporation of digital technologies and solutions in managing todays' global Supply Chains. The application of Blockchain, Artificial Intelligence (AI) and Internet of Things (IoT) technologies are at the center of the current research. In addition, the department is responsible for the transfer of terrestrial logistics solutions to the space exploration sector.

Axel joined Fraunhofer in 2013 after working for nearly 15 years in the Procurement and Supply Chain area. His professional activities during this time include various international assignments in Switzerland, Russia, and South America.

In addition, until 2022, Axel held an Associate Professorship at St. Petersburg State University in St. Petersburg, Russia and was heading the Research Center for International Logistics CIL.

Prior to his professional career Axel Schulte gained an engineering degree and a doctorate degree in Business and Economics, both at the Technical University of Munich.

Metaverse

Shared Manufacturing

Patrick Stuckmann-Blumenstein, Larissa Krämer, Dominik Bons, Patrick Keitzl, and Eugen Burov

1 Introduction

The manufacturing industry has recently faced obstacles, such as increased competition, shorter life cycles, and crises, leading to uncertainty. With the emergence of the internet and affordable automation devices, companies can analyze processes in detail and collect a vast amount of data, often referred to as the new oil due to its value for future technologies. In this context, cooperation is essential, as companies often need help in tackling these challenges. High investments in machinery, uncertainty regarding the market demand, shorter innovation cycles and location-specific risks such as supply difficulties, political problems or natural disasters force traditional manufacturing processes to undergo a paradigm shift. For example, during the COVID-19 crisis manufacturers relying on overseas suppliers experienced considerable difficulties to overcome supply shortages [1].

Additionally, sustainability has become vital in recent years due to its potential to address environmental concerns by reducing resource consumption, minimizing pollution, and improving long-term economic viability. Adopting sustainable practices helps manufacturers meet changing regulatory requirements and consumer

---

P. Stuckmann-Blumenstein (✉) · L. Krämer · D. Bons · P. Keitzl
TU Dortmund, Dortmund, Germany
e-mail: patrick.stuckmann@tu-dortmund.de

L. Krämer
e-mail: larissa.kraemer@tu-dortmund.de

D. Bons
e-mail: dominik.bons@tu-dortmund.de

P. Keitzl
e-mail: patrick.keitzl@tu-dortmund.de

E. Burov
Fraunhofer IML, Dortmund, Germany
e-mail: eugen.burov@iml.fraunhofer.de

expectations and contributes to a sustainable industrial landscape. These factors have led to the reorganization of procurement priorities, in which delivery reliability and the ecological footprint have become more important in addition to costs. Logistics is crucial in fulfilling these goals and relies on data availability and information technology. The efficiency of logistics processes also depends on the availability and capabilities of the technology.

The emergence of Web3 technology allows to close the gap between the virtual and physical world, enabling real-time visualization and management of logistics processes. Collected data, such as AI-controlled processes, can be analyzed or provided for further usage. However, conventional systems cannot keep up with the necessary depth of collaboration due to their centralized approaches and resulting dependencies. Web3 offers a promising approach to fulfil the requirements as an enabler for decentralized and collaborative logistics processes [2]. One form of collaborative and decentralized logistics process is balancing resources in value-added networks summarized under the term shared manufacturing.

The shared manufacturing paradigm is one manifestation of shared resource utilization, illustrated in Fig. 1. It is a platform-based approach that reverberates across the entire production chain, primarily influencing business-to-business-related business models [3]. This chapter provides a condensed overview of shared manufacturing in Web3. First the understanding of shared manufacturing in the context of the sharing economy is examined, followed by a discussion of the role of Web3 in the field of sharing approaches and an explanation of various stages of development. In the following sections, areas of application in the field of supply chain management and intralogistics are presented. The chapter concludes with a brief exploration of emerging trends, limitations and a discussion of prospects.

2 Understanding Shared Manufacturing in the Context of the Sharing Economy

The paradigm of shared production is a subset of the sharing economy and was first described in 1990 [4, 5]. Although the origin of the sharing economy term is unclear, it is often attributed to Bootsmann and Rogers in 2010 [6]. Unlike the traditional concept of sharing, the sharing economy explicitly embodies a non-zero-sum game and emphasizes the commercial exchange of goods and services coordinated via platforms without any transfer of ownership [6]. The cooperative use of production resources is based on platforms on which users can assume the role of consumers, prosumers (both producers and consumers), or producers [5]. Participants can benefit from the economies of scale and greater flexibility offered by the ecosystem. Analogous to biological ecosystems, different actors work together in these ecosystems to produce different products using local resources and recycling [6].

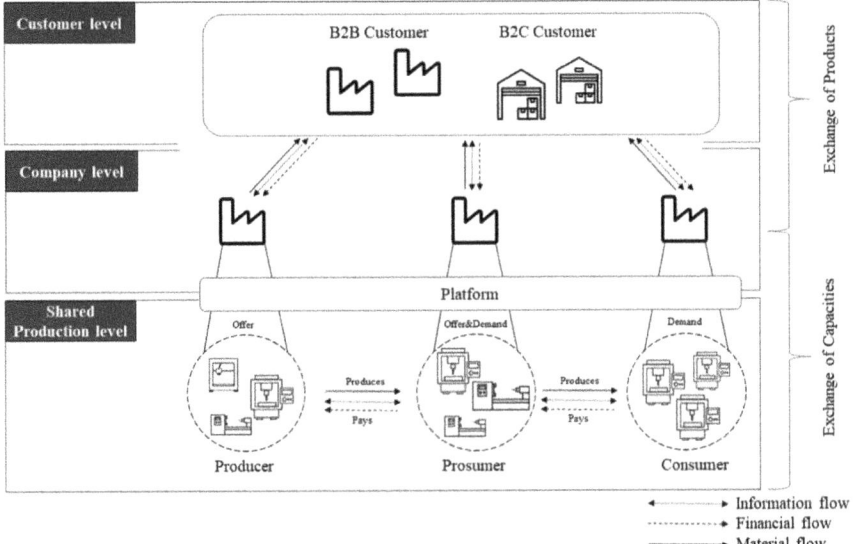

Fig. 1 Shared manufacturing in a nutshell

These principles enable small and medium enterprises (SMEs) to compete with global players in a decentralized and flexible manner [7]. Consequently, digital platforms need to be integrated into business processes to achieve a high degree of automation and to build trust among participants through visual information and key performance indicators. The digital integration of machines using sensors and actuators is a prerequisite for participation in a shared manufacturing network. Machines can be integrated either by direct connection to the platform or indirectly through the integration of an operator.

The following section focuses on the technological requirements for shared manufacturing and briefly overviews Web3, the metaverse, and blockchain in this context.

3 Web3, Metaverse and Blockchain as Potential Spaces for Sharing Approaches

Implementing shared manufacturing concepts successfully in a company requires information technologies to create data management infrastructures and suitable working environments. The decentralized basis of shared manufacturing increases the need for an infrastructure that meets the requirements for transparency, data provision and data privacy. As a result, the terms Web3, metaverse, and blockchain are often mentioned among possible solutions.

Blockchain is a digital, decentralized, and distributed ledger that logs transactions chronologically, aiming to create permanent and tamper-proof records [8]. This technology relies on cryptographic methods, game theory, and graph theory to achieve this tamper resistance and to guarantee a trustworthy transaction history. All participants can view the transaction log, providing transparency and building trust in the peer-to-peer network. Blockchain networks are decentralized with real-time access, utilizing a consensus mechanism to ensure node agreement and prevent data tampering. In addition, smart contracts, i.e., computer programs embedded in blockchain networks, automatically execute a predetermined code when certain conditions are fulfilled [9]. Smart Contracts ensure the timely execution of activities and the automation of processes. Consequently, distributed ledger technologies such as the blockchain can provide the basis for data management within the sharing economy [10].

With the rising importance and popularity of blockchain, the topic of decentralization comes more into focus. The term Web3 is increasingly used in the literature and represents further development of today's internet. The central thesis is that the centralization of the internet will not comply to social standards in the long term. An interoperable system that benefits all users rather than individual operators should be created [11]. Following Gan et al. [12], the following Table 1 outlines the different phases of web development and their main differences.

The first applications on the Internet, which defined the Web 1.0 era, were made up of static pages on which visitors could only perform essential functions such as reading and clicking [12]. As Web 2.0 emerged, it marked a more interactive development that allowed users to create content actively. Whether the user data is provided voluntarily or not, it initially belongs to the respective Web 2.0 platforms, enabling revenue maximization without any corresponding benefit for the users [13]. In contrast to Web 2.0, in Web 3.0 data sovereignty is transferred to the users, including the power to determine the dissemination of their data [13]. This is made possible by the decentralized hosting of Web 3.0 applications in a peer-to-peer network. This generation of the web enables better networking and processing of data [14]. Web3 emerges as the latest evolutionary phase of the Internet, evolving in parallel with Web 3.0. Web 3.0 and Web3 may both stem from the Semantic Web, which essentially aims to enhance data interoperability and comprehension among

Table 1 Differences among stages of webs based on [12]

Web	Architecture	Characteristics	Benefit distribution
Web 1.0	centralized	Host-generated content, host-generated authority	Platform monopoly
Web 2.0	centralized	User-generated content, host-generated authority	Profit-sharing (plattforms and users)
Web 3.0	decentralized	User-generated content, user-generated authority	Peer-to-peer
Web3	decentralized	User-generated content, user-generated authority	Smart contract

machines, but they diverge significantly in their core principles and applications [12]. While Web 3.0 primarily revolves around the concept of smart input terminals and decentralized hosting in a peer-to-peer network, Web3 emphasizes decentralized governance utilizing blockchain technology, with a focus on economic elements like non-fungible tokens and value exchange. Additionally, Web3 places greater emphasis on trust, relying on stable reputation systems to ensure reliability, whereas Web 3.0 prioritizes data ownership by users and utilizes mature distributed technologies to uphold decentralized thinking.

As Web3 continues to redefine the landscape of the Internet with its emphasis on decentralized governance and blockchain technology, it sets the stage for the next frontier: the Metaverse, a virtual shared space where users can interact in real-time, transcending traditional boundaries of space and identity [15]. Transitioning to the industrial sphere, the concept of the Industrial Metaverse emerges, wherein it functions as a system of digital twins depicting various industrial scenarios, offering unparalleled opportunities for simulation, collaboration, and innovation within industrial contexts. In the industrial metaverse, humans can interact with each other and with machines to make adjustments in production or to simulate and optimize industrial processes [15]. The users should be able to work and communicate in a virtual environment similar to their actual industrial environment, as it is more convenient and cost-effective. This environment, which aims to map entire value creation networks digitally, offers significant opportunities for the sharing economy, particularly in collaborative process control and joint product development. The industrial metaverse also offers the opportunity to integrate expertise for problem-solving or training regardless of location [16]. It is also possible to share production expertise in the industrial metaverse as part of the sharing economy.

4 Application Areas of Shared Manufacturing

Given the dynamic development in production and logistics, the convergence of both fields is an attractive environment for technological and operational developments. As a result, new forms of cooperation and business transactions will become possible. Although this has positive aspects, such as facilitating the short-term balancing of demand peaks in shared manufacturing, it also entails additional risks that affect more than one network partner. An example of this is the emergence of ad hoc supply chains during the COVID-19 pandemic [17], which combines the rapid adaptation of products and production resources with the fast, demand-driven distribution of these products. The continuous progress of crisis adoption requires a reconsideration of the planning and execution of production and logistics, with the transition from a "single company paradigm" to a "collaborative network paradigm" [6, 18]. This evolving collaboration model also requires a new management approach (see Chapter 1.3) which involves establishing network rules or robust and sustainable network governance [15]. Decentralized technologies are the primary enablers for these developments. The integration of both aspects makes it possible to create a networked production and logistics area encompassing all facets from production to network level, as all data, processes, and events pass through these levels.

4.1 Supply Chain Management Perspective

As value chains have become increasingly complex, new challenges have arisen. Companies are confronted with a lack of transparency and trust. Collaboration along the supply chain and in ecosystems is essential for companies to address these hurdles and increase their resilience to crisis scenarios.

Web3 serves as a tool for SME to participate in decentralized production ecosystems on a global scale. Conventional platforms are usually based on centralized structures, which leads to dependencies between participants. As a result, these platforms are often controlled by larger companies that have the necessary market influence to implement their systems successfully. In contrast, decentral systems based on peer-to-peer systems, supported by a trusted ledger, enable users to gain control over their data. The availability of relevant data is a prerequisite for automation and autonomy and forms the basis for shared manufacturing [19]. Collaboration is no longer limited to the value creation process but can include vertical and horizontal connections within networks, culminating in an autonomous production ecosystem [19]. This perspective has resulted in the emergence of blockchain, International Data Spaces (IDS), and digital twins. Blockchain, a decentralized and reliable layer, promises to ensure data consistency and transparency between ecosystem participants. The blockchain-based system is complemented by IDS, which serves as a secure and standardized framework for data exchange [20]. Digital supply chain twins are federated to monitor processes and machines using data provided by blockchain or IDS. Integrating IoT devices and cyber-physical systems (CPS) allows each product to be traced from its origin through each stage of the supply chain.

This transparency guarantees the authenticity of products and facilitates monitoring stock levels in real-time, reducing the risk of stock-outs and overstocking. Products, machines, or intellectual property ownership can be tokenized, creating a tradable and traceable link between data and assets [20]. Machine or process data can be verified by the data associated with the token, simplifying the certification process. Smart contracts can automatically execute the agreements set out in the contract based on the token's information. The digital linkage of information and legal consequences ensures that agreements are validated and checked even in trustless environments.

This information linkage can be applied in insurance, for example, in the case of transportation or the provision of a service agreed in advance, increasing transparency and enabling the automatic processing of insurance claims [21]. As a digital product passport, such real-time evidence can be provided and consolidated with other production details, including the ecological footprint or proof of quality. In conjunction with all data across the production life cycle, this also paves the way for a circular economy [22]. In shared manufacturing, the simplification of cross-border value networks depends on the management of cross-border transportation based on a valid (production) history.

The decentralized structure of Web3 and shared manufacturing offers a feasible solution to reduce the impact of crisis situations. Through the network balancing effects of an ecosystem, manufacturer can fall back on alternative production sites in case of temporary challenges. Possible challenges could include delivery interruptions, staff shortages, or machine breakdowns. This enables cost savings and the reduction of emergency stocks while maintaining high resilience in crises. Cooperation between different regions helps offset the impact of weather-related and political crises on production and ensures that supply guarantees are maintained. Effectively functioning ecosystems can create location-related advantages and possibly offset associated disadvantages. This also includes energy bottlenecks that may arise from the growing dependence on renewable energies [23]. Shared manufacturing can play a role in balancing electricity grids by providing access to supra-regional, transnational or even cross-continental resources [6].

Illustrated in Fig. 2 is a manifestation of a decentralized capacity exchange, where multiple manufacturers are integrated into an ecosystem. Each manufacturer is linked to decentralized services through an individual Eclipse Dataspace Connector (EDC) and a Blockchain Connector (BCC). Platforms can also connect with these decentralized services, eliminating a central dependency on a single platform. Additionally, the central systems of the manufacturer (ERP, PDM, MES) can engage directly with the ecosystem, lowering entry barriers for other participants to join the network. The example shown was designed to meet the specific requirements of production networks, consisting of standardized machines for additive manufacturing and CNC or turning processes. Integrating a file storage system such as IPFS, which is similar to blockchain, enables the provision of templates or the inclusion of end users who do not have an EDC connector.

4.2 Shared Manufacturing from an Intralogistics Perspective

From an intralogistics perspective, the shared manufacturing concept enables several companies to work together within a single production space. Companies in joint production are often unfamiliar without a shared common basis for trust. They may even be competitors or have conflicting interests, resulting in a hesitancy to share production-specific data. This is counteracted by the need for more data in manufacturing to effectively utilize a shared manufacturing system by predicting over or under-capacity. These intertwined challenges create opportunities for fraudulent behavior and, at the very least, lead to coordination problems between the relevant players. Besides, as a result of shared manufacturing, the variety of product designs in manufacturing can increase while the number of units per variant decreases [24].

The challenges of shared manufacturing require a production system that is characterized by a high degree of flexibility. A viable solution to this is a matrix production system in which workstations are configured to perform at least two different work

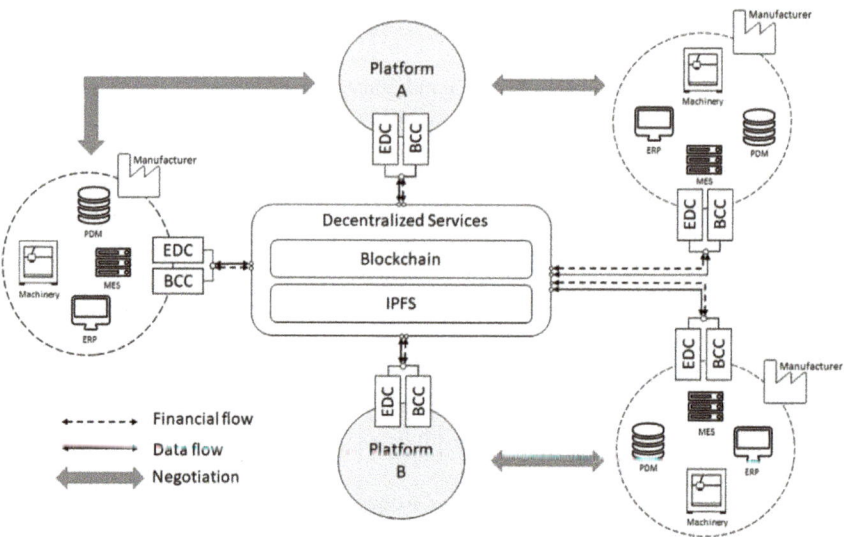

Fig. 2 Manifestation of a decentral capacity exchange

steps and operate with variable cycle times [25]. Thus, matrix production systems offer high flexibility regarding throughput times, routing alternatives, and the ability to accommodate rush orders, delays, and breakdowns [26]. Combining the flexibility of a matrix production system with the efficiency of an assembly line results in a hybrid production approach. The Cyber-Physical Matrix Production System (CPMPS) as the highest level of maturity incorporates both concepts and uses a digital twin [24]. These systems, shown exemplarily in Fig. 3, can visualize and simulate processes before execution [27].

The digital twin acts as the virtual representation of a physical system and ensures syn-chronization between the virtual and physical world through a bidirectional flow of information [28]. Visualization and simulation functions are used to seamlessly exchange data between the production and stakeholders. The effectiveness of the digital twin-based system depends heavily on the accuracy and validity of the data provided by the physical sys-tem. Automated processes, digitized data management, and the integration of financial processes are mandatory to replace manual procedures

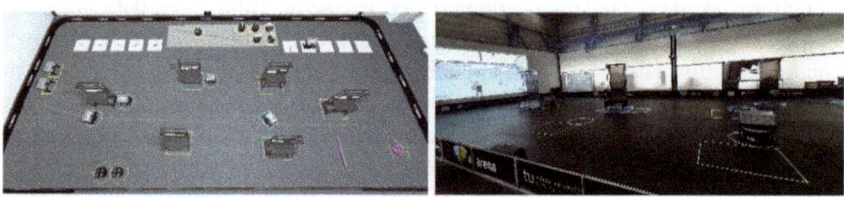

Fig. 3 The digital twin of a hybrid production connected to a blockchain with the virtual and physical production system in a research laboratory [27]

and achieve end-to-end trans-parency for an efficient material flow [29]. Web3, with its decentralized and transparent concept, serves as an enabler for the end-to-to-end transparency and validity of the data. In the case of pay-per-use business models, costs can be broken down to individual process levels via micropayments. Applying the cost-by-cause principle means that costs can be visualized in the digital twin and used for billing or further processing. Integrating resources such as workstations, robots, or machines within a production system into a decentralized system enables processes to be tracked precisely, resulting in lower overhead costs and more accurate billing.

Web3 technologies, such as blockchain, ideally incorporate a digital twin to achieve high transparency and automation. Figure 4 displays a framework for integrating a digital twin and blockchain [27]. The framework combines the different levels of blockchain with the dimensions of a digital twin to create a holistic digital twin with process and data integrity [30].

The financial transactions within a blockchain framework can be processed through the native cryptocurrency or payment tokens, offering benefits such as lower gas fees and higher transaction speed. In contrast to native cryptocurrency coins, tokens are more versatile [31]. This versatility enables functionalities in shared production, like the usage as an incentive for stakeholders to adhere to specific network policies. Tokens can enable access to a shared manufacturing network as a whole or to specific products and services within the network. For example, they can allow the reservation of machines for a specific operation. In shared manufacturing, utility or reward tokens can also play a role for an operational reputation mechanism that ensures the legitimacy of users and the reliability of processes. For example,

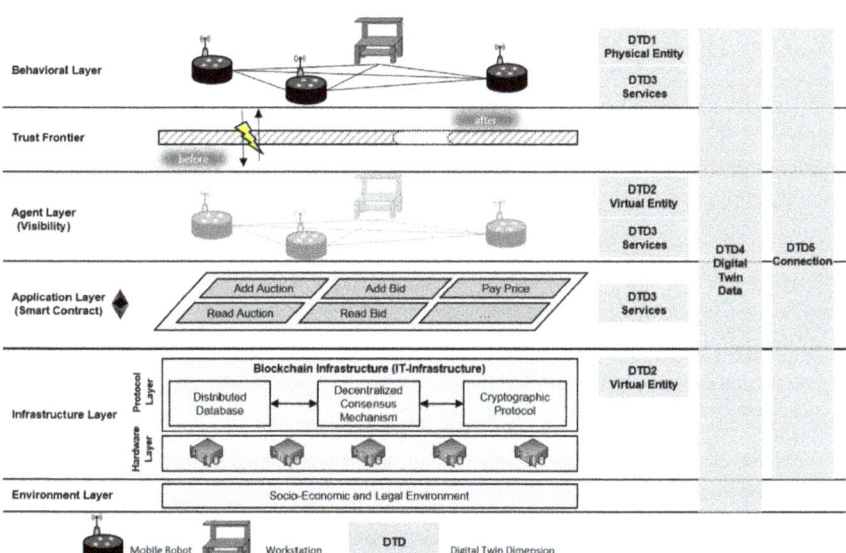

Fig. 4 Framework for designing a blockchain-based digital twin for CPPS [30]

tokens can be distributed for completed production tasks and correctly executed payments. This approach incentivizes reliable and trustworthy participants in shared manufacturing and correct behavior. Shared manufacturing usually involves SMEs that can jointly own a machine via an asset-backed token. This token can also increase transparency regarding the origin of a product, as all components can be recorded with the token, enabling seamless transfer from one machine to another during the manufacturing process [32].

5 Discussion

Web3 and the concept of shared manufacturing, which is characterized by decentralization and automation, are closely linked. In the context of shared manufacturing, Web3 technologies facilitate the development of autonomous peer-to-peer value creation networks. In particular, blockchain, as a transparent and trusted technological backbone, promotes collaboration in fields lacking trust. Tokens and smart contracts are two essential tools for automation based on valid data. For instance, tokens can also be used to establish access between centralized and decentralized platforms and facilitate the transition to Web3. Challenges that hinder the successful implementation of shared manufacturing are usually the initial investment, a lack of standardization, and concerns about loss of data control [5]. Web3, however, has the potential to overcome these hurdles.

Emerging Web3 and shared manufacturing trends will increasingly focus on pay-per-use models, leading to new business models and governance considerations [33]. However, this development depends on the level of process digitalization and accessibility. Cyber-physical (production) systems will play a key role in sharing approaches and traceability [33]. Sustainability and a growing need for resilience will accelerate shared manufacturing. With the gradual digitalization of processes and the increasing complexity of supply chains, the shift towards a sharing economy in manufacturing is already on its way but still needs further contributions.

References

1. A. Cherrafi, A. Chiarini, A. Belhadi, J. El Baz, A. Chaouni Benabdellah, Digital technologies and circular economy practices: vital enablers to support sustainable and resilient supply chain management in the post-COVID-19 era. TQM **34**(7), 179–202 (2022). https://doi.org/10.1108/TQM-12-2021-0374
2. A. Murray, D. Kim, J. Combs, The promise of a decentralized internet: what is Web3 and how can firms prepare? Bus. Horiz. **66**(2), 191–202 (2023). https://doi.org/10.1016/j.bushor.2022.06.002
3. A. Klarin, Y. Suseno, A state-of-the-art review of the sharing economy: scientometric mapping of the scholarship. J. Bus. Res. **126**, 250–262 (2021). https://doi.org/10.1016/j.jbusres.2020.12.063
4. E. Brandt, A vision for shared manufacturing. Mech. Eng. CIME **112**(12), 52–56 (1990)

5. C. Yu, X. Xu, S. Yu, Z. Sang, C. Yang, X. Jiang, Shared manufacturing in the sharing economy: concept, definition and service operations. Comput. Ind. Eng. **146**, 106602 (2020). https://doi.org/10.1016/j.cie.2020.106602
6. P. Stuckmann-Blumenstein, D. Bons, N. Große, L. Benkhoff, Problems and potentials of shared manufacturing in the context of industrial ecosystems: a bibliometric analysis, in *Hawaii International Conference on Systems Science (HICSS)* (2024)
7. R. Stammes, E. Burov, T. Ludwig, T. Gürpinar, Strategic realignment of medium-sized companies due to distributed ledger technologies in supply chain management (2022)
8. M. Lange, S.C. Leiter, R. Alt, Defining and delimitating distributed ledger technology: results of a structured literature analysis, in *Lecture Notes in Business Information Processing, Business Process Management: Blockchain and Central and Eastern Europe Forum*. ed. by C. Di Ciccio et al. (Springer International Publishing, Cham, 2019), pp.43–54
9. B.K. Mohanta, S.S. Panda, D. Jena, An overview of smart contract and use cases in blockchain technology, in *2018 9th International Conference on Computing, Communication and Networking Technologies (ICCCNT)*, (Bangalore, 2018), pp. 1–4
10. A. Rejeb, J.G. Keogh, S.J. Simske, T. Stafford, H. Treiblmaier, Potentials of blockchain technologies for supply chain collaboration: a conceptual framework. IJLM **32**(3), 973–994 (2021). https://doi.org/10.1108/IJLM-02-2020-0098
11. J. Zheng, D.K.C. Lee, Understanding the evolution of the internet: Web 1.0 to Web3.0, Web3 and Web 3. SSRN J. (2023). https://doi.org/10.2139/ssrn.4431284
12. W. Gan, Z. Ye, S. Wan, P.S. Yu, Web 3.0: The Future of Internet, in *Companion Proceedings of the ACM Web Conference 2023*, (Austin TX USA, 2023), pp. 1266–1275
13. F.A. Alabdulwahhab, Web 3.0: the decentralized web blockchain networks and protocol innovation, in *2018 1st International Conference on Computer Applications & Information Security (ICCAIS)*, (Riyadh, 2018), pp. 1–4
14. D. Sheridan, J. Harris, F. Wear, Cowell, Jerry, Jr, E. Wong, A. Yazdinejad, Web3 Challenges and Opportunities for the Market (2022)
15. Z. Zheng et al., Industrial metaverse: connotation, features, technologies, applications and challenges, in *Communications in Computer and Information Science, Methods and Applications for Modeling and Simulation of Complex Systems*, eds. by W. Fan, L. Zhang, N. Li, X. Song, (Springer Nature Singapore, Singapore, 2022), pp. 239–263
16. H. Wang et al., A survey on the metaverse: the state-of-the-art, technologies, applications, and challenges. IEEE Internet Things J. **10**(16), 14671–14688 (2023). https://doi.org/10.1109/JIOT.2023.3278329
17. J. Müller, K. Hoberg, J.C. Fransoo, Realizing supply chain agility under time pressure: Ad hoc supply chains during the COVID-19 pandemic. J. Ops. Manage. **69**(3), 426–449 (2023). https://doi.org/10.1002/joom.1210
18. D. Bons, P. Keitzl, H. Schulz, P. Stuckmann-Blumenstein, T. Gürpinar, S. Brüning, A taxonomy for the formation of enterprise blockchain consortia. Blockchain Cryptocurreny **1**, 18–35 (2023)
19. M. ten Hompel, M. Schmidt, Silicon economy: logistics as the natural data ecosystem, in *Springer eBook Collection, Designing Data Spaces: The Ecosystem Approach to Competitive Advantage*, eds. by B. Otto, M. ten Hompel, S. Wrobel, 1st edn., (Springer International Publishing; Imprint Springer, Cham, 2022), pp. 263–278
20. W. Prinz, T. Rose, N. Urbach, Blockchain technology and international data spaces, in *Springer eBook Collection, Designing Data Spaces: The Ecosystem Approach to Competitive Advantage*, eds. by B. Otto, M. ten Hompel, S. Wrobel 1st edn., (Springer International Publishing; Imprint Springer, Cham, 2022), pp. 165–180
21. L. Bader, J. C. Burger, R. Matzutt, K. Wehrle, Smart contract-based car insurance policies, in *2018 IEEE Globecom workshops (GC wkshps): Proceedings: Abu Dhabi, UAE, 9-13 December 2018* (Abu Dhabi, United Arab Emirates, 2018), pp. 1–7
22. V. Gallina, B. Gal, Á. Szaller, D. Bachlechner, E. Ilie-Zudor, W. Sihn, Reducing remanufacturing uncertainties with the digital product passport, in *Lecture Notes in Mechanical Engineering, Manufacturing Driving Circular Economy: Proceedings of the 18th Global Conference on Sustainable Manufacturing, October 5–7, 2022, Berlin*, eds. by H. Kohl, G. Seliger,

and F. Dietrich, 1st edn. (Springer International Publishing; Imprint Springer, Cham, 2023), pp. 60–67
23. V.V. Prabhu, Services for competitive and sustainable manufacturing in the smart grid. Serv. Orientat. Holonic Multi-Agent Manuf. Control. **402**, 227–240 (2012). https://doi.org/10.1007/978-3-642-27449-7_17
24. A. Hellmich et al., Umsetzung von cyber-physischen Matrixproduktionssystemen (2022)
25. M. Schönemann, C. Herrmann, P. Greschke, S. Thiede, Simulation of matrix-structured manufacturing systems. J. Manuf. Syst. **37**, 104–112 (2015). https://doi.org/10.1016/j.jmsy.2015.09.002
26. C. Hofmann, N. Brakemeier, C. Krahe, N. Stricker, G. Lanza, The impact of routing and operation flexibility on the performance of matrix production compared to a production line, in *Advances in Production Research: Proceedings of the 8th Congress of the German Academic Association for Production Technology (WGP), Aachen, November 19–20 2018*, ed. By R. Schmitt (Springer International Publishing AG, Cham, 2019), pp. 155–165
27. L. Krämer, P. Kaiser, J. Kajewski, M. Roidl, Reversing the digital twin – smart-contracting in hybrid production, in *2023 IEEE International Conference on Omni-layer Intelligent Systems (COINS)* (Berlin, Germany, Jul. 2023), pp. 1–6
28. W. Kritzinger, M. Karner, G. Traar, J. Henjes, W. Sihn, Digital Twin in manufacturing: a categorical literature review and classification. IFAC-PapersOnLine **51**(11), 1016–1022 (2018). https://doi.org/10.1016/j.ifacol.2018.08.474
29. D.A. Wuttke, C. Blome, M. Henke, Focusing the financial flow of supply chains: an empirical investigation of financial supply chain management. Int. J. Prod. Econ. **145**(2), 773–789 (2013). https://doi.org/10.1016/j.ijpe.2013.05.031
30. L. Krämer, N. Große, P. Stuckmann-Blumenstein, R. Ahlbäumer, M. Henke, M. ten Hompel, *Designing a Blockchain-Based Digital Twin for Cyber-Physical Production Systems* (publishing, Hannover, 2022)
31. P. Freni, E. Ferro, R. Moncada, Tokenization and blockchain tokens classification: a morphological framework, in *2020 IEEE Symposium on Computers and Communications (ISCC)* (Rennes, France, 2020), pp. 1–6
32. L. Krämer, J. Rutinowski, J. Endendyk, R. Vaut, M. Roidl, On Blockchain-based Token Usage in Cyber-Physical Production Systems (2022)
33. T. Gürpinar et al., Blockchain technology in supply chain management—a discussion of current and future research topics, in *Springer eBook Collection, Science and Technologies for Smart Cities: 7th EAI International Conference, SmartCity360°, Virtual Event*, vol. 442, 2–4 Dec 2021, *Proceedings*, ed 1st ed. by S. Paiva et al. (Springer International Publishing; Imprint Springer, Cham, 2022), pp. 482–503

Patrick Stuckmann-Blumenstein is a research associate at the Chair of Enterprise Logistics at TU Dortmund University. He holds a Master's degree in Industrial Engineering from TU Dortmund and has a strong background in the Dortmund logistics ecosystem. Patrick has worked on the "Blockchain Europe" project and is currently contributing to the "TECON" project, which focuses on the tokenized exchange of construction data. His research interests include digital supply chain twins, blockchain technology, artificial intelligence, resilience, and simulation. Patrick is also an associate member of the Research Training Group "Adaption Intelligence of Factories in a Dynamic and Complex Environment" (GRK2193).

Larissa Krämer is a research associate at the Chair of Material Handling and Warehousing at TU Dortmund University. She holds a Master's degree in industrial engineering and logistics from Dortmund. With deep roots in the logistics ecosystem of Dortmund, she has extensive experience and understanding of the region's logistical challenges and innovations. Her research primarily focuses on blockchain technology, digital twins in manufacturing, and simulation modeling of material flow systems. Larissa has actively contributed to the "Blockchain Europe" project, leveraging her expertise to advance blockchain applications in logistics. She is also a member of the Research Training Group "Adaption Intelligence of Factories in a Dynamic and Complex Environment" (GRK2193).

Dominik Bons is a research associate at the Chair of Enterprise Logistics at TU Dortmund University. He holds a Master's degree in Mechanical Engineering from TU Dortmund and has a strong background in the Dortmund logistics ecosystem. Dominik has worked on the "Blockchain Europe" project and is currently contributing to the "SKALA" project, which focuses on the interplay of Blockchain technology and artificial intelligence in logistics. His research interests include distributed ledger technologies and artificial intelligence and their impact on supply chain risk management and resilience.

Patrick Keitzl is a research associate at the Chair of Enterprise Logistics at TU Dortmund University. He holds a bachelor's degree in industrial engineering from TU Dortmund and has extensive experience in the logistics ecosystem in Dortmund. Patrick has contributed to the "Blockchain Europe" project. He is currently involved in the "SKALA" project, which focuses on developing AI and blockchain solutions for automation and autonomy in value-creation networks. His research interests include the efficient use of digital twins, blockchain technology, and artificial intelligence in supply chain management.

 **Eugen Burov** is a researcher at the Fraunhofer IML in Dortmund. He holds a Master's degree in Industrial Engineering. His scientific and practical background focuses on planning logistics processes and blockchain.

Virtual Workspaces and Collaboration

Erik Jarne Prinz and Cedric Muschick

1 Introduction

The digital landscape has undergone an extraordinary metamorphosis in recent years, with the metaverse captivating the collective imagination of individuals, industries, and the academic community. This transformation marks a momentous leap in the way we engage with virtual environments, transcending traditional boundaries and ushering in a new era of innovation. At the nexus of this technological revolution and the evolving nature of work, the metaverse is poised to redefine the very concept of virtual workrooms and collaboration.

This chapter embarks on a comprehensive exploration of the ever-evolving metaverse, its far-reaching implications, and its consequent impact on the reshaping of virtual work environments. In an era where work is increasingly characterized by distributed teams, remote interactions, and digital ecosystems, the metaverse offers a promising solution. As we delve deeper into this landscape, our investigation will encompass the nascent technologies, the intricate dynamics of digital societies, and the convergence of the virtual and physical realms. The central objective of this study is to shed light on the transformative potential of the metaverse in enhancing collaborative processes within virtual workspaces, thus heralding a groundbreaking epoch in remote work and digital collaboration.

However, this transformation is not without its challenges. Security, privacy, and the need for effective digital infrastructure are paramount concerns that must be addressed as we venture into this uncharted territory. Furthermore, the blurring of boundaries between work and personal life necessitates careful consideration of the psychological and societal implications of this transition.

The metaverse represents not only a technological shift but a fundamental reimagining of how we connect, create, and collaborate. Its emergence has not only blurred

E. J. Prinz (✉) · C. Muschick
AVANTECH, Euskirchen, Germany
e-mail: contact@aitanastrategicadvisors.de

the lines between the physical and digital worlds but has also opened up a realm of opportunities and challenges previously uncharted. This paper serves as a comprehensive guide to navigating this uncharted territory. As we venture further into the depths of the metaverse, we anticipate uncovering the keys to unlocking the full potential of this transformative landscape for the betterment of virtual workrooms and collaborative endeavors. This journey promises to be as intriguing as it is vital for individuals, organizations, and societies alike, as they adapt to and capitalize on the metaverse's revolutionizing force.

2 Transforming the Traditional Office

The traditional office, once an emblematic center of professional life, is now standing at the threshold of a profound metamorphosis within the emerging metaverse. This transformative journey illuminates the evolving nature of workspaces and their pivotal role in the ever-changing landscape of the digital age. Historically, the traditional office has been synonymous with a physical, centralized location where individuals convene to carry out professional tasks. This conventional model has been characterized by its geographical and temporal constraints, often necessitating the daily commute and adherence to fixed working hours. While it has been the cornerstone of productivity for decades, this traditional paradigm is now under scrutiny in the wake of the metaverse's advent.

The metaverse, with its promise of interconnected digital realms, offers a compelling alternative to the conventional office. It liberates work from the confines of a specific physical location, enabling professionals to seamlessly transition between virtual and physical spaces. This shift entails a reimagining of the workspace as an immersive, dynamic environment where individuals can collaborate, innovate, and communicate, irrespective of geographical boundaries. The convergence of virtual reality (VR), augmented reality (AR), and advanced communication technologies within the metaverse empowers professionals to transcend the limitations of traditional office settings. Virtual meetings, collaborative projects, and shared workspaces become more immersive and interactive, enhancing productivity and fostering a sense of presence.

The transformation of the traditional office within the metaverse is emblematic of the profound changes shaping the future of work. It represents a departure from the rigid confines of the physical office to a more flexible, interconnected, and digitally augmented workspace. As professionals adapt to this evolving landscape, they must grapple with new opportunities and challenges, all while recognizing the transformative potential it holds for the way we define and engage with work.

3 The Evolution of Virtual Environments

The transformation of virtual environments provides a compelling narrative, underscoring their progressive development and their critical role within the emerging metaverse. As we stand at the cusp of this transformative era, it is imperative to examine the historical continuum of virtual environments, from their nascent origins to their contemporary prominence. The genesis of virtual environments can be traced back to the rudimentary computer simulations and text-based Multi-User Dungeons (MUDs) of the 1970s. These formative experiments laid the groundwork for the concept of collaborative online spaces, albeit in a limited and textual form. The 1990s saw the ascent of 3D virtual worlds exemplified by platforms like Second Life. Here, users could craft avatars, engage in social interactions, and cultivate digital communities. This marked a pivotal shift toward more immersive and shared digital realms. The convergence of virtual environments with gaming and social media platforms brought forth a notable integration of virtual reality (VR) and augmented reality (AR). This integration significantly enhanced the realism and interactivity of virtual experiences. COVID-19 accelerated the adoption of virtual environments as indispensable tools for remote work and collaboration. Platforms like Zoom and Webex became essential, demonstrating their capacity to facilitate professional engagement within an increasingly remote work landscape. This evolutionary journey finds its apex in the metaverse. The metaverse seamlessly amalgamates virtual environments into an expansive, interconnected digital domain, facilitating fluid transitions between occupational, recreational, educational, and social realms.

The metaverse, as an amalgamation of diverse virtual environments, promises to redefine the manner in which individuals engage in work, education, social interaction, and innovation. It transcends the boundaries of mere gaming, interpersonal interactions, and professional collaboration, offering a comprehensive array of experiences and applications. Within the metaverse, users can collaboratively work on projects, gain knowledge, entertain themselves, and conduct business affairs, all within a cohesive digital space. It marks the next chapter in the continuous evolution of virtual spaces, offering unprecedented opportunities for exploration, interaction, and growth within this intricate digital landscape.

4 The New Age of Virtual Workrooms

The conventional concept of workrooms, typically comprising physical office spaces with desks, cubicles, and meeting rooms, has undergone a radical reimagining. In the metaverse era, workrooms transcend physical confines and are redefined as digitally mediated, collaborative environments. These spaces are facilitated by augmented and virtual reality technologies, enabling professionals to interact, communicate, and create in an entirely new dimension. The metaverse empowers individuals and teams to congregate within virtual workrooms, regardless of their geographic dispersion.

This virtual togetherness is made possible through avatars, digital replicas of individuals, and spatial audio that fosters a realistic sense of presence. Virtual workrooms, equipped with interactive whiteboards, shared screens, and 3D models, provide dynamic settings for brainstorming, project management, and knowledge exchange. Furthermore, these digital workrooms extend beyond traditional work-related tasks. They are adaptive and versatile, fostering creativity, networking, and knowledge sharing. Professionals can move seamlessly between focused, task-oriented work and informal, social interactions, simulating the serendipitous encounters that often occur in physical office spaces.

Nonetheless, the transition to the new age of virtual workrooms is not devoid of challenges. The potential for isolation, distraction, and disconnection is palpable. Maintaining a balance between work and personal life becomes increasingly complex as the boundaries blur within the metaverse. It enables professionals to transcend geographical boundaries, fosters dynamic collaboration, and redefines the work environment as an immersive, interactive digital space. As we navigate the opportunities and challenges presented by this new age of virtual workrooms, we must remain cognizant of the profound changes it brings to the landscape of work and collaboration, as discussed in the context of the evolving traditional office.

5 Collaboration in the Metaverse

In the present landscape, collaboration within the metaverse is already a reality. Within this environment, the possibilities for collaboration are virtually limitless. Professionals can collaborate on projects, share and manipulate 3D models, and even engage in virtual trade shows or conferences. Learning is transformed into an interactive experience through collaborative virtual classrooms, and artistic expression is enhanced through collaborative digital art installations.

Looking ahead, the metaverse presents a rich landscape of upcoming possibilities for collaboration. One of the most promising aspects is the potential for interdisciplinary collaboration. Professionals from diverse fields can seamlessly collaborate within shared virtual spaces, fostering innovation and the cross-pollination of ideas. These digital realms facilitate not only work-related tasks but also informal social interactions, fostering the serendipity that often leads to creative breakthroughs. Traditionally, physical distances have often hindered collaboration, imposing limitations on the ability to work with individuals and teams from different parts of the world. However, the Metaverse offers a transformative solution to this issue. Physical distances are no longer insurmountable obstacles; participants can interact as if they were in the same room, even when separated by vast geographical expanses. This capacity has far-reaching implications for various domains, such as global businesses, research initiatives, and cross-cultural exchanges. Teams can be composed of members from diverse continents, each contributing their unique perspectives and expertise. Geographical barriers no longer impede the assembly of highly diverse and international teams. This not only enriches collaborative efforts but also fosters a greater understanding of different cultures, which can lead to more innovative and well-rounded solutions to complex challenges.

It ensures that talent and expertise from every corner of the world can be harnessed, fostering a dynamic and interconnected global collaborative landscape. This is poised to be a defining feature of the Metaverse's impact on how we collaborate and innovate in the digital age. As we navigate this evolving landscape, we must remain mindful of the transformative impact it has on how we collaborate in the context of work, learning, and innovation, as discussed in the preceding texts.

6 Privacy, Security, and Digital Infrastructure

The emergence of the metaverse as a burgeoning digital landscape designed to facilitate work and collaboration has ushered in an era replete with opportunities and challenges, most notably those related to privacy, security, and digital infrastructure. As professionals venture into the metaverse to engage in work and collaborative endeavors, a complex tapestry of issues emerges, with existing concerns and future challenges intertwined.

In the context of work and office spaces within the metaverse, privacy issues assume heightened significance. The metaverse's immersive nature blurs the boundaries between personal and professional domains, increasing the vulnerability of individual data and behavioral patterns to exposure. The tracking, profiling, and ownership of data in this digital realm pose ongoing challenges, underscoring the necessity of robust data protection measures, especially concerning confidential documents and sensitive work-related information.

Security vulnerabilities within the metaverse represent a present and immediate concern for digital offices. The expansive digital terrain of the metaverse offers an array of opportunities for cyber threats, including security breaches, unauthorized access to virtual workspaces, and data hacking. The integrity of confidential information and sensitive communications is at risk, necessitating the implementation of stringent cybersecurity measures. Ensuring the secure exchange of documents and data within the metaverse's office spaces is paramount. Digital infrastructure readiness, including network stability, bandwidth allocation, and hardware capabilities, emerges as a critical concern in the metaverse's work context. The widespread adoption of the metaverse for professional purposes hinges upon a robust digital infrastructure that can support the demands of virtual offices. Addressing the digital divide across various regions and demographics is an essential step in ensuring equitable access to these virtual workspaces.

Looking to the horizon, the metaverse raises the specter of upcoming challenges for the workplace. One pressing issue is the lack of a comprehensive legal framework tailored to the metaverse. Existing regulations may inadequately address issues like data protection, intellectual property rights, and liability, creating a legal vacuum.

To navigate this challenge, there is an urgent need for the formulation of metaverse-specific laws, particularly in the context of digital offices where the sharing of confidential documents is a daily practice. Ensuring the authenticity of identities within the metaverse and the accurate representation of individuals in virtual workspaces is an issue on the horizon. Emerging technologies such as blockchain hold promise in addressing this challenge, although their full implementation remains to be realized. Trust within the workplace ecosystem becomes a paramount concern. This encompasses the credibility of digital assets and the reliability of information exchanged within virtual office spaces. Addressing these challenges requires a multifaceted approach. Collaboration between policymakers and legal experts is essential for establishing a regulatory framework specific to the metaverse, taking into account the unique attributes of this digital environment, especially within digital offices where confidential documents are at the forefront.

Investing in enhanced security measures is a prerequisite to protect sensitive work-related information. Cybersecurity measures, including encryption, multi-factor authentication, and security protocols, must be a priority to safeguard the secure exchange of documents and data in virtual office spaces. Interdisciplinary collaboration between experts from various domains, including law, technology, and ethics, is essential for formulating a comprehensive approach to metaverse governance within digital offices. Collaboration between academia, industry, and governmental entities is indispensable in shaping a metaverse that supports secure professional work. Infrastructure investment is a fundamental step to ensure equitable access to the metaverse, particularly in the context of virtual offices. Governments and private sectors must invest significantly in digital infrastructure, addressing the digital divide, augmenting network capacity, and enhancing hardware accessibility.

These challenges encompass existing concerns related to data protection and cybersecurity as well as emerging issues associated with identity verification and the credibility of digital assets. To conquer these challenges and create a secure and innovative metaverse for professional work, a proactive, interdisciplinary approach is imperative, addressing the unique needs and characteristics of digital offices where the sharing of confidential documents and information is a core aspect of daily operations.

7 Summary

The Metaverse represents a profound shift in the landscape of work and collaboration, offering innovative solutions to age-old challenges. Its transformative potential is underscored by its capacity to transcend geographical barriers, uniting individuals from diverse corners of the world in shared virtual workspaces. This newfound inclusivity fosters global collaboration, where physical distances are no longer limiting factors. Moreover, the integration of cutting-edge technologies, such as haptic feedback and enhanced avatars, promises to further enhance the sense of presence and the ability to collaborate effectively in virtual spaces. These advancements contribute to a collaborative experience that closely mirrors physical face-to-face interactions, promoting nuanced communication and shared experiences. However, this transformation also brings forth a unique set of challenges, including concerns related to privacy, security, and digital infrastructure, particularly in the context of digital offices where confidential documents and sensitive information are exchanged regularly. Addressing these challenges necessitates a proactive, interdisciplinary approach, including the formulation of metaversespecific regulations, the implementation of robust cybersecurity measures, and the promotion of digital literacy.

As the Metaverse continues to evolve, it offers a vision of the future where meaningful global collaboration is limited only by the extent of our imagination. It is a realm where professionals, researchers, and individuals from all walks of life can come together to work, create, and innovate in a space that knows no geographical boundaries, offering a promising path toward a more connected and inclusive global collaborative landscape.

The Role of Blockchain and Distributed Ledger Technologies in the Industrial Metaverse

Orhan Küpeli, Alexander Grünewald, Tan Gürpinar, Max Schwarzer, and Austin King

1 Introduction

In today's rapidly evolving digital world, the idea of the metaverse is gaining traction, capturing the interest of both tech enthusiast individuals and enterprises. Within this shifting landscape, distributed ledger technologies are emerging as an important backbone infrastructure. This integration of distributed ledger technologies and particularly blockchain solutions into the metaverse hold immense promise, especially in the context of Web3 aligned decentralization, authentication, and user empowerment.

Blockchain solutions offer promising frameworks and applications for redefining digital interactions and transactions in the metaverse and in enabling industrial metaverse approaches to be used in enterprise networks. By leveraging decentralized IT architectures and immutable ledger capabilities, blockchains hold the potential to instill trust and transparency in virtual environments, essential for fostering robust and sustainable ecosystems. Furthermore, blockchains facilitate secure transactions and asset ownership, empowering users with greater control over their digital identities and assets. This aspect aligns closely with the principles of Web3, which prioritize user sovereignty and data ownership. As such, the integration of blockchain into the metaverse not only enhances security and authenticity but also aligns with the overarching goals of the Web3 paradigm.

Our chapter seeks to explore these synergies, analyzing how blockchain technology can enhance the functionality and resilience of industrial metaverse platforms.

O. Küpeli (✉) · A. Grünewald
Fraunhofer IML, Dortmund, Germany
e-mail: kuepeli.orhan@outlook.de

T. Gürpinar · A. King
Quinnipiac University, Hamden, USA

M. Schwarzer
TU Dortmund, Dortmund, Germany

By identifying and addressing potential weaknesses in existing approaches, we aim to elucidate the different roles and transformative potential of blockchain within the context the metaverse context. Therefore, we conduct a literature review exploring the relationship and answering the following question:

In which application areas are blockchain and other distributed ledger technologies meaningfully utilized in industrial metaverse approaches?

In the review, we will focus the industrial metaverse which is differing from the consumer-oriented metaverse in its utilization for industrial contexts and can be subdivided into various application areas. In Table 1, the industrial metaverse has been categorized into different domains and elucidated through various practical examples. It illustrates how corporates are already experimenting with the industrial metaverse and for what reason they are implementing it. To build upon this, subsequently, the role of distributed ledger technologies will be analyzed in depth and by means of 95 selected scientific papers on the topic.

The analysis focuses on elucidating how blockchain and distributed ledger technology are referenced in the context of the metaverse and on identifying the various practical approaches in which they are utilized. The results show that blockchain technology is frequently associated with two main and multiple other domains: firstly, the establishment of a transparent and trustworthy environment, and secondly, the secure management of user's content and data.

As the metaverse continues to evolve, its applications extend beyond the realm of social interactions, giving rise to the concept of the industrial metaverse. Table 1 serves as a comprehensive overview of exemplary approaches within the industrial metaverse space, designed to differentiate it from the traditional metaverse primarily centered around social dynamics. In order to delineate the distinctive features and key components of the industrial metaverse, this table provides insights into various current industry projects. By focusing on these innovative approaches, we aim to offer a nuanced understanding of the industrial metaverse, exploring its potential impact and significance in the context of this book chapter.

To the end of this chapter, we will explore the potential business models associated with the metaverse. As technology continues to evolve, it is essential for businesses to understand how they can leverage the metaverse to enhance their operations and engage with customers in novel ways. We will delve into various business models and elucidate how companies can effectively integrate the metaverse into their strategies to drive innovation and foster growth.

2 Literature Research

The chapter bases on a comprehensive literature research as introduced in the introduction. The main goal is to identify which role the blockchain and distributed ledger technologies have in the industrial metaverse. Through that, we can identify and build upon crucial domains and areas. Later on, the identified areas will be addressed

Table 1 Industrial metaverse specification and use cases

Industrial metaverse specification	Organization	Type of use
Analytics	Tampere University	70% of visual inspections, and 65% of routine HVAC management tasks replaced with analytics driven automation [1]
	GTR	Optimization of departure and arrival schedules, predictive maintenance, and early anomaly detection for equipment and systems [1]
	Renault	Joint production and estimated $330 million in production cost savings by 2025 [2]
	BMW	Production process and product design optimization using real-time data [3]
	Tesla	Predictive equipment maintenance, and product quality control [3]
Digital twins	Equinor	Virtual prototype tests to optimize product selection [4]
	Space Perspective	Product design optimization related to supplier selection [1]
	Michelin	Digital business model used to monitor KPI performance [3]
Simulations	Equinor	Sustainable decision making related to product selection [4]
	BMW	30% reduction in production planning times [5]
	Space Perspective	Thermal environment tests for spacecraft interior [1]
	Intel	Virtual collision, and on-road driving tests [6]
	Michelin	$11 million logistics cost savings [3]
	Hyundai	Physical resource optimization activities conducted through virtual monitoring and simulations [3]
Augmented and virtual reality	Bimbo	Enhanced productivity, employee augmentation, 10–15% reduction in annual capital expenditures [4]
	Lowe's	Inventory management, customer service, and employee augmentation improved through augmented reality connectivity [2]
	Toyota	Use of Microsoft HoloLens 2 to improve employee training, efficiency, and inter-departmental collaboration [3]

separately and explained in more detail. Figure 1 shows the identified areas and the corresponding number of papers.

The findings from the analysis of the role of blockchain indicate that this technology is mainly employed for data security and protection. This leads to enhanced transparency in metaverse operations, particularly concerning the traceability of financial transactions [7]. As blocks are generated, interconnected, and

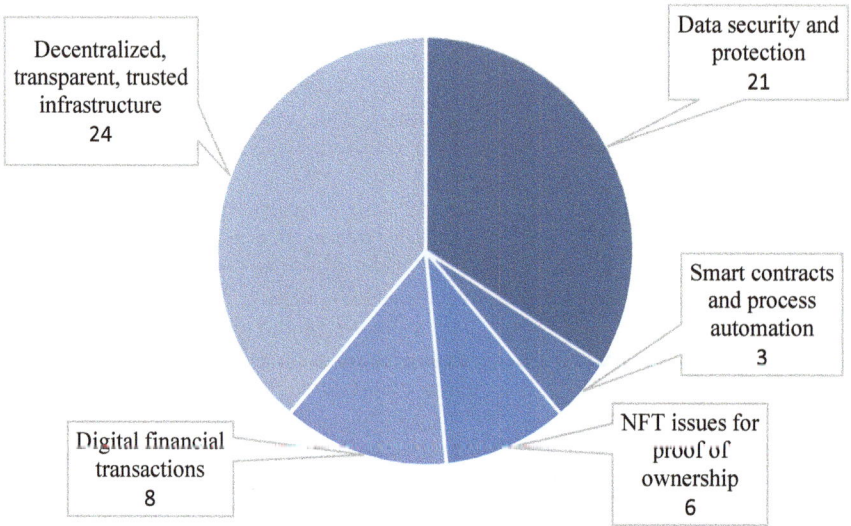

Fig. 1 Role of blockchain and distributed ledger technologies in the metaverse

interlinked, manipulating transactions involving avatars, purchases, and services in B2C scenarios becomes challenging [8]. Consequently, blockchain solutions offer B2B scenarios to improve industrial metaverse concepts such as secure identity management and access control of digital twins in smart factories [9]. Additionally, they streamline payment systems, enabling transactions in crypto-currencies and the exchange of digital assets between enterprises. A pivotal component is the utilization of blockchain-based non fungible token (NFT) by which objects can be attributed to their respective owner [10]. This leads to the possibility to proof ownership and trade of virtual properties [11]. Furthermore, smart contracts automate processes within the metaverse, reducing errors and expediting transactions [12]. In promoting transparent decision-making, blockchains facilitate tamper-proof voting and governance processes. Asset tokenization in the metaverse allows for creating and managing digital assets for various business purposes, as seen in energy trading systems where energy can be represented as a digital asset [13]. Moreover, blockchains incentivize data sharing and collaboration between enterprises, fostering innovation [14].

2.1 Financial Payments and Token Utilization

The industrial metaverse, contributing to the convergence of industrial IT and operational technology to combine physical and virtual worlds, is expected to have a market opportunity between 8 and 13 trillion USD [15]. Fields of application vary broadly, depending on context, involved participants, and use case. Similarly, the technological foundation to operationalize respective use cases may vary, depending on the requirements and goals. One idea is to implement Web3 technologies, including blockchain technology, as a key enabler to facilitate payments, handling of assets, or the usage of smart contracts [7]. Immanent characteristics of blockchain technology, such as immutability and decentralization, especially come into play in an inter-organizational context with multiple (unknown) parties involved [16]. Additionally, tokens allow for including a (native) medium of exchange, store of value, or voting/governance mechanism. In general, different types of tokens can be distinguished, e.g., utility tokens, asset tokens, and payment tokens. Definitions and subsets of respective categories may overlap and/or vary slightly. However, the key message is that blockchain-based tokens may fulfill a broad set of requirements in the industrial metaverse. "Asset tokens are crypto tokens linked to physical or digital assets (financial and non-financial). They can represent security and investment tokens as well as digital registries. […] Payment tokens are crypto tokens used for making digital payments; the term mainly refers to cryptocurrencies, a specific form of digital currency built upon ledgers. […] Tokens that provide a certain utility to users (such as access rights, membership rights, or identification and authentication), or that serve as rewards, fall into the category of utility tokens" [17, pp. 7–8].

Besides the previously mentioned advantages of including blockchain-based tokens, disadvantages may remain depending on the technology setup. Governance questions related to decision rights, accountability, and roles and responsibilities in an inter-organizational setup need to be assessed and carefully implemented as they drive incentives and scalability prospects. From a legal perspective, dispute resolution and enforceability of rights may vary depending on the applicable jurisdiction. Last but not least, the technology allows for various modes of implementation (e.g., public-permissionless vs. private-permissioned setup), which need to be designed to fulfill minimum enterprise requirements related to security, profitability, and sustainability [17].

2.2 Record Keeping and Data Security

In blockchain technology, record-keeping and data security play a significant role. The innovative ledger technology ensures transparent and immutable records and establishes robust mechanisms for safeguarding data integrity, fundamentally reshaping conventional information security paradigms. This will be important for realizing the success of a decentralized industrial metaverse that lacks data storage and validation [18]. Moreover, given that the metaverse collects a wide range of data, ensuring the data´s quality is imperative to prevent potential stakeholders from losing trust in the metaverse [19].

Blockchain generally represents a database that a multitude of participants can operate. Through the egalitarian participation of all involved entities, blockchain technology constitutes a peer-to-peer network. Transactions sent to another participant's public key are stored under a hash value along with the corresponding information, including timestamps. A hash value consists of a predetermined length of records

Fig. 2 Structure and connectivity from blocks within the blockchain [21]

that encrypt a transaction, thereby preventing the alteration of input data or rendering it impossible to modify retrospectively. This leads to using blockchain technology in an untrustworthy environment, where transactions cannot be changed afterward [20]. However, it remains feasible to identify the associated transaction using the hash value and verify its correctness or accuracy. Every entry in the blockchain is allocated a hash value, significantly influencing the overall data control process within the blockchain. Following this, blocks are created by consolidating these transactions. This makes manipulation or alterations of existing data within the blockchain quite challenging. As a hash includes a timestamp, transactions can be reconstructed chronologically, ensuring a certain degree of data security. Figure 2 shows how multiple hash values create a block and that multiple blocks are connected to each other.

As this structure doesn't need a third party, users can verify and review every transaction independently and inexpensively [22]. This enables the possibility of validating and storing data on a decentralized basis within the industrial metaverse [18]. In case of industrial usage, various approaches are already being explored, where data protection is decentralized but the data is encrypted, allowing sensitive data continue to be protected. This involves, for example, collecting large amount of data, as with IoT devices [23]. This is quite crucial, as data collection represents an extremely valuable and important resource, with Big Data sets being a critical component for manufacturing and production systems [7]. Furthermore, blockchain and distributed ledger technology can significantly protect those data within the metaverse ecosystem.

2.3 Blockchain as Infrastructure in the Industrial Metaverse

In addition to its use cases for financial payments, tokenization, and data protection, industrial metaverse approaches can also be set up in a decentralized manner utilizing blockchain solutions as a backbone infrastructure. Here, companies converge and find a common ground to exchange data, co-produce goods and services, and collaborate on shared digital twins [24]. The utilization of blockchain technology as a foundational infrastructure in this case is primarily due to its intrinsic technical attributes that cater to the essential requirements of an ecosystem [25]. Blockchain's decentralized and immutable ledger capabilities offer a trusted and secure environment where diverse entities can interact seamlessly [26]. By leveraging blockchain's features, the industrial metaverse can establish a transparent, trustworthy, and interoperable framework for multi-party collaboration [24]. Smart contracts facilitate the automated execution of agreements, enabling seamless interactions and transactions within the network of entities.

The integration of blockchain as a backbone technology in industrial metaverse approaches itself enables further use cases. One such use case involves supply chain monitoring, where the distributed infrastructure facilitates information sharing for real-time tracking and traceability of goods across the involved production and distribution networks. This transparency ensures authenticity, reduces counterfeiting, and enhances trust among participating stakeholders. Additionally, blockchain-based digital twins serve as virtual tamper-proof replicas, offering a secure and synchronized platform for collaborative design, simulation, and testing, thereby streamlining product development and innovation [27]. Moreover, blockchain's role in decentralized identity management ensures data sovereignty and privacy while enabling secure access control across the metaverse ecosystem [28].

Figure 3 shows the seamless integration of blockchain technology into the metaverse, with distinct puzzle pieces symbolizing the synergistic connection between these two components. It summarizes the topics and shows, how the industrial metaverse can benefit from the blockchain, which addresses the weaknesses of the metaverse. The first puzzle piece underscores the foundational role of blockchain as the backbone of the metaverse infrastructure, aligning with principles such as data sovereignty and distributed record keeping. Here, blockchain technology facilitates immutable ledger capabilities and decentralized identity management. Transitioning to the second puzzle piece, our focus shifts to the paramount significance of digital assets and payments within the metaverse, where blockchain acts as a catalyst for payments and tokenization. The third puzzle piece highlights aspects of data safety and continuous record-keeping within the metaverse. Blockchain contributes significantly to ensuring data privacy and continual data recording through its data control and security features. Collectively, these puzzle pieces showcase how blockchain technology can not only seamlessly integrate into the metaverse but also contribute to neutralizing potential weaknesses, establishing a robust foundation for secure and efficient metaverse utilization.

3 Business Model Implications for Enterprises

Predictions suggest that the metaverse will become a major industry in the coming years. The emerging metaverse technologies may still be in their infancy, but they are already opening up strategic opportunities beyond virtual reality (VR). The metaverse will impact the physical world in the future by transforming or extending physical

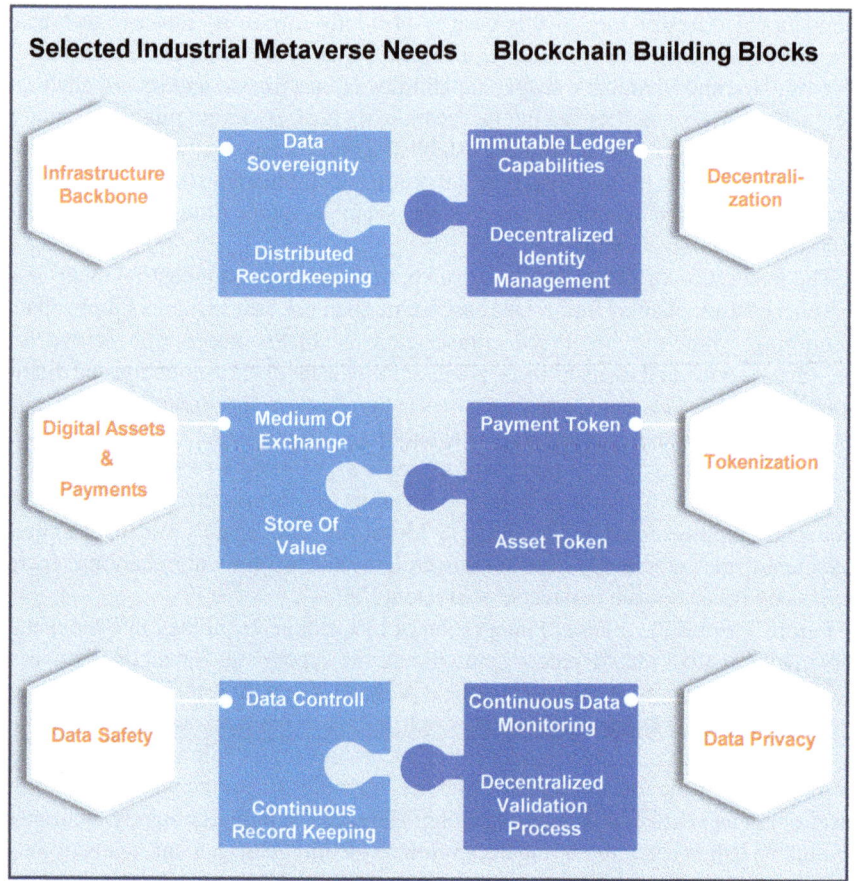

Fig. 3 Connection between industrial metaverse and blockchain technology

activities into a virtual environment. A virtual ecosystem is created in which the physical world merges with the digital world. This level of interaction is the foundation for new innovative business models.

Business models, in general, describe how a company creates, delivers, and captures value [29]. About the metaverse, a business model thus describes how companies carry out economic activities in the virtual world by offering virtual goods and services. The metaverse thus opens new economic opportunities by creating digital assets and using new models for value exchange [30].

Some companies, especially from the gaming industry, lifestyle brands, social media, and technology companies, have already launched metaverse projects and entered the digital world. In the first five months of 2022, companies invested over 120 billion US dollars in the metaverse, representing massive growth compared to the 57 billion US dollars invested in 2021 [31]. J.P. Morgan, for example, is the first bank that has established a store in the metaverse. They forecast a $1 trillion market opportunity and have targeted virtual real estate [32]. While real estate investments and online gaming partnerships are often companies' first steps into the metaverse, some of the fashion industry's most prominent players have also looked to NFTs to generate new revenue streams. For example, Adidas sold 30,000 items as NFTs in the metaverse worth $23 million within minutes of going on sale [33]. The participation of companies in the digital world can be of immense importance to them and offers promising opportunities.

Companies consider the consumer-oriented metaverse a chance to move their products and services into the virtual world to create an improved customer experience and thus achieve better customer relationships [34]. In contrast to the real world, a store in the metaverse is open non-stop and can be reached from any location. Even now, virtual goods are sold that are only available in the metaverse or can only be purchased through the metaverse in the real world [35]. For example, VR gives customers a haptic product presentation, allowing products to be displayed far more realistically than conventional catalogs and websites [36]. This offers consumers a unique shopping experience. At the same time, this new direct way of communication also generates a huge amount of data that companies can use to collect valuable information about consumer behavior [35, 36]. Although the metaverse is still in its infancy, Gartner [32] predicts that by 2026, 25% of people will be active in it to work, shop, educate themselves, use social media, and find entertainment.

The metaverse is also increasingly being discussed in the industry, albeit still as an abstract vision. While the consumer-oriented metaverse is person-centered in its origins, the industrial metaverse focuses on industrial applications and uses. In an industrial context, it enables (real-time) communication with physical objects, processes, and environments in a virtual environment or overlays the natural environment with digital content. The industrial metaverse focuses on use cases in companies geared towards real-world challenges and business requirements. It enables a range of use cases for industry, from training and education under simulated conditions to construction, maintenance, repair, and virtual product testing. Technology companies such as Siemens, BMW, and Deutsche Bahn have already tested such use cases of the industrial metaverse [37].

The possible applications of the industrial metaverse are wide-ranging and strengthen companies' existing business models by enabling process improvements and increasing efficiency. They also open opportunities to expand or develop new digital business models [38]. By 2030, Gartner predicts that industrial applications in the metaverse will already be in the early mainstream of implementation, offering considerable potential for groundbreaking changes in various industries [32].

Equivalent to the consumer-oriented metaverse, production companies can use the industrial metaverse to showcase themselves and their products and services and create a virtual experience for potential customers. The industrial metaverse enables sectors like the automotive industry to address real problems digitally. With virtual sensors that predict equipment failures, autonomous trucks that optimize logistics, and collaborative robots that increase productivity, the metaverse offers transformative use cases [39, 40]. In product development, for example, new product designs can be realized virtually and planned, designed, simulated, and tested safely with the customer. Similarly, implementing new production systems and the production ramp-up of product series can be tested digitally without interrupting production. Virtual learning processes can also accelerate the introduction process of physical production. Employees can acquire the necessary skills in the metaverse, gain practical experience, and share best practices across locations or worldwide. This makes access to talent and training independent of location. The industrial metaverse also provides starting points for new digital business models, e.g., as a virtual (trading) platform on which production companies bring their customers, partners, and suppliers together in a marketplace model. The metaverse can also act as an interface for digital data spaces and offer production companies the opportunity to monetize the data and the information products they generate [37].

Entrepreneurs recognize that the metaverse offers new opportunities in the virtual and physical world. Today's technical possibilities have already led to the development of the first business models, although companies are currently focusing on selling services or products in the metaverse. In the future, this fusion of two worlds will also be increasingly used in various ways within industry as the so-called industrial metaverse. On the one hand, we see entrepreneurs in the physical world introducing virtual experiences, transactions, and artifacts and merging their physical world with the metaverse, e.g., virtual fashion [37, 41], virtual jewelry [42], or the recording of movements with haptic feedback [41, 43]. On the other hand, the fields of application of the industrial metaverse demonstrate transformative possibilities that can revolutionize work processes and create added value for companies.

4 Conclusion and Future Outlook

In this chapter, we present a review on the role of blockchain and distributed ledger technologies in the industrial metaverse. A number of 95 papers have been analyzed and presented to distinguish several application scenarios. In conclusion, the integration of decentralized technologies into the industrial metaverse holds significant promise for revolutionizing various aspects of modern industry as they play a pivotal role in enabling financial payments, token utilization, data security, and infrastructure development within industrial metaverse approaches.

By leveraging blockchain technology, industrial players can facilitate secure and transparent financial transactions, implement tokenization mechanisms for asset management and governance, and ensure robust data security and integrity across decentralized networks. The immutable and decentralized nature of blockchain ledgers provides a solid foundation for building trust and facilitating seamless interactions among multiple stakeholders within the metaverse. Furthermore, blockchains serve as a foundational infrastructure for the industrial metaverse, offering a decentralized framework for data sharing, collaboration, and value creation. Through the integration of blockchain, companies can explore innovative use cases such as supply chain monitoring, digital twin management, and decentralized identity solutions, thereby enhancing operational efficiency, transparency, and trustworthiness throughout the value chain. The opportunities presented by the industrial metaverse are reshaping traditional business models, driving digital transformation, and paving the way for new avenues of economic growth and innovation. As industry leaders and entrepreneurs embrace the potential of the metaverse, they are poised to unlock new revenue streams, improve customer experiences, and drive unprecedented levels of efficiency and productivity.

Even though real-world implementation projects on both enterprise blockchains and industry metaverse approaches are still in concept stages and a lot of technical, organizational, human, and legal challenges still need to be addressed, these technologies continue to evolve and mature steadily and will undoubtedly play a central role in shaping the future of industrial innovation and economic prosperity.

Future research activities should prioritize the trilemma between decentralization, scalability, and security of blockchain services for the industrial metaverse. Also, integrating blockchain services into existing systems while considering the interplay to other technologies such as AI, particularly in federated learning approaches should be considered. Therefore, also the interoperability of multiple blockchain solutions is crucial for seamless transactions in the broader industrial ecosystem. Finally, future studies should address the social, ecological and economic sustainability of blockchain solutions and their powered industrial metaverse platforms.

References

1. Siemens Knowledge Hub, Launching Siemens Xcelerator. [Online]. Available: https://www.youtube.com/watch?v=NxLkgcI5RfY. Accessed 16 Feb 2024
2. N. Kshetri, The Economics of the industrial metaverse. IT Prof. **25**(1), 84–88 (2023). https://doi.org/10.1109/MITP.2023.3236494
3. A. Meige, R. Eagar, The Industrial Metaverse: Making the Invisible Visible to Drive Sustainable Growth, (2023)
4. Microsoft, The Industrial Metaverse. [Online]. Available: https://www.youtube.com/watch?v=wAlcX7QaWkc. Accessed 16 Feb 2024
5. Z. Lin, P. Xiangli, Z. Li, F. Liang, A. Li, Towards metaverse manufacturing: a blockchain-based trusted collaborative governance system. Int. Conf. Blockchain Technol. (ICBCT) **4**, 2022 (2022)
6. Intel Corporation, Intel Tech Helping Design Prototype Fusion Power Plant. [Online]. Available: https://www.intc.com/news-events/press-releases/detail/1632/intel-tech-helping-design-prototype-fusion-power-plant. Accessed 18 Feb 2024
7. D. Mourtzis, J. Angelopoulos, N. Panopoulos, Blockchain integration in the era of industrial metaverse. Appl. Sci. **13**(3), 1353 (2023). https://doi.org/10.3390/app13031353
8. T. Huynh-The et al., Blockchain for the metaverse a review. Futur. Gener. Comput. Syst., 401–419 (2023)
9. T. Gürpinar, J. Kampfhues, M. Austerjost, J. Maaßen, M. Henke, F. Yildirim, Blockchain technology as the backbone of the internet of things—an introduction to blockchain devices to blockchain devices, in *Conference on Production Systems and Logistics*, no. 3 (2022)
10. D. Chalmers, C. Fisch, R. Matthews, W. Quinn, J. Recker, Beyond the bubble: Will NFTs and digital proof of ownership empower creative industry entrepreneurs? J. Bus. Ventur. Insights **17**, e00309 (2022). https://doi.org/10.1016/j.jbvi.2022.e00309
11. B. Guidi, A. Michienzi, Social games and Blockchain: exploring the Metaverse of Decentraland, in *International Conference on Distributed Computing Systems Workshops*, pp. 199–204 (2022)

12. P. Que, Y. Zeng, F. Gao, The current situation and prospect of the development of metaverse technology, in *2022 4th International Conference on Applied Machine Learning (ICAML)*, Changsha, China, pp. 1–5 (2022). https://doi.org/10.1109/ICAML57167.2022.00089.
13. M. Moniruzzaman, A. Yassine, R. Benlamri, Blockchain and metaverse for peer-to-peer energy marketplace: research trends and open challenges, in *ICTMOD* (2022)
14. Y. Cui, H. Idota, M. Ota, Reforming supply chain systems in Metaverse, in *The 9th Multidisciplinary International Social Networks Conference*, Matsuyama Japan, pp. 39–43 (2022). https://doi.org/10.1145/3561278.3561289
15. Citi, Metaverse and Money: Decrypting the Future. [Online]. Available: https://www.citigroup.com/global/insights/citigps/metaverse-and-money_20220330. Accessed 29 Nov 2023
16. M. Schwarzer, T. Gürpinar, M. Henke, To join or not to join?—A framework for the evaluation of enterprise blockchain consortia. Front. Blockchain (2022)
17. R. Beck, A.B. Pedersen, J. Schwiderowski, Token classification and taxonomy framework. Inf. Syst. Front. (2023)
18. D.M. Doe, J. Li, N. Dusit, Z. Gao, J. Li, Z. Han, Promoting the sustainability of blockchain in Web 3.0 and the metaverse through diversified incentive mechanism design. IEEE Open J. Comput. Soc. **4**, 171–184 (2023). https://doi.org/10.1109/OJCS.2023.3260829
19. S. Mishra, H. Arora, G. Parakh, J. Khandelwal, Contribution of blockchain in development of metaverse, in *2022 7th International Conference on Communication and Electronics Systems (ICCES)*, Coimbatore, India, pp. 845–850 (2022). https://doi.org/10.1109/ICCES54183.2022.9835986
20. A.G. Gad, D.T. Mosa, L. Abualigah, A.A. Abohany, Emerging trends in blockchain technology and applications: a review and outlook. J. King Saud Univ. Comput. Info. Sci. **34**(9), 6719–6742 (2022). https://doi.org/10.1016/j.jksuci.2022.03.007
21. S. Schacht, C. Lanquillon (eds.), *Blockchain und maschinelles Lernen: Wie das maschinelle Lernen und die Distributed-Ledger-Technologie voneinander profitieren* (Springer Vieweg, Berlin, Heidelberg, 2019)
22. A.S. Rajasekaran, M. Azees, F. Al-Turjman, A comprehensive survey on blockchain technology. Sustain. Energy Technol. Assess. **52**, 102039 (2022). https://doi.org/10.1016/j.seta.2022.102039
23. R. Patan, R.M. Parizi, Securing Data Exchange in the Convergence of Metaverse and IoT Applications (2023)
24. O. Küpeli, T. Gürpinar, Towards a definition of the industrial metaverse applied in context of the blockchain and Web3 ecosystem, in *Blockchain and Cryptocurrency Congress (B2C' 2023)* (2023)
25. A. Vaghani, T. Gürpinar, N. Große, A taxonomy characterizing blockchain-empowered services for the metaverse, in *Blockchain and Cryptocurrency Congress*, pp. 64–66 (2022)
26. N. Große, D. Leisen, T. Gürpinar, R.S. Forsthövel, M. Henke, M. ten Hompel, in *Evaluation of (De-)Centralized IT technologies in the fields of Cyber-Physical Production Systems* (Publishing, Hannover), 2020
27. K. Salah, R. Jayaraman, M. Uddin, Blockchain for digital twins recent advances and future research challenges. IEEE Netw. **34**(5) (2020)
28. J. Gelhaar, T. Gürpinar, M. Henke, B. Otto, Towards a taxonomy of incentive mechanisms for data sharing in data ecosystems, in *Twenty-fifth Pacific Asia Conference on Information Systems*, 2021
29. A. Osterwalder, Y. Pigeuner, *Business Model Generation: A Handbook For Visionaries, Game Changers, and Challengers* (2010)
30. A. Grünewald, T. Gürpinar, C. Culotta, A. Guderian, Archetypes of blockchain-based business models in enterprise networks. Inf. Syst. E-Bus. Manage. (2024)
31. T. Elmasry et al., Value Creation in the Metaverse: The Real Business of the Virtual World (McKinsey & Company), 2022.
32. J. Wiles, What is a metaverse? And should you be buying in?. [Online]. Available: https://www.gartner.com/en/articles/what-is-a-metaverse. Accessed 29 Nov 2023

33. Adidas, NFTs are here: welcome to the Metaverse. [Online]. Available: https://www.adidas.de/en/metaverse. Accessed 29 Nov 2023
34. A. Cibi, Relativity of metaverse in business model: a conceptual analysis. Contrib. Environ. Sci. Innov. Bus. Technol., 9–17 (2023)
35. T. Benny, Demystifying metaverse in business: a conceptual study. Contrib. Environ. Sci. Innov. Bus. Technol., 1–8 (2023)
36. S. Periyasami, Perisayamy, A. Prince, Metaverse as future promising platform business model: case study on fashion value chain. MDPI, 527–545 (2022)
37. Bitkom, Industrial Metaverse: Use Cases, Mehrwerte und Potenziale für den Wirtschaftsstandort Deutschland, 2023. [Online]. Available: https://www.bitkom.org/sites/main/files/2023-09/bitkom-leitfaden-industrial-metaverse.pdf
38. V. Krishna, Investigating the effects of the metaverse on business models. J. Adv. Manage. Eng. Sci. **2023**, 60–72 (2023)
39. P. Bhattacharya et al., Towards future internet: the metaverse perspective for diverse industrial applications. Mathematics **11**(4), 941 (2023). https://doi.org/10.3390/math11040941
40. C.-H. Chu, D.K. Baroroh, J.-K. Pan, S.-M. Chen, An exemplary case of industrial metaverse: engineering product demonstration using extended reality technologies. Int. J. Precis. Eng. Manuf. Smart Tech. **1**(2), 243–250 (2023). https://doi.org/10.57062/ijpem-st.2023.0038.
41. A. Williams, Nike Sold an NFT Sneaker for $134,000, *The New York Times*, 26 May 2022. [Online]. Available: https://www.nytimes.com/2022/05/26/style/nike-nft-sneaker.html. Accessed 29 Nov 2023
42. K. Youde, Jewellers and watchmakers grapple with when to enter the metaverse: Luxury brands study the possibilities while some warn of danger of being left behind, *Financial Times*, 08 Jul 2022. [Online]. Available: https://www.ft.com/content/13aa0026-b559-4c03-af76-96c5d61d2f4f. Accessed 29 Nov 2023
43. Teslasuit, Meet our Haptic VR Suit and Glove with Force Feedback, *Teslasuit*, 02 Mar 2022. [Online]. Available: https://teslasuit.io/. Accessed 29 Nov 2023

Orhan Küpeli obtained his Master of Science in Industrial Engineering from Technical University Dortmund, Germany. He started working at the Fraunhofer Institute for Material Flow and Logistics as a working student and is involved in research on Blockchain and Web3 topics to empower decentralized data ecosystems. Additionally, Orhan Küpeli is part of a research team working on industrial metaverse applications and exploring its integration with blockchain or distributed ledger solutions.

Alexander Grünewald is a research associate at the Fraunhofer Institute for Material Flow and Logistics IML in Dortmund, focusing on technology and innovation management. He completed both his Bachelor's and Master's degrees in Industrial Engineering at the Technical University of Dortmund. His primary research interests lie in the field of blockchain technology. In his research, he investigates the extent to which blockchain technology can act as an enabler for sustainable economic practices. He examines the impact of blockchain technology on business models and its economic, ecological, and social implications for value creation networks. Alexander Grünewald is particularly committed to technology transfer into practice. In his research, he demonstrates how blockchain technology can be effectively applied. Through his work, he aims to combine theoretical insights with practical solutions to promote sustainable and innovative business models.

Tan Gürpinar is an Assistant Professor of Business Analytics & Information Systems at Quinnipiac University, Connecticut, US, with a focus on the impact of blockchain solutions on sustainability and profitability. He also holds an affiliation with the Fraunhofer Institute for Material Flow and Logistics, Germany where blockchain research findings are transferred into the practice to empower decentralized data ecosystems in various industries. He is an Editorial Board Member at the Blockchain & Cryptocurrency Journal and part of the Academic Advisory Body of the International Association of Trusted Blockchain Applications. Before becoming an Assistant Professor, he has been a blockchain team lead at Dortmund University of Technology and a co-founder of the associated blockchain PhD group. Tan Gürpinar's expertise lies in in blockchain governance and deploying interdisciplinary approaches to seamlessly integrate blockchain solutions into enterprise networks. His involvement in the Web3 Compendium reflects his commitment to advancing the understanding of blockchain's synergies with emerging technologies within the Web3 framework, aiming to translate theoretical insights into actionable strategies for transformative impact.

Max Schwarzer is a Tech Strategy Consultant working for the consulting company Accenture. His key areas of expertise include future technologies and Tech M&A in the Financial Services industry. Before his current role, he helped clients from the automotive and logistics industry with research and development of web3 technologies, digital assets, as well as IT and platform ecosystem governance. His primary research interest includes blockchain governance as a key topic, specifically the design of governance mechanisms for enterprise blockchain solutions in an inter-organizational context. He is fascinated about the parallels to Game Theory and theoretical possibilities of implementing governance with blockchain technologies. Max is a PhD student at the University of Dortmund. He holds a Masters degree from the University of Regensburg and a Bachelors degree from the University of Bayreuth in Economics.

Austin King holds a Master of Science in business analytics and a Bachelor of Science in management from Quinnipiac University, Connecticut, United States. He is conducting research in the field of blockchain technology, particularly cryptography, and focuses on identifying the challenges and benefits of cryptographic encryption.

Potentials and Applications of the Industrial Metaverse Using the Example of Synthetic Data Generation

Oliver Petrovic, Josefine Monnet, Petar Tesic, Yannick Dassen, and Werner Herfs

1 The Industrial Metaverse

Since Facebook Inc. announced its reorientation and renaming to Meta Platforms Inc. in 2021, referring to the Metaverse as the next evolution of the internet, the term, which originated in the early 1990s, has become a megatrend permeating all areas of life. Although there are various interpretations, they all share the common goal of merging the virtual and physical worlds [1] (Fig. 1). Technologies to achieve this vision, such as Digital Twins, VR/AR, and IIoT, already exist. However, the holistic approach of the Metaverse and developments in areas like Artificial Intelligence, graphic computing power, and Mixed Reality open up significant new potentials beyond the current Industry 4.0 approaches [2].

In addition to the opportunities in the consumer field, where applications in social media or entertainment are driving new forms of communication and media consumption, the potential in the industrial environment is immense. This manifestation of the Metaverse is called the Industrial Metaverse, and its development offers great potential for improving the productivity, sustainability, and resilience of companies and networks [3]. As a result, the technology provides significant opportunities to address current challenges such as climate change, volatile supply chains, and skilled labor shortages, and to disruptively change the way of collaboration, engineering, and process design in the productive environment.

Technological Foundations

The core idea of the Industrial Metaverse can be described as the merging of physical elements with their virtual representation. Consequently, real-time interfaces between physical assets (e.g., machines and production systems) and the virtual

---

O. Petrovic (✉) · J. Monnet · P. Tesic · Y. Dassen · W. Herfs
RWTH Aachen, Aachen, Germany
e-mail: o.petrovic@wzl.rwth-aachen.de

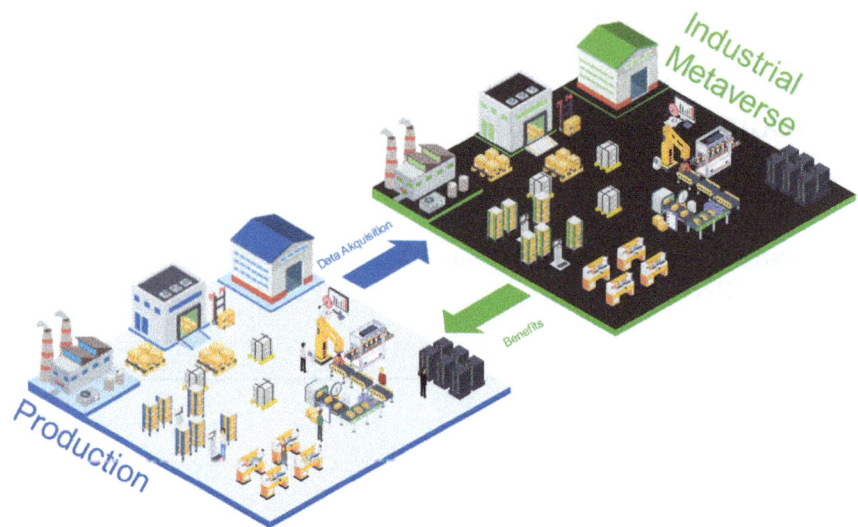

Fig. 1 The Industrial Metaverse as an immersive fusion of the physical and virtual worlds

environment are required to ensure the coherence of both worlds. Additionally, immersive user interfaces for interacting with virtual representations and enabling Mixed Reality approaches, where reality is overlaid with virtual information, must be available. For the Industrial Metaverse to operate as a heterogeneous platform, not only must the vertical interfaces to reality be standardized, but also the horizontal interfaces between services and applications within the virtual platform.

The core of the actual platform is a central integrated 3D model, which, when enriched with real, contextualized production data, becomes a Digital Twin [4]. Numerous functionalities are built on this model, creating added value and being combinable into specific applications. These functionalities can be primarily categorized into collaboration and simulation. This overarching classification of the components of the Industrial Metaverse is illustrated in Fig. 2.

To implement and further develop such a comprehensive system in the industrial environment, numerous essential enabling technologies must be combined [6]. Among these, the following technologies are particularly noteworthy:

The **Industrial Internet of Things (IIoT)** plays a central role in the Industrial Metaverse by providing the foundation for networking and communication between machines, systems, and plants. By utilizing standardized interfaces and communication protocols essential for Industry 4.0, the IIoT enables efficient data collection and transmission. The integration of platform-internal data models and formats, such as Universal Scene Description (USD) formats, enhances this efficiency by enabling a unified representation of complex scenes and objects in 3D visualizations. These formats support consistency and interoperability within the Industrial Metaverse by providing a common language for describing product data and process information.

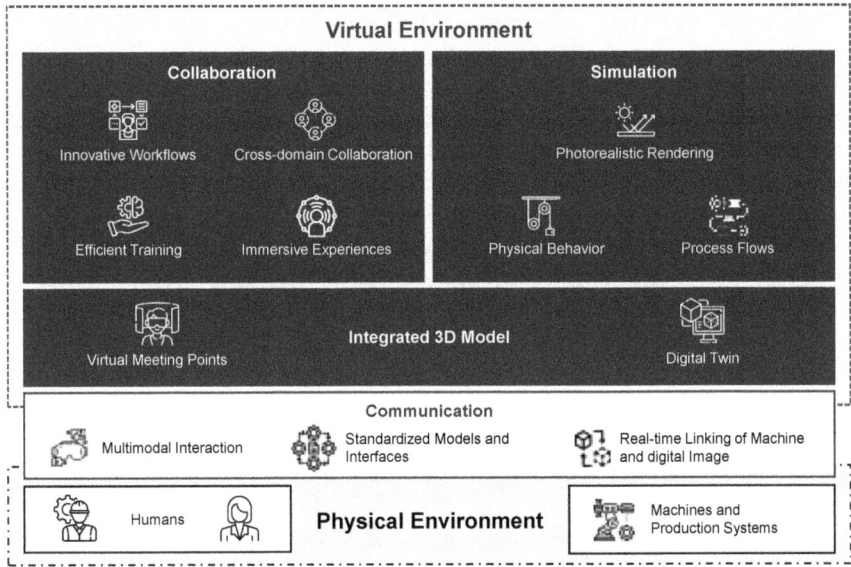

Fig. 2 Functionalities of the industrial metaverse based on [5]

This facilitates the integration of various systems and technologies, enables seamless collaboration, and promotes the development of new applications and services.

The introduction of real-time, low-latency **communication technologies like 5G and 6G** is crucial for the realization of the Industrial Metaverse. They enable real-time data exchange, which is essential for applications such as remote maintenance, immersive AR/VR experiences, or dynamic process adjustments. Improved connectivity promotes the seamless integration of physical and digital worlds, forming the foundation for advanced industrial applications and services in the Industrial Metaverse.

Edge and Cloud Computing play a central role in the Industrial Metaverse by providing the necessary computing power for graphically intensive applications and enabling the operation of digital twins in the cloud. These technologies allow companies to conduct complex simulations and analyses without investing in expensive local hardware. Specifically, the edge computing component brings processing closer to the data generation site, significantly reducing latency, which is critical for applications such as real-time monitoring and control of production processes. Cloud computing complements this by offering scalable resources that enable the storage and analysis of vast amounts of data and provide a platform for hosting digital twins. Additionally, as-a-service offerings enhance the accessibility of Metaverse functionalities by making advanced technologies like VR/AR, AI, and IoT available to companies of all sizes. This combination not only supports more efficient resource utilization and innovation but also enables a new level of interactivity and collaboration in the Industrial Metaverse.

In the Industrial Metaverse, **Artificial Intelligence (AI)** plays a central role, particularly through the use of advanced technologies such as Large Language Models (LLMs), Generative AI (GenAI), Foundation Models, and Transformer models, which open up new potentials. LLMs and Transformer models are used in the Industrial Metaverse for understanding natural language and immersive user interaction, while GenAI and Foundation Models offer new possibilities for creating synthetic data and generating realistic simulation scenarios.

Die Fortschritte im Ray-Tracing und die Effizienzsteigerung moderner GPUs haben **fotorealistisches Rendering** im Industrial Metaverse zu einem unverzichtbaren Werkzeug gemacht, das weit über die Generierung von Bild-Datensätzen für Vision-Applikationen hinausgeht. Diese Technologie ermöglicht es, immersive Erfahrungen zu schaffen, die Nutzern eine nahezu reale Interaktion mit virtuellen Umgebungen und Produkten bieten. Sie revolutioniert Bereiche wie Engineering, Fabrikplanung und Produktentwicklung, indem sie eine präzise Visualisierung und Analyse komplexer Systeme in einer virtuellen Welt ermöglicht, wodurch der Entwicklungsprozess beschleunigt, die Zusammenarbeit gefördert und Kosten reduziert werden. Zudem leistet fotorealistisches Rendering einen entscheidenden Beitrag zur Mitarbeiterqualifizierung, indem es interaktive, realitätsnahe Trainingsumgebungen schafft, die das Lernen und die Vorbereitung auf reale Arbeitsbedingungen verbessern.

Advances in ray tracing and the increased efficiency of modern GPUs have made **photorealistic rendering** an indispensable tool in the Industrial Metaverse, extending far beyond the generation of image datasets for vision applications. This technology enables the creation of immersive experiences that provide users with almost real interactions with virtual environments and products. It revolutionizes areas such as engineering, factory planning, and product development by enabling precise visualization and analysis of complex systems in a virtual world, thus accelerating the development process, promoting collaboration, and reducing costs. Additionally, photorealistic rendering makes a significant contribution to employee training by creating interactive, realistic training environments that improve learning and preparation for real-world working conditions.

VR and AR technologies enable immersive experiences by immersing users in virtual environments or integrating digital information into their real surroundings. These technologies offer significant advantages for the industry by enabling interactive training, complex product designs, and maintenance procedures through visual overlays, which enhance accessibility and understanding. In the Industrial Metaverse, VR and AR play a central role in creating an intuitive and immersive interface between humans and the virtual world, revolutionizing product development and maintenance.

Blockchain technology is essential in the Industrial Metaverse for secure, transparent transactions and data authenticity. It enables smart contracts for automated interactions between systems and supports new business models such as DeFi and the tokenization of industrial goods. Additionally, blockchain ensures the data integrity of digital twins, promoting a decentralized, interoperable infrastructure in the Industrial Metaverse, which is crucial for connected, autonomous industrial ecosystems.

Application Fields, Benefits and Potentials

The Industrial Metaverse opens up advanced application fields for the industry throughout the entire product lifecycle, as illustrated in Fig. 3. These can be divided into three primary categories: empowering people, efficient engineering, and smart automation.

First, the Metaverse enhances human work through immersive and simulative technologies. Immersive learning environments promote virtual qualification, while simulation platforms, networked with real production systems, enable remote maintenance and control of physical assets. Advanced monitoring solutions and assistance systems, which go beyond traditional dashboards, are realized through graphically enriched work instructions.

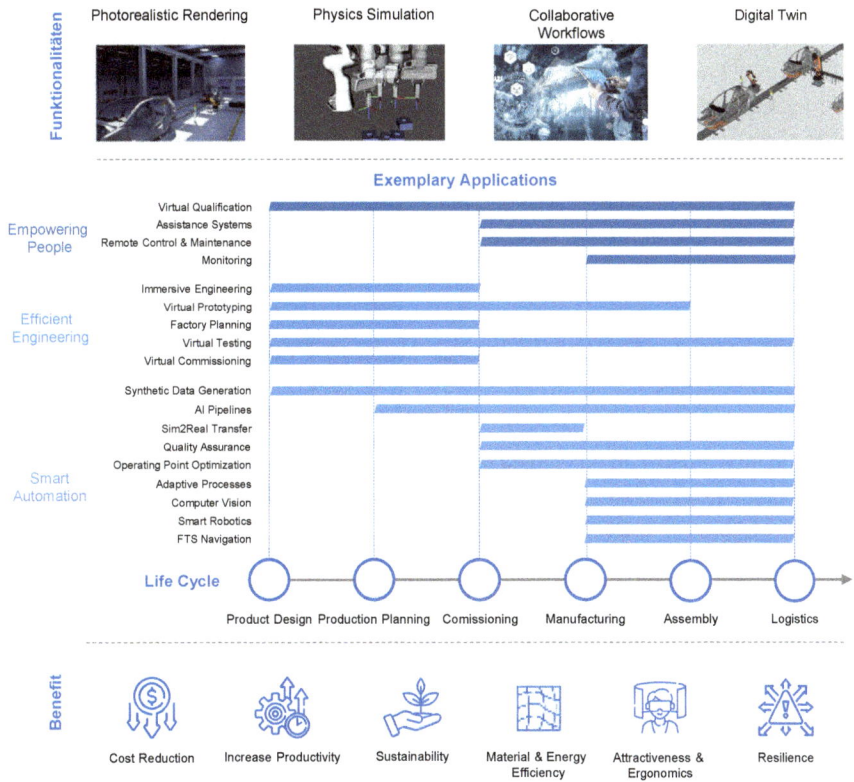

Fig. 3 Applications of the industrial metaverse along the lifecycle [5, 7]

Second, novel engineering approaches revolutionize the product development cycle. The integration of CAx, simulation, and planning tools with a centralized data base allows for iterative testing of virtual prototypes regarding functionality and manufacturability. This enables the parallelization of previously sequential engineering processes and significantly accelerates the production ramp-up, resulting in substantial economic benefits. Additionally, the simulation of production under various scenarios allows for in-depth analyses of efficiency, resilience, and sustainability.

Third, the Metaverse serves as a catalyst for intelligent automation processes in production. By generating synthetic data early in the lifecycle, AI applications can be trained more effectively. Cognitive robotics, machine vision for quality control, optimization of production facilities, and intelligent navigation of autonomous transport systems are examples of how the Metaverse optimizes implementations in intelligent automation.

In this contribution, the aforementioned potentials are specifically examined using the example of synthetic data generation for AI applications in production. The high effort required for generating and labeling image datasets for training often makes the introduction of ML algorithms uneconomical. This effort increases exponentially with a high variety of components and requires continuous retraining with product adjustments. At this point, the generation of synthetic data has the potential to significantly accelerate the adoption of AI-based systems by generating the necessary dataset through simulation tools and automatically labeling it [8]. The opportunities and challenges arising from the use of these synthetic image data will be discussed in the following section using two typical use cases.

2 Case Study—Synthetic Image Data in Production Applications

Synthetic image data can be generated in unlimited quantities and with desired variance through the creation of photorealistic renderings. However, they only represent a model of the real system. The difference between real and synthetically generated data often results in lower performance of the trained AI model on real data compared to the synthetically generated data. Overcoming this effect, known as the "Sim2Real Gap," requires the most accurate possible replication of both the product and the environment. The recent emergence of Metaverse technologies has had a significant impact on simulation technologies for industrial applications. Modern simulation platforms allow companies to virtually build complex processes and test them under real conditions and challenges before implementing them in the physical world. NVIDIA's Omniverse is a platform within the context of the industrial Metaverse that aims to further close the Sim2Real Gap through the efficient use of

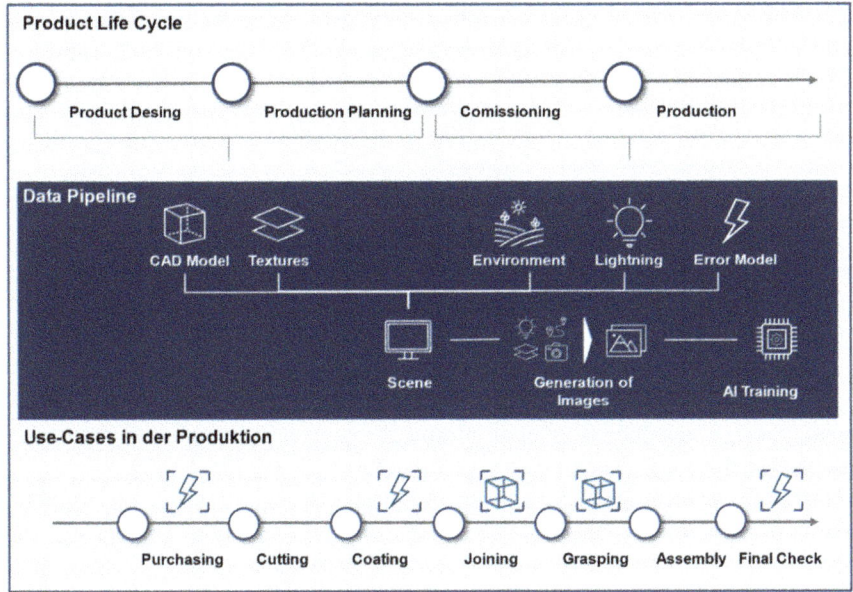

Fig. 4 Data pipeline for generating synthetic data across the product lifecycle based on [10, 11]

modern graphics cards [9]. The targeted variation of the simulation model's properties during data generation, also known as domain randomization, can make AI systems particularly robust for Sim2Real transfer.

The process of generating a two-dimensional image from a three-dimensional model is called rendering and involves converting 3D model data, which includes geometries, textures, lighting, and shading information, into pixel or vector data [10, 11]. The information required for generating synthetic data in an industrial environment comes from various phases of the product lifecycle, as illustrated in Fig. 4.

The CAD files originate directly from the engineering of the product and production, thus forming the basis for the rendering process. The choice of a specific material is made as part of the product design but may vary depending on suppliers or previous processes that alter the surface texture, such as grinding or polishing. The replication of the environment is also relevant. Environmental objects, which appear in the camera's field of view either directly or through their shadows, must be replicated in the simulated environment. Additionally, it is crucial to recreate the lighting conditions as realistically as possible. Various light sources are available in the simulation environment, which need to be selected and aligned accordingly.

Within production, numerous opportunities arise for the use of AI-based systems. Identifying suitable use cases requires consideration of both organizational factors (e.g., the provision of workpieces) and economic factors (cost–benefit analysis), as well as an assessment of technical feasibility. Classic application fields for AI-based image processing systems include, for example, object recognition/localization systems or quality assurance tools.

2.1 Object Recognition/Localization

Object recognition and localization in image processing is a core area of computer vision and a key factor for many applications such as autonomous driving, robotics, facial recognition, medical imaging, and security systems. ML-based image processing for object recognition and localization can be divided into different task types. These tasks, illustrated in Fig. 5, differ in the depth and degree of visual understanding and the resulting amount of information generated during the evaluation of an image [12].

The ability to identify objects and determine their location in space plays a crucial role in robotics, as it enables robots to react flexibly and autonomously to their environment. These visual informations can serve as the basis for intelligent path planning and autonomous handling operations, for example [13].

The rapid advancements in deep learning technologies, along with the increasing computational power of graphics processors, have led to the development of numerous robust and high-performance object localization systems in recent years [13]. A common characteristic of these systems is the high amount of training data required, which must be available in the form of annotated images of the objects

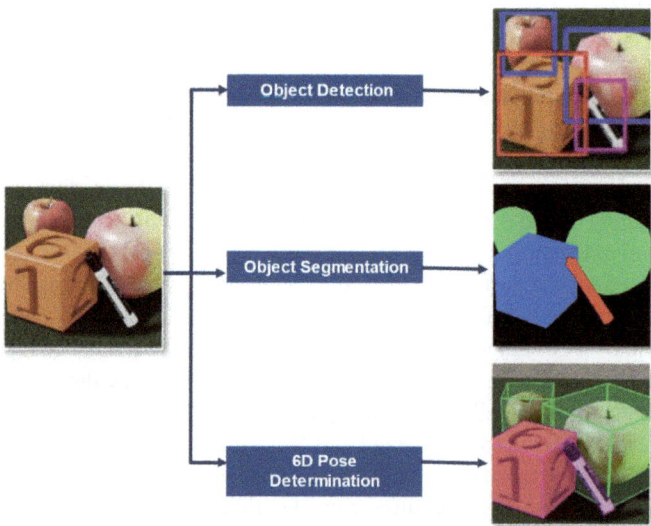

Fig. 5 Object detection, object segmentation and pose estimation

to be localized. Manually capturing these images is usually a time-consuming and costly endeavor, as the dataset must include all possible edge cases and the collected images must be manually labeled.

The above-described pipeline for generating synthetic data offers an effective alternative for creating this training dataset. Tools of the industrial Metaverse enable photorealistic rendering of objects while varying the properties of the virtual environment (lighting, background, etc.). Through this so-called "domain randomization" during dataset generation, it can be ensured that the AI trained based on the simulation can be successfully transferred to real images. The rendering of the dataset takes only a fraction of the time required for manual capture. Additionally, synthetic data generation provides high-quality annotations that are not contaminated by human inaccuracy and judgment, enabling precise object pose estimation [14].

2.2 Quality Inspektion

Another example is automated quality control for detecting defects. Currently, a large proportion of inspections are performed manually, which entails a high level of effort [15]. Quality inspections occur after certain processing steps or sometimes only at the end of production, meaning defective parts pass through the entire value chain only to be rejected during the final quality check. According to the rule of ten, an undetected defect at each stage of the value chain increases the quality costs, making early detection economically beneficial [16]. An automated quality inspection that runs parallel to the process would be desirable, as it would reduce lead times, save personnel costs, and ensure a consistent quality standard.

When considering whether to replace or supplement an existing system with an AI-based system, various factors must be taken into account. Figure 6 provides an overview. First, the current level of automation needs to be examined. This includes evaluating the tools used and the personnel effort involved. Additionally, the reliability of the current inspection process must be compared to the requirements. Furthermore, the current inspection duration must be related to the overall lead time. The value creation process between the cause of the defect and its detection must be monetarily evaluated in terms of early detection. Lastly, defect rates and the impacts of respective defects (scrap or rework of defective products) need to be included in the evaluation process.

After the organizational and economic assessment, an evaluation of the technical feasibility and the effort required for implementing such a system follows. Various indicators, which are outlined in Fig. 6, play a role in this assessment. The modeling is based on the data mentioned earlier and illustrated in Fig. 4. This includes both CAD data of the components and environmental models. Modeling occurring quality defects often requires a very detailed replication of textures. There are numerous pre-modeled surface textures in libraries that are integrated into the Metaverse. If the relevant texture is not available there, this – like the creation of environmental

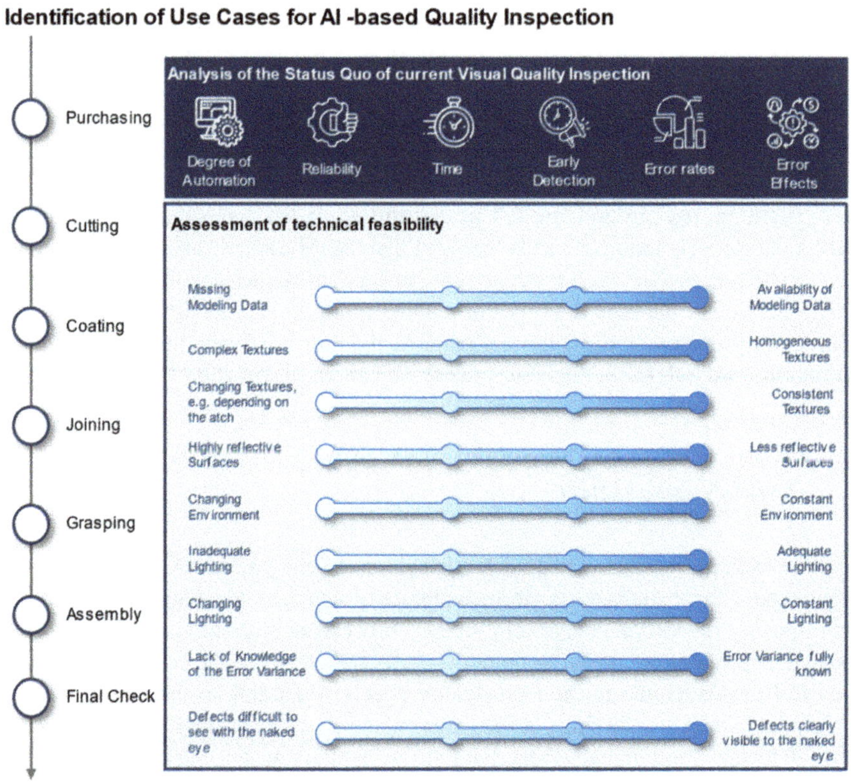

Identification of Use Cases for AI-based Quality Inspection

Fig. 6 Identification of use cases for AI-based quality inspection

models – entails additional effort. Complex textures are more challenging to replicate than homogeneous textures, and variations in textures, such as those caused by changing material batches, require greater dataset variance and thus result in higher modeling effort. Highly reflective surfaces complicate the detection of surface defects. In this case, lighting plays a particularly crucial role. A constant and adequate lighting setup is desirable. Depending on the component and defect, an individual lighting concept must be decided upon. Knowledge of relevant defects, including their variations in shape, type, and dimension, forms the basis for generating the necessary data. Generally, a good indicator is that defects easily visible to the naked eye can be well replicated and detected by AI. However, each case must be evaluated individually.

3 Conclusion and Outlook

The concept of the Industrial Metaverse offers numerous possibilities to fundamentally change workflows in the manufacturing industry in the near future. Beyond the use of synthetic data for training artificial intelligence, as discussed in this article, the potentials of the Industrial Metaverse can be identified at all stages of the product lifecycle. Shorter innovation cycles through collaborative product design and virtual prototyping, faster implementation of new processes and systems through virtual commissioning, and innovative training concepts in virtual reality, as well as the ability to visualize production processes in three dimensions to an unprecedented extent and present them intuitively, are just some of the conceivable implications. In particular, the innovative integration of functionalities that were previously often strictly separated creates entirely new workflows that save time and resources. In this sense, the Industrial Metaverse represents an evolutionary extension of Industry 4.0, further blurring the boundaries between the physical and digital worlds and laying the foundation for the future of industrial production.

As this contribution shows, we are already on a good path to realizing the vision of the Industrial Metaverse in some areas. Despite these advances, companies still face several challenges in implementing the Industrial Metaverse. A 2023 study highlights high implementation costs, lack of expertise, and integration issues with existing systems as major barriers [7]. However, these problems are not insurmountable: the cost factor tends to decrease over time as the technology matures and spreads, and the lack of know-how can be addressed in the long term with targeted training and further education measures or in the short term by purchasing expertise through consulting services. Crucial, however, is the evaluation of the added value that the technologies provide for specific business contexts.

Before such steps are initiated, however, the question should be asked as to whether and how the Industrial Metaverse technologies deliver concrete added value for the company in question. It will therefore be our task in the near future to continue to critically analyze the possibilities of industrial metaverse technologies with regard to the generation of added value in companies and to make further developments where necessary. Analogous to the possibilities of using synthetic data for quality control presented above, the focus should be on the symbiosis of technology and real use cases. The seamless integration of existing systems, the reduction of the gap between simulation and reality and an improved information network are core areas for future research. At the same time, the role of humans in this system must not be lost sight of. Modern possibilities for human–machine interaction, such as augmented reality, virtual reality and chatbots based on large-language models such as ChatGPT, offer a wide range of options for visualizing content and implementing an exchange between humans and industrial metaverse.

However, it must be taken into account that, in addition to any lack of expertise in using devices and tools, employee acceptance must also be addressed. Innovative technologies only work if the people who work with them know how to use them correctly and have a certain level of trust in the technology.

A first step towards realizing the desired vision is being taken in parallel by several German automotive companies [13, 14, 17, 18] and Siemens AG [15, 19] in cooperation with NVIDIA. The Omniverse platform developed by NVIDIA is being used, which uses the Universal Scene Description (USD) file exchange format to seamlessly integrate a wide variety of programs for the development of three-dimensional environments [14]. By using correspondingly powerful hardware, three-dimensional, photorealistic environments can be built and used for various purposes. To further develop the USD format and promote greater interoperability between 3D tools and data, the "Alliance for OpenUSD" was founded at the beginning of August 2023 [16, 20].

These examples show that collaborative efforts are already being made in various places to overcome the current inhibiting factors in the implementation and application of the industrial metaverse. However, the future task remains to test the limits of the technologies, identify meaningful use cases and design practical, participatory implementation options. This requires ongoing cooperation between research institutes and technology companies in order to make the vision of the Industrial Metaverse a reality.

References

1. L.-H. Lee et al., in *all one needs to know about metaverse: a complete survey on technological singularity*. Virtual Ecosystem, and Research Agenda (2021). https://doi.org/10.48550/arXiv.2110.05352
2. Plattform I4.0: Industrial Metaverse, Impulspapier (2023). https://www.plattform-i40.de/IP/Redaktion/DE/Downloads/Publikation/Industrial_Metaverse.pdf?__blob=publicationFile&v=19
3. C. Brecher et al., in *Sustainable Production-as-a-Service, Conference Paper* (2023). https://doi.org/10.24406/publica-949
4. e.V. Bitkom, Industrial metaverse—use cases. Merhwerte und Potenziale für den Wirtschaftsstandort Deutschland. https://www.bitkom.org/sites/main/files/2023-09/bitkom-leitfaden-industrial-metaverse.pdf
5. O. Petrovic, Y. Dassen, C. Brecher, Potenziale des Industrial Metaverse—Konvergenz von der Simulation bis zur Realität, 2023. https://www.industrie-management.de/sites/industrie-management.de/files/IM_05-2023-Petrovic.pdf
6. https://assets.new.siemens.com/siemens/assets/api/uuid:260d1ead-250b-45fb-b32f-caa9924c1ddd/The-Emergent-Industrial-Metaverse.pdf
7. https://www2.deloitte.com/us/en/insights/industry/manufacturing/industrial-metaverse-applications-smart-factory.html
8. O. Petrovic, D.L.D. Duarte, W. Herfs, Generating synthetic data—using a knowledge-based framework for autonomous productions, in 2023 *IEEE/ASME International Conference on Advanced Intelligent Mechatronics (AIM)*, ed. (IEEE, 28–30 June 2023), pp. 1086–1093
9. NVIDIA: What Is Isaac Sim?—Omniverse IsaacSim latest documentation. https://docs.omniverse.nvidia.com/isaacsim/latest/overview.html. Lasted checked 04 Nov 2023
10. J. Peddie, in *Ray Tracing: A Tool for All. (Springer eBooks Computer Science*, ed. 1st edn. (Springer, Cham, 2019), p. 7–27
11. K. Lehn, M. Gotzes, F. Klawonn, in *Introduction to Computer Graphics. Using OpenGL and Java. (Undergraduate Topics in Computer Science)*, 3rd edn. (Springer International Publishing; Imprint Springer, 2023), p. 1–13

12. O. Russakovsky, J. Deng, H. Su, J. Krause, S. Satheesh, S. Ma, Z. Huang, A. Karpathy, A. Khosla, M. Bernstein, A.C. Berg, L. Fei-Fei, ImageNet large scale visual recognition challenge (2014)
13. Z. Zou, K. Chen, Z. Shi, Y. Guo, J. Ye, Object detection in 20 years: a survey, in *Proceedings of the IEEE*, vol. 111(3), p. 257–276 (2023)
14. Z.-Q. Zhao, P. Zheng, S.-T. Xu, X. Wu, Object detection with deep learning: a review. IEEE Trans. Neural Netw. Learn. Syst. **30**(11), 3212–3232 (2019)
15. J.M. Rožanec, P. Zajec, E. Trajkova, B. Šircelj, B. Brecelj, I. Novalija, P. Dam, B. Fortuna, D. Mladenić, Towards a comprehensive visual quality inspection for industry 4.0. IFAC-PapersOnLine **55**(10), 690–695 (2022). https://doi.org/10.1016/j.ifacol.2022.09.486
16. https://www.sixsigmablackbelt.de/fehlerkosten-10er-regel-zehnerregel-rule-of-ten/
17. BMW Group: BMW Group und NVIDIA heben virtuelle Fabrikplanung auf die nächste Ebene. Internet-adresse. https://www.press.bmwgroup.com/deutschland/article/detail/T0329569DE/bmw-group-und-nvidia-heben-virtuelle-fabrikplanung-auf-die-naechste-ebene?language=de. Zuletzt aufgerufen am 10 Sep 2023
18. D. Shaprio, Manufactured in the metaverse: mecedes-benz assembles next-gen factories with NVIDIA omniverse. Internetadresse. https://blogs.nvidia.com/blog/2023/01/03/mercedes-benz-next-gen-factories-omniverse/
19. Siemens: Siemens and NVIDIA partner to build the Industrial Metaverse. Internetadresse. https://www.sie-mens.com/global/en/company/insights/siemens-and-nvidia-partner-to-build-the-industrial-metaverse.html. Zuletzt aufgerufen am 29 Feb 2024
20. Alliance for OpenUSD: The Alliance for OpenUSD (AOUSD): Shaping the Future of 3D Technology. Internetadresse. https://aousd.org/the-alliance-for-openusd-aousd-shaping-the-future-of-3d-technology/. Zuletzt aufgerufen am 29 Feb 2024

Oliver Petrovic holds the position of Chief Engineer and Head of Department for Automation and Control Technology at the Laboratory for Machine Tools WZL at RWTH Aachen University. His department is intensively researching and advancing topics such as robotics, Industrial IoT, Artificial Intelligence and human-technology interaction. Thus, a very broad field of excellence in the automation and digitalization of industrial production is being advanced.

Before leading the department, Oliver Petrovic spent many years applying his expertise in the fields of robotics, AI and industrial assembly in numerous research and industrial projects. He disseminates this expertise to a broad community in lectures, standardization committees and conferences https://www.wzl.rwth-aachen.de/.

Tokenization

Introduction to Tokenisation

Lukas Wagner

1 Introduction

Even before the publication of the Bitcoin whitepaper in 2008 and the further developments of what is now commonly referred to as blockchain technology (or, even more broadly, distributed ledger technology (DLT)), there had been different attempts to establish digital representations of physical assets in order to facilitate transactions with these assets. With the development of DLT over time, a wide range of use cases, frameworks and implementations has emerged under the concept of what has since been coined tokenisation. Today, some estimate that tokenised digital securities in the amount of $ 4 to $ 5 trillion could be issued by 2030 [8], while others even project more than $ 16 trillion or 10% of global GDP to be tokenised by that time, highlighting considerable potential in relation to currently illiquid assets [17]. This chapter will provide an introduction to the concept of tokenisation (Sect. 2), its history (Sect. 3) and more recent developments in the form of different legal frameworks (Sect. 4). The chapter will then provide a high level overview over the process of tokenisation (Sect. 5), and discuss tokenisation's benefits and current shortcomings (Sect. 6), before closing with an outlook (Sect. 7).

2 Tokenisation as a Concept

As a concept, and in the capital markets context discussed in this chapter, tokenisation commonly refers to the creation of a digital representation of something of value [5, 20]. Such digital representation can refer to what are often called 'Real World

---

L. Wagner (✉)
General Counsel, NYALA Digital Asset AG, Berlin, Germany
e-mail: l.wagner@nyala.de

Assets',[1] but also purely digital benefits or services, or individual aspects of replaceable physical goods, such as (in the case of the Blockstream Mining Note) a certain amount of hash rate employed for the mining of bitcoin [24]. In a frequently used differentiation, such tokens[2] would be considered so-called extrinsic crypto assets, as they derive their value from an object outside of the DLT system, in contrast to intrinsic crypto assets that are not linked to anything beyond the crypto asset itself [1]. How such link between the digital representation and the underlying value is achieved, however, varies in detail, and will be set out further below (Sect. 4).

3 History

Even before the current use of the concept of tokenisation in a capital markets context as set out in this chapter, tokenisation had already been used in data security. There, it refers to the process of replacing sensitive information, such as credit card or transaction data, with a randomly generated token to avoid risks for customers due to data breaches [29]. Somewhat similarly, in machine learning, tokenisation refers to the practice of converting a sequence of text into smaller parts, the tokens. This is aimed at helping machines understand human language by breaking a text down into bite-sized pieces that are then easier to analyse [4].

In the interpretation as set out above under Sect. 2, on the other hand, the concept was discussed in the sphere of the so-called Cypherpunks [15], even ahead of the creation of Bitcoin in 2008–9 as the first implementation of what is now commonly referred to as blockchain technology. One early such discussion referred to digital, cryptographically secured property titles [27].

Finally, after the inception of Bitcoin, there were attempts to represent value in token form beyond the first blockchain network's native currency bitcoin. The first widely discussed such project was Mastercoin [33], followed by further variations including the concept of Colored Coins [3]. However, the programmability of Bitcoin is limited to a relatively narrow scripting language. As several early users found this functionality to be too limited, Ethereum was created as a Turing-complete[3] smart contracting platform. It also included one of the first more elaborate descriptions of the concept of a token in a blockchain environment [7], which was developed even further over time with the creation of the ERC-20 token standard [31] as well as later iterations, including standards for tokenised securities specifically [19, 23]. From there, more and more alternative blockchain platforms emerged, creating a lot of

---

[1] Often abbreviated as RWA; however, this acronym should be avoided for real word assets, as it has been used for decades in the financial context to refer to risk-weighted assets, in particular in the context of capital requirements [6].

[2] On a technical level, the term coin is commonly used to refer to the native currency of a DLT system, such as bitcoin (BTC) in the case of Bitcoin, or Ether (ETH) in the case of Ethereum, while units issued on top a DLT system are referred to as tokens [2].

[3] After [30].

hype around the potential of tokenisation, leading up to the idea of the 'tokenization of everything' [21].

Even more recently, the term tokenisation has also been used to refer to proprietary solutions in the financial sector, which, while borrowing some characteristics from public permissionless blockchains, are permissioned and very centralised as regards the interactions between the holder and their ledger entry. While some legal frameworks account for these structural differences, such as the German eWpG (see Sect. 4.3.3), these more centralised approaches are often also conflated with the original idea of public permissionless tokens that are mainly controlled by their owner, leading to considerable confusion.

4 Legal Frameworks

4.1 Overview

As tokens in tokenisation (as understood in this chapter) refer to objects outside of the DLT system, the question arises how to link the token in question to the underlying value. Over time, different legal approaches have been developed for this. Initially, issuers had attempted to provide a mere contractual link in the terms of issue or similar documentation. As this approach proved unsatisfactory due to legal uncertainty, e.g. in terms of negotiability and in cases of insolvency of issuers and custodians, certain jurisdictions introduced dedicated legal frameworks, often equating DLT-based securities to traditional paper-based securities. More recently, novel approaches have been developed to provide even wider negotiability beyond what a traditional paper-based security would provide, and potentially making it easier to encompass value not captured as such under traditional legal concepts [32].

4.2 (Contractual) Security Token

The first wave of fund raising based on DLT was the boom of Initial Coin Offerings (ICOs) in 2017–8. Projects issued a token that often had a rather loose connection to any proceeds to be generated by the project. Frequently, the token was meant to be used as a means of payment on a platform or system yet to be built. At the moment of fund raising, there was thus no direct consideration associated with it [26].

Out of these ICOs, Security Token Offerings (STOs) developed as a more structured form of raising money, closer to traditional securities [18]. While such security tokens typically have a closer link to the revenues generated by a project than the indirect effect of a project's (future) payment token gaining in (expected) value, the link between the token and the project was still merely contractual and dependent upon the token's documentation (if any).[4] This put investors in STOs at a disadvantage compared to investors in traditional (paper-based) securities, as the mere contractual link between the token and the contractual claim against the project can break in several circumstances (e.g. vice of consent during a transfer, or inheritance), and there are no provisions enabling a bona fide acquisition of future acquirers in good faith. Furthermore, the mere contractual nature of the claim against the project has certain disadvantages in case of insolvencies of the issuer or intermediaries, where the security token, unlike a physical object such as a paper certificate, cannot clearly be separated from the insolvency estate [26].

4.3 Statutory (Broad) Tokenisation Frameworks

4.3.1 Overview

Against the backdrop of these disadvantages of mere contractual security tokens, more and more jurisdictions have introduced dedicated statutory frameworks for tokenisation over time to provide legal certainty for both issuers and investors.

4.3.2 Liechtenstein TVTG

One of the first such legal frameworks was the Token and Trusted Technology Service Provider Act (*Token- und VT-Dienstleister-Gesetz* – TVTG, also known as the Blockchain Act) of Liechtenstein, which was introduced in 2019. It is relatively broad in scope, and not only encompasses the tokenisation of claims as well as membership rights in corporations, but also of property in physical objects. For the verification that such physical objects actually exist and the respective property rights are enforceable, the TVTG introduces the role of a regulated entity, the so-called physical validator [12]. While this framework appears very wide and flexible in terms of scope, and while 27 companies have been registered under the framework since its inception [13], the adoption in concrete projects still seems limited, which might also be due to the limited relevance of the jurisdiction in question.

---

[4] A similar approach has been taken in particular by so-called neo brokers offering what they call 'fractionalised shares'. These instruments are not actual stocks of the company in question but derivative contracts that represent a claim on the economic performance of the underlying company. Recently, this concept has been extended into so-called stock tokens, which are meant to represent such derivative contracts in token form [22, 25].

4.3.3 German eWpG

As one building block in a series of DLT-related legal frameworks, Germany also introduced an Electronic Securities Act (*Gesetz über elektronische Wertpapiere* – eWpG) in 2021. Over time, its scope has been extended: while it only covered bearer bonds initially, it now also applies to fund units and registered shares.

In contrast to some legal frameworks for digital securities in other countries, which do not clearly account for the structural differences between traditional securities held in book-entry form and tokens that are directly held by the investors, the eWpG clearly distinguishes what it calls central register securities (registered in a collective entry – *Sammeleintragung*) from so-called crypto securities (registered in an individual entry – *Einzeleintragung*). Electronic central register securities in that sense follow the traditional logic of paper-based securities held in book-entry form. They are held by investors via a chain of custody relationships, leading back to the top-level custody account, which is held for all investors in the name of either a central securities depository (CSD) or a custody bank. Where the account and register are held by a CSD, central register securities are automatically recorded in book-entry form and thus tradable on a trading venue [14].

Crypto securities (registered in an individual entry), on the other hand, are directly held by the investor in the form of a token. This has consequences for the custody of the crypto security, which can be held in self-custody without the involvement of a custody bank or other intermediary.

Since the inception of the eWpG, more than 1.5 million central register securities and more than 190 crypto securities have been issued.[5]

4.3.4 Further Initiatives in EU Member States

More recently, and in light of the EU's DLT Pilot Regime (Regulation (EU) 2022/858 of 30 May 2022 on a pilot regime for market infrastructures based on distributed ledger technology – DLTR), which is applicable since March 2023 and aims at creating market infrastructure for DLT-based securities, other EU Member States have also introduced frameworks for digital securities. Two examples are Italy's decree law no 25 of 17 March 2023 (*Disposizioni urgenti in materia di emissioni e circolazione di determinati strumenti finanziari in forma digitale e di semplificazione della sperimentazione FinTech* – DL Fintech) and Spain's law 6/2023 on Securities Markets and Investment Services (*Ley 6/2023 de los Mercados de Valores y de los Servicios de Inversión* – LMVSI). Since their inception, the national regulators have also published implementing regulations as well as guidance as regards the licence required for the issuance of digital securities under these frameworks [9, 10]. However, while the first licences have been issued, many companies are still

---

[5] As of July 2025; see, for central register securities, Clearstream's Public Security Terms at https://public-security-terms.clearstream.com/, and for crypto securities BaFin's list of crypto securities at https://www.bafin.de/ref/19659410.

in the process of applying for their respective licences and of implementing the requirements.

4.4 Promise of a Reward

Beyond these explicit and detailed dedicated legal frameworks, there have been more innovative attempts to make use of the legal concept of a promise of a reward to achieve even further negotiability than would be the case for traditional paper-based securities. This concept provides for a binding promise (and a resulting claim) in case a reward for undertaking an act was offered by means of a notice by publication.[6] Under this concept, the claim that is to be tokenised can be linked in the token documentation to the holder of the token returning the same token to the issuer. As the requirement for the claim is merely the holder returning the token, it is, as a general principle, irrelevant how the token holder came into possession of the token, which provides for more comprehensive protection of transactions, as an acquirer of the token is exposed to very limited risks of his acquisition being disputed later when redeeming the token for the underlying asset. This framework is currently used in Germany to de-facto tokenise shares in limited liability companies (*Gesellschaften mit beschränkter Haftung* – GmbHs), which cannot be tokenised under the eWpG at this stage [28]. However, as a novel and somewhat unorthodox solution, this approach might be less suited for traditional investors.

4.5 Dedicated Frameworks for Specific Token Types

In addition to these securities-related legal frameworks, further dedicated frameworks for specific token types have been developed over time, which will not be discussed in detail in this chapter. The EU's Markets in Crypto-Assets Regulation (Regulation (EU) 2023/1114 of 31 May 2023 on markets in crypto-assets – MiCAR), for instance, provides a framework for the issuance of stablecoins. Stablecoins are pegged to the value of one or more official currencies, or one or more other assets, and such pegging is supposed to ensure a stable value. Under MiCAR, they can be issued in the form of so-called asset-referenced tokens and e-money tokens.

Furthermore, the European Commission has now proposed a regulation for the issuance of a digital euro by the European Central Bank (Proposal for a regulation on the establishment of the digital euro, COM(2023) 369 final) as a so-called Central Bank Digital Currency (CBDC). Other jurisdictions such as China, the Bahamas or Nigeria have already launched such CBDCs, raising concerns, inter alia in relation to the privacy of its users and excessive control due to a potential programmability of the digital money [16].

---

[6] See, e.g., Sec. 657 of the German Civil Code (*Bürgerliches Gesetzbuch* – BGB).

5 High-Level Overview Over the Tokenisation Process and Life Cycle

For context, the following section shall give a high-level overview over the wider tokenisation process and life cycle, with a focus on the tokenisation of a bearer bond under the German eWpG (see Sect. 4.3.3 above). For a better overview, the description may skip some minor technical steps. Details may also vary under other legal frameworks, and even for the issuance of other assets under the eWpG.

- **Sourcing**: Where parties beyond the issuer itself are involved, the investment may be sourced via platforms such as crowdfunding platforms or similar service providers.
- **Documentation**: Once an investment has been identified, the legal documentation needs to be drafted, including the terms of issue and further information documents, such as a prospectus (where required) and/or a securities information sheet.
- **Registration**: Once the terms of issue are finalised, they are recorded with the registrar as the basis for the token.
- **Token Creation**: Tokens are then created on the chosen blockchain protocol under the International Securities Identification Number (ISIN) of the security, which link to the registered terms of issue.
- **Creation of Investor Wallets**: In case investors do not yet have a wallet/address on the chosen blockchain protocol, wallets for these investors are created, potentially involving a custodian.
- **Distribution of Tokens**: The tokens are distributed to the wallets/addresses of the initial investors.
- **Secondary Market Trading**: Once distributed to the initial investors, these can trade the securities onwards, either OTC/bilaterally or on a trading venue.
- **Servicing/Corporate Actions**: As investors are holders of the tokens and thus have a wallet/address on the blockchain network, the servicing of the securities could be done via an on-chain cash representation on the same blockchain network. Similarly, corporate actions could be dealt with on- or off-chain using the respective private key relating to the investor's address. However, these processes are currently for the most part still completed via traditional means, i.e. bank transfers and conventional (digital) means of communication.
- **Redemption**: Upon maturity of the security, it is redeemed with the issuer against payment of the principal, and the respective token is deleted/marked as redeemed.

6 Potential Benefits of Tokenisation

6.1 Overview

Proponents of tokenised securities, such as the commentators estimating $ 4 to $ 16 trillion in tokenised assets by 2030 as cited above, refer to a number of benefits of tokenisation over traditional paper-based securities, which will be discussed in this section.

6.2 Increased Efficiency

Natively digital securities in token form enable much higher levels of automation throughout the security life cycle, inter alia through the programmability of the token itself. This can increase the efficiency of both the initial issuance as well as primary and secondary markets, with settlements close to real time (depending on the block time of the chosen infrastructure), compared to the settlement cycles of one or more days in traditional settlement infrastructures.[7] The automation of and rules-based interactions with programmable tokens might also render certain misuses of the securities technically impossible, thereby making roles of previously required intermediaries, such as clearing houses, obsolete.

However, such notable efficiency gains presuppose that the DLT logic of direct, peer-to-peer interaction between participants is followed through to a significant extent. In practice, established players along the value chain are often more comfortable with a setup that mirrors their previous arrangements more closely for the time being. This often leads to solutions that mimic legacy infrastructure to a greater extent than would be necessary, with solutions that are ultimately more complex than what would be technically and legally required. On top of this, the mandatory involvement of state regulators or similar entities under the specific legal framework might also lead to increased efforts and costs. An example is the continued requirement of a fairly manual publication of every issuance of a crypto security under the German eWpG in the German Federal Gazette, which costs close to EUR 30 per issuance.[8] Finally, even where participants would be willing to adopt new solutions, some of these potential efficiency gains cannot yet be realised due to a lack of adoption of the necessary infrastructure by market players or challenges regarding the interoperability of different blockchain systems.

---

[7] In traditional market terminology, T + x is used to express the timeframe between the trade and its settlement, e.g. T + 1 referring to a timespan of one day between the trade day and the settlement day [11].

[8] As part of the Financing for the Future Act (*Zukunftsfinanzierungsgesetz*) of 11 December 2023, this requirement is set to be abolished, but only with effect from 1 November 2025.

6.3 Cost Savings

Where it can be achieved however, such reduction of the intermediaries required and involved in the process, combined with a higher degree of automation due to digitalisation and standardisation of processes, can lead to significant operational cost savings. It can lead to even further indirect cost savings due to a higher capital efficiency, as funds are not tied up for days while a transaction is awaiting clearance (see [5]).

At the same time, establishing these new infrastructures and integrating them into existing systems often requires relatively high upfront investments in IT but also the surrounding processes, which need to be amortised before cost savings of the setup can materialise. Potential issuers with relatively low volumes will therefore have to wait a considerable time before any net cost savings can be seen.

6.4 Increased Liquidity/Democratisation of Investments

The efficiency gains and reduced costs enable the tokenisation of assets that could not previously be issued as securities, increasing the liquidity of these assets. In turn, the technology also enables the fractionalisation of an investment into very small ticket sizes, which opens up the market to investors that could not traditionally invest in it. However, even where the smaller ticket sizes should open up the market to new investors, some asset classes might still remain inaccessible for retail investors for regulatory reasons (see [5]).

6.5 More Compliant and More Auditable Processes

Finally, the high degree of automation and programmability, which renders non-compliance technically impossible and the whole process publicly auditable on the blockchain, will lead to more compliant and more easily auditable processes. In the future, this might render certain legal requirements obsolete. As the systems could be accessed by regulators directly via API where they require certain information, this might also include the reporting requirements financial institutions have to fulfil, which are currently very extensive. However, that presupposes a regulator that is tech-savvy enough to make use of these possibilities, which is unfortunately not everywhere the case at the moment.

7 Outlook

As has been shown, the tokenisation of securities is a promising development for the democratisation of finance. At the same time, many frameworks and technological solutions are still in the process of being developed and adopted. It might therefore still take some time before the concept and its implementations can show its full potential.

References

1. Allen&Overy, *Germany to Set Out Civil Law Treatment of Crypto-Assets* (2022).
2. A.M. Antonopoulos, G. Wood, *Mastering Ethereum: Building Smart Contracts and DApps* (O'Reilly Media, Sebastopol, 2018)
3. Y. Assia, V. Buterin, L. Hakim, M. Rosenfeld, R. Lev, *Colored Coins whitepaper* (2013). https://docs.google.com/document/d/1AnkP_cVZTCMLIzw4DvsW6M8Q2JC0lIzrTLuoWu2z1BE/edit. Accessed 18 Jul 2025
4. A.A. Awan, *What is Tokenization?* (2023). https://www.datacamp.com/blog/what-is-tokenization. Accessed 18 Jul 2025
5. A. Banerjee, I. De Bode, M. de Vergnes, M. Higginson, J. Sevillano, *Tokenization: A digital-asset déjà vu* (2023). https://www.mckinsey.com/industries/financial-services/our-insights/tokenization-a-digital-asset-deja-vu. Accessed 18 Jul 2025
6. Basel Committee on Banking Supervision, *RBC: Risk-based Capital Requirements RBC20: Calculation of Minimum Risk-based Capital Requirements* (2023). https://www.bis.org/basel_framework/chapter/RBC/20.htm?inforce=20230101&published=20201126. Accessed 18 Jul 2025
7. V. Buterin, *Ethereum: A Next-Generation Smart Contract and Decentralized Application Platform* (2014). https://ethereum.org/content/whitepaper/whitepaper-pdf/Ethereum_Whitepaper_-_Buterin_2014.pdf. Accessed 18 Jul 2025
8. Citigroup, *Money, Tokens, and Games* (2023). https://www.citigroup.com/global/insights/citigps/money-tokens-and-games. Accessed 18 Jul 2025
9. CNMV, *Preguntas y Respuestas frecuentes sobre los Instrumentos Financieros basados en Tecnologías de Registros Distribuidos (TRD)* (2024). https://www.cnmv.es/DocPortal/Fintech/FAQ_IFbasadosTRD.pdf. Accessed 18 Jul 2025
10. Consob, *Regolamento sull'emissione e circolazione in forma digitale di strumenti finanziari* (2023). https://www.consob.it/documents/1912911/1950567/reg_consob_2023_22923.pdf. Accessed 18 Jul 2025
11. Finra, *Understanding Settlement Cycles: What Does T+1 Mean for You?* (2024). https://www.finra.org/investors/insights/understanding-settlement-cycles. Accessed 18 Jul 2025
12. Government of Liechtenstein, *Report and Application Concerning the Creation of a Law on Tokens and TT Service Providers (Tokens and TT Service Provider Act; TVTG) and the Amendment of Other Laws* (2019). https://www.naegele.law/files/Downloads/2019-07-12_BuA_TVTG_en_full_report.pdf. Accessed 18 Jul 2025
13. Government of Liechtenstein, *Stellungnahme der Regierung an den Landtag des Fürstentums Liechtenstein zu den anlässlich der ersten Lesung betreffend die Abänderung des Token- und VT-Dienstleister-Gesetzes (TVTG) und weiterer Gesetze aufgeworfenen Fragen* (2023). https://bua.regierung.li/BuA/pdfshow.aspx?nr=116&year=2023. Accessed 18 Jul 2025
14. C. Heise, *Now also in Electronic Form: Securities* (2021). https://www.bafin.de/ref/19620014. Accessed 18 Jul 2025
15. E. Hughes, *A Cypherpunk's Manifesto* (1993). https://www.activism.net/cypherpunk/manifesto.html. Accessed 18 Jul 2025

16. Human Rights Foundation, *CBDC Tracker* (2025). https://cbdctracker.hrf.org/home. Accessed 18 Jul 2025
17. S. Kumar, R. Suresh, D. Liu, B. Kronfellner, A. Kaul, *Relevance of On-chain Asset Tokenization in 'Crypto Winter'* (2022). https://www.bcg.com/publications/2022/relevance-of-on-chain-asset-tokenization. Accessed 18 Jul 2025
18. T. Lambert, D. Liebau, P. Roosenboom, Security token offerings. Small Bus. Econ. **59**, 299–325 (2022). https://doi.org/10.1007/s11187-021-00539-9
19. J. Lebrun, T. Malghem, K. Thizy, L. Falempin, A. Boudjemaa, *ERC-3643: T-REX - Token for Regulated Exchanges* (2021). https://eips.ethereum.org/EIPS/eip-3643. Accessed 18 Jul 2025
20. M. Ong, *Back To Basics: Tokenization Explained* (2023). https://www.forbes.com/sites/forbesfinancecouncil/2023/12/20/back-to-basics-tokenization-explained/. Accessed 18 Jul 2025
21. S. Radocchia, *How the Blockchain Could Help the Digital Lives of Musicians and Artists* (2018). https://hackernoon.com/how-the-blockchain-could-help-the-digital-lives-of-musicians-and-artists-7bd7b2566cdd. Accessed 18 Jul 2025
22. Robinhood, *Private Stock Giveaway Terms* (2025). https://cdn.robinhood.com/assets/robinhood/legal/private_stock_giveaway_promotion_terms_and_conditions_eu.pdf. Accessed 18 Jul 2025
23. J. Shiple, H. Marks, D. Zhang, *ERC-1450: A Compatible Security Token for Issuing and Trading SEC-Compliant Securities* (2018). https://eips.ethereum.org/EIPS/eip-1450. Accessed 18 Jul 2025
24. Sicos Securities, *MERJ Pre-Listing Statement for the Listing of the Blockstream Mining Notes* (2022). https://stokr.io/blockstream-mining/terms. Accessed 18 Jul 2025
25. P. Smith, *Robinhood's Private Stock Tokens Lure Investors, Draw Scrutiny* (2025). https://www.bloomberg.com/news/articles/2025-07-12/robinhood-s-private-stock-tokens-lure-investors-draw-scrutiny. Accessed 18 Jul 2025
26. Swiss Federal Council, *Legal Framework for Distributed Ledger Technology and Blockchain in Switzerland* (2018). https://www.newsd.admin.ch/newsd/message/attachments/55153.pdf. Accessed 18 Jul 2025
27. N. Szabo, *Secure Property Titles with Owner Authority* (1998). https://nakamotoinstitute.org/library/secure-property-titles. Accessed 18 Jul 2025
28. Tokenize.it, *Legal setup* (2025). https://www.tokenize.it/documentation/legal-setup. Accessed 18 Jul 2025
29. Trustcommerce, *Where Did Tokenization Come From?* (2017). https://web.archive.org/web/20170224055805/, https://trustcommerce.com/blog/where-did-tokenization-come-from/. Accessed 18 Jul 2025
30. A. Turing, On computable numbers, with an application to the Entscheidungsproblem, in *Proceedings of the London Mathematical Society*, vo. 2–42, no. 1 (1937), p. 230
31. F. Vogelsteller, V. Buterin, *ERC-20: Token Standard* (2015). https://eips.ethereum.org/EIPS/eip-20. Accessed 18 Jul 2025
32. L. Wagner, M. Esmer, *Comments on Draft Unidroit Principles* (2023). https://www.unidroit.org/wp-content/uploads/2023/03/W.G.8-Doc.-4-Public-Consultation-Comments-with-Annexes-1.pdf. Accessed 18 Jul 2025
33. J.R. Willet, *MasterCoin—A Second-Generation Protocol on the Bitcoin Blockchain for Creating and Trading New Currencies* (2012). https://github.com/bitsblocks/mastercoin-whitepaper/blob/master/index.md. Accessed 18 Jul 2025

Lukas Wagner heads the legal team at NYALA. NYALA provides technology for regulated crypto financial services such as crypto custody and crypto securities registry. At NYALA and in his previous role, Lukas was involved in the issuance of some of the first DLT-based electronic securities, including the first tokenisation of a stock corporation under the German Electronic Securities Act (eWpG). Prior to that, he worked at international law firms, advising companies including fintech businesses on financial regulatory issues, among other things, and before that at a start-up focused on AI solutions to combat money laundering. He studied law in Munich, Mainz and Rome, and law and finance in Oxford (MSc). After his legal studies, he completed his legal traineeship in Frankfurt (Main), Hong Kong and Brussels.

ERC Token Standards Powering NFTs: An Overview

Lorenz Raphael Lehmann

1 Introduction

Non-fungible tokens (NFTs) have emerged as a significant and transformative new form of ownership in the digital landscape [1]. These unique digital tokens have evolved from the realm of cryptocurrency, including the first which was Bitcoin [2]. Unlike Bitcoin, which is a fungible token, meaning it can be divided into smaller units [2], each NFTs is inherently unique and inherently not capable of being split into smaller parts [3]. NFTs are made possible through the power of smart contracts on a blockchain [3]. These smart contracts are self-executing code snippets with the terms of the execution written into the code itself [4]. They enable the creation and validation of NFTs, ensuring that each token is distinct from one another [3]. This feature of uniqueness is what sets NFTs apart, making them non-interchangeable or non-fungible [1]. Each NFT carries specific information that makes it both distinct from other tokens and verifiable, providing a level of security and ownership [1].

The versatility of NFTs is evident in their wide range assets they can represent [1]. From digital art and gaming items to virtual real estate, NFTs have enabled whole new industry to emerge [1, 5]. They serve as a key to owning a variety of assets that were previously difficult to tokenize or authenticate [5]. Provable uniqueness and ownership have caused some NFTs to hold monetary value. In an internet environment where duplication and piracy are widespread, NFTs provide a mechanism to authenticate and prove ownership of digital assets [1]. User can do with their NFTs as they like, which led to a speculation bubble in 2021–2022 where rare NFTs were sold for significant amounts of money [6].

L. R. Lehmann (✉)
RWTH Aachen, Aachen, Germany
e-mail: lorenz.lehmann@rwth-aachen.de

In an internet environment where duplication and piracy are rampant, NFTs provide a mechanism to authenticate and prove ownership of digital assets [5]. This aspect is particularly vital for natively digital items, where proving ownership has traditionally been challenging [1].

2 Historic Development of NFTs

2.1 The First NFTs Created

The history of NFTs begins with the inception of a unique digital asset on a not typical known blockchain platform. In 2014 the first known NFT named Quantum was minted on the Namecoin blockchain. Created by Kevin McCoy, Quantum represented a piece of generative art stored fully on the blockchain [7].

At the time of Quantum's creation, there was no established token standard for NFTs, leading the design of the NFT up to the creator [7]. Namecoin, the blockchain on which Quantum was minted, had a unique requirement: users needed to renew their digital items every 250 days to retain ownership [8]. This requirement later led to complications, particularly highlighted when Quantum was sold in 2021 for $1.47 million, resulting in an ownership dispute [8]. Despite being the first known NFT to be created, Namecoin and its NFTs did not capture the attention of the mainstream consumer market [6].

Following these early developments, other projects began to explore the potential of NFTs. One notable example was Spells of Genesis developed by EverdreamSoft [9]. This project was built on top of the Bitcoin blockchain and represented the first known instances of gaming NFTs. Spells of Genesis was a trading card game that gave players genuine ownership of their digital assets, distinguishing it from traditional online gaming models [9].

The Ethereum blockchain, which would become a central hub for NFT development, saw significant milestones with projects like CryptoPunks and CryptoKitties [10]. Launched in 2017 by Larva Labs, CryptoPunks was an art project that comprised the first significant NFT art collection [10]. Initially, these tokens were minted as ERC-20 tokens, a standard designed for fungible tokens, as NFT-specific standards were not developed yet [11, 12].

In the same year CryptoKitties, created by Dapper Labs, launched and quickly gained significant popularity [13]. It was one of the first games on the Ethereum blockchain and played a crucial role in the development of the NFT ecosystem [13]. The game's success and the unique challenges it presented were instrumental in inspiring the creation of the ERC-721 token standard [11]. ERC-721 is a standardized token model specifically designed for NFTs, addressing limitations of previous used standards and paving the way for a wide array of NFT-based applications [11].

2.2 Ethereum Revolutionising the NFT Landscape

Ethereum's role in the NFT landscape is a pivotal chapter in early success of NFTs development. Ethereum's core innovation over other blockchains like Namecoin lies in the inner workings, particularly the Ethereum Virtual Machine (EVM) [14]. Using high level programming languages like Solidity, Ethereum was able to create more sophisticated and complex contract logics, vastly expanding the potential applications of blockchain technology [14]. One of the big improvements was the advancements and simplification of NFT ownership transfers [11, 12]. This improvement was critical in enhancing the usability and accessibility of NFTs, allowing for a more seamless and user-friendly experience. In the realm of digital assets, where proof of ownership and transferability are of upmost importance, this advancement played a crucial role in promoting the broader adoption of NFTs [13].

Perhaps the most significant advancement achieved by Ethereum was the establishment of a clear and standardized token model for NFTs. This began with the introduction of Ethereum Improvement Proposal 721 (EIP-721) in January 2018 [11]. EIP-721 was ground-breaking as it provided a standardized framework for NFT creation, enabling consistent and predictable behaviour of these digital assets across all applications and platforms [15]. This standardization was crucial for developers and creators, as it reduced complexity in NFT creation and integration [11]. This interoperability was a key factor in unlocking more functionalities for NFTs, allowing them to interact in a broader ecosystem of decentralized applications (dApps) [15]. As a result, NFTs could be used in a wide range of contexts, from digital art and collectibles to gaming and decentralized finance (DeFi) [15].

Ethereum's powerful smart contract capabilities combined with the establishment of standardized token models, have been instrumental in transforming the NFT landscape [16]. In the years following, the NFT landscape experienced significant growth in users and value. This growth can be tracked by the NFT trading volume, which in February 2023 alone saw 1.65 billion$ worth of NFT trades on various marketplaces on the Ethereum blockchain [6].

3 ERC Token Standards

An ERC (Ethereum Request for Comment) token standard is an integral component of the Ethereum blockchain ecosystem, playing a crucial role in the development and operation of its diverse applications [11]. Understanding the ERC token standard requires an appreciation of the underlying structure and processes of the Ethereum network.

Ethereum was launched on July 30, 2015 and operates using the Ethereum Virtual Machine (EVM) [17]. The EVM is a powerful and flexible computing environment. A key feature of Ethereum is its ability to execute custom code, which forms the basis of smart contracts. Smart contracts are programs stored on the Ethereum blockchain that can execute automatically when certain conditions are met. These contracts are essentially Ethereum accounts with which users can interact, allowing for a wide range of decentralized applications [17].

The development and enhancement of the Ethereum blockchain are governed by Ethereum Improvement Proposals (EIPs) [11, 12]. EIPs are formal suggestions made by community members for improving the Ethereum network. They fall into three categories: Standards, Meta and Informational. Standards EIPs propose standards to the Ethereum protocol, by which the community can agree upon. While not directly changing the inner workings of the blockchain, EIP standards have significant impact on how smart contracts are written [14]. Meta EIPs on the other hand, focus on altering the process or rules governing how Ethereum functions. Informational EIPs provide guidance and information on Ethereum-related topics without suggesting changes to the network's functioning [17].

The process of proposing an EIP is democratic and open to anyone [11, 12]. Proposals are submitted through the Ethereum Improvement Proposal repository on GitHub [18]. Once an EIP is proposed, it undergoes a rigorous review process, involving discussion and debate within the Ethereum community. This process includes stakeholders such as developers, node operators and users. The decision-making regarding EIPs is achieved through a consensus mechanism, ensuring that changes are agreed upon by a broad spectrum of the community [18]. After reaching consensus, feedback is incorporated and developers begin implementing the changes before the official release [11, 12].

In the context of ERC token standards, the term ERC is often used interchangeably with EIP that fall under the Standards category [11, 12]. ERC token standards are fundamental in ensuring interoperability and consistency across various applications built on Ethereum, serving as a fundamental feature for innovation and development within the Ethereum ecosystem.

4 Smart Contracts

Smart contracts represent a fundamental technology in the blockchain ecosystem, particularly within the Ethereum network. They are essentially self-executing contracts, with the terms of the agreement between parties directly written into lines of code. These contracts reside and operate on the blockchain, ensuring transparency, security and immutability [4].

At the core of smart contract functionality is the principle that they automatically execute when predefined conditions are met. This autonomous execution eliminates the need for intermediaries, reducing time and potential for disputes. Solidity, a high-level, statically-typed programming language, is predominantly used for writing smart contracts on Ethereum. However, other languages like Viper also exist [8]. The choice of programming language is flexible as the written code is eventually compiled into low-level bytecode. This bytecode is what the Ethereum Virtual Machine (EVM) executes, allowing for the diverse range of programming languages, adding to the decentralised nature of Ethereum [4].

Once deployed on the blockchain, a smart contract is immutable, which means it cannot be updated or changed. This immutability is good and bad at the same time: it guarantees the integrity and permanence of the contract but also means that any flaws or bugs in the code are permanent and can be exploited [4].

A smart contract comprises data, functions, and event logs. Data can take the form of either persistent storage, memory or environmental variables. Functions within a smart contract are designed to either retrieve ('View' functions, require no EVM call) or update information (requires EVM call). Functions can be categorized as public or private, where private functions are only visible within the contract code they are defined in and public functions can be called externally. Internal functions take the private functions one step further, allowing to be inherited into other smart contracts, while remaining only callable within the code. And external functions are similar to public functions except that the program within can't call this function [4].

Event logs in smart contracts play a crucial role in recording occurrences. While they have no direct effect on the transaction's outcome, they are an import tool for tracking and monitoring contract activities. Smart contracts can emit events and write logs to the blockchain, which can then be processed by frontend applications [4].

Lastly, smart contract interfaces are essential for interactions within the blockchain ecosystem. An interface is a collection of function definitions without implementation. A so to speak blueprint of functions that a contract must have. It enforces a defined set of properties and functions on a contract, making it crucial for interacting with other smart contracts. By declaring an interface, a contract can interact with other contracts and call functions in another contract without needing access to the entire codebase. This mechanism is particularly important in complex blockchain applications where multiple contracts interact with each other [7, 16].

Smart contracts empower a wide array of applications and must be well understood to understand the details of the ERC token standards empowering NFTs.

5 ERC721

The EIP-721, initially submitted in 2017, represents the single biggest advancement in the standardisation of NFTs on the Ethereum blockchain. EIP-721, conceptualized by Dieter Shirley of Dapper Labs, lays out a specific token standard for NFTs, which are unique digital assets distinguishable from each other. This proposal has led to the creating of a unified and standardized approach to the creation, issuance and interaction with NFTs on EVM chains [11].

EIP-721 introduces a set of rules and functions that a smart contract must implement to be compliant with the standard, ensuring consistent behaviour across different implementations and dApps. The standard defines a minimum interface, IERC721, which is necessary for a contract to be considered ERC721-compliant. This interface includes essential functions and events that facilitate the tracking and transferring of individual tokens [11].

Due to the high cost of persistent storage on the Ethereum, EIP-721 has a practical approach to the storage of information about NFTs, such as art, attributes or other specifications. Instead of storing this data directly on the blockchain, the standard utilizes Uniform Resource Identifiers (URIs) to reference the data. This approach means that most NFTs which represent the ownership of an Art for example, include only a link to an IPFS (InterPlanetary File System) server or other internet location where the actual image or asset is stored. NFTs on Ethereum are typically representations of ownership rather than direct stores of the asset's data, which is often offloaded off-chain through various data providers, reducing transaction costs [11].

To be ERC721 compliant, a contract must implement several key functions as defined by the IERC721 interface. These include [11]:

- *BalanceOf(owner):* a view function to return the balance of a specified address
- *OwnerOf(tokenId):* another view function to identify the owner of a specific token ID
- *SafeTransferFrom(from, to, tokenId) & transferFrom(from, to, tokenId):* which are functions for transferring tokens
- *Approve(to, tokenId), getApproved(tokenId), setApprovalForAll(operator, _ approved) & isApprovedForAll(owner, operator):* allows the owner of the NFT to approve another address to transfer this specific NFT on their behalf.

The standard also specifies critical event logs such as [11]:

- *Transfer(from, to, tokenId):* logs a transfer of ownership
- *Approval(owner, approved, tokenId) & ApprovalForAll(owner, operator, approved):* logs an approval for transfer on someone else behalf

EIP721 also includes optional interfaces, IERC721Metadata and IERC721Enumerable, which add further functionalities. The IERC721Metadata interface provides additional identification for the NFT involved, including functions like [11]:

- *Name():* defines the name of the NFT collection
- *Symbol():* defines the ticker of the NFT collection
- *TokenURI(tokenId):* defines the location of the asset the NFT represents

The IERC721Enumerable focuses on enhancing the ability to enumerate tokens and requires the following additional functions [11]:

- *TotalSupply():* a view function that returns the total number of NFT that can exist in this collection
- *TokenOfOwnerByIndex(owner, index):* a view function to get the tokenId numbers an owner address owns
- *TokenByIndex(index):* a view function to get the tokenId numbers based on the index.

Important to note that the IERC721 standard requires the import of the IERC165 interface, which includes interface support via the function **supportsInterface(interfaceId)**. This requirement ensures that ERC721 tokens are compatible with a broader ecosystem of smart contracts and applications on Ethereum [11].

6 ERC1155

EIP-1155 was introduced by blockchain gaming platform Enjin in 2018 and is also a NFT token standard. EIP-1155, also known as the Multi Token Standard, addresses and improves several limitations found in ERC721. One of the key advancements of EIP-1155 is its efficiency in handling batch transfers, a feature missing in ERC721. In contrast to ERC721, which requires a separate transaction for each NFT transfer, EIP-1155 allows for the batch transfer of multiple tokens in a single transaction, significantly reducing transaction costs of the transaction [12].

EIP-1155 introduces the core interface IERC1155, which is mandatory for compliance with the standard. Additionally, there are optional interfaces, IERC1155MetadataURI and IERC1155Receiver, which provide extended functionalities. The IERC1155 interface includes the following functions [12]:

- *BalanceOf(account, id):* This view function provides the balance of a specific token for a specific account
- *BalanceOfBatch(accounts, ids):* This view function extends the balance query to handle multiple accounts and token IDs simultaneously
- *SetApprovalForAll(operator, approved):* This function enables or disables an operator to manage all of the caller's tokens
- *IsApprovedForAll(account, operator):* This view function checks if an operator is approved to manage all tokens of an account
- *SafeTransferFrom(from, to, id, amount, data):* This function safely transfers a specific amount of tokens of a single ID from one account to another
- *SafeBatchTransferFrom(from, to, ids, amounts, data):* An extension of the transfer function, this allows for the safe transfer of multiple token types in a single transaction.

In addition to these functions, IERC-1155 interface defines several event logs to ensure traceability of transactions [12]:

- *TransferSingle(operator, from, to, id, value):* This event is emitted for single token transfers
- *TransferBatch(operator, from, to, ids, values):* This event is emitted for batch token transfers
- *ApprovalForAll(account, operator, approved):* This event indicates the approval status change of an operator for an account
- *URI(value, id):* This event provides information about where the metadata for a specific token ID is stored.

The inclusion of these functions and events in EIP-1155 allows for a more versatile and efficient handling of tokens on the Ethereum blockchain. It enables the creation of both fungible and non-fungible tokens within the same contract, a flexibility not offered by ERC721. This multi-token approach is particularly beneficial for applications in gaming and collectibles, where a variety of item types and quantities can be managed more effectively. [11, 12].

7 Inscriptions

The concept of inscriptions, represents a recent development in the blockchain space, also regarding NFTs. This innovation utilizes the calldata of a blockchain transaction to inscribe data directly onto the blockchain in a cost-effective manner. This approach marks a significant departure away from smart contract token standards [19].

The genesis of this concept occurred on the Bitcoin blockchain, a platform with no fundamental smart contract functionality. Through the Ordinals protocol, introduced in December of 2022, it became possible to inscribe arbitrary content (a string with content) onto individual satoshis, the smallest unit of a Bitcoin. This technique effectively turns each inscribed satoshi into a unique digital artifact, akin to an NFT. These Bitcoin-native digital NFTs are unique because the data is inscribed directly onto the Bitcoin blockchain, leveraging its inherent security and immutability [19].

The concept of Ordinals and inscriptions is not limited to Bitcoin and can be applied to other blockchains. The underlying principle is to use the blockchain's native transfer feature to store the URI strings directly onchain. Then using an offchain third party to calculate the state root and verify the true ownership. This approach differs from the more common method seen in Ethereum's ERC token standards, where the URI are stored within a smart contract and the blockchain transacts and checks the valid state [19].

References

1. Q. Wang et al., Non-fungible token (NFT): Overview, evaluation, opportunities and challenges (2021). arXiv preprint arXiv:2105.07447
2. S. Nakamoto, Bitcoin: A peer-to-peer electronic cash system. Satoshi Nakamoto (2008)
3. U.W. Chohan, Non-fungible tokens: blockchains, scarcity, and value, in *Non-Fungible Tokens*. (Routledge, 2021), pp. 1–11
4. Z. Zheng et al., An overview on smart contracts: challenges, advances and platforms. Futur. Gener. Comput. Syst. **105**, 475–491 (2020)
5. M. Nadini et al., Mapping the NFT revolution: market trends, trade networks, and visual features. Sci. Rep. **11**(1), 20902 (2021)
6. V. Hategan, Dead NFTs: the evolving landscape of the NFT market. DappGambl 2023
7. M. Xu et al., When quantum information technologies meet blockchain in web 3.0. IEEE Netw. **38**(2), 255–263 (2023)

8. T.T. Ochoa, Non-fungible tokens (NFTS) and copyright law. St. Clara High Technol. Law J. **40**(1), 1 (2024)
9. U.W. Chohan ed. Non-fungible Tokens: Multidisciplinary Perspectives (Taylor & Francis, 2024)
10. J.N. Wang et al., Dissecting returns of non-fungible tokens (NFTs): evidence from CryptoPunks. N. Am. J. Econ. Financ. **65**, 101892 (2023)
11. E. William et al., *Erc-721 non-fungible token standard*. Ethereum Improvement Protocol, EIP-721 (2018)
12. R. Witek et al., *Eip-1155: Erc-1155 multi token standard*. Ethereum Improvement Protocol, EIP-1155 (2018)
13. L. Ante, Non-fungible token (NFT) markets on the Ethereum blockchain: temporal development, cointegration and interrelations. Econ. Innov. New Technol. **32**(8), 1216–1234 (2023)
14. Y. Fu et al., Evmfuzz: differential fuzz testing of ethereum virtual machine. J. Softw. Evol. Process. **36**(4), e2556 (2024)
15. A. Konagari et al., NFT marketplace for blockchain based digital assets using erc-721 token standard, in *2023 International Conference on Sustainable Computing and Smart Systems (ICSCSS)* (IEEE, 2023)
16. M. Mazur, E. Polyzos, Non-fungible tokens (NFTs). The Elgar Companion to Decentralized Finance, Digital Assets, and Blockchain Technologies. (Edward Elgar Publishing, 2024), pp 280–297
17. M. Di Angelo, G. Salzer, Tokens, types, and standards: identification and utilization in Ethereum. in *2020 IEEE International Conference on Decentralized Applications and Infrastructures (DAPPS)* (IEEE, 2020)
18. V. Buterin et al. *Ethereum Improvement Proposals*. GitHub, 27 Aug (2024)
19. J. Messias et al., The Writing is on the Wall: Analyzing the Boom of Inscriptions and its Impact on EVM-compatible Blockchains (2024). arXiv preprint arXiv:2405.15288

Lorenz Raphael Lehmann Serves as the Research Lead at growthepie, a role that underscores his expertise in the web3 space and his commitment to tokenizing the future. He completed dual master's degrees in Chemical and Energy and Power Engineering from RWTH Aachen University in Germany and Tsinghua University in China, providing him with a solid educational foundation that spans two continents. Driven by a vision to level the playing field and enable equal access to financial technology for everyone, Lorenz is predominantly passionate about infrastructure through Layer 2 scaling solutions and prediction market designs. His commitment to web3 extends beyond building; he is also dedicated to educate the public on blockchain technology, believing that knowledge empowers community transformation. Lorenz's interest in web3 began during his university years when he founded the Aachen Blockchain Club. This student initiative at RWTH Aachen focuses on broadening the understanding of blockchain technology among students. Under his leadership as president, the club thrived, helping to establish a prominent web3 scene in Aachen.

Utilizing Tokenized Real-World Assets in DeFi

Lorenz Raphael Lehmann

1 Introduction

The landscape of digital finance has undergone a significant transformation in recent years, primarily due to the combination of blockchain technology and digital assets [1]. This change can be best observed by total number and trading volume of digital coins or cryptocurrencies, which have experienced massive growth [2]. Through the use of token standards on smart contract-enabling blockchains like Ethereum anyone can create their own digital asset in minutes [3, 4]. This evolution has not only expanded the cryptocurrency landscape but also opened up the way for Decentralized Finance (DeFi) [1].

DeFi enables complex financial instruments to operate fully onchain [5]. This includes lending, borrowing, saving, trading and derivatives [5]. As most blockchains are inherently permissionless, DeFi democratized access to financial services to allow anyone to engage with the financial market without the need of traditional gatekeeping entities [1]. However, the rise of cryptocurrencies and DeFi platforms has not been without criticisms. Sceptics argue that cryptocurrencies are essentially valueless, as they are not backed by anything real [6]. This criticism highlights a challenge for the blockchain space: All assets apart from blockchain native assets remain outside the blockchain ecosystem, meaning not onchain [6].

Through a process called tokenization a trend is emerging into bringing real-world assets (RWA) onchain. According to DefiLlama, as of February 2024, there are $4.4 billion worth of RWA tokenized and put onchain [2]. This process involves converting the rights to an asset into a digital token that exists within the blockchain. This development is significant as it bridges the gap enabling a pathway for RWA to be integrated into the blockchain ecosystem, thereby enhancing their utility and accessibility [7]. The process of bringing RWA onto the blockchain is not without its

---

L. R. Lehmann (✉)
RWTH Aachen University, Aachen, Germany
e-mail: lorenz.lehmann@rwth-aachen.de

challenges. It requires the addition of a trusted third party to oversee the issuance and redemption of these tokenized assets, which introduces another layer of complexity and trust to the system [7, 8].

2 Background

Decentralized Finance (DeFi) refers to financial tools that are built in a permissionless way on the blockchain. Biggest differentiator to traditional financial tooling provided by e.g. banks is that it is more transparent and accessible due to its permissionless nature [1]. The term refers to not only a single application or platform but rather encompasses all protocols built on top of smart contract-capable blockchains that offer financial services to end-users. These services range from lending and borrowing platforms to complex financial instruments such as derivatives and earning yield, operating in a decentralized manner [1]. The principle of DeFi is to create a system where financial operations do not rely on centralized entities or gatekeepers, but instead are executed peer to peer between participants facilitated by smart contracts [10]. This setup ensures that all transactions are transparent, rules are clearly defined and trust lies within the code [10].

Tokenized real-world assets (RWA) are tokens representing ownership of diverse physical or traditional financial assets [7]. The tokenization process involves creating a token transferrable onchain with transferrable ownership e.g. whoever holds the token in their wallet, is the owner of the RWA asset. This process makes them bring RWA onchain and to be used on a blockchain [11]. With RWA can include for example cash, commodities, equities and bonds [12].

2.1 Getting RWA Onchain

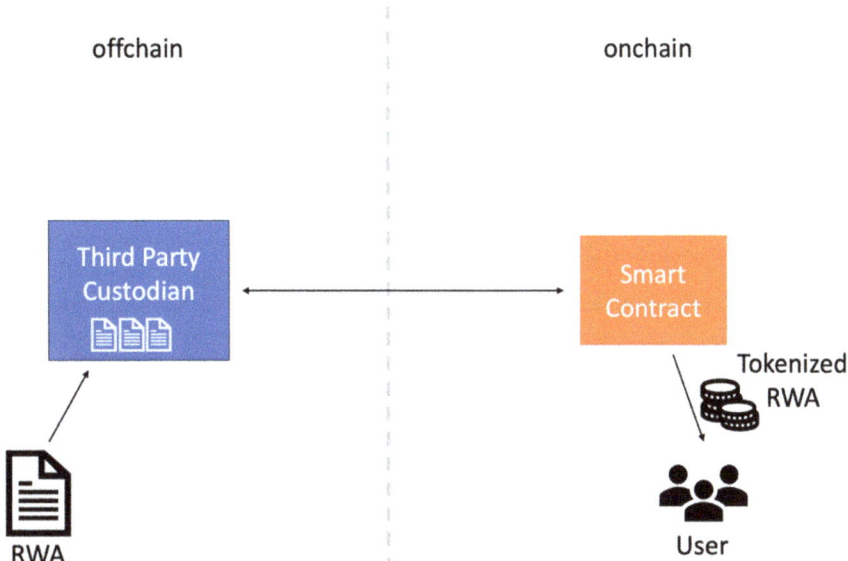

As RWA are fundamentally offchain, there is a delicate process required in bringing RWA onchain and into the DeFi ecosystem. The process includes thinking about how the asset should be tokenized and being mindful of the additional trust assumptions [11]. Generally, the process involves selecting the right RWA, specifying the tokenization specifics and choosing a trusted third party. The process described here disregards other steps that might be applicable with the process of tokenizing RWA, such as legal, compliance and technical barriers [1].

Before bringing a RWA onchain it is important to select the correct RWA for tokenization. This decision is fundamental as it dictates the asset's suitability for the blockchain environment and determines the specific regulatory and technical requirements needed for successful tokenization. Fundamentally blockchain data is open, so bringing confidential assets to the chain might not be suitable [10].

RWA can be tokenized in different ways, so in this second step choosing the right characteristics of the digital equivalent is essential. This involves determining whether the asset will be represented as a fungible token, e.g. through an ERC-20 token standard or as a non-fungible token (NFT) through an ERC-721 or ERC-1155 standard [3, 4]. The specifications chosen here can have large effects on the asset's functionality and interoperability within the blockchain ecosystem.

The most critical step is choosing a custodian that is trusted who will manage the issuance and redemption mechanisms of the tokens, ensuring the underlying RWA will back the digital asset. A new level of trust is attributed to this third party, therefore a common practice for them is to provide users with a proof of reserves, as it guarantees redeemability and maintains the system's credibility and trust [11].

The motivation to bring RWA onchain often stems from the desire to leverage DeFi's functionalities. The successful onchain integration of RWA contributes to the expansion of the DeFi sector, providing a more diversified, inclusive and efficient financial environment [12].

3 RWA Onchain

For those reasons integrating tokenized RWA into DeFi onchain comes with benefits. Through the blockchains immutable ledger, a greater level of trust and a reliable source of ownership are established [13]. It eliminates discrepancies and facilitates atomic settlements, which is the act of transferring assets and payments simultaneously, streamlining the entire trading process and minimizing settlement risks [13].

Self-executing, autonomous smart contracts that power DeFi automate a significant portion of a trading process [14]. This automation reduces the need for traditional intermediaries thereby cutting down fees and removing any human error or interference. The result is a more streamlined, cost-effective and lower-risk transaction process, which in the end saves the users a lot of money when dealing with RWA [12].

RWA tokenization brings a high level of transparency to RWA due to the open nature of transaction data on public blockchains. Ownership activity and transaction data are publicly accessible in real-time through the blockchain ledger, allowing anyone to verify the authenticity and ownership of the asset. Such high level of transparency mitigates any instances of dealing with ownership disputes [15].

Another advantage of integrating RWA into DeFi is through the potential to enable liquidity for assets that were previously considered illiquid. This process can be facilitated by a methodology known as fractionalization, which entails dividing a larger asset into smaller more manageable shares that can be individually bought and sold. Moreover, as the selling process is significantly simplified, the otherwise costly and time-consuming sale of an RWA in the real world becomes almost frictionless [12].

3.1 RWA in DeFi

While DeFi has grown to 80b$ of value onchain [2], the sector remains relatively small compared to traditional financial services [9]. This disparity is due to the barriers to entry in onchain DeFi, as blockchains are largely isolated from the real world, limiting DeFi to cryptocurrencies, which represent only a fraction of global assets (Fig. 1). Real-world assets (RWA) could address this limitation by bringing a broader range of assets on-chain, potentially creating synergies that expand the scope and utility of DeFi [1].

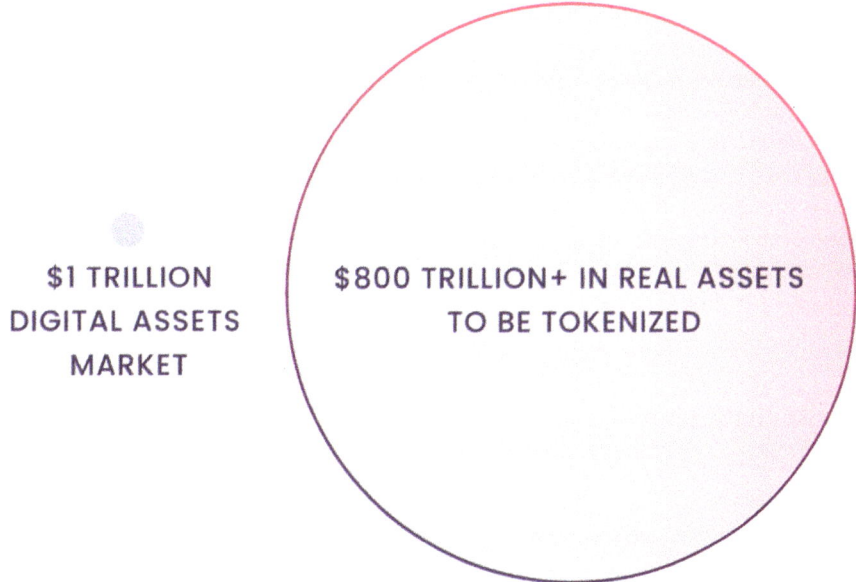

Fig. 1 Market size comparison (not to scale) [9]

Once the RWA is tokenized and onchain, there are endless use cases and applications to leverage in DeFi [7]. Building ontop of the permissionless nature of blockchain anyone can create their own DeFi protocol that uses RWA and the space is rapidly growing. The use cases highlight some of the best way to utilise this through applications [1].

The most widely used application is trading, which builds upon harnessing blockchain's innate efficiencies, such as instant settlement and transparency [7]. Trading RWA onchain compared to tradition mechanisms is not only more frictionless but also accessible to a broader audience [16]. The permissionless nature of blockchains democratizes access, breaking down the barriers that have historically restricted participation in asset markets.

Lending against your assets is an age-old practice that dates back to the first pawn shops in China over 3000 years ago [17]. In the contemporary DeFi ecosystem, this translates to enabling asset holders to leverage their possessions as collateral for loans, often at rates significantly more favourable than those offered by traditional financial institutions [18]. For instance, leveraging a tokenized high-value asset like a luxury watch can facilitate substantial loans with reasonable interest rates, illustrating the tangible benefits of RWA tokenization in DeFi lending platforms [10].

Getting offchain yield onchain is a significant driver behind the tokenization of RWAs, aiming to bring the yields from traditional treasury assets onto the blockchain [19]. This shift addresses the growing demand for treasury yields within the DeFi space, allowing users to transition from non-interest-bearing stablecoins to RWAs that offer attractive onchain yields. Initiatives by entities such as stUSDT or Ondo

Finance exemplify this trend, having surpassed over $1.5 billion worth of RWAs on-chain to capitalize on rising interest rates in treasury instruments [20].

Fractionalization stands out as a transformative use case, enabling the division of high-value RWAs into more manageable and accessible segments [21]. This process democratizes investment opportunities, allowing individuals to overcome the cost barriers and engage with markets previously beyond their reach [1]. By investing in fractions of an asset, participants can share in the financial rewards and contribute to the economic growth associated with diverse assets like SME bonds or overdue invoices [21]. This not only enhances the liquidity of traditionally illiquid assets but also allows investors to spread risk more effectively.

4 Challenges and Risks

4.1 Legal

The legal landscape surrounding tokenization of RWA presents a complex array of considerations. Regulations and law for blockchain technology varies a lot across different countries and areas. This variation often results in certain RWAs being classified as illegal in certain countries if they fail to comply with local laws and regulation. This significantly affects the global uniformity and adoption of tokenized assets. The fear of legal trouble hinders investor participation, as potential lawsuits associated with non-compliance create significant monetary damage [22].

One of the innovative solutions offered by blockchain technology is the ability to encode complex compliance requirements into the tokens themselves at the protocol layer. This means that the tokens' smart contract logic can enforce regulatory compliance, such as Know Your Customer (KYC) verifications, by permitting transactions only from addresses that have been previously approved or verified. This built-in compliance mechanism can mitigate legal risks, making it easier for issuers to adhere to the necessary legal frameworks for investors to engage with tokenized RWA [22].

4.2 Types of Risk

The tokenization of Real-World Assets (RWA) introduces several types of risks that stakeholders and users must carefully consider to ensure the stability and integrity of the DeFi ecosystem. Understanding these risks is crucial for both issuers and holders of tokenized assets.

4.2.1 Issuer Risk

The interaction between the third-party issuer of the tokenized RWA and the blockchain represents a significant point of vulnerability [7]. Token holders are inherently reliant on the third-party for the redemption of their assets. This dependency introduces a risk where a sudden inability of the issuer to redeem the asset could lead to abrupt repricing of the asset, potentially triggering liquidations in lending markets [7]. If the tokenized asset cannot be sold to cover outstanding debts, so called "bad debt" will be created, negatively affecting the overall stability and liquidity of the whole DeFi lending platform [23].

4.2.2 Fractionalization Risk

The process of fractionalizing RWAs poses its own set of challenges. There is a risk that if an asset is divided among too many holders, it may become practically irredeemable [21]. This fragmentation can lead to a disparity between the asset's market value and its intrinsic value, causing it to trade at a discount [24]. Such a scenario affects the utility and efficiency of the asset within the DeFi ecosystem, as the perceived value of the fractionalized pieces may significantly deviate from their actual worth [23].

4.2.3 Ownership Risk

The coexistence of digital and physical forms of an asset can lead to complex ownership disputes [21]. For instance, if a tokenized piece of art is stolen and subsequently recovered by the original owner, questions regarding the ownership of the digital tokens representing the artwork arise [22]. Typically, legal systems will favour the physical asset's owner, rendering the digital tokens worthless. This scenario underscores the importance of clear legal frameworks and robust mechanisms to resolve disputes [21].

5 Conclusion

The integration of Tokenized RWA into the DeFi ecosystem marks a significant evolution in the landscape of digital finance. This article explored the different aspects of RWA tokenization, from its inception and legal considerations to the diverse risks involved in participating in DeFi applications. The potential of tokenized RWAs to revolutionize financial markets, enhance liquidity, reduce costs and democratize access to investment opportunities is immense. However, this potential comes with its own set of challenges including regulatory compliance, issuer reliability, and the complexities of ownership and fractionalization.

The legal and operational frameworks surrounding tokenized RWAs are still in development, varying significantly across different jurisdictions, which poses a substantial challenge to global adoption. Furthermore, the reliance on third-party issuers introduces issuer risk, showcasing the importance of robust trust mechanisms. Despite these challenges, the benefits of integrating RWA into DeFi offer compelling reasons for users to continue to invest and push this space forward.

As we move forward, it is crucial for stakeholders in the DeFi ecosystem to collaborate in addressing these challenges. By creating a regulatory environment that supports innovation while ensuring investor protection and market integrity, the full potential of tokenized RWA can be realized. Continued research will be needed to navigate the complexities of tokenization and harnessing its full potential to create a more inclusive and efficient financial landscape.

References

1. B. Sriman, S. Ganesh Kumar, Decentralized finance (defi): the future of finance and defi application for Ethereum blockchain based finance market, in *2022 International Conference on Advances in Computing, Communication and Applied Informatics (ACCAI)* (IEEE, 2022)
2. DefiLlama, *DefiLlama*, www.defillama.com. Accessed July 2024
3. E. William et al., Erc-721 non-fungible token standard, in *Ethereum Improvement Protocol, EIP-721* (2018)
4. R. Witek et al., Eip-1155: Erc-1155 multi token standard, in *Ethereum Improvement Protocol, EIP-1155* (2018)
5. J.R. Jensen, V. von Wachter, O. Ross, An introduction to decentralized finance (DeFi). Compl. Syst. Inf. Model. Quarter. **26**, 46–54 (2021)
6. N. Carter, L. Jeng, DeFi protocol risks: the paradox of DeFi, in *Regtech, Suptech and Beyond: Innovation and Technology in Financial Services* (Riskbooks, forthcoming Q3, 2021)
7. B. Notheisen, J.B. Cholewa, A.P. Shanmugam, Trading real-world assets on blockchain: an application of trust-free transaction systems in the market for lemons. Bus. Inform. Syst. Eng. **59**, 425–440 (2017)
8. R. Garg, *Blockchain for Real World Applications* (Wiley, 2023)
9. Real-World Assets, *Polymesh*, https://polymesh.network/real-world-assets. Accessed June 2024
10. K. Qin et al., *CeFi vs. DeFi---Comparing Centralized to Decentralized Finance*. arXiv preprint arXiv:2106.08157 (2021)
11. S. Aramonte, W. Huang, A. Schrimpf, DeFi risks and the decentralisation illusion. BIS Quarter. Rev. **6** (2021)
12. Exploring Fractional Ownership of Real-World Assets in DeFi. Polytrade Blog, https://blog.polytrade.finance/defi/exploring-fractional-ownership-of-real-world-assets-in-defi/. Accessed June 2024
13. C. Schinckus, The good, the bad and the ugly: an overview of the sustainability of blockchain technology. Energy Res. Soc. Sci. **69**, 101614 (2020)
14. Z. Zheng et al., An overview on smart contracts: challenges, advances and platforms. Fut. Gener. Comput. Syst. **105**, 475–491 (2020)
15. J. Sunny, N. Undralla, V. Madhusudanan Pillai, Supply chain transparency through blockchain-based traceability: an overview with demonstration. Comput. Ind. Eng. **150**, 106895 (2020)
16. V. Buterin, Ethereum: platform review. Opportun. Chall. Priv. Consort. Blockchains **45**, 1–45 (2016)

17. Y. Peng, *The Chinese Banking Industry: Lessons from History for Today's Challenges* (Routledge, 2007)
18. J. Xu, N. Vadgama, From banks to DeFi: the evolution of the lending market, in *Enabling the Internet of Value: How Blockchain Connects Global Businesses* (2022), pp. 53–66
19. N. Walton, *Yield Generation Using Decentralized Financial (DeFi) Applications* (2022)
20. L. Ante, I. Fiedler, E. Strehle, The influence of stablecoin issuances on cryptocurrency markets. Financ. Res. Lett. **41**, 101867 (2021)
21. E. Vitelaru, L. Persia, Fractional vehicle ownership and revenue generation through blockchain asset tokenization. Transp. Telecommun. **24**(2), 120–127 (2023)
22. L. Efimova, O. Sizemova, D. Chub, Digital financial assets: concept and legal nature. BRICS LJ **11**, 32 (2024)
23. Qin, Kaihua, et al. "An empirical study of DeFi liquidations: Incentives, risks, and instabilities." Proceedings of the 21st ACM Internet Measurement Conference. 2021.
24. M. Aquilina, J. Frost, A. Schrimpf, Decentralized finance (DeFi): a functional approach. J. Financ. Regul. **10**(1), 1–27 (2024)

Tokenisation of Tangible Assets

Markus Fehn

1 Introduction

Blockchain technology has revolutionised the financial sector, leading to increased digitalisation and automation. By implementing these strategies, the efficiency of processes can be improved, resulting in faster transactions and fewer intermediaries involved in the process chain. In the near future, these developments have the potential to lower the expenses associated with issuing and managing financial products, particularly those that are standardised.

In this context, understanding the significance of smart contracts is crucial for fully grasping the advantages they offer. These contracts and instructions, stored in blocks, initiate purely digital processes, allowing for near real-time settlement.

In the near future, we can anticipate a rise in the issuance, storage and management of financial products in a purely digital form on the blockchain. Ever since Germany implemented the Electronic Securities Act (eWpG), products on the blockchain have been given the same status as traditional products under the law.

Tokenising real-world assets will have a significant impact as it introduces a fresh approach to generating value from assets that are typically difficult to sell. Take the case of a numismatist or philatelist who wants to make their rare coin or stamp collection more accessible. They can achieve this by dividing ownership of the assets into fractions and offering them to interested buyers worldwide. Additionally, they can choose to transfer the physical assets to a museum, ensuring their display and safekeeping. By utilising this strategy, collectors can keep a portion of their assets while also increasing their financial flexibility.

As an illustration, artists have the ability to produce videos on social media and sell NFTs that grant usage rights for marketing and commercial purposes. This allows

---

M. Fehn (✉)
Chartered Investment: Technology and Innovation, e-Sec GmbH, Dusseldorf, Germany
e-mail: Markus.Fehn@chartered-investment.com

© The Author(s), under exclusive license to Springer Nature Switzerland AG 2025
W. Prinz and D. Trauth (eds.), *Tokenizing the Future*,
https://doi.org/10.1007/978-3-031-91405-8_32

them to generate income from their creative work without having to share profits with centralised platforms.

Tokenisation also allows for different methods of funding infrastructure and creates new financing possibilities for small and medium-sized enterprises (SMEs) through decentralised finance (DeFi) channels.

2 Brief History of Blockchain Technology in the Financial Sector

Distributed ledger technology (DLT)[1] is the foundation for applying blockchain technology in the financial industry. This technology enables the recording of transactions in a decentralised manner. DLT is different from the traditional, centrally managed ledger.

The concept involves documenting transactions in a standardised format within decentralised databases, which can then be accessed by authorised parties. Just like any other decentralised databases, synchronised versions of the classic accounting journal can be found in various storage locations.

When blockchain technology was first introduced in the cryptocurrency world, it caught the attention of major players in the financial industry. They quickly realised the potential of using blockchain as a tool for managing distributed ledgers. In 2016, the requirements for functional ledgers on the blockchain became clear. This aims to document transactions between users in digital payment and business transactions, eliminating the need for a central authority to validate each transaction.

The process of legitimisation is accomplished through the utilisation of blockchain technology. This involves the creation of a new block or token that is derived from the transaction data and then added to the existing blockchain. With the current status easily accessible to all participants, thanks to cryptographic encryption, there is always a reliable ledger.

Practical procedures have been developed in practice, especially in the realm of cryptocurrencies like Bitcoin. These advancements have not only facilitated other applications in the financial industry, but have also made it possible to issue bonds on the blockchain.

The Ethereum blockchain saw the launch of its first structured product on 16 March 2018,[2] marking a significant milestone in the world of blockchain technology. This GBP bond offered capital protection and was linked to the performance of the FTSE 100 Index. This product employed a cutting-edge platform connected to the Ethereum blockchain to streamline pricing, issuance and administration. This ground-breaking

---

[1] https://www.bafin.de/SharedDocs/Veroeffentlichungen/EN/Fachartikel/2016/fa_bj_1602_blockchain_en.html.

[2] https://chartered-investment.com/fileadmin/user_upload/opus/PressReleases/World_First_Blockchain_Structured_Product_release_final.pdf.

product was the pioneer in being tested within the regulatory sandbox of the UK Financial Conduct Authority (FCA).

3 Definitions

3.1 What is the Meaning of Real-World Assets?

Illiquid assets or commodities with a tangible, intangible, or real value are referred to as "real-world assets" (RWAs). Examples of these assets include real property, machinery, raw materials (including gold and other precious metals), intellectual property such as copyrights and patents, and certain types of contracts. The RWA category also encompasses additional rights, such as carbon credits. Real-world assets encompass a variety of financial assets that are not commonly traded or classified as securities. Examples include invoices from the trade in goods, bundled personal loans, or mortgages. In general, it's a term that lacks clarity and is currently understood in various ways depending on the context. It is important to approach this status with a critical eye, as the potential advantages of blockchain technology can only be fully realised if there is clear definition of terms and a widespread understanding in the market. It seems unrealistic to expect that the new Web3 world will become a reality tomorrow, especially when every market participant has a different understanding of the term RWA.

These RWAs or illiquid assets can be tokenised on the blockchain, making them easily tradable and accessible. Several service providers currently offer the tokenisation of real-world assets through various platforms and protocols.

3.2 What is the Meaning of Financial Assets?

Real-world assets encompass tangible items and certain intangible assets, while financial assets represent monetary values in a broad sense. These are cash and equivalent values:

Accounts Receivable and Credit Balances, including bank balances, amounts owed by buyers and customers, and various types of loans;

Securities, including shares, bonds and their derivatives like options and futures, as well as other tradable financial instruments;

Fund Units, which represent ownership in a diverse range of real and financial assets;

Insurance Products, including life and pension insurance, as well as other types of insurance policies;

Pensions and Annuities in the form of entitlements to future pension payments or pension benefits;

Foreign Exchange (currencies) traded on the foreign exchange markets.

3.3 What is the Meaning of the Term "Crypto Securities"?

The German Act on Electronic Securities (eWpG): defines crypto securities as follows: "A crypto security is an electronic security that is entered in a crypto securities register."[3] In essence, a crypto security is a specific type of electronic security where the issuer enters the information in an electronic securities register, rather than issuing a physical securities certificate.

The role of the crypto securities registrar[4] is quite unique. When investors want to buy or sell a crypto security, they need to place their order through a user-friendly interface. Ownership of the crypto security is transferred when the entry for the holder is modified. As part of the transfer process, it is necessary for the holders to prove their identity to the crypto securities registrar. Furthermore, the registrar is legally obligated under the German Money Laundering Act.

It is crucial to have clear and easily understandable terms and conditions for issuing and investing. It is crucial for investors to have access to these terms and conditions whenever they need them. Due to this requirement, the crypto securities registrar is responsible for securely storing them as a permanent electronic document. These documents should be easily accessible on the internet and appropriately linked in the register. Furthermore, it is important to document and sequentially number any modifications made to the terms and conditions of issuance. This responsibility places a crypto securities registrar at the centre of attention from a technical and regulatory perspective. Service providers who have a solid understanding of the practical application of blockchain technology have a distinct competitive advantage in this field.[5]

3.4 A Comparison of the Various Blockchains Utilised in the Financial Sector

Ever since the *Bitcoin blockchain* was introduced in 2009, numerous blockchain platforms have surfaced, with some fading away over time. As time has passed,

---

[3] https://www.gesetze-im-internet.de/ewpg/eWpG.pdf.

[4] https://www.bafin.de/DE/Aufsicht/FinTech/Geschaeftsmodelle/DLT_Blockchain_Krypto/Kryptowertpapierregisterfuehrung/Wertapierregister_node.html.

[5] https://chartered-investment.com/fileadmin/user_upload/opus/PressReleases/World_First_Blockchain_Structured_Product_release_final.pdf.

developers have expanded their focus beyond cryptocurrencies to explore other applications of blockchain technology. They have created blockchains that aim to tokenise values and streamline the process of creating smart contracts. Several blockchains are becoming increasingly significant in the financial industry or are well on their way to doing so[6]:

Ethereum: After Bitcoin initially gained popularity in the first phase of blockchain applications in the financial sector, Ethereum[7] later emerged as the preferred platform for tokenisation in financial and asset transactions. Ethereum serves as a versatile framework for developing a variety of applications, with a special focus on facilitating the creation of financial products and their tokenisation.

Hyperledger Fabric: The Linux Foundation maintains this distributed ledger platform as an open source initiative. It serves as the foundation for the IBM Blockchain Platform and other related initiatives. Hyperledger Fabric[8] serves as a versatile framework for creating corporate applications, with a particular focus on the financial industry.

Polygon is a platform that enables the creation of blockchain applications that seamlessly integrate with Ethereum. Polygon[9] utilises sidechains to process transactions outside of the Ethereum blockchain, resulting in enhanced speed and reduced costs. This platform provides a software development kit (SDK) and has its own native token called MATIC.

Quorum[10] is a blockchain platform that JP Morgan utilises for a range of purposes, such as tokenising securities, monitoring supply chains and streamlining contracts.

Avalanche[11] is a blockchain platform that offers compatibility with Ethereum and is designed for a variety of applications. It operates its own token, known as AVAX.

Tezos[12] is a blockchain platform and coin that provides a sophisticated smart contract language called Michelson.[13]

Ripple serves as a blockchain platform primarily designed for facilitating cross-border currency transactions. Additionally, it serves as a framework for various financial applications.

---

[6] https://geekflare.com/blockchain-platforms-for-finance-applications/.
[7] https://en.wikipedia.org/wiki/ethereum.
[8] https://www.ibm.com/topics/hyperledger.
[9] https://polygon.technology/.
[10] https://consensys.net/quorum/.
[11] https://www.avax.network/.
[12] https://en.wikipedia.org/wiki/tezos.
[13] https://www.michelson.org/.

3.5 The Purposes of Blocks and Tokens on the Blockchain

When it comes to blockchain technology, there are two distinct concepts for recording information and transactions:

Block: The block serves as a fundamental, self-contained entity within the blockchain. The data or transaction data collected in a block is securely encrypted, ensuring its utmost protection. These blocks are seamlessly interconnected within the blockchain, following a set of well-defined rules.

Every block in a blockchain includes a reference to the block that came before it, forming a chain of blocks known as a blockchain. By referencing the previous block, the data in the blockchain is safeguarded, ensuring its integrity and security.

Token: In blockchain technology, a token serves as a digital representation of assets or other values. Tokens are generated and distributed on a blockchain platform.

Tokens serve various purposes and can be utilised in different ways. Currently, there are several significant types of tokens: utility tokens, security tokens, payment tokens (including stablecoins). Security tokens play a crucial role in the financial sector as they represent ownership shares in various assets. Security tokens operate as digital securities. In contrast, stablecoins are seen as a type of cryptocurrency that is tied to a stable asset, like a fiat currency or a commodity.

In a nutshell: Blocks serve as the foundation of a blockchain, housing transactions and data. On the other hand, tokens act as digital assets within a blockchain, serving various purposes. Blocks serve as structural components, while tokens represent values in a digital format.

4 Legal Status

4.1 The Significance of the Electronic Securities Act and Its Influence on the Tokenisation of Assets

The German Electronic Securities Act (eWpG) of 10 June 2021 has opened up opportunities for a diverse array of crypto securities. In addition to this law, the legislator has also introduced a new financial service that now requires a licence: the crypto securities register. The eWpG enables the elimination of the previously necessary written certificate for issuing securities, replacing it with the option to enter the securities in an electronic securities register. From the issuer's point of view, crypto securities provide significant benefits when compared to traditionally issued securities:

Agile Processes: Agile processes have revolutionised the way issuing works, making it faster and more cost-effective for simply structured, high-volume products. One of the key changes is the elimination of the central securities depository, among other things.

Automated Management: Register updates can be easily streamlined through the use of algorithms, while smart contracts offer the convenience of automating tasks like dividend and coupon payments.

Transparency and Security: The use of tamper-proof recording systems such as blockchain ensures the utmost security and transparency of transactions.

A crypto bond is issued through entry in a crypto securities register and necessitates a registrar. According to Section 16 (2) eWpG, this entity is considered to be whoever the issuer designates as such for the holder. To comply with Section 32 German Banking Act (KWG), the registrar must obtain a licence to maintain a crypto securities register.

The Electronic Securities Register Ordinance (eWpRV) complements the eWpG by requiring the register to be maintained on a secure recording system. For optimal organisation, it is essential to record the data in a chronological order and ensure its protection against any unauthorised deletion or alteration.

This recording system operates as a decentralised association, ensuring the integrity and authenticity of each entry by distributing control rights to participating units. DLT systems, like blockchain, are well-suited for this purpose.

When developing a crypto security, the registrar needs to input several essential details into the crypto securities register. When it comes to bonds, all the necessary details are provided. This includes information about the issuer and holder, the volume and amount of the issue, as well as the interest rate and maturity of the crypto securities. Furthermore, it is essential to thoroughly document the unique attributes of the recording system and the cryptographic procedures employed.

4.2 The Impact of the German Electronic Securities Act on the Legal Landscape of Other Nations

Many legislators in various countries are closely observing the German eWpG, and in certain jurisdictions, preliminary guidelines are already being implemented in accordance with German law.

In September 2020, *Switzerland* implemented a law that governs the utilisation of blockchain and distributed ledger technology in the financial sector. This law became effective on 1 August 2021.[14]

Singapore has implemented a range of laws and regulations pertaining to digital tokens and cryptocurrencies. In January 2020, the Payment Services Act (PS Act)

---

[14] https://www.sif.admin.ch/sif/en/home/documentation/press-releases/medienmitteilungen.msg-id-84035.html.

was implemented, making it mandatory for all crypto companies operating in the country to register and obtain a licence.[15]

The laws currently under consideration in the UK primarily centre around cryptocurrencies. However, the regulations do not specifically address digital tokens.[16]

Laws and regulations are currently being developed in nearly all *European Union member states*.

5 Technology

5.1 Comparing the Securitisation of Assets: Tokens Versus Certificates

Records of property ownership could potentially be traced back to the ancient Sumerian civilisation, where they were first documented using cuneiform script around the third millennium BC.[17] Deeds have always held immense significance across various cultures, especially when it comes to documenting ownership. This is particularly true in societies where sign-orientated languages are prevalent. The act of drawing up deeds has become a crucial legal process, ensuring that ownership is properly recorded and documented.

Share Certificates serve as official proof of ownership for shares in a company.

Bond Certificates serve as official documentation that verifies an individual's ownership of bonds in either a company or government.

Land Registry Extracts are official documents that provide confirmation of real property ownership.

In Germany, it is a legal requirement for documents to be in writing, as stated in Section 126 German Civil Code (BGB), unless there are specific laws that state otherwise. In certain situations, especially in the financial sector, it is feasible to issue these documents electronically rather than in paper form. However, similar to traditional paper documents, it is crucial to store securities in a secure and tamper-proof manner. In the past, these actions were typically converted into digital versions of the original paper documents.

The German legislator has now established regulations with the eWpG, ensuring that purely digital forms of recording hold the same legal weight as deeds, as long as they adhere to the specified regulations. Notably, blockchain and distributed ledger technology are explicitly acknowledged.

---

[15] https://cryptovalleyjournal.com/focus/legal-and-compliance/the-global-state-of-crypto-regulation/.

[16] https://www.finanznachrichten.de/nachrichten-2023-11/60635285-grossbritannien-auf-dem-weg-zur-krypto-regulierung-ein-neues-zeitalter-fuer-digitale-waehrungen-303.htm.

[17] https://en.wikipedia.org/wiki/cuneiform_law.

Since the 1980s, documents in the financial industry have transitioned to digital form, just like how written records in the form of deeds have been replaced by digital storage in many areas of the economy and society. The emergence of blockchain and distributed ledger technology has presented exciting possibilities for streamlining digitalisation, enhancing transparency and bolstering security.

A security token replaces the certificate in digital documentation, serving as a reliable alternative. Security tokens are a way to digitally represent ownership in assets like shares, bonds and investment funds. These tokens can be issued and traded on the blockchain, making it easier to manage and transfer ownership.

5.2 The Benefits of Tokenisation in the Financial Sector

Security tokens in the financial sector have numerous advantages compared to traditional transaction documentation concepts:

Global Access: With security tokens securely stored in the blockchain, they open up the world of trading and transferring securities, making it easier and more liquid.

Automated Compliance: By utilising smart contracts in the blockchain, compliance rules and adherence to regulations can be automated, resulting in a trading process that is both transparent and efficient.

Fractionalised Ownership: Security tokens can be divided into smaller units, allowing individuals with limited capital to access investment opportunities. This fractionalised ownership opens doors for a wider range of people to participate in investments.

Tokenisation is a valuable tool that enhances the efficiency of presenting real-world assets on investors' balance sheets. It has the potential to decrease liquidity requirements and streamline the collateralisation process. Furthermore, the enhanced transparency helps bridge information gaps, greatly improving the likelihood of increased liquidity and, consequently, tradability.

Furthermore, the blockchain has the capability to seamlessly integrate all the necessary elements of traditional securities and transactions, whether they are backed by physical or digital certificates, making it incredibly convenient and efficient. Smart contracts are crucial in this context.

5.3 The Significance of Smart Contracts in the Tokenisation of Value

Blockchain platforms can truly be considered as a complete alternative to traditional financial data and transactions when they incorporate smart contracts in the form of security tokens. When engaging in securities trading, the parties involved enter into contracts that precisely outline their rights and obligations.

When it comes to a bearer bond, the essential part of the bond agreement not only outlines that the issuer will transfer a specific share to the buyer in exchange for an agreed payment, but also specifies the conditions related to the duration and interest payment. These types of agreements can be represented as smart contracts using tokenisation.

Every security token that represents an asset on the blockchain must also include a smart contract. These contracts are designed to be user-friendly, with coded programmes that automatically execute when specific conditions are met. For instance, they can be set to trigger when a share's dividend is paid.

From a technical standpoint, smart contracts are crafted in specialised languages that have been designed specifically for tokenising values or are well-suited for this purpose.

5.4 Approach to the Tokenisation of Real-World Assets

In essence, the process of tokenising real-world assets is quite similar to tokenising financial assets. In both scenarios, the value is symbolised by generating a security token on the blockchain, making it easily exchangeable. There is a significant distinction, not in the technical aspect, but rather in the financial aspect. In the real world, assets are always backed by depositing tangible or intangible assets as collateral.

The process of tokenising real-world assets involves transforming an illiquid asset into digital tokens, which can be easily traded on a blockchain platform. The process may differ based on the asset and platform, but there are several fundamental steps that are typically followed when tokenising real-world assets:

Legal and Regulatory Review: Ensuring compliance with all legal requirements and regulations is essential. It is important to ensure compliance with the conditions of the eWpG, the prospectus requirement and other relevant regulations pertaining to the specific asset and its trading market.

Token Structuring: The tokens are structured by converting the assets into digital form and representing them with smart contracts on a blockchain. Tokens can be issued in proportion to the asset's size, ensuring that each token represents a share or unit of the actual asset.

Choosing the Right Platform and Creating Tokens: When it comes to selecting a blockchain platform, it's important to consider factors like security, scalability, transaction costs and functionality. Smart contracts are used to create tokens on the chosen blockchain platform.

Managing Custody and Trading: After the tokens are created, they need to be securely stored in a wallet and made available for buying and selling on a reliable trading platform. It is important for these platforms to adhere to the laws and regulations of the country they operate in.

6 Application Example

6.1 An Illustrative Instance of the Tokenisation Process Applied to Real-World Assets

There are few assets that hold the same level of authenticity and value as precious metals, with gold being a prime example. Although the reputation of gold as a reliable hedge against inflation has waned in recent times, bars and coins remain highly sought after, particularly among individual investors.

In December 2023, Chartered Investment introduced a new investment product called "Responsibly Sourced Gold—Series II and III" through a public offering. This product stands out because it provides investors with the option to choose between a traditional and a tokenised version. This unique feature makes a significant contribution to bridging the gap between conventional capital markets and Web3 offerings.

Chartered Investment, along with its partners, has created a unique concept that offers investors the opportunity to subscribe to a certificate in both traditional form (ISIN: CH1305317765) and tokenised form (ISIN: LU2718166007). You can easily obtain both securities through a public offer. The electronic security adheres to the stringent regulations of the eWpG.

As the crypto registrar[18] of the tokenised product, e-Sec, a Chartered Investment subsidiary, has been providing this service under the provisional approval of the German Federal Financial Supervisory Authority (BaFin)[19] since 10 December 2021. It successfully issued the first crypto security, called "ZSquare Venture Opportunity Crypto Security",[20]—in accordance with the strict rules of the eWpG on 8 December 2021.

The "Responsibly Sourced" product is issued and managed as tokens on the Ethereum blockchain. The associated token holds a "smart contract" that encompasses all the bond's features and the mutually agreed terms and conditions between the issuer and the holder.

By utilising Twin-ISIN and conducting a public offering, the world's most significant precious metal becomes accessible to investors through various investment channels. This creates a distinct advantage in today's market. Both versions of the

---

[18] http://www.ewpg.de/mitgliedsunternehmen/.

[19] https://chartered-investment.com/media/news/deutschlands-erste-emission-eines-kryptowertpapiers-gemaess-ewpg/.

[20] https://www.finanznachrichten.de/nachrichten-2021-12/54709737-deutschlands-erste-emission-eines-kryptowertpapiers-gemaess-ewpg-chartered-investment-gruendet-tochtergesellschaft-E-sec-gmbh-E-sec-fuehrt-kryptowertp-007.htm.

product are backed by physical assets. Investors have the flexibility to select their preferred access route to participate in the price development of gold.

A significant distinction from the current supply of gold tokens is the creation of a publicly accessible security. This security is not only fully backed by gold, but it also exists in both the traditional and the new blockchain-based world. This development marks an important stride towards harmonising these two realms.

7 Outlook

7.1 Potential Prospects for the Tokenisation of Tangible Assets

The potential for revolutionising the creation, trading and management of real assets is evident in the tokenisation process. The reason for this is that the underlying technology provides a multitude of advantages—from enhanced liquidity and global accessibility to the democratisation of investments.

Nevertheless, it is evident that the fate of tokenisation relies on regulatory advancements, technological progress and the embrace from investors and institutions.

7.2 Exploring the Complexities of Tokenising Tangible Assets

There is no significant difference in the risks associated with converting tangible assets into digital tokens compared to other aspects of issuing, trading and managing crypto securities.

There is no need to worry about newly created or changed legal frameworks making it harder or obstructing the tokenisation of assets. All relevant global jurisdictions have already enacted corresponding laws or are currently developing them, ensuring a smooth process. Nevertheless, there is a potential concern that the legal framework may not be adequately synchronised or promptly synchronised, thereby impeding international trade of crypto securities in particular.

There is a certain level of risk associated with the widespread adoption of blockchain technology, as its reputation has been tarnished by the volatility of cryptocurrencies in recent times. Furthermore, there are ongoing concerns about the long-term viability of blockchains that operate on the proof-of-work principle.

The future of tokenisation of real-world assets hinges on finding solutions in key areas. These solutions will play a crucial role in ensuring the continued global success of this innovative concept.

Digital Money—CBDCs and Stablecoins: A Central Bank Digital Currency (CBDC) is a state-issued digital currency, typically built on blockchain or distributed ledger technology. It aims to modernize payments, improve financial inclusion, and enhance system efficiency. While CBDCs remain under development in many countries, stablecoins—privately issued digital tokens pegged to fiat currencies—have gained significant traction. Regulated stablecoins now play a key role in enabling secure, fast, and programmable settlement for tokenised real-world assets. Both CBDCs and stablecoins offer pathways to digital money on-chain, helping ensure seamless transactions in increasingly tokenised financial markets.

Interoperability has been a challenge in the financial industry when it comes to blockchain applications. These applications have been developed separately and with limited connections to traditional financial products. Furthermore, the process of tokenising assets continues to occur on various technical platforms with different technologies. This hinders the desired compatibility between systems.

Standardisation and clear interfaces will play a crucial role in addressing the challenges that impact the entire world of blockchain technology. It is highly desirable to have standards in place for payment processing, data transfer and data exchange that allow for seamless transactions between different systems and platforms.

Government bodies, research institutions and entities in the financial sector are all actively pushing forward with developments related to these standards.

Legal Framework: A global legal framework is essential for ensuring transparent and secure global trading in blockchain-based products, including tokenised real-world assets. Put simply, if the legal frameworks in different countries are designed to enable reliable transactions across borders.

7.3 Exploring the Expansion of Markets Through the Tokenisation of Real-World Assets

Germany's efficient implementation of the eWpG, in accordance with the EU Commission's guidelines for crypto securities, has positioned Germany and other European countries as global pioneers in the tokenisation of various assets, including real assets.

It's worth noting that crypto bonds issued in Germany, Luxembourg, the UK and other European countries have the advantage of obtaining an ISIN and being treated similarly to traditional products. This gives them a competitive edge.

Simultaneously, key players in the global financial industry are actively involved in the tokenisation of tangible assets and are dedicating significant resources to blockchain and distributed ledger technology. The pace of development is impressive, with progress clearly visible and a rapidly expanding range of tokenised products of all kinds.

7.4 The Importance of Tokenising Tangible Assets in the Future

In the near future, the tokenisation of real-world assets will likely become the preferred method for issuing corresponding securities. This is due to the numerous advantages it offers over traditional processes, such as faster transactions, increased accessibility, improved transparency and enhanced security.

Markus Fehn is Head of Strategy and Innovation at Chartered Investment and has more than 15 years of professional experience in Financial Services. His core expertise ranges across digital finance and capital markets while he is well-versed in digital transformation at the same time. As part of this, he has developed deep knowledge of technologies with the potential to disrupt entire markets, among them blockchain technology and generative artificial intelligence. Academically, Markus holds a Master of Business Administration (MBA) from the WHU—Otto Beisheim School of Management. In the past, Markus was the architect of important milestones in blockchain transactions. In 2018, the first Structured Product using blockchain was launched by Chartered Investment with Markus' contribution. Moreover, the first Crypto Security under eWpG was realized by Markus and the Chartered Investment team. Although there are many challenges to make a blockchain business case fly today, Markus is deeply convinced that the advantages in the long run will prevail. This is true for both, blockchain as well as AI technology.

Engagement Reimagined: Translating Psychological Ownership into Token-Based Engagement Models

Lea Horn

1 Introduction

Since the advent of Web3 as a technological and cultural concept, ownership has taken on a meaning beyond physical possession. The emergence of blockchain as the technological backbone facilitating Web3 and the principles underlying the decentralized web, such as democratization, participation, and co-creation, have sparked a reconfiguration of interactions around digital assets and inter-community relationships. As scholars and practitioners begin to understand this emergent landscape, it is essential to identify the motivations, benefits, and potential pitfalls associated with the concept of ownership. Digital collectibles, decentralized autonomous organizations (DAOs), and social tokens based on smart contracts are changing how individuals and collectives perceive, enact, and distribute different forms of ownership. At the core of this are participatory mechanisms that can increase the democratization of digital spaces and change how value is allocated and stored. By leveraging the unique engagement capabilities of Web3 technologies, actors can foster deeper, more transparent, and interactive connections, turning passive consumers into active stakeholders and co-creators. Brands of non-digital products can likewise take advantage of these characteristics if they succeed in creating a meaningful link between participative methods and their products.

Building on broad secondary research with literature in consumer behavior, developmental psychology, social identity theory, and the philosophy of technology, this chapter examines why and how Web3 enhances psychological mechanisms of ownership that drive consumer engagement. It then presents a structured framework of the Web3-enabled building blocks of ownership to offer a set of tokenized engagement models that entities can draw from when designing their Web3 initiatives. The goal

---

L. Horn (✉)
German Association for the Digital Economy (BVDW), Berlin, Germany
e-mail: leahorn.contact@gmail.com

is to allow the reader to take full advantage of the opportunities offered by Web3, to envision broader systems underpinned by co-creation, and to anticipate potential challenges.

2 Ownership with Web3: What Has Changed?

There is a shift in the way consumers want to interact with brands, demanding more personalized experiences, more value, and a broader selection of offers to choose from, going beyond mere transactions and one-fits-all products and services [1]. The introduction of Web3 technologies both causes and shapes this shift by challenging traditional notions of transactional relationships and digital asset management. At the heart lies the principle of psychological ownership, a deeply personal sense of ownership over an asset, which in the context of Web3 can extend beyond the physical to include digital assets and experiences. Web3 technologies redefine what it means to 'own' something in the digital world. There are four core motivations that drive psychological ownership: Self-identity, affiliation, efficacy and effectance, and value creation. They provide the foundation for our framework's Building Blocks (Fig. 1) and will, therefore, be described in more detail below.

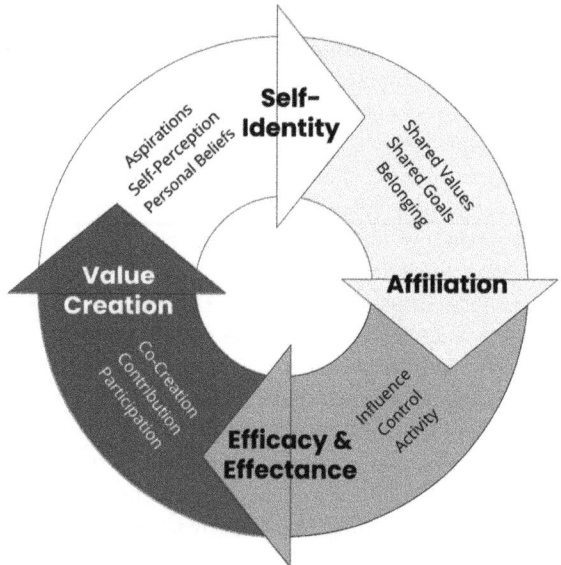

Fig. 1 The Building Blocks Framework for Ownership-Driven Engagement

2.1 The Invention of Ownership and People's Desire to Own

We differentiate two distinct yet interrelated concepts that often intersect in our understanding of possession, asset management, and control: legal ownership and psychological ownership.

Legal ownership refers to the formal, recognized right or title to an object, property, or idea codified by law. As the Web3 infrastructure continues to evolve, it aims to provide a decentralized, verifiable record of digital transactions and ownership on the blockchain, securing digital assets such as digital collectibles, cryptocurrencies, and other digital goods, as mentioned above. It is worth noting, however, that this legal recognition and protection can vary and is still evolving in many parts of the world.

In contrast to that, psychological ownership is rooted in one's feelings or perceptions about an object, idea, or property and does not depend on legal titles. It is driven by personal investment, emotional attachment, or the perception of control or association with the object or idea in question. People invest money, time, and emotion in acquiring digital goods, whether it is a piece of digital art, a domain on a decentralized network, or a virtual plot of land in a metaverse. This investment can foster a sense of attachment, pride, and perceived control, even if the digital asset does not have a tangible, physical form.

As described by Karahanna et al. [2], the concept of psychological ownership is underpinned by three core motivations: self-identity, affiliation, and efficacy and effectance. In the course of this chapter, we will introduce a fourth pillar that can enhance ownership motivations through the application of Web3 concepts and technologies: value creation.

2.1.1 Self-identity

Self-identity refers to one's understanding of oneself as a distinct and continuous entity, encompassing aspects such as personal beliefs, values, experiences, and attributes. According to Karahanna et al. [2], self-identity plays a pivotal role in the context of psychological ownership as one of the core factors influencing one's attachment to both material and non-material possessions. When individuals claim ownership of an object or idea, it often serves as an extension of their self-concept, reflecting who they are or who they aspire to be. For example, a person might buy a book not just for its content but because it resonates with their intellectual proclivities, thereby reinforcing their self-image as a learner. Similarly, ownership of intellectual property can be a manifestation of one's creativity and innovation, making it an integral part of their self-identity. Through this lens, possessions, whether tangible or intangible, are not just external entities. Instead, they become interwoven with, reinforce, and validate one's sense of self and personal narrative. This deep connection between possessions and self-identity is one reason why people often feel a deep sense of attachment and pride in what they own.

2.1.2 Affiliation

Affiliation is another key element in the development of psychological ownership [2]. It touches on the deeply rooted human desire to belong and identify with larger entities or groups. Ownership is, in many ways, a manifestation of this innate desire. When individuals own material objects or intangible assets such as ideas or intellectual property, it often symbolizes belonging to a particular group, culture, or ideology. For example, owning an artwork can link an individual to a community of art enthusiasts, while owning a patent can link someone to a community of innovators in a particular field. In addition, shared ownership or mutual investment in an idea or value fosters a sense of communal belonging, emphasizing shared goals and collective identity. By doing so, ownership not only satisfies personal needs but also strengthens social bonds, underlining the role of belonging. This intertwining of personal and collective identity through the medium of ownership emphasizes the profound influence of affiliation.

2.1.3 Efficacy and Effectance

At its core, the human desire to own is not merely about possession but about exerting control and influence over one's environment. Efficacy relates to the belief in one's ability to produce desired outcomes through one's actions. Owning an object or an idea grants a person the power to control, modify, or utilize it, reinforcing their sense of personal efficacy. Effectance motivation, on the other hand, pertains to the inherent desire to interact with and have an effect on one's environment. The act of acquiring and maintaining ownership satiates this motivation, as it provides tangible evidence of one's capacity to influence the external world. Together, efficacy and effectance contribute to the deep human need for competence and mastery that ownership, whether of material items or intellectual property, effectively satisfies [2].

2.1.4 Value Creation

The motivations that are traditionally connected to psychological ownership are three-fold: self-identity, affiliation, and efficacy and effectance. Beyond these psychological motivations, humans have a strong and intrinsic desire to create new meaning, either as tangible objects or intangible ideas. Web3 is particularly characterized by co-creation and participation, where customers can actively contribute to the value chain. In addition, Web3 and its related technologies can enhance existing products, assets, infrastructure and the like in ways that increase their perceived value. Therefore, we add value creation as the fourth core pillar of psychological ownership. In the broadest sense, value creation describes the creation or formation of something new. On the one hand, it can refer to the creation of value out of what previously had no value, such as an asset that becomes endowed with value. On the other hand, it can refer to the direct creation of something, such as the design of a product, where

Web3 allows for a verifiable and permanent attribution to the creator. When individuals create value—whether by contributing to DAOs or submitting design ideas to create digital collectibles —they inherently develop a deeper sense of ownership. Their investments, both in terms of time and financial resources, are not just for utilization but for actual enhancement and growth of the digital ecosystem. This act of adding value and witnessing the tangible results of one's contributions solidifies one's connection to the asset, reinforcing the feeling that they are not just users or spectators but integral stakeholders. In this way, the creation of value in the decentralized digital space reinforces the inherent desire for ownership in a unique way because contributing effectively enhances the feeling of ownership.

3 How Web3 and Blockchain Technologies Shape Ownership

In the traditional digital world, ownership and transactions are overseen by central authorities, be they banks, companies, or other institutions. This centralized model has inherent vulnerabilities, such as power in the hands of with a non-transparent record of transactions. Blockchain technologies emerge as a fundamental change from this paradigm, bringing a decentralized approach to data management and ownership.

At its core, a blockchain is a publicly viewable digital record [3] of transactions that provides the technological infrastructure to facilitate and implement the vision of a user-empowered, decentralized internet postulated by Web3. These transactions are encrypted and linked together in a way that ensures the integrity and immutability of the data. In order to carry out these transactions, a consensus mechanism is needed. It ensures that all participants in a distributed network agree on a 'single source of truth' of the stored data. With that, it guarantees the network's resistance to fraudulent activities, such as the distribution of counterfeit assets, thereby fostering trust among decentralized entities. Non-Fungible Tokens (NFTs) are a type of blockchain-based digital asset commonly used to represent ownership of digital collectibles. . In essence, NFTs are digital certificates recorded on a blockchain that serve as proof of ownership [3]. This proof of ownership can cover real-world assets, such as fine art or real estate, and digital assets, such as any form of digital file or music royalties. The latter, in particular, is a novelty, as previously, it was not possible to distinguish the actual owner and rights holder of a digital asset from someone who had merely downloaded a copy of said asset [4]. Provided these fundamentals, NFTs are said to hold the potential to "redistribute market power" [5]. Although the digital realm is more prevalent in the context of Web3, we will consider various forms of tangible and intangible assets in the course of this discussion to outline possible adaptations of psychological ownership for engagement models with Web3.

3.1 How Digital Assets Create Value

In the context of Web3 and NFTs in particular as the repository for storing digital assets, the question of value and added value creation has often been raised, with critics of the movement arguing that the potential benefits of Web3-based solutions are artificial, do not add measurable value to existing solutions, and do not offer alternative solutions to problems that have not yet been solved by other technological or philosophical approaches [6]. In addition, the perceived lack of tangible results on business-critical issues, coupled with a diffuse array of impactful and scalable use cases, may support the belief that Web3-enabled processes do not have a significant enough impact. In fact, it can, in the worst case, even add complexity to existing processes, which, in turn, would subtract value from them. The question is, therefore, in what form and in what context do Web3 technologies deliver the added value that brands and other trademark holders are looking for, and when do they not?

For consumers who are more actively engaged with a brand, storing entries on the blockchain enables three attractive characteristics. First, they can provide everlasting proof of one's intellectual and creative contribution to a product or the community. Imagine a customer winning the design competition for a product design that hits the stores. Before, while there was a form of recognition, for example, as part of the product launch campaign or press coverage accompanying the design competition, this recognition of contribution faded quickly. Now, this integral contribution to a brand can be inscribed onto a blockchain with a visual representation in the form of an NFT certificate for the customer to digitally access at all times and forever. Second, customers can receive shares or royalties of the purchases made from a product that they co-designed. The NFT that commemorates their initial contribution then serves as proof of the rights holder's entitlement to those royalties. Third, the information stored on the blockchain, such as the metadata associated with the NFT, often contains information about the nature of the asset and the specific time and date of its creation or tokenization. Thus, from an idealistic point of view, NFTs relate to moments in time [7]. Web3's underlying technology allows these moments in time—history— to become tradable [7], which in turn can give them both financial and emotional value. According to Steve Kaczynski and Scott Duke Kominers [4], "[the users'] decision to embrace the NFTs quite literally imbues those NFTs with their meaning and establishes their initial value" (p. 7).

Building on this, Pierce et al. [8] highlight that NFTs, hence, allow for users, consumers, fans, and others to establish "an emotional connection between themselves and their past" (p. 13), rendering it an intrinsically motivated, highly personal and integral experience in shaping one's identity, which in turn allows digital assets to be perceived as valuable.

4 Assessing Suitable Approaches to Creating Token-Based Engagement Models

In the next section, we will outline how the insights gained can be applied as token-based engagement models. The means by which a brand chooses to address any or all of the drivers that foster a sense of "mineness," resulting in more engaged, loyal consumers, ultimately depend on how much influence and control a brand is willing to give them over the brand, its strategy, its product design, or the overarching ecosystem. To help brands make these decisions, we created the Media Matrix for Tokenization (Fig. 2). This tool maps out different Web3 products and concepts, and shows how each one taps into the four main psychological drivers that promote a feeling of ownership.

For example, offering profile-picture NFTs (PFPs) as a form of expression of self-identity, brands, and trademark holders can play with this unique aspect of avatar identities for digital worlds while building a community around digital collectibles that can be collected, showcased and traded. Other Web3 products rely on co-creation, where the lines between producers and consumers can become increasingly permeable. The idea of involving consumers directly in the design process has its appeal, as it allows brands and trademark holders to receive direct feedback that helps them tailor their products and experiences to the user.

Yet, these entities need to carefully consider their commitment to user involvement in the design and implementation of their products because value-creating contributions or decision-making rights can be directly traced and linked to an asset holder. It is, therefore, important to avoid throwing around terms such as 'co-creation' without actually involving the respective asset holders to ensure that well-intentioned initiatives do not lead to negative backlash if delivered poorly. While Web3 may enable a greater sense of psychological ownership, there remains a critical discourse about

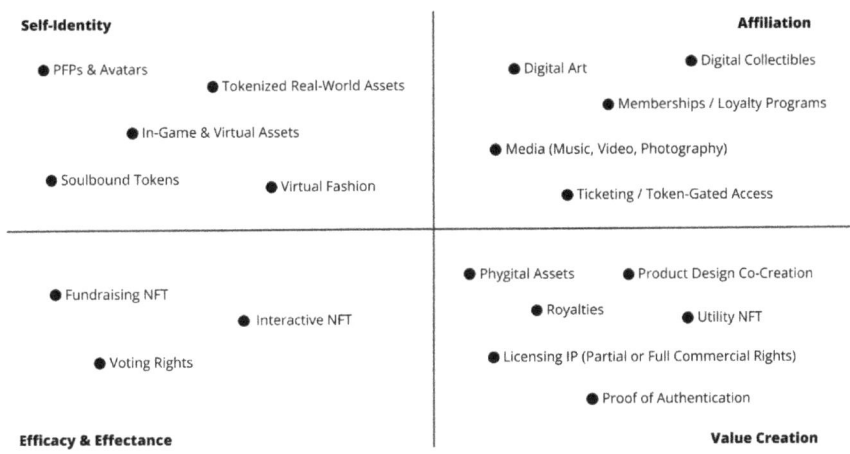

Fig. 2 Media Matrix for Tokenization

the expectations fueled by entities in terms of perceived ownership versus what is ultimately delivered in terms of ownership opportunities. To circumvent this, staying authentic to a brand's DNA when adopting Web3 strategies and setting expectations early on about what form of value, benefit, or involvement consumers can expect is pivotal to successfully leveraging Web3 engagement models for consumer engagement.

To illustrate how this translates into practice, we will present three use cases that are related to existing examples, their objectives, and the token-based engagement strategies used to achieve these objectives based on the Building Blocks Framework for Ownership-Driven Engagement (Fig. 1) and the Media Matrix for Tokenization (Fig. 2).

4.1 Use Case: A Web3-Native Sports Brand Growing Revenue Streams

A Web3-native brand in the sports industry aimed to (a) facilitate the financial support of young athletes' journeys to a professional career in sports and (b) increase revenue to sustain the business's operations. They did so by using smart contracts to execute fundraising activities geared towards investing in emerging sports athletes to support their sports careers. By harnessing NFTs as proof of contribution and fundraising, the NFT holders were entitled to receive varying shares of the athlete's prize money depending on the amount of their initial contribution.

This company has been able to successfully engage its consumers and members because it taps into all four of the Building Blocks. First, consumers who interact with this brand are likely to see themselves as savvy investors who want to grow a portfolio of assets. Investing in emerging athletes thus allows them to enhance their self-identity, as they are able to build a sports asset portfolio. Second, holding an NFT that proves their contribution to the career of a specific athlete automatically makes them a part of their team. Hence, this brand succeeds at fostering a strong community that people want to be affiliated with. Third, by sponsoring an athlete, the NFT holder plays an active part in the progress of this individual, tapping into the pillar of efficacy and effectance, which is characterized by being able to actively impact the external environment. Lastly, owning an NFT as proof of one's initial contribution is tied to financial rewards and pay-outs in prize money shares.

With multiple six-figures in prize money having been distributed back to NFT holders, this example of a token-based incentive for increased engagement created new value and financial gain for community members.

4.2 Use Case: Media and Entertainment Brands Forging More Direct Bonds with Fans

This use case has emerged across a wide range of media and entertainment companies, such as musicians and film studios. Known for their extensive libraries of digital-first intellectual property and high levels of intermediary involvement, companies in this industry are looking to (a) provide fans with new and interactive ways to engage with their favorite IPs and (b) create a more direct connection with fans beyond the traditional entertainment consumption experience through intermediaries like music labels or theater chains. They achieved this by offering digital collectibles in the form of NFTs that represent ownership over an asset related to an IP—be it collectible cards of specific movie characters that unlocked exclusive experiences or a share of the royalties generated from future streams of a song, effectively giving fans a more tangible stake in an artist's or IP holder's success.

Initiatives to create more compelling experiences through more personalized forms of entertainment have led to positive consumer and fan sentiment by appealing to the desire to affiliate with a community or fan base of shared interests (affiliation). Moreover, especially when connected to royalties or licensing rights, these initiatives also allow for those who hold such an NFT to benefit from value creation in the form of financial gains through asset appreciation and distribution of said royalties or shares.

However, while both the examples from 4.1 and 4.2 highlight how Web3 can enable a sense of psychological ownership by allowing consumers and fans to feel a sense of control, agency, and belonging over digital assets and communities, there remains a critical discourse about the expectations set by brands and IP owners in terms of what is ultimately delivered. All of these initiatives have not been without their critics. A key point of contention was that NFT holders were not given the actual licensing rights to an IP, but rather the prospect of earning a small share of all royalties generated, for example, from song streams. Furthermore, the often very modest (royalty) share came under scrutiny, particularly when considering the initial cost of acquiring an NFT, which represents shares in prize money or song stream revenues. Critics pointed out that the financial return for NFT holders may not be substantial, raising questions about the fairness and value of such an investment for fans. This emphasizes the importance of expectation management and thoughtful, detailed communication of what consumers can expect when using psychological ownership motives to foster engagement.

4.3 Use Case: A Sneaker Brand Leveraging Phygitals to Merge Physical and Digital Products

The third example highlights the innovative approach to consumer engagement taken by a global retailer and footwear brand that became known and loved by consumers across all demographics and age groups for their sneakers and sports apparel. The brand's objective was twofold: First, to provide its loyal customers with new ways to engage and be rewarded tapping into the Building Blocks of self-identity and affiliation. Second, to expand its product portfolio with digital offerings to increase revenue streams, which ties into the Building Block of value creation and influencing the brand's value chain. The brand leveraged several of the media formats outlined in Fig. 2 to build a holistic, integrated engagement program. To do so, it harnessed the potential of soulbound tokens and digital collectibles, while also inviting consumers to participate in the co-creation of physical product designs. A cornerstone of this initiative was the launch of a unique 'phygital' product: an exclusive pair of sneakers paired with an NFT symbolizing ownership. This dual offer not only rewarded existing customers with new experiences and deeper brand loyalty, but also attracted Web3 natives into the brand ecosystem, converting them into both advocates and paying customers. Importantly, those different forms of media have not been offered all at once, but have been rolled out sequentially. This has enabled the brand to test each form of tokenization in relation to its engagement goals and iterate with each new launch.

The execution of this Web3 project roadmap was diligent, with the steady introduction of new touchpoints that were seamlessly integrated into the brand's narrative and product line. Communication strategies spanned both Web2 and Web3 channels, ensuring broad reach and engagement. As a result, the brand not only generated multiple seven-figure revenue streams, but also saw a significant uplift in customer sentiment, as evidenced by the positive online coverage of the project. In addition, the initiative had a tangible impact on customer lifetime value (CLV), with a noticeable increase in the number of returning customers eager to take advantage of their exclusive access to certain products. This case illustrates how to integrate Web3 concepts and technologies into traditional business models, fostering deeper customer relationships and driving sustainable, measurable growth.

5 Future Outlook and Conclusion

Brands need to be top of mind with their users, customers, and fans to build loyalty, drive repeat business, and gain a competitive edge in markets where capturing and retaining consumer attention is vital. In the future, as these technologies mature and become more integrated into mainstream applications, it is likely that there will be a proliferation of innovative engagement models that leverage the unique capabilities of Web3 and its related technologies, infrastructure, and values. Brands, trademark and IP holders, and other organizations that adeptly navigate this new landscape will find themselves at the forefront of a more democratized, participatory, and co-creative digital economy. This will not only enhance the consumer experience but also foster deeper, more meaningful connections between organizations and their stakeholders. Understanding why consumers want to interact with and be connected to brands and their ecosystems helps to design engagement initiatives that are authentic, meaningful and add value. Psychological ownership mechanisms can be a starting point to achieve this.

However, with great potential comes the need for responsibility and foresight. Actors venturing into this space must be mindful of the ethical, legal, and social implications of these emerging technologies. As we embrace this new era of digital ownership and consumer engagement, it is crucial to ensure that these advances lead to more equitable, transparent, and inclusive digital spaces. The journey to Web3 represents not just a technological shift, but a philosophical and cultural one, offering a canvas for innovation, collaboration, and reimagined value creation and exchange in our increasingly digital world.

References

1. S. Suherlan, M.O. Okombo, Technological innovation in marketing and its effect on consumer behaviour. Technol. Soc. Perspect. (TACIT) **1**(2), 94–103 (2023). https://doi.org/10.61100/tacit.v1i2.57
2. E. Karahanna, S.X. Xu, N. Zhang, Psychological ownership motivation and use of social media. J. Market. Theor. Pract. **23**(2), 185–207 (2015). https://doi.org/10.1080/10696679.2015.1002336
3. A. Robertson, *NFT Makers Are Trying to Build the Next Disney* (2021). https://www.theverge.com/22785051/nft-collectibles-intellectual-property-decentralized-disney. Accessed 13 Nov 2023
4. S. Kaczynski, S.D. Kominers, *How NFTs Create Value* (2021). https://hbr.org/2021/11/how-nfts-create-value. Accessed 19 Nov 2023
5. D. Chalmers, C. Fisch, R. Matthews, W. Quinn, J. Recker, Beyond the bubble: will NFTs and digital proof of ownership empower creative industry entrepreneurs? J. Bus. Ventur. Insights (2022). https://doi.org/10.1016/j.jbvi.2022.e00309
6. A. Park, M. Wilson, K. Robson, D. Demetis, J. Kietzmann, Interoperability: our exciting and terrifying Web3 future. Bus. Horiz. **66**(4), 529–541 (2023). https://doi.org/10.1016/j.bushor.2022.10.005
7. K. Reichert, *Krypto-Kunst* (Wagenbach, Berlin, 2021). ISBN 978-3-8031-3711-1
8. G. Brown, J.L. Pierce, C. Crossley, Toward an understanding of the development of ownership feelings. J. Organ. Behav. **35**(3), 318–338 (2014). https://doi.org/10.1002/job.1869

 **Lea Horn** is a seasoned technology executive with over a decade of experience in digital transformation. She has held senior positions at leading technology companies including Meta, where she developed digital sales and marketing solutions for national and international markets. As an entrepreneur and strategic consultant, she specializes in emerging technologies including Web3, Blockchain, and AI, with particular focus on customer experience innovation. Since 2023, Horn has served on the Immersive Experiences working group board of the German Association for the Digital Economy (BVDW), where she advises organizations and Government institutions on integrating immersive technologies into their digital strategies.

NFT—Non-Fungible Tokens

Diana Dabboussi-Gürman

1 Introduction

In recent years, a growing trend has emerged within Web3: Non-Fungible Tokens (NFTs). They have evolved into a multi-functional tool used in a variety of industries and business contexts, starting with digital artwork and collectibles [1]. NFTs can be used to authenticate not only virtual assets such as digital artwork or collectibles, but also physical objects [2]. By 2021, NFTs, specifically in the form of Profile Picture Collections (PFPs), experienced a significant increase in value. Many users saw these JPEGs as an expression of their digital identity, creating communities parallel to the evolution of the Web3. The concept of NFTs is based on an Ethereum token standard that uniquely identifies each token with specific attributes, serving as an individual identifier for virtual or digital objects [3]. Notably, the artist "Beeple" gained attention by selling a digital artwork at a Christie's auction for approximately $69.3 million [4]. This has impacted both the decentralized application (DApps) market and the Web3 space. The blockchain technology ensures trust within the network, forming the backbone of this technology. The growing demand have made NFTs extremely attractive to businesses, with companies from almost every industry participating [4].

The origins of NFTs are linked to the digital art scene. However, their influence extends far beyond the present day. Companies are increasingly recognizing the potential of this technology to develop innovative business models and achieve efficiencies. Besides profitability, companies are exploring NFTs for several reasons. They reach new audiences, strengthen existing customer relationships, deliver new experiences, and enhance brand reputation by being at the forefront of a new generation of technology [3]. However, the development of the NFT ecosystem is in its early stages. For newcomers, there is a risk of getting lost in the rapid evolution, as there

---

D. Dabboussi-Gürman (✉)
Fraunhofer Institute for Applied Information Technology FIT, Sankt Augustin, Germany
e-mail: diana.dabboussi-guerman@fit.fraunhofer.de

© The Author(s), under exclusive license to Springer Nature Switzerland AG 2025
W. Prinz and D. Trauth (eds.), *Tokenizing the Future*,
https://doi.org/10.1007/978-3-031-91405-8_34

are few systematic summaries available. This work describes the technical foundations of NFTs and presents examples of their integration into business contexts. The goal of this chapter is to analyze the role of NFTs in the industry and illustrate their potential for business applications.

2 Foundations of NFTs

2.1 Non-Fungible Tokens

Non-fungible tokens (NFTs) represent a blockchain-based mechanism for issuing verifiable digital certificates of authenticity. They are increasingly used to represent ownership, access rights, or provenance of both digital and physical assets ranging from media content and intellectual property to supply chain data, educational credentials, and product authenticity records. NFTs are defined by their uniqueness and indivisibility, in contrast to cryptocurrencies like Bitcoin, which are fungible and interchangeable by design. This distinctiveness enables not only immutable and transparent recordkeeping, but also the creation of programmable, interoperable assets that support a wide range of decentralized applications an essential component in the evolving architecture of Web3.

Technical Aspect

This section introduces the technical components of an NFT, which form the base of a functional NFT system. Ethereum is the most used blockchain platform for NFTs, providing a secure environment for executing smart contracts [5]. Blockchain technology provides a decentralized and tamper-resistant infrastructure for storing and verifying a wide range of information related to NFTs, including provenance, authenticity, transaction history, metadata, and programmable logic. This transparency and immutability form the basis for trust in the integrity, uniqueness, and functional value of an NFT. Each NFT has a unique identifier that is created using cryptographic hashing algorithms and is stored on the blockchain. An NFT is irreplaceable because each represents its own value and cannot be divided. NFTs are based on the implementation of smart contracts. The term "Smart Contract" may seem misleading at first, as it encompasses more than just "contracts". A smart contract is a computer protocol that automates the verification, processing, or execution of programmable conditions and rules. They are also based on blockchain technology and utilize a combination of cryptography and computer code. Originally introduced by Szabo, they are designed to facilitate, verify, or enforce digital negotiations [6]. While a traditional contract is a written agreement between parties monitored by an authority, Smart Contracts go beyond this framework. They enable parties to exchange without a trusted third party and play a central role in the functionality and management of NFTs, including aspects such as buying, selling, transferring, and defining properties.

The blockchain address and the transaction are fundamental concepts in cryptocurrency. A wallet address is a unique identifier for a user to send and receive assets, like a bank account for spending assets in a bank. It consists of a fixed number of alphanumeric characters generated from a pair of public and private keys. These keys, known as public and private keys, are used to verify the owner's identity.

- **Public key**: This key serves as the address to which the NFT can be sent. It is publicly available.
- **Private key**: This key is strictly confidential and is used to control access to the wallet. Only the person with the private key can transfer or control the NFT.

To securely store and manage the keys, a digital wallet is required. Transactions can be executed through the wallet by interacting with NFT marketplaces, for example. Trading NFTs and the associated transaction fees also require the use of cryptocurrencies on blockchain platforms. These transaction costs depend on the respective blockchain network. The combination of these factors allows users to maintain control over and interact with their NFTs. Each NFT contains unique data such as metadata, token identifiers, and smart contract references that distinguishes it from all other tokens. Ownership of an NFT is recorded on the blockchain and corresponds to control over a specific token address, cryptographically signed by the issuer. While the underlying digital asset (e.g., an image or file) may be copied, such duplication does not confer ownership, which is defined solely by the blockchain token record.

Within an NFT workflow there are several components, including the server side, the user side, the transaction information (Tx Info) and the internal and external storage (see Fig. 1). On the server side, the NFT itself has a unique identifier, called the Token ID, which identifies the NFT and the smart contract address, including the creation date. This address contains the rules and conditions that define the NFT, such as transferability or specific functions written in the smart contract. Any transaction that changes the ownership or other properties of the NFT is visible in the transaction history which is displayed in the transaction information. Metadata, such as title, creator, description, date, offers an overview by providing information for users and must be added by the creator. Due to high transaction costs, the raw data of an NFT is stored in external storage such as IPFS, while only the smart contract is stored on the blockchain [7]. On the user side, the user can interact within these systems and submit proof of ownership or confirmations within the workflow of the NFT systems, for example using digital wallets.

In summary, the technical aspects of an NFT, as well as the features and functions, are secured on the blockchain. However, to ensure interoperability, the ERC (Ethereum Request for Comments) standards are important. These standards provide a framework for the smart contract and management of NFTs.

ERC standards are defined in Ethereum Improvement Proposals (EIPs). An EIP is a formal proposal for changes, improvements, or extensions to the Ethereum platform. These proposals are created by the Ethereum community, including the definition of new standards for tokens and smart contracts, such as the ERC standards. The ERC standards themselves are numbered proposals within the EIPs, specifying the

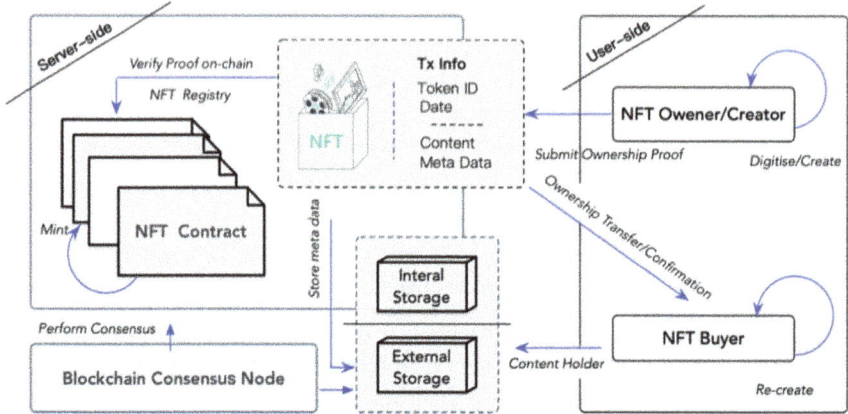

Fig. 1 Non-fungible token (NFT): workflow system [23]. *Note* Adapted from: "Non-fungible token (NFT): Overview, evaluation, opportunities and challenges." by Wang, Q., Li, R., Wang, Q., & Chen, S., 2021, p. 6. https://doi.org/10.48550/arXiv.2105.07447

exact requirements and specifications for functions or token types [8]. They play a fundamental role in the definition and standardization of tokens and smart contracts, serving as a guide for developers and enabling the interaction of various decentralized applications and tokens. Applying these standards facilitates the development process, promotes acceptance, and integrates NFTs into various application areas. ERC-721 [9] and ERC-1155 [10] define the structure and behavior of NFTs. Table 1 illustrates them based on their specific applications and features.

While ERC-721 and ERC-1155 are the standards for Ethereum, other blockchains may use different standards or protocols for the implementation of NFTs.

NFT Types

Various types of NFTs are suitable for specific use cases. In the following, the types of NFTs that represent the diversity of the technology will be explained.

Table 1 ERC-standards

ERC-standard	Application	Features
ERC-721	*Standard of NFTs*	Unique and non-fungible; ideal for individual assets; features for creation, transfer, and management of NFTs; unique identifier and metadata
ERC-1155	*Standard for multiple fungible and non-fungible tokens*	Creates both fungible and non-fungible tokens in the same smart contract; reduces transactions; supports "semi-fungible" tokens; enables efficient batch transfers; flexible for dApps with different token types

POAPs (Proof of Attendance Protocol): POAPs are based on the smart contract by POAP Inc. Users create these NFTs to allow participants to prove their attendance at specific events. POAPs are automatically distributed to participants based on predefined criteria once they claim these tokens for example at events [11].

Soulbound Tokens: They consist of smart contracts that ensure uniqueness and immutability. As the name suggests, these tokens are tied to the owner and non-transferable [12].

Dynamic NFTs: The technology requires logic for changing the content within the smart contracts, for example, to alter the representation of the NFT. The dynamics can be triggered by external data sources, interfaces, or predefined events [13]. They represent a type of interactive NFTs that process user inputs and modify the token.

Multilayered NFTs: These NFTs expand on traditional NFTs by integrating multiple levels of content, properties, or rights. Controlled by smart contracts, they enable complex digital assets with different access levels, ownership shares, and the possibility of combining with other NFTs.

3 NFT—Use Cases

NFTs allow digital assets to be managed and customized to meet individual business needs, providing a flexible way to enable digitalization across various industries. Use cases span multiple sectors, including for example art and supply chain management. In the art world, NFTs enable artists to directly monetize their digital creations while ensuring provenance and authenticity. Most of the industries use NFTs for exclusive content and fan engagement, while supply chains use them to improve transparency and traceability.

When considering the integration of NFTs, it is important to assess the specific needs of the business and develop a tailored approach. One major application of NFTs is in the context of loyalty programs. By leveraging NFTs, companies can create more engaging and personalized loyalty experiences by offering unique rewards and benefits that resonate with their customers. This chapter explores the potential of NFTs to transform traditional loyalty programs, providing insights into their implementation and the value they can bring to both companies and their customers.

Example: Loyalty Program

Loyalty programs are one of the most effective tools to increase the visibility of a company and build relationships between a business and its customers [14]. A study by Experian shows that over 70% of companies see a return on investment in loyalty programs [15]. Consumers often seek discounts and ways to acquire free products and exclusive offers. NFTs have gained importance in these systems as they can build a community around brands, as seen in collections from well-known NFT projects such as Nike, Starbucks and others. They impact the incentive structure of Web2 loyalty programs through the integration of digital identity and ownership. Customers are actively involved in business development in the digital age. Changes brought about by Web3 have prompted companies to renew their marketing communication. Companies use NFTs to create rewards for their customers, including exclusive access to products, limited editions, or digital memories represented as NFTs. This creates a sense of exclusivity that can increase customer loyalty.

To implement a loyalty program with NFTs, smart contracts must be appropriately developed. The concept for this can vary. For instance, the smart contract identifies NFT owners and tracks their loyalty points. It is programmed to automatically award loyalty points based on predefined actions such as purchasing products or participating in specific activities. Loyalty points can be rewarded in the form of additional NFTs, exclusive content, or discounts for future purchases. This automation ensures the effectiveness of a loyalty program. Simultaneously, a smart contract can send notifications about new offers and exclusive events to NFT owners. When designing NFT-based loyalty programs, it is essential to strike a balance between Web3-native users who are already familiar with blockchain technologies and Web2 users, who may be new to decentralized systems. The following are two case studies to provide a general understanding of implementing such programs.

Case: Nike

Nike's NFT initiative illustrates a strategic shift toward digital co-creation and consumer engagement. The company's goal was to leverage NFTs not just as digital assets, but as tools to build community, foster brand loyalty, and unlock new revenue models. To achieve this, Nike acquired RTFKT, a pioneer in digital sneakers, securing both creative talent and technical infrastructure. Building on that foundation, Nike launched the ".SWOOSH" platform, which enables users to design and trade virtual products for use in games and virtual experiences. This collaborative model includes revenue-sharing opportunities, positioning users as creative partners rather than passive consumers [16]. The first collection a series of digital "Air Force 1" shoes priced at $19.82 per box marked a controlled market entry. Complementing this, Nikeland, Nike's virtual environment, was introduced to deepen engagement and foster interaction [17]. With over $185 million in NFT revenue in a single year, Nike's phased approach by comprising acquisition, platform development, product launch, and community integration and illustrates a structured entry into the NFT space. This strategy yielded multiple benefits: increased digital engagement, the development of a co-creative brand ecosystem, and the expansion of Nike's presence into virtual and gamified environments. Through the.SWOOSH platform, Nike positioned itself not only as a product provider but as an enabler of user-generated content, aligning with emerging trends in participatory digital culture. From a strategic standpoint, this project contributed to both short-term monetization and longer-term brand differentiation in digital markets. The initiative also offered a testing ground for new forms of value exchange between brands and consumers, potentially informing future business models based on digital ownership and collaboration [16]. However, several analytical considerations arise. The long-term viability of NFT-based engagement remains subject to evolving user behavior, regulatory developments, and technological infrastructure. Additionally, while early financial metrics are promising, further research is needed to assess the impact of such initiatives on brand equity, customer retention, and the scalability of co-creation models. As a case study, Nike's implementation provides a valuable reference point for examining how established brands adapt to decentralized, digitally-native environments.

Case: Starbucks

In 2023, Starbucks launched Starbucks Odyssey, an invitation-only Web3 loyalty program integrating NFT technology [16]. The program enables members to earn digital "stamps" through activities such as quizzes and store visits. In its initial testing phase, the first NFT drop the Siren Collection, priced at $100 per two-stamp pack sold out within 18 minutes, temporarily overwhelming the website due to unexpected demand [18]. Secondary market activity quickly followed, with some NFTs trading for over $2,000, suggesting early signals of customer enthusiasm and perceived value [16]. By integrating NFTs into its loyalty ecosystem, Starbucks aims to augment traditional rewards with tokenized, tradable assets that offer both intrinsic and collectible value. This aligns with a broader trend of brands leveraging blockchain-based mechanisms to deepen engagement and personalize the customer experience.

Loyalty programs based on NFTs represent a significant shift in how companies conceptualize customer engagement. Rather than offering static rewards, NFTs allow for dynamic, personalized incentives such as limited digital collectibles, exclusive access, or gamified participation. These digital assets are secured on the blockchain, ensuring transparent ownership and tradability. Smart contracts further enhance these systems by automating reward issuance based on predefined customer behaviors such as purchases, event participation, or social engagement. This automation streamlines program management and opens up new opportunities for cross-platform and interoperable rewards. Case studies such as Nike's.SWOOSH platform and Starbucks Odyssey demonstrate how established brands are experimenting with NFTs to create differentiated loyalty experiences. While both initiatives show potential for increased customer interaction and new revenue streams, their long-term effectiveness will depend on sustained user engagement, seamless integration into existing ecosystems, and the ability to balance the needs of digitally fluent users with those new to Web3 technologies. As blockchain applications mature, loyalty programs are likely to evolve from transaction-based systems to relationship-based ecosystems. NFTs provide a framework for this shift by enabling unique, verifiable, and flexible rewards. However, future success will rely on careful program design, user education, and adaptability to regulatory and technological developments.

Example: NFT Authenticity Certificates

Another function of NFTs is serving as digital certificates. In this context, NFTs act as proofs for specific information or transactions [19]. Like traditional certificates or credentials, NFTs provide authenticity and trust, but in a decentralized and transparent manner. By certifying the authenticity of digital or physical objects, NFTs attract consumers by assuring them of the uniqueness and verifiability of their possession enhancing perceived value and ownership [19]. A notable application is in the education sector, where academic degrees, certificates, or professional qualifications can be represented as NFTs on a blockchain [20, 21]. This allows employers to verify the authenticity of credentials, while graduates retain control over their digital certificates. Such systems reduce administrative overhead and offer a tamper-resistant solution for credential validation. In the context of sustainable supply chains, NFTs can also function as digital product passports. These digital passports serve as digital records that store key information about a product's origin, production conditions, and sustainability attributes. For example, farmers can use NFTs to certify the origin and ecological standards of their products [22, 23]. Consumers, in turn, can trace the entire supply chain when purchasing food, ensuring alignment with sustainability criteria. This application not only promotes transparency and accountability, but also addresses growing consumer demands for ethically sourced and environmentally responsible products.

In summary, the application of NFTs extends beyond loyalty programs to various sectors, offering innovative solutions for authentication and transparency [24, 25]. These use cases in education and supply chains demonstrate the broader potential of NFTs to streamline verification processes, reduce administrative burdens, and align with evolving consumer values in the digital economy [26].

Design Decisions: NFT Project

In designing NFT projects, uniqueness and authenticity take center stage. To successfully analyze NFT projects, several factors need to be considered [27–29]. The following outlines design decisions and critical factors based on already successful NFT projects.

- The project's community strategy particularly on platforms like Twitter and Discord offers insights into engagement and reach. Metrics such as the number of active Discord members help assess the project's social traction and community strength.
- The roadmap focuses on strategic planning over time what the project intends to achieve, including launch phases, key milestones, and expansion goals.
- The composition and visibility of the team are key factors in evaluating the credibility and success potential of an NFT project.
- The clarification of ownership rights is a crucial design consideration. It is important to determine whether NFT holders receive full commercial rights, limited usage rights, or no rights beyond possession. The scope of ownership directly impacts the perceived value and utility of the NFTs, and therefore plays a central role in the project's long-term appeal and legal clarity.
- The economic model focuses on value creation and distribution how the project is financially structured, including revenue sources and stakeholder incentives.

Potential Caution, Including

- Projects without an active community may encounter difficulties.

- A lack of presence on social media platforms can indicate a lack of visibility.
- The roadmap should be critically assessed for lack of coherence or unrealistic targets.
- Visual or thematic similarities to existing projects can compromise the originality of a project.
- A lack of transparency within a team can create mistrust and affect the success of the project.

This approach provides a structured framework for evaluating the strengths and weaknesses of NFT projects and can support more informed decision-making. The following steps are recommended as guiding considerations when approaching such a project:

Step 1: Project Planning

This involves defining goals, target audiences, and technical requirements for the NFT project. Comparing blockchain providers leads to the selection of the appropriate blockchain. Use cases, components, and objects are also defined. Precise target audience definition and selection of suitable communication channels are crucial. Additionally, risk analysis and resource planning should inform fundamental decisions for the NFT project [27].

Step 2: NFT Creation

The focus is on the conception of smart contracts and the design of NFTs. The concept includes the conditions and rights that the NFT should encompass. This involves creating added value, or utilities, for buyers, whether through exclusive content, special access rights, or other benefits.

Step 3: NFT Implementation

This step involves the technical implementation of smart contracts and blockchain infrastructure. In this phase, the architectural overview must be documented and adapted to the company's processes. NFT integration into actual business processes is also carried out, and user interfaces are developed.

Step 4: Monetization of NFTs

This step focuses on the monetization of NFTs by exploring various revenue streams. Businesses get the insights into the economic perspective to understand and maximize the financial benefits of NFT implementation, where an effective marketing strategy is essential [19].

Step 5: Project Scaling and Growth

In the final step, data on the project's performance and usage are collected and analyzed. Based on these insights, adjustments, expansions, or updates can be made to ensure that the NFT project continues to meet business requirements [19]. This phase enables responding to changes in the company or the blockchain market. In summary, for a successful NFT project, clear goals and careful blockchain selection are recommended. Since, the conception and design of NFTs are complemented by technical implementation and seamless integration into business processes [19, 30]. The focus is on monetization, followed by continuous analysis for adjustments and scaling. This approach ensures effectiveness and adaptability to future requirements.

4 Conclusion

In conclusion, NFTs represent a foundational innovation within the broader blockchain ecosystem, combining decentralized ownership, verifiability, and programmable logic through smart contracts. This chapter introduced the technical and conceptual foundations of NFTs, including token standards (such as ERC-721 and ERC-1155), the role of smart contracts, and the importance of ownership rights and interoperability. These elements form the basis for a wide range of emerging applications across sectors. Beyond their technical structure, NFTs enable novel use cases that extend far beyond digital art. Companies are exploring NFT-based solutions for loyalty programs, digital memberships, authenticity certificates, educational credentials, and sustainable supply chain tracking. These examples demonstrate the versatility of NFTs as tools for enabling transparency, efficiency, and new forms of interaction between stakeholders. In particular, the ability of NFTs to serve as digital certificates or digital product passports shows their potential in supporting traceability, data integrity, and consumer trust. To implement NFT projects successfully, companies should follow a structured approach that includes clear goal definition, audience identification, technical planning, and integration into broader business processes. Critical success factors include not only robust technical design but also meaningful value propositions that resonate with end users. Projects must also navigate challenges such as regulatory uncertainty, technological complexity, and limited public understanding of blockchain systems.

Overall, NFTs offer a flexible and powerful digital infrastructure for a wide range of applications. Their effectiveness, however, depends on careful planning, responsible use, and the ability to deliver real, demonstrable value. As the field evolves, NFT projects will need to move beyond experimentation and align with practical needs, industry standards, and long-term strategic goals. Only then can NFTs realize their potential as a transformative element in the digital economy.

References

1. A. Park, J. Kietzmann, L. Pitt, A. Dabirian, The evolution of nonfungible tokens: complexity and novelty of NFT use-cases. IT Professional **24**(1), 9–14 (2022)
2. S. Dupont, I. Schuiling, *How Can Brands Effectively Utilize NFTs as a Marketing Tool to Increase Customer Acquisition, Build Customer Engagement, and Establish Long-Term Loyalty?*

3. A. Notaro, All that is solid melts in the ethereum: the brave new (art) world of NFTs. J. Vis. Art Pract. **21**(4), 359–382 (2022)
4. A. Park, J. Kietzmann, L. Pitt, A. Dabirian, The evolution of nonfungible tokens: complexity and novelty of NFT use-cases. IT Professional **24**(1), 9–14 (2022). https://doi.org/10.1109/MITP.2021.3136055
5. G. Wood et al., Ethereum: a secure decentralised generalised transaction ledger. Ethereum Project Yellow Paper **151**, 1–32 (2014)
6. N. Szabo, Smart contracts: building blocks for digital markets. EXTROPY: J. Trans. Thought **18**(2) (1996)
7. J. Benet, *Ipfs-Content Addressed, Versioned, p2p File System* (2014). arXiv preprint arXiv:1407.3561
8. M. Becze, H. Jameson et al., *Ethereum Improvement Proposals*. Retrieved from https://eips.ethereum.org/
9. E. William, S. Dieter, E. Jacob, S. Nastassia, *Eip-721: Erc-721 Non-Fungible Token Standard* (2018). Retrieved from https://eips.ethereum.org/EIPS/eip-721
10. R. Witek, C. Andrew, T. Philippe, B.E. James, S. Ronan, Eip-1155: Erc-1155 multi token standard, in *Ethereum Improvement Protocol, EIP-1155* (2018). Retrieved from https://0xjac.github.io/EIPs/EIPS/eip-1155
11. Poap, *The Bookmarks of Your Life* (2023). Retrieved from https://poap.xyz/
12. E.G. Weyl, P. Ohlhaver, V. Buterin, *Decentralized Society: Finding web3's Soul* (2022). SSRN 4105763
13. S. Allen et al., *NFTs for Art and Collectables: Primer and Outlook* (2022)
14. H. Sahin, A.O. Kusakci, B. Mbowe, The effects of frequent flyer programs in the airline industry on customer loyalty. Herit. Sustain. Dev. **3**(2), 130–47. https://doi.org/10.37868/hsd.v3i2.69
15. J. Barbara, *Wie man Kunden mit einem effektiven Treueprogramm an seinen Onlineshop bindet* (2018). https://news.creativestyle.de/kunden-mit-einem-treueprogramm-an-seinen-onlineshop-binden
16. M. Faithfull, *In Pictures: How Nike, Starbucks, Adidas and Gucci Are Jumping into NFTs* (2023, April 19). https://www.chargedretail.co.uk/2023/04/19/luxury-sportswear-nfts/
17. The Amazing Ways Nike Is Using The Metaverse, Web3 and NFTs. (2022, June 1). Accessed 14 May 2023. https://www.forbes.com/sites/bernardmarr/2022/06/01/the-amazing-ways-nike-is-using-the-metaverse-web3-and-nfts/?sh=4042f2cb56e9
18. M. Garfinkle, *Starbucks Odyssey NFT's Sell Out in Minutes, Reselling for Thousands. Entrepreneur* (2023, March 10). https://www.entrepreneur.com/business-news/starbucks-odyssey-nfts-sell-out-in-minutes-reselling-for/447520
19. O. Ali, M. Momin, A. Shrestha, R. Das, F. Alhajj, Y.K. Dwivedi, A review of the key challenges of non-fungible tokens. Technol. Forecast. Soc. Chang. **187**, 122248 (2023)
20. P. Khati, A.K. Shrestha, J. Vassileva, Student certificate sharing system using blockchain and nfts, in *International Congress on Blockchain and Applications* (Springer Nature Switzerland, Cham, 2023, July)
21. S. Kolvenbach, R. Ruland, Grather, W., Prinz, W., *Blockchain 4 Education* (2018)
22. A. Chang, N. El-Rayes, J. Shi, Blockchain technology for supply chain management: a comprehensive review. FinTech **1**(2), 191–205 (2022)
23. A. Musamih, I. Yaqoob, K. Salah, R. Jayaraman, M. Omar, S. Ellahham, Using NFTs for product management, digital certification, trading, and delivery in the healthcare supply chain. IEEE Trans. Eng. Manage. (2022)
24. F. Chiacchio, D. D'Urso, L.M. Oliveri, A. Spitaleri, C. Spampinato, D. Giordano, A non-fungible token solution for the track and trace of pharmaceutical supply chain. Appl. Sci. **12**(8), 4019 (2022)
25. M. Nadini, L. Alessandretti, F. Di Giacinto, M. Martino, L.M. Aiello, A. Baronchelli, Mapping the NFT revolution: market trends, trade networks, and visual features. Sci. Rep. **11**(1), 20902 (2021)
26. R.B. Dos Santos, N.M. Torrisi, R.P. Pantoni, Third party certification of agri-food supply chain using smart contracts and blockchain tokens. Sensors **21**(16), 5307 (2021)

27. J. Huang, Y. Shan, Y. Wang, Design and implementation of NFT-based system. Highlight. Sci. Eng. Technol. **32**, 1–6 (2023)
28. D. Lanza, From the algorithm to the new art collector. Design, development and launch of an innovative NFT collection, in *International Conference on Design and Digital Communication* (Springer Nature Switzerland, Cham, 2023)
29. V. Patel, *How to Build Trust in nft Communities: An Analysis of Strategies to Build Trust and Commitment in nft Projects* (2022)
30. Q. Wang, R. Li, Q. Wang, S. Chen, *Non-Fungible Token (NFT): Overview, Evaluation, Opportunities and Challenges* (2021). arXiv preprint arXiv:2105.07447
31. B. Jarmul, *Wie man Kunden mit einem effektiven Treueprogramm an seinen Onlineshop bindet* (2018). https://news.creativestyle.de/kunden-mit-einem-treueprogramm-an-seinen-onlineshop-binden

Diana Dabboussi-Gürman is a research associate at the Fraunhofer Institute for Applied Information Technology (FIT). With a Bachelor of Science in Business Informatics and a Master of Science in Media Informatics, she is currently pursuing her doctorate in Computer Science at RWTH Aachen University, where she focuses on the development of Web3 applications. In her current research, she analyzes the potential of digitization through Web3 technologies and manages blockchain projects with focus on NFTs. For her PhD, she is developing an ecosystem for identifying transactions based on blockchain technology and researching the impacts of such systems.

The goal is to help companies recognize the potential of emerging technologies by making NFT fundamentals accessible and developing practical use cases that highlight how blockchain and NFTs can open new opportunities for digital assets and future business models. Viewed as transformative tools, Web3 technologies like NFTs have the capacity to redefine digital value creation and ownership structures. A solid understanding of these technologies will be crucial for shaping future business models and advancing digital markets.

On the Role of Tokenization for Pursuing Environmental Sustainability

Vincent Schaaf, Jonathan Lautenschlager, Tobias Guggenberger, Marc-Fabian Körner, Jens Strüker, and Nils Urbach

1 Introduction

Tackling the pressing challenges of climate change and environmental degradation, the Paris Agreement and the United Nations' Sustainable Development Goals (SDGs) stand as pivotal frameworks guiding global efforts [1]. The Paris Agreement, an accord within the United Nations Framework Convention on Climate Change (UNFCCC), represents a collective commitment to limit global warming to well below 2, preferably to 1.5 °C, compared to pre-industrial levels. This goal is intricately aligned with the broader ambitions of the UN's SDGs. Both these global

---

V. Schaaf · J. Lautenschlager (✉) · T. Guggenberger · M.-F. Körner · J. Strüker · N. Urbach
Branch Business and Information Systems Engineering, Fraunhofer FIT, Sankt Augustin, Germany
e-mail: jonathan.lautenschlager@fit.fraunhofer.de
URL: https://www.wi.fit.fraunhofer.de/

V. Schaaf
e-mail: vincent.schaaf@fit.fraunhofer.de
URL: https://www.wi.fit.fraunhofer.de/

T. Guggenberger
e-mail: tobias.guggenberger@fit.fraunhofer.de
URL: https://www.wi.fit.fraunhofer.de/

M.-F. Körner
e-mail: marc-fabian.koerner@fit.fraunhofer.de
URL: https://www.wi.fit.fraunhofer.de/

J. Strüker
e-mail: jens.strueker@fit.fraunhofer.de
URL: https://www.wi.fit.fraunhofer.de/

N. Urbach
e-mail: nils.urbach@fit.fraunhofer.de
URL: https://www.wi.fit.fraunhofer.de/

commitments not only symbolize the need for international cooperation but also highlight the indispensable role of environmental sustainability in securing a viable and equitable future for all [2]. As nations and organizations strive to align their strategies and operations with these goals, the importance of innovative, sustainable solutions fostering effective digital emission tracing between nations and organizations has never been more necessary [3]. For instance, providing interoperable end-to-end information flows that can link sustainability data such as carbon sources and sinks while keeping end consumers and 'businesses' sensitive data confidential could drastically reduce carbon emissions across (international) borders [4, 5].

To that end, researchers have specifically proposed blockchain-based solutions with their ability to provide cryptographic verifiability and to unite different stakeholders on a single decentralized and neutral data-sharing platform to address the need for confidential end-to-end information flows [6, 7]. Hence, blockchain emerges as a promising approach, offering novel solutions to persistent challenges in achieving SDGs. While blockchain initially gained attention as the underlying technology for Bitcoin, it has rapidly evolved beyond its cryptocurrency origins. As an immutable infrastructure and distributed ledger technology, it also takes on an essential role in the implementation of sustainable and decentralized automated supply chains and tracing systems [8], the realization of decentralized (machine) identity management solutions [3], or particularly within the emergence of Decentralized Finance (DeFi), a paradigm shift that redefines financial interactions without relying on centralized intermediaries [9–11].

When considering the need for end-to-end information flows and its corresponding need for interoperable data exchanges between organizations, blockchain technology stands as an enabler by making use of its multifold characteristics of simultaneously advancing decentralization [8], fostering transparency [7, 12], enhancing data confidentiality, facilitating interoperability across diverse platforms, and enabling the innovative use of digital assets and tokenization [13, 14]. Regarding this, public and private organizations benefit from those blockchain characteristics as its features may foster the achievement of SDGs by providing a solution for a decentra lized end-to-end information flow [15].

As we delve deeper into the synergistic potential of blockchain and Web3, the concept of 'real-world' tokenization emerges as a critical component in bridging the digital and physical realms [14]. Tokenization in general describes the concept of representing an asset, object or any arbitrary information on a blockchain. 'Real-world' to kenization, i.e., tokenizing assets and objects outside of the blockchain, represents an integration that is essential for onboarding tangible assets into the blockchain and Web3.0 ecosystem, thereby expanding the scope of these technologies beyond virtual transactions to encompass real-world applications [3]. Particularly in environmental sustainability, numerous individual solutions at this intersection are beginning to surface while each is offering unique approaches to address sustainability challenges [16, 17].

Hence, we aim to provide a comprehensive overview of the current state of relationship between blockchain-based tokenization and environmental sustainability.

We explore innovative ways on how the amalgamation of blockchain and tokenization can enhance environmental sustainability, including the use of tokenized assets on blockchain networks for representing environmental assets like carbon credits, renewable energy certificates, or natural resources. Moreover, we outline how tokenization has the potential to increase transparency, accountability, verifiability, and efficiency while maintaining data confidentiality in managing environmental resources within sustainability solutions.

Therefore, this book chapter is structured as follows. First, we give a overview of challenges and requirements within the current processing of sustainability data. Afterwards, we investigate two promising areas and their potential of tokenization for sustainability solutions and measures. First, we explain how tokenization can support the process of traceability and verifiability of sustainability data, and second, we investigate which role tokenization can play when considering using it for sustainability assets. Last, we give an outlook on the challenges that tokenization of sustainability data and assets faces and point out the measures different stakeholders from research and industry need to take to address these challenges and maximize the potentials of tokenization for global sustainability.

2 Requirements and Challenges with Current Sustainability Data

Trustworthy "sustainability data" is crucial for the pertinent implementation of regulation frameworks such as ESG (Environmental, Social, and Governance). Sustainability data primarily encompasses CO_2 and other greenhouse gas (GHG) emissions data alongside broader ESG metrics [18, 19]. These data types are integral for evaluating and managing the environmental impacts of various entities and activities. They serve as key indicators for measuring progress toward achieving the SDGs and provide a mechanism for fine-grained carbon accounting, among other things. Traceability and verifiability of end-to-end information flows allow stakeholders to access and understand the CO_2 emissions associated with their processes and enable sustainability adaptive actions. Furthermore, it ensures comprehensive disclosure to both relevant stakeholders and legislative authorities [4].

However, immediate access to such detailed and verifiable sustainability data is often not feasible as their disclosure is typically occurring ex-post. This lack of immediate access to sustainability data results in data on CO_2 emissions possessing insufficient temporal granularity to effectively encourage adaptive behaviours. Additionally, instruments like the European Guarantees of Origin (GOs) and European Union Emissions Trading System (EU-ETS) certificates, which account for substantial quantities of Megawatt hours (MWh) of electricity and tons of CO_2, lack the necessary resolution to engage smaller stakeholders in CO_2-adaptive decision-making processes [20]. Hence, the existing manual and non-automated reporting process lack practicality for scaling data processing to fine granularity regarding time and quantity, hindering CO_2-adaptive control in business processes and consumption. This also limits stakeholders' ability to verify their own or their products' carbon footprints effectively and precisely [4, 16, 21].

Thus, one of the major requirements for an end-to-end information flow and, thus, for a continuous carbon emissions accounting process is the provision of reliable and authentic sustainability data [22, 23]. To date, the absence of data verifiability and the resulting deficit of trust have been significant barriers, hindering the implementation of continuous exchange and processing of sustainability data within globally spanning supply chains [22, 23]. In these mostly competitive market segments, however, it is also not feasible to establish a single source of truth that measures, reports, and verifies sustainability data. This is due to the complexity of the worldwide trade system but also the absence of a single trusted entity that could potentially consolidate and exploit sensitive business data. Therefore, decentralized systems are seen as a suitable solution since they do not rely on the existence of a single centralized authority but manage to establish trust and verifiability in a distributed network of many stakeholders without incurring the risk of unintentionally disclosing sensitive business data to individual third parties [24].

In the context of recording and processing sustainability data in supply chains, sensitive business information is often used to issue certificates that represent certain (mostly positive) actions and their underlying data. The goal of those certificates is to provide uniqueness, traceability, and verifiability features of sustainability data through supply chains. Further reliability and impact are achieved through the uniqueness of the information, which allows the definite traceability of the data to its origin. The most prominent examples are certificates for networked carbon markets and particularly carbon offsets, of which some, however, do not use unique but fungible certificates that do not allow for the tracing of the data to its origin [25, 26]. Besides the verifiability and reliability of the data itself, the uniqueness of the certificates and their redemption plays a crucial role. In the example of the carbon certificates, it is not only important that the certificate includes the correct amount of carbon emissions but that this certificate can only be used once and not multiple times [27]. This can enable the sustainability data or certificate to be used as an asset as it is uniqueness and digital ownership are ensured. Besides sustainability certificates, the class of sustainability assets also includes other sustainability-propelling investments, such as renewable energy production units.

Tokenization can solve both aspects as it can make the flow of data traceable and verifiable but also prevent the malicious or double-spending of certificates [14]. Generally, the token often serves as a digital representation of a physical good, recording important information about the product, including the carbon emissions generated during its production. Accordingly, we want to investigate how practitioners can utilize the concepts of tokenization to enhance the uniqueness, traceability, and verifiability of sustainability data and moreover, how sustainability assets can benefit from tokenization and the new opportunities associated with it.

3 Tokenization for Traceability and Verifiability of Sustainability Data

The traceability and verifiability of sustainability data is often a global task that needs to be performed in complex cross-organizational and international supply chains. As highlighted before, these are rarely contexts where there is a single entity that has the capability to trace sustainability data worldwide, or that is trusted by all stakeholders. Thus, decentralized systems that can establish neutrality and reliability without relying on a central party are required. Blockchains are particularly well suited as they can ensure system availability and integrity even under a certain percentage of malicious actors. Additionally, they provide chronological and tamper-proof storage of data and a reliable and verifiable automation of processes through clearly defined rules in the form of smart contracts. With these characteristics, blockchains enable digital ownership as an accessible and tamper-proof transaction record, allowing for verification of ownership and preventing double spending.

In the blockchain context, tokens are used as digital representations of a potentially unlimited variety of things. The most prominent category is cryptocurrencies which are either tokens natively integrated into the blockchain, e.g., bitcoin or Ethereum, or currency of other applications issued on top of existing blockchains, e.g., the AAVE [28] or MKR [29, 30] tokens issued by the respective protocols on the Ethereum blockchain. Furthermore, tokens can also be used to represent "real-world" assets or certificates [31], e.g., event tickets [32]. Therefore, tokens can provide a bridge between the outside world and the blockchain ecosystem [29, 30], making the tokenized assets tangible on the blockchain.

The approach of tokenizing sustainability certificates or the underlying data has already been widely suggested [10, 10, 16, 25, 33, 34]. In this context, tokenization provides the following advantages for sustainability data:

Immutability: Once the certificate is tokenized, i.e., the data has been onboarded, the blockchain data structure and the decentralized consensus ensure that the information cannot be tampered [35].

Traceability: As blockchain provides a complete ledger of all transactions, the tokenization of certificates and the forwarding on the blockchain afterwards enable the traceability of the certificates and their underlying data, e.g., through a supply chain, to the data source [4].

Double-spend resistance: When tokenizing certificates, blockchain ensures that one certificate cannot be forwarded to multiple parties or be redeemed multiple times, as the blockchain ledger enables the detection of previous spending [36].

Not depending on a single authority: All the features listed above can also be achieved with traditional, centralized information systems. In these systems, however, ensuring immutability and double-spend resistance and providing traceability is dependent on a single system authority. Blockchain can provide these advantages as well as the availability and integrity of the overall system in a decentralized setting, i.e., not depending on a single party but on an honest majority of the governing parties [37].

While the tokenization of sustainability certificates offers several benefits, there are also some challenges and hurdles that need to be addressed:

Distinguishability for storing unique metadata: Using fungible tokens for sustainability data allows for the tokenization of quantitative data, such as the amount of carbon emissions produced or saved in a certain process. While this information is already valuable, and these systems enable the simple forwarding and netting of these tokens, they do not allow for the inclusion of unique metadata, such as time and origin of production. However, enabling this end-to-end traceability and connecting carbon sources and sinks could further improve the impact of a sustainability data system. To achieve distinguishability and the ability to store unique metadata, the usage of nun-fungible tokens (NFTs) is required as they can be created as unique objects, distinguishable from similar tokens, and can store arbitrary additional information [4, 13, 32].

Fractionalizability to enable fine-granularity: However, NFTs typically come with the practicability drawback that they can only be transferred. In many supply chains, for example, certificates need to be split up, e.g., when one energy plant supplies multiple factories, the certificate with information on the energy production of a given time needs to be split into smaller fractions. To enable the forwarding of certificates in a flexible and fine-granular way, the tokens need to provide fractionalizability, i.e., it has to be possible to break NFTs into arbitrary fractions or combine fractions while ensuring parity between the initial tokens and the sum of all fractions [4].

Data privacy for sensitive user and corporate data: Blockchains are a pseudonymous but transparent system per default. Such systems, however, are prone to de-anonymization if the information available in—or outside of the system is sufficient to recognize patterns and establish linkages between on-chain pseudonymous and real-world entities [38, 39]. A fine-granular tracing system would most certainly provide enough detailed information to, on the one hand, establish such linkages and, on the other hand, for the information to harm trade secrets and user protection and violate data protection laws such as the EU's general data protection regulation (GDPR) [24]. Thus, sensitive data, e.g., the electricity consumption data of a household or an industrial plant that would be visible in a carbon emission tracing system in the electricity supply chain, cannot be stored on a transparent ledger. Although typically this transparency enables the key advantages of blockchain in the immutability and traceability of information and the prevention of double spending, there are other ways to retain these properties while ensuring data privacy. The most prominent approach is the utilization of zero-knowledge proofs (ZKPs) [40, 41]. ZKPs are a cryptographic primitive that allows proving and verifying the correctness of a statement or a computation without the need to disclose the underlying information. With this, the rules of the blockchain, i.e., information cannot be altered and tokens cannot be double-spend, while keeping the information itself private [4, 24, 42].

Scalability for fine-granular datapoints: The fine-granularity of the data, in combination with the extensiveness of many supply chains, leads to a larger number of newly captured data that needs to be tokenized and many certificates that need to be transferred among the supply chain. For example, when looking at a system to trace carbon emissions in the German electricity market in a 15-min time resolution, there are 40 million households consuming, resulting in at least 40 million transactions every 15 min, resulting in roughly 45 thousand transactions per second. This number of transactions cannot be handled by a single system, especially not by a blockchain, but requires a sensible segmentation that splits the overall system into more manageable subsystems but does not cause too many interactions between these different subsystems. Such segmentation can then be used by blockchain scalability solutions such as ZK-Rollups. They leverage the potential of ZKPs to outsource transactions to a more centralized second blockchain but prove the correctness of the aggregate of all transactions to the main blockchain with a ZKP. Thus, they can take transaction load from the main blockchain but still maintain the guarantees regarding data integrity that are provided by the decentralized main blockchain [7].

Verifiable and trustworthy data sources: While one of the main ideas of a blockchain is to guarantee the integrity and availability of its data, these security guarantees are only limited to the information that is stored on the blockchain [11, 43, 44]. In contrast, blockchains are unable to verify information outside of the network. This kind of information, however, is what underlies all sustainability data, as it consists of carbon emission, other greenhouse gases or many other measurements from the "realworld". The problem of reliable and verifiable access to information outside the blockchain is better known as the oracle problem and has been widely discussed in the literature [45]. Some approaches rely on the reporting of the information by a trusted party that is given permission to do so, while other approaches apply decentralization to the oracles as well and use data that is agreed upon through a consensus mechanism with multiple oracles.

Alternatively, for data acquired through physical measurement devices like smart meter gateways, which monitor electricity production and consumption, there exists a different approach to enhance the trustworthiness of oracle data. This approach leverages concepts known in digital identities to equip smart metering devices with machine identities issued by trustworthy institutions. Every identity is based on a public–private key pair that is stored in a hardware secure module that ensures the physical binding of the identity to the respective device. By reporting data with the signature of the respective key and the identity information attached, all users can verify the authenticity of the information, its origin, and the fact that this device was installed or maintained by a trusted institution [3].

4 Tokenization of Sustainability Propelling Assets

The financial sector is one of the most dominant application areas of blockchains today. In its context, blockchain enables the merging of settlement and clearing of financial assets and thus can represent efficient infrastructure for settlement and clearing of payments and other more complex financial transactions. Therefore, the usage of blockchain as an infrastructure for the financial system is being explored by banks and other corporations in the financial sector but is also recognized and accounted for by regulators and legislators. The EU, for example, has established an explicit regulatory framework around cryptocurrencies and other blockchain-based assets with the "Markets in Crypto-assets Regulation" (MiCa R) [46] and has further introduced a regulatory sandbox for blockchain-based settlement and clearing of financial assets [47].

Besides the transformative impact on the traditional financial system, blockchain also enabled the emergence of a totally new, decentralized financial ecosystem (DeFi). "DeFi is a decentralized financial system that enables financial services and instruments to be offered and used without the need for intermediaries as the system is based on public blockchains and smart contracts" [10]. DeFi can provide an open, efficient, and trustless financial system that can replicate existing financial products but can also be the origin of new financial services through its automatability and composability [48]. Although DeFi still faces several challenges regarding scalability, security, and adoption, it is widely seen as a promising new paradigm of financial systems [49].

As mentioned above, sustainability certificates are not only used to trace sustainability data and prove it to third parties but they are also often treated as tradable assets, e.g., carbon offsets. Against this backdrop, the tokenization of sustainability certificates and their underlying data cannot only ensure verifiability and integrity, as pointed out in the previous section, but it can also represent an asset that is natively integrated into the blockchain-based financial infrastructure. With this, these assets can be integrated into more open and efficient financial markets, both in the traditional financial system and in the newly arising decentralized financial system. Easier access and more efficient processes can increase their desirability as financial assets and the liquidity and efficiency of the whole market.

The most advanced markets for the tokenization of sustainability certificates are voluntary carbon markets (VCMs) that facilitate the issuance and trading of voluntary carbon credits (VCCs). These markets and the credits traded on top usually follow a certain lifecycle [50]: First is the general project design, followed by registration at an institution that certifies carbon standards such as Verra [51] or Gold Standard [52]. Afterwards, the monitoring, reporting, and verification (MRV) processes that were defined during the project design and certification are carried out. Based on the verified data, the VCCs are issued on the market, where they can then be transacted and retired by the final buyer who wants to claim the environmental benefit associated with the credits. VCMs in DeFi, such as Klima [53] or Toucan [54], have adopted one or more carbon standards and facilitate the issuance, trading, and retirement of carbon credits from projects certified under this standard in a decentralized way on public and permissionless blockchains. Platforms like Senken are more geared towards the traditional financial system and directly facilitate the issuance and purchase of VCCs [55].

Besides sustainability certificates, tokenization can also be applied to other sustainability propellant projects, such as the expansion of renewable energy sources. By tokenizing these projects and the assets produced, they can be opened to more investors, providing increased liquidity and leading to more efficient markets. Especially, renewable energy sources that introduce distributed energy production can benefit from new funding and investing opportunities.

While the tokenization of sustainability assets could foster environmental sustainability by providing access to open and more efficient capital markets, it currently still faces several challenges. Most of these challenges are closely related to the challenges associated with the tokenization for traceability and verifiability of sustainability data outlined in the previous section. The oracle problem that is linked to the verifiability of the initial information is crucial, for example, for the verification of carbon offset projects, their authenticity, and their factual impact. Furthermore, blockchain-based financial systems, especially DeFi, still face several challenges regarding the scalability of the technology itself and the adoption of these systems by regulators, corporations, and consumers [10]. However, the tokenization of sustainability assets itself could provide DeFi with another asset class, decreasing its correlation and dependency on the cryptocurrency market and advancing the adoption of DeFi. Thus, DeFi and tokenized financial assets could mutually benefit from each other.

5 Conclusion

In conclusion, the intersection of blockchain technology and environmental sustainability, particularly through the lens of tokenization, presents a novel paradigm for managing and verifying sustainability data. The tokenization of sustainability certificates and underlying information holds significant potential to revolutionize the way we handle environmental data while building trust, offering a decentralized approach that ensures traceability and verifiability without the need for a central authority. This approach aligns with the growing global emphasis on transparent and accountable environmental governance.

However, this innovative approach is not without its challenges. Among these issues, the emphasis on privacy is substantial, especially because sustainability data frequently includes confidential information. The inherent transparency and distributed structure of blockchain technology might result in the exposure of confidential corporate data or sensitive information about individual consumption patterns. Such exposure could have far reaching implications, ranging from competitive disadvantages for businesses to privacy violations for individuals. Hence, striking a balance between transparency and privacy is essential, necessitating the utilization of sophisticated cryptographic solutions like zeroknowledge proofs to enable privacy-preserving transactions and data exchange on public blockchains [24, 41].

The persistent oracle problem regarding the authenticity of the information onboarded to the blockchain system also poses a significant challenge. It highlights the potential disconnection between the physical world and the blockchain's digital ledger. Reliable and accurate data input is the backbone of the blockchain's credibility, particularly in environmental sustainability contexts where data accuracy and reliability is essential. Hence, the integration of trustworthy and verifiable data sources, possibly through sophisticated machine-verifiable devices or trusted digital reporting mechanisms, is crucial to ensure the verifiability of the data being tokenized. In this context, ensuring the reliability and accuracy of data inputs is crucial, as blockchain's immutability means any errors or falsifications in initial data entry persistently exist throughout the system. This characteristic of blockchain demands rigorous standards for data entry, end-to-end verification processes, and ongoing monitoring to maintain the system's integrity, which need to be investigated in further research [4, 10].

Besides these challenges, the tokenization of sustainability assets, encompassing both certificates and projects or infrastructure, also has further potentials. Especially, it can pave the way for their integration into new, more open, and efficient blockchain-based systems and markets. This integration is particularly promising within the context of DeFi enhancing traditional financial sectors by employing blockchain infrastructure. While these sectors are still maturing, both technologically and in terms of widespread adoption, sustainability assets could potentially serve as a pivotal technology for practical diffusion. This application could spearhead a new wave of adoption, demonstrating the practical utility and benefits of blockchain technology in a field with significant societal and environmental impact.

Ultimately, the symbiotic relationship between sustainability, blockchain, and particularly tokenization is promising to create a mutual benefit. On the one hand, blockchain technology, particularly through tokenization, offers a robust framework for managing sustainability data with enhanced transparency, accountability, and efficiency. On the other hand, the integration of sustainability assets into blockchain platforms could catalyze further development and adoption of these technologies, offering a concrete, impactful application that addresses global climate challenges. As the field continues to evolve, it will be critical to continue to address current challenges, particularly around privacy and authenticity, in order to fully realize the potential of this convergence to promote a more sustainable future.

References

1. United Nations, *The Paris Agreement* (3 April 2015). Available at: https://www.un.org/en/climatechange/paris-agreement. Accessed: 3 Apr 2024
2. United Nations, in *The-Sustainable-Development-Goals-Report 2023: Special edition: Towards a Rescue Plan for People and Planet* (2023). Available at: https://unstats.un.org/sdgs/report/2023/The-Sustainable-Development-Goals-Report-2023.pdf. Accessed: 20 Nov 2023
3. M. Babel et al., Vertrauen durch digitale Identifizierung: Über den Beitrag von SSI zur Integration von dezentralen Oracles in Informations systeme. HMD Praxis der Wirtschaftsinformatik **60**(2), 478–493 (2023). https://doi.org/10.1365/s40702-023-00955-3
4. M. Babel et al., Enabling end-to-end digital carbon emission tracing with shielded NFTs. Energy Inform. **5**(S1), 21 (2022). https://doi.org/10.1186/s42162-022-00199-3
5. J. Strüker et al., *Decarbonisation Through Digitalisation: Proposals for Transforming the Energy Sector* (2021)
6. T. Guggenberger, A. Schweizer, N. Urbach, Improving interorganizational information sharing for vendor managed inventory: toward a decentralized information hub using blockchain technology. IEEE Trans. Eng. Manage. **67**(4), 1074–1085 (2020). https://doi.org/10.1109/TEM.2020.2978628
7. J. Sedlmeir, F. Völter, J. Strüker, The next stage of green electricity labeling: using zero-knowledge proofs for blockchain-based certificates of origin and use. ACM SIGENERGY Energy Inform. Rev. **1**(1), 20–31 (2021)
8. B.-J. Butijn, D.A. Tamburri, W.-J. van Heuvel, Blockchains: a systematic multivocal literature review. ACM Comput. Surv. (CSUR) **53**(3), 1–37 (2020)
9. A. Bechtel et al., The future of payments in a DLT-based European economy: a roadmap, in *The Future of Financial Systems in the Digital Age: Perspectives from Europe and Japan*. Springer, Singapore (2022), pp. 89–116
10. V. Gramlich et al., A multivocal literature review of decentralized finance: current knowledge and future research avenues. Electron. Mark. **33**(1), 11 (2023)
11. T. Guggenberger, M. Kuhn, B. Schellinger, in *Insured? Good! Designing a Blockchain-Based Credit Default Insurance System for DeFi Lending Protocols* (MENACIS2021, 2021). Available at: https://aisel.aisnet.org/menacis2021/8
12. M. Baza et al., Privacy-preserving blockchain-based energy trading schemes for electric vehicles. IEEE Trans. Veh. Technol. **70**(9), 9369–9384 (2021)
13. L. Arnold et al., Blockchain and initial coin offerings: blockchain's implications for crowdfunding, in *Business Transformation Through Blockchain*, vol. I, ed. by H. Treiblmaier, R. Beck (Springer International Publishing, Cham; Imprint, Palgrave Macmillan, 2019), pp. 233–272. Available at: https://link.springer.com/https://doi.org/10.1007/978-3-319-98911-2_8
14. A. Sunyaev et al., Token economy. Bus. Inf. Syst. Eng. **63**(4), 457–478 (2021)
15. M.C. Lacity, S.C. Lupien, *Blockchain Fundamentals for Web 3.0* (University of Arkansas Press, 2022)
16. A. Al Sadawi et al., A comprehensive hierarchical blockchain system for carbon emission trading utilizing blockchain of things and smart contract. Technol. Forecast. Soc. Chang. **173**, 121124 (2021)
17. M. Schletz et al., Blockchain and regenerative finance: charting a path toward regeneration. Front. Blockchain **6**, 13 (2023). https://doi.org/10.3389/fbloc.2023.1165133
18. P. Ahi, C. Searcy, An analysis of metrics used to measure performance in green and sustainable supply chains. J. Clean. Prod. **86**, 360–377 (2015)
19. A. Qorri, Z. Mujkić, A. Kraslawski, A conceptual framework for measuring sustainability performance of supply chains. J. Clean. Prod. **189**, 570–584 (2018)
20. R. Watanabe, G. Robinson, The European Union Emissions Trading Scheme (EU ETS). Clim. Policy **5**(1), 10–14 (2005). https://doi.org/10.1080/14693062.2005.9685537
21. F. Knirsch et al., Decentralized and permission-less green energy certificates with GECKO. Energy Inform. **3**, 1–17 (2020)

22. M.-F. Körner et al., Accelerating sustainability in companies: a taxonomy of information systems for corporate carbon risk management, in *ECIS 2023 Research Papers* (2023). Available at: https://aisel.aisnet.org/ecis2023_rp/248
23. M.-F. Körner et al., Digital carbon accounting for accelerating decarbonization: characteristics of IS-enabled system architectures, in *Proceedings of the 29th Americas Conference on Information Systems (AMCIS)* (2023)
24. J. Sedlmeir et al., The transparency challenge of blockchain in organizations. Electron. Mark. **32**(3), 1779–1794 (2022)
25. A. Jackson et al., Networked carbon markets: permissionless innovation with distributed ledgers?, in *Transforming Climate Finance and Green Investment with Blockchains* (Elsevier, 2018), pp. 255–268
26. L. Pigeolet, A. van Waeyenberge, Assessment and challenges of carbon markets. Braz. J. Int'l L. **16**, 74 (2019)
27. M. Platt et al., Information privacy in decentralized applications, in *Trust Models for Next-Generation Blockchain Ecosystems* (2021), pp. 85–104.
28. AAVE Liquidity Protocol, *AAVE Liquidity Protocol* (2023). Available at: https://aave.com/. Accessed: 22 Nov 2023
29. M. Brennecke et al., The de-central bank in decentralized finance: a case study of MakerDAO, in *Proceedings of the 55th Hawaii International Conference on System Sciences, Hawaii International Conference on System Sciences: Hawaii International Conference on System Sciences* (2022). https://doi.org/10.24251/HICSS.2022.737
30. M. Brennecke et al., The human factor in blockchain ecosystems: a sociotechnical framework, in *Wirtschaftsinformatik 2022 Proceedings* (2022). Available at: https://aisel.aisnet.org/wi2022/finance_and_blockchain/finance_and_block-chain/3
31. T. Guggenberger et al., Kickstarting blockchain: designing blockchain-based tokens for equity crowdfunding. Electron. Commer. Res. 1–35 (2023). https://doi.org/10.1007/s10660-022-09634-9
32. F. Regner, A. Schweizer, N. Urbach, NFTs in practice: non-fungible tokens as core component of a blockchain-based event ticketing application, in *Proceedings of the 40th International Conference on Information Systems (ICIS)*, Munich, Germany (2019). Available at: https://eref.uni-bayreuth.de/52509/
33. Dena, *Blockchain in the Integrated Energy Transition: Dena Multi-Stakeholder Study* (2019). Available at: https://www.dena.de/fileadmin/user_upload/dena-Studie_Blockchain_Integrierte_Energiewende_EN.pdf. Accessed: 22 Nov 2023
34. L. Einhellig, J. Strüker, *Blockchain in the Integrated Energy Transition: Dena—Multistakeholder Study. Study Findings (dena)* (2019). Available at: https://www.dena.de/fileadmin/user_upload/dena-Studie_Blockchain_Integrierte_En-ergiewende_EN.pdf. Accessed: 18 Dec 2023
35. Z. Zheng et al., An overview of blockchain technology: architecture, consensus, and future trends, in *2017 IEEE International Congress on Big Data Congress 2017: 25–30 June 2017, Honolulu, Hawaii, USA: Proceedings, 2017 IEEE International Congress on Big Data (BigData Congress). Institute of Electrical and Electronics Engineers, Honolulu, HI, USA, 6/25/2017–6/30/2017* (IEEE, Piscataway, NJ, 2017), pp. 557–564
36. S. Nakamoto, *Bitcoin: A Peer-to-Peer Electronic Cash System* (2008). Available at: https://bitcoin.org/bitcoin.pdf. Accessed: 22 Nov 2023
37. G. Fridgen et al., *Cross-Organizational Workflow Management Using Blockchain Technology—Towards Applicability, Auditability, and Automation* (2018). Available at: https://scholarspace.manoa.hawaii.edu/items/b868becb-3554-47db9e0d-7f34030b3780
38. A. Biryukov, D. Khovratovich, I. Pustogarov, *Deanonymisation of clients in Bitcoin P2P network* (2014). Available at: https://arxiv.org/pdf/1405.7418.pdf
39. V. Gramlich et al., Decentralized finance nobody knows you are a dog, in *Proceedings of the 57th Hawaii International Conference on System Sciences (HICSS)* (Honolulu, 2024). Available at: https://eref.uni-bayreuth.de/id/eprint/87878/
40. S. Goldwasser, S. Micali, C. Rackoff, in *The knowledge complexity of interactive proof-systems*, in *Proceedings of the Seventeenth Annual ACM Symposium on Theory of Computing, the*

Seventeenth Annual ACM Symposium. Association for Computing Machinery; ACM Special Interest Group on Algorithms and Computation Theory, Providence, Rhode Island, United States, 5/6/1985–5/8/1985 (ACM, New York, 1985), pp. 291–304. https://doi.org/10.1145/22145.22178

41. M. Principato et al., *Towards Solving the Blockchain Trilemma: An Exploration of Zero-Knowledge Proofs* (2023)
42. EY Blockchain, *EYBlockchain/nightfall_3: A Mono-Repo Containing an Optimistic Version of Nightfall*, 22 November (2023). Available at: https://github.com/EYBlockchain/nightfall_3. Accessed: 22 Nov 2023
43. T. Guggenberger et al., A structured overview of attacks on blockchain systems, in *Proceedings of the 25th Pacific Asia Conference on Information Systems (PACIS)*. Dubai, United Arab Emirates (2021). Available at: https://eref.uni-bay-reuth.de/id/eprint/66899/
44. V. Schlatt et al., Attacking the trust machine: developing an information systems research agenda for blockchain cybersecurity. Int. J. Inf. Manage. **68**, 102470 (2023). https://doi.org/10.1016/j.ijinfomgt.2022.102470
45. M.D. Sheldon, Auditing the blockchain oracle problem. J. Inf. Syst. **35**(1), 121–133 (2021). https://doi.org/10.2308/ISYS-19-049
46. European Commission, *Regulation of the European Parliament and of the Council on Markets in Crypto-assets, and Amending Directive (EU) 2019/1937*, 26 November (2020). Available at: https://eur-lex.europa.eu/legal-content/EN/TXT/?uri=CELEX%3A52020PC0593. Accessed: 26 Nov 2023
47. European Commission, *Launch of the European Blockchain Regulatory Sandbox*, 26 November (2023). Available at: https://digital-strategy.ec.europa.eu/en/news/launch-european-blockchain-regulatory-sandbox. Accessed: 26 Nov 2023
48. F. Schär, *Decentralized Finance: On Blockchain and Smart Contract-Based Financial Markets* (2021), p. 103. Available at: https://research.stlouisfed.org/publications/review/2021/02/05/decentralized-finance-on-blockchain-and-smart-contract-based-financial-markets
49. V. Gramlich et al., *Decentralized Finance (DeFi): Foundations, Applications, Potentials, and Challenges* (2022). Available at: https://www.fit.fraunhofer.de/content/dam/fit/de/documents/Whitepaper/DeFi/English_Version_2022_07_15.pdf. Accessed: 26 Nov 2023
50. R. Mills, *How to Build a Trusted Voluntary Carbon Market* (Rocky Mountain Institute, 2 September 2022). Available at: https://rmi.org/how-to-build-a-trusted-voluntary-carbon-market/. Accessed: 27 Nov 2023
51. Verra, *Home–Verra* (2023), 27 Nov 2023. Available at: https://verra.org/. Accessed: 27 Nov 2023
52. Gold Standard Org., *Goldstandard* (2023). Available at: https://www.goldstandard.org/. Accessed: 18 Dec 2023
53. D.A.O. Klima, *KlimaDAO | Public Good for Our Planet*, 27 November (2023). Available at: https://www.klimadao.finance/. Accessed: 27 Nov 2023
54. Toucan, *Welcome to Toucan* (2023). Available at: https://docs.toucan.earth/toucan/introduction/overview. Accessed: 18 Dec 2023
55. Senken, *Senken—Authentic Carbon Credit Portfolios* (2023). Available at: https://www.senken.io/de. Accessed: 18 Dec 2023

Vincent Schaaf is a postdoctoral researcher at the Branch Business and Information Systems Engineering of Fraunhofer FIT and the Frankfurt University of Applied Sciences. Additionally, he is a PhD Student at the University of Bayreuth. His research focusses on blockchain, digital identities, and other applied cryptography and their interplay and symbiotic usage for building trustless information systems that enable data verifiability and privacy. Thereby, Vincent Gramlich, both conducts fundamental research on these technologies and their affordances and challenges as well as designs practical frameworks and applications, especially in the energy and financial sector. His works have been published in a broad array of academic journals and conferences such as Electronic Markets (EM), Energy Informatics (EnIcs), ACM Symposium on Applied Computing (SIGAPP), or Journal of Banking Law and Banking (JBB).

Jonathan Lautenschlager is a postdoctoral researcher at the Frankfurt University of Applied and currently pursuing his PhD in Information Systems at FIM Research Center, University of Bayreuth, and the Branch Business and Information Systems Engineering of the Fraunhofer FIT. In his research, Mr. Lautenschlager works on blockchain technology, decentralized identity management and related fields. Thereby, Jonathan Lautenschlager, both conducts fundamental research on these technologies, their affordances, and challenges as well as designs practical frameworks and applications to overcome barriers for practical diffusion. He received his M.A in Media Economics with specialization in media informatics. His research has been published in Business and Information Systems Engineering (BISE), Electronic Markets (EM), Government Information Quarterly (GIQ), Communications of the Association for Information Systems (CAIS) and the International Conference on Wirtschaftsinformatik (WI).

Tobias Guggenberger is a postdoctoral researcher at the University of Bayreuth and the Branch Business and Information Systems Engineering of Fraunhofer FIT. He holds an M.Sc. in business administration and a Ph.D. in information systems engineering. His research primarily concentrates on the design and management of information systems, leveraging emerging technologies like artificial intelligence (AI) and blockchain. In particular, his work is notably centered on exploring the impact of decentralization and distributed decision-making on the collaborative and value-creation practices of firms. His scholarly contributions have appeared in several renowned conference proceedings, including the European Conference on Information Systems (ECIS) and the International Conference on Information Systems (ICIS). His research has also been published in prestigious international journals such as IEEE Transactions on Engineering Management (IEEE

TEM), Electronic Markets (EM), and Computers and Industrial Engineering.

Marc-Fabian Körner is a postdoctoral researcher at the Branch Business and Information Systems Engineering of Fraunhofer FIT and the Fraunhofer BlockchainLab. He is also a habilitation candidate at the University of Bayreuth in the field of Information Systems (IS). Additionally, he is affiliated with the Research Center FIM, where he leads a research group on Digital Sustainability and Decarbonization. His research focusses on Green IS and related topics in energy informatics, data ecosystems and cross-organizational collaboration, as well as the application of Web3-technologies in private and public organizations. Marc-Fabian is an active reviewer, associate editor, and track chair for various journals and conferences. Within various funded projects, he also advises several organizations on digitalization- and technology-related issues, e.g., in the context of data governance and sustainability. His interdisciplinary work has been published in journals including Business and Information Systems Engineering (BISE), Electronic Markets (EM), Applied Energy (APEN), and the European Journal of Operational Research (EJOR). Marc-Fabian also regularly presents his work at international conferences.

Jens Strüker is a professor of Information Systems and Digital Energy Management at the University of Bayreuth, deputy head of the Business and Information Systems Engineering department of the Fraunhofer Institute for Applied Information Technology (FIT), director of the FIM Research Institute for Information Management as well as co-director of the Fraunhofer Blockchain Lab in Bayreuth. As a habilitated business informatics specialist and economist, his research focuses on real-time energy markets, CO_2 tracking, and supporting web.3 3technologies such as IoT, AI, data spaces, and distributed ledger technologies.

Nils Urbach is Professor of Information Systems and Digital Business as well as Director of the Research Lab for Digital Innovation and Transformation at the Frankfurt University of Applied Sciences, Germany. Furthermore, he is Director at the FIM Research Center for Information Management and the Branch Business and Information Systems Engineering of Fraunhofer FIT, as well as Co-Founder and Director of the Fraunhofer Blockchain Lab. Nils Urbach has been working in the fields of digital innovation and transformation for several years. His work has been published in several academic journals such as Information Systems Research (ISR), the Journal of Strategic Information Systems (JSIS), the Journal of Information Technology (JIT), MIS Quarterly Executive (MISQE), Information and Management (I&M), and Business and Information Systems Engineering (BISE). He advises several companies on digitalization issues and regularly appears as a speaker on this topic.

Legal

Crypto Art and Intellectual Property

Su-Zeong Fröhlich and Kerstin Gold

1 Introduction

1.1 The Rise of Digital Art and the Birth of Crypto Art

Digital Art, as an intersection between art and technology and a facet of new media art, has redefined our understanding of creative expression. This introduction illuminates the complex evolution of Digital Art that laid the foundation for the emerging phenomenon of Crypto Art.

The roots of Digital Art trace back to the 1960s. Artists began to experiment with emerging technologies, marrying traditional artistic methods with the capabilities of the computer (Fig. 1). There are myriad expressions of Digital Art, from algorithm-driven pieces to interactive installations [1]. However, capturing the history of Digital Art presents a unique challenge though. Its relatively nascent timeline, coupled with sporadic documentation and constant transformation, makes it difficult to maintain a clear and consistent overview [2]. For example, many of its pivotal works and writings have not been archived consistently [3].

Digital Art has been a vibrant yet underrecognized segment in the art world [4]. While digital creative practices began flourishing with the proliferation of personal computers and the internet in the 1990s, Digital Art faced a unique challenge stemming from the digital format itself: it lacks physical 'originals' and is easily replicable which makes ownership and trade more conceptual than tangible. This hindered its recognition and valuation in the traditional art market, where uniqueness and ownership are crucial. Finally, the commercialization of Digital Art shifted drastically with the advent of Non-Fungible Tokens (NFTs) in the mid-2010s [5].

S.-Z. Fröhlich (✉) · K. Gold
Berlin, Germany
e-mail: szfroehlich@posteo.de

Fig. 1 Vera Molnar "Hommage à Monet" (1983), 20 × 25 cm, plotter drawing. Courtesy by the artist and DAM projects

In the contemporary art market, Digital Art is experiencing a shift in recognition and investment, as evidenced by its inclusion in prestigious collections and exhibitions at major institutions such as the Museum of Modern Art (MoMA) [6], Whitney Museum of American Art in New York, the Los Angeles County Museum of Art (LACMA) [6], and the Centre Georges Pompidou [7] and Musée d'Orsay in Paris. While the latest Art Basel Report for 2023 shows a decrease in overall spending on Digital Art from the previous year [8], the market for art-related NFTs has seen significant cumulative growth over the past years [9], and Gen Z collectors have emerged as leading spenders in Digital Art [8, 10]. These patterns suggest that Digital Art is not only gaining traction among younger collectors but is also positioned to maintain and possibly increase its relevance in the future art market.

The introduction of NFTs has certainly made a game-changing impact on Digital Art [8, 11]. By creating a system of ownership and authenticity for digital works, NFT technology has bridged a crucial gap [12]. This advancement has not only added tangible value to Digital Art but also facilitated its integration into the conventional art canon. NFTs provide a way to distinguish original digital works from copies [13], offering a sense of scarcity and uniqueness previously unattainable in the digital realm. As a result, artists, collectors, and the broader art community have begun to acknowledge and appreciate Digital Art as a legitimate and valuable form of creative expression [14].

The newfound acceptance and validation of Digital Art has laid the groundwork for Crypto Art, which has seen a surge in popularity and exploration [15]. Artists are now increasingly viewing blockchain technology not just as a tool for certifying and trading art but also as a creative medium in its own right. This perspective shift is leading to innovative uses of blockchain beyond the creation of NFTs, thereby pushing the boundaries of what art can be in the digital age.

In this chapter, we will refer to Crypto Art as a type of "Digital Fine Art," based on Christine Paul's definition of "Digital Art" in her corresponding seminal book. Other formats, such as 'Profile Pic (PFP) NFT-Collectibles' (e.g., Bored Ape Yacht Club, one of their most notable representatives), are not considered in this chapter. While they represent a new and intriguing phenomenon in digital graphics, PFP-NFT-Collectibles are acquired for different collecting motivations. These motivations encompass branding enhancement, financial investment, and speculation. Additionally, such collectibles typically provide an array of utilities, including but not limited to, serving as admission tokens granting access to exclusive events. The primary impetus for collecting Crypto Art on the other hand lies in the appreciation of the artwork itself, rather than extrinsic factors [16]. Additionally, NFT-Collectibles target a different audience and operate under distinct market and marketing mechanisms, which differ from those of the traditional art market. Therefore, we believe it is sensible to make a distinction between these two applications of NFTs for the purposes of this chapter. Consequently, the task of accurately defining this new art phenomenon enabled through blockchain technology is challenging. Although terms like "art-related NFT" [17] or "NFT art" [18] are frequently used, these terms may overemphasize the underlying technology. While Blockchain is being utilized as a new creative medium of our time, similar to how a brush represents a fundamental tool in traditional art creation, it is important to recognize 'Crypto' as an overarching term. In our view, 'Crypto' serves best as the umbrella term that encapsulates this innovative use of technology in the art world. "Crypto Art" [19] a term yet to be universally agreed upon but, from our perspective, most likely characterizes this new art direction, which notably began with Jennifer and Kevin McCoy's NFT "Quantum" in 2014, marking the genesis of Crypto Art [20].

The origins of Crypto Art can be traced back even further in time to a period when the desire to own digital assets first emerged among tech enthusiasts or so-called "techies" [21]. "Pepe the Frog" became widely known as one of the first internet memes that could be collected, and thus evolved into a symbol of the movement, rather than a piece of art in the classical sense [22]. Assessing whether it qualifies as 'Fine Digital Art' exceeds the scope of this chapter, but it is nevertheless a noteworthy mention. Eventually, Crypto Art truly catapulted into mainstream awareness in 2021 with the remarkable auction sale of Mike Winkelmann's (known as Beeple) piece, 'Everydays: The First 5000 Days,' at Christie's [23]. This art NFT sold for an astonishing $69 million, becoming the most expensive digital artwork ever sold and placing Mike Winkelmann among the top three most expensive living artists. This event not only legitimized NFTs in the art world but also introduced them to a broader audience.

1.2 Classification of Crypto Art

In the twenty-first century, the Digital Art landscape has undergone a dramatic shift due to the advent of NFTs. Through blockchain-based tokens, notions of ownership, trading, and valuation in the digital world have been redefined [15]. There are numerous subcategories of art NFTs, both content-wise—such as Generative Art, Digital Painting, Photography, and many more—and technically, differentiated into on-chain and off-chain NFT art as well as dynamic-NFTs. They all use blockchain technology in different ways and thus offer distinctive attributes.

What distinguishes Generative Art from the other classifications mentioned here is its innovative fusion of creativity and technology, which uses the power of algorithms to create sophisticated and ever-changing artworks. Generative Art represents a specific form within the spectrum of Crypto Art that particularly benefits from or makes extensive use of blockchain technology. Artists use creative coding to generate varied and evolving art each time a collector mints a piece. This process turns Generative Art into unique and immersive experiences, elevating the concepts of 'originality' and 'uniqueness' to new heights. Each minted piece carries an element of surprise, enhancing the unpredictability and offering a distinct, personalized asset to collectors. The advancement of Generative Art can be attributed to artists' growing recognition of blockchain as a dynamic creative medium, allowing them to both utilize and challenge the technology, resulting in entirely new artistic expressions. Notable platforms that are supporting and nurturing this phenomenon, are the Generative Art platforms Artblocks and fx(hash) [24] (Fig. 2).

In on-chain art, the blockchain itself is harnessed as an artistic medium, while off-chain NFTs function as "digital twins" that aren't cryptographically secured [25]. Dynamic-NFTs, exemplified by Beeple's "Human One", signify a progression by facilitating interactive and mutable content [26]. They are revolutionizing the art domain, contrasting traditional art by being modifiable in real-time, even post-acquisition.

Furthermore, NFTs can be linked to artworks that are unaffordable for the majority of people and subsequently fractionalized, allowing for collective ownership of a piece of art by many [27]. This approach was taken by Particle with Banksy's 'Love Is In The Air', for example [28].

In conclusion, the trajectory of Digital Art—from its inception in the 1960s to the NFT revolution in 2020—showcases a compelling confluence of creativity, technology, and market dynamics. With this historical insight, both art aficionados and novices can better appreciate and gauge contemporary trends and prospective potentials.

Fig. 2 Elsif, "Slice of Live #1" (2023), Generative NFT, on fx(hash)

2 Common Blockchain Protocols in the Crypto Art Ecosystem: An Overview

Ethereum (ETH): As a trailblazer in the NFT realm, Ethereum remains the platform of choice for many artists. Esteemed platforms such as OpenSea, Rarible, and Foundation are built on Ethereum. However, high transaction fees can pose barriers to emerging artists seeking entry.

Polygon (previously Matic): As a Layer-2 solution tailored for Ethereum, Polygon offers expedited transactions at reduced fees. It has garnered acclaim within the art community by circumventing Ethereum's limitations, such as high transaction fees and low transaction speed, while preserving compatibility with its infrastructure.

Tezos (XTZ): Emphasizing environmental sustainability in contrast to Proof-of-Work blockchains like Ethereum, Tezos has attracted a plethora of artists. Hic et Nunc, Objkt and fx(hash) stand out as platforms leveraging Tezos for NFT purposes amongst creators.

Solana: With its commendable transactional speed and affordability, Solana is increasingly pulling a significant number of artists and developers towards it. Platforms such as Magic Eden and Metaplex serve as representatives for Solana-centric marketplaces. Its distinctive consensus mechanism, Proof of History, facilitates brisk transactions.

Cardano (ADA): Despite its nascent foray into the NFT and Smart Contract sectors, there's already a burgeoning roster of NFT projects developing on the Cardano blockchain.

Flow: Conceived by the creators of CryptoKitties, from its very early stage, Flow has been specifically designed for NFTs and decentralized applications (dApps). NBA Top Shot, one of the most prosperous NFT platforms, operates on Flow.

Lastly, it is worth noting that Bitcoin, the inaugural and most renowned blockchain network, is now venturing into the NFT space with Bitcoin-based ordinal inscriptions, known as '**Ordinals**', introduced to the public in January 2023 [29].

By late July 2023, total ordinal inscriptions on the Bitcoin blockchain exceeded 18.5 million, showcasing a wide variety of content types and signaling a robust and diverse participation from the community. This growth trajectory underscores the growing interest and potential of Bitcoin-based ordinals in reshaping the Digital Art landscape and offering new forms of cultural expression [30]. The significance of the Bitcoin network is growing within creative spheres, gaining importance for both artists and collectors.

3 How NFTs Challenge the Traditional Art Market

3.1 Mechanisms of the Traditional Art Market

The traditional art market is a complex ecosystem with distinctive players and mechanisms: artists create unique works that are mostly sold through intermediaries like galleries or auction houses. Collectors, both individual and institutional, drive market demand which is not solely determined by the aesthetic value of the works. It is a market frequently viewed as an elite and complex one, considered by many to be inaccessible and opaque. This market's dynamics are influenced by a variety of external factors, including value principles, exclusivity, reputation, and speculative forces, rather than solely by the intrinsic beauty of the art as its own piece. For collectors, acquiring a coveted artwork is often challenging, even if there are no budgetary limitations. Galleries and art advisors wield significant gatekeeping power, often prioritizing placement of important works in esteemed private collections over selling to the highest bidder. This intermediary role of galleries and advisors adds another layer of complexity and exclusivity to the art acquisition process.

Predominantly featuring physical and typically unique artworks, their distribution spans galleries, auctions, and dealerships. The valuation process is inherently subjective, deeply influenced by experts, critics, and curators. Digitization, however, has brought more transparency to this world—initially through galleries showcasing their pieces with prices online [31], introducing "Click & Buy" options, and subsequently magnified through blockchain technology. The perceived value of an artwork transcends its physical manifestation, being significantly shaped by its contextualization, art historical ties, and expert endorsements.

In the traditional art market, there is a strong symbiotic interplay between artists and esteemed institutions—while celebrated artists seek these platforms, institutions favor established names. Artists initiating their careers in prestigious venues tend to have steadier trajectories. There is a dominance of key European and North American institutions, suggesting a concentrated flow of art commerce in these regions. Through a network-centric ranking methodology, a correlation can be observed between institutional prestige and the economic worth of artworks they house [32].

Conclusively, the art domain emerges as a labyrinth of relationships and hierarchies, where an artist's reputation and affiliations, along with the collector's connections and network, are critical for success [32]. It also highlights a high level of dependency for artists and gated access for collectors. This underlines the imperative for emerging artists to meticulously cultivate and expand their professional networks for market success, and for collectors to navigate the nuanced web of intermediaries and gatekeepers to acquire desired artworks.

3.2 Crypto Crash ≠ Crypto Art Crash

While the cryptocurrency market is known for its volatility, a downturn in this sector does not necessarily signal a decline in Crypto Art, as noted in the Survey of Global Collecting in 2022 by Art Basel and UBS [33] and the Art + Tech Report 2023 [16]. The relationship between Crypto Art and traditional cryptocurrencies like Bitcoin and Ether is a distinct one. As highlighted in the paper by K. Vasan et al., the NFT market operates with its own set of dynamics, independent of cryptocurrency market fluctuations [34]. This is primarily because collectors of Digital Art often state that the volatility of cryptocurrencies does not impact the perceived or personal value of their art collections. Their motivation for purchasing is predominantly driven by the artistic value and emotional resonance of the artwork (Fig. 3).

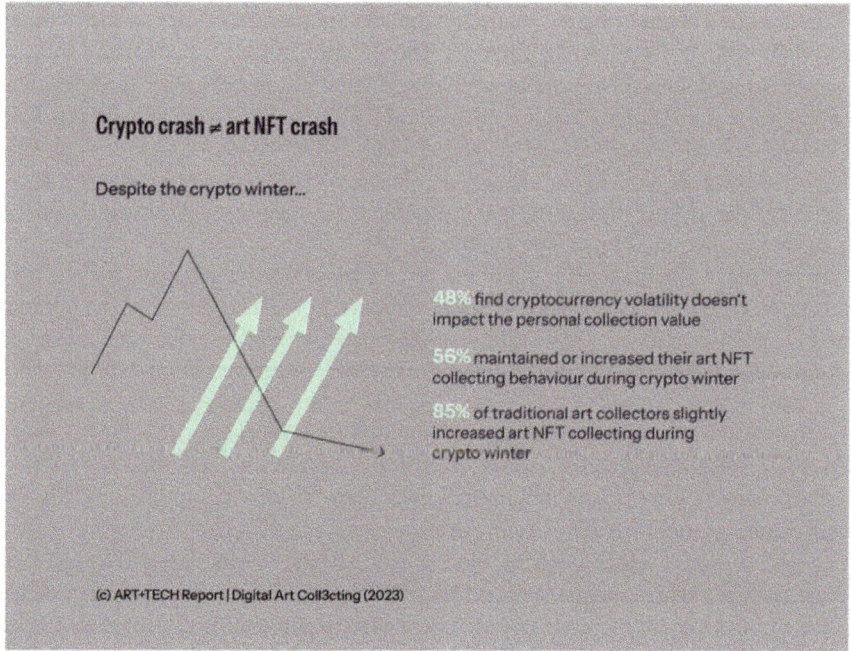

Fig. 3 Crypto crash ≠ crypto art crash [16]

3.3 How NFTs Are Challenging Existing Art Market Paradigms

The advent of Web3 technologies, including NFTs, has enabled a democratization of the art market across multiple dimensions. Firstly, there is an increased accessibility as artists can offer their works directly to a worldwide audience, circumventing physical barriers and costly intermediaries. Secondly, the inherent transparency of blockchain provides an indisputable public record, enhancing traceability of provenance and authenticity. Lastly, the paradigm shift facilitates artists' direct engagement with collectors, potentially ensuring more equitable financial reciprocation. As NFTs reshape the Digital Art landscape, the repercussions vary for its primary stakeholders: artists gain augmented autonomy, securing royalties from secondary sales [9], and expanding their reach unfettered by physical constraints; collectors enjoy the security and clarity afforded by the blockchain while accessing a more diverse artwork repertoire and engaging in direct artist-collector-relationships; intermediaries, on the other hand, face challenges as conventional roles wane but simultaneously, novel roles burgeon, particularly in the digital domain. The forthcoming chapter will delve deeper into these nuanced changes for each stakeholder—artists, collectors and intermediaries—setting the stage for a comprehensive exploration in subsequent discussions.

4 What Does Crypto Art Mean for Artists? The Renaissance of Digital Art in the Age of Blockchain Technology

The advent of Non-Fungible Tokens (NFTs) has catalyzed a shift in the valuation and appreciation of Digital Art. Long-standing Digital Artists, such as Kenny Schachter (Fig. 4), Tyler Hobbs [9], Mike Winkelmann [35], Harm van den Dorpel, Olive Allen [36], Gretchen Andrew, and the artist duo Looping Lovers, comprising Philipp Ries and Thomas Mayer, find themselves at the forefront of this paradigm shift, witnessing a renewed esteem for their craft. This has enabled NFT artists to succeed outside the traditional gallery structure [27]. Take Beeple, for instance, whose digital collage, 'Everydays: The First 5000 Days' (2021), became famous when it was sold as an NFT at Christie's for $69.3 million in March of 2021 [37]. This transformative moment is not just about valuation but is emblematic of a broader evolution in this space. Blockchain and NFTs have not only emerged as innovative financial tools but also as novel creative mediums, ushering in unprecedented art forms and augmenting traditional artistic endeavors with unique utilities. Moreover, this era signals a redistribution of power, granting artists enhanced self-determination and autonomy [38, 39]. From ensuring immutable attribution and ownership of data to fostering a newfound individual agency and independence that bypasses the conventional gallery system, artists are now in an empowered position. Furthermore, the direct artist-collector relationship is a significant and unique development for artists. They can now maintain constant, direct contact with collectors, nurturing future collecting interest with repeated purchases and increasing the level of participation in the artistic process. Notably, they have the theoretical ability to engage directly in primary sales and gain from secondary market royalties, though the long-term viability of this model continues to be a subject of discussion. Concurrently, the fusion of blockchain technology, artificial intelligence (AI), and the Web3 concept is drawing an influx of emerging artists, championing a more democratized access to the art world [40]. However, these advancements also have their drawbacks. Artists are having to take on various roles, including project management, community coordination, and administrative tasks, echoing the trends seen among Instagram artists. This multitasking can negatively impact the creative process. Additionally, the ease with which anyone can now create and directly sell art online has led to an excessive proliferation of artworks, making it challenging to discern high-quality pieces. Consequently, it is not surprising that we are beginning to see the emergence of curatorial initiatives on platforms aimed at offering guidance and curation [16].

Fig. 4 Kenny Schachter "Scam Likely" (2019), digital print

5 What Does Crypto Art Mean for Collectors? An Era of Artistic Diversity and Digital Ownership

The rise of Crypto Art and blockchain technology has significantly disrupted the conventional art world. This shift towards a Web3 context is not only attracting new collectors but also enhancing the popularity of on-chain art in particular, creating a novel ecosystem that facilitates easier participation. This marks the onset of a new era characterized by artistic diversity and digital ownership. In recent years, we have noticed signs of democratization in the art market. Notably, the introduction of novel technologies and platforms has lowered barriers to entry, both for artists and collectors. There has also been a notable increase in both the influence and decision-making power of collectors, as well as the emergence of an entirely new category of collectors. Crypto Art has allowed for new buyer groups to enter the space who would otherwise not have gained access [16, 37, 41]. This shift calls into question the necessity of traditional gallery access for the collector side and has ushered in an era marked by more accessible price points and heightened transparency. This evolution is drawing a fresh cohort of collectors, characterized by their youth, gender diversity, and innovative perspectives, making the art collecting landscape more inclusive and varied.

Additionally, Crypto Art distinguishes itself through its direct artist-collector interface, a paradigm shift from traditional art dealings. The absence of intermediaries like galleries has rendered the buying experience more intimate and genuine. This evolution of the artist-collector relationship, previously influenced by platforms like Instagram where collectors could send a direct message to artists, has further deepened in the Web3 era. Platforms like Discord and X, formerly Twitter, offer a more expanded and interactive space for engagement, unlike the more unidirectional interaction on Instagram [42, 43]. The shift to active, real-time interaction on these new platforms significantly enhances the engagement and connection between artists and collectors.

Another marked difference in this era is the added value accompanying an artwork—often coined as "Utility." Acquiring a piece of art is no longer just about the piece itself but might include added benefits or experiences. Moreover, collectors are now not just investing in an art piece but are becoming part of a like-minded community. The sense of belonging, shared experiences, and the entire ecosystem around the art piece intensifies the collector's association to the piece and the community as a whole.

Furthermore, the advent of blockchain technology is transforming notions of ownership. The emergence of digital ownership paves the way for innovative forms of collecting and possessing art, like collective ownership and fractionalization. However, as with any revolution, there are inherent challenges. A pertinent question arising is the extent of influence collectors should wield over art production.

Lastly, while NFTs suggest the possibility of gamification like introducing virtual galleries, art games, and experiences, it is vital to note that the significance of these elements is not as pronounced in the art domain as it is in the NFT and collectibles arena [16] (Fig. 5).

6 What Does Crypto Art Mean for Intermediaries? Navigating the Challenging Landscape of Curation and Distribution

The art world is undergoing a transformative shift with the emergence of NFT, compelling traditional galleries to reevaluate their roles. The rise of Crypto Art and the democratized means of purchasing and collecting it pose significant challenges for galleries, especially in attracting new, diverse collector segments including the younger generation. Traditional galleries, which have long been pivotal in the art market, are now grappling with their relevance in a market with new, diverse consumers entering the field.

Despite these challenges, some specialized NFT galleries like Superchief Gallery are successfully carving out a niche, highlighting that the evolution of galleries is not about their demise but about adaptation. This new era does not mark the end of curation or contextualization [44]. On the contrary, there is a growing demand for

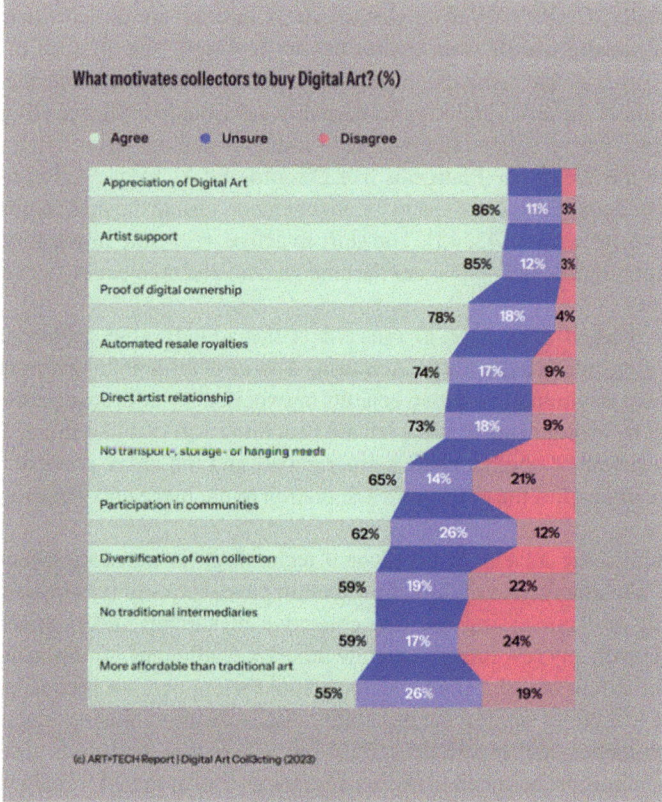

Fig. 5 What motivates collectors to buy digital art? (%) [16]

these services despite Web 3's core concept of decentralization. Crypto Art platforms such as Rarible, Nifty Gateway, Foundation, and Objkt are addressing this need by offering curated experiences, signifying the enduring necessity of galleries in providing guidance and expertise in the art selection process.

Furthermore, the rise of online communities has introduced a new dynamic in the art world. This chapter aims to draw attention to a question: are communities emerging as the new intermediaries? The answer is emphatically affirmative [45]. As elucidated in the Art + Tech Report 2023, communities have indeed become essential players in the Digital Art collecting world. They wield substantial influence over various facets of the art market, including the trajectories of artist careers, the formation of opinions, collecting decisions, and even shaping the art market as influential curatorial and sales-determining entities. They are becoming a new kind of intermediary in the art world. This shift towards community-driven curation and influence represents a significant departure from traditional art market mechanisms, heralding a more inclusive and democratized art landscape (Figs. 6 and 7).

Crypto Art and Intellectual Property 539

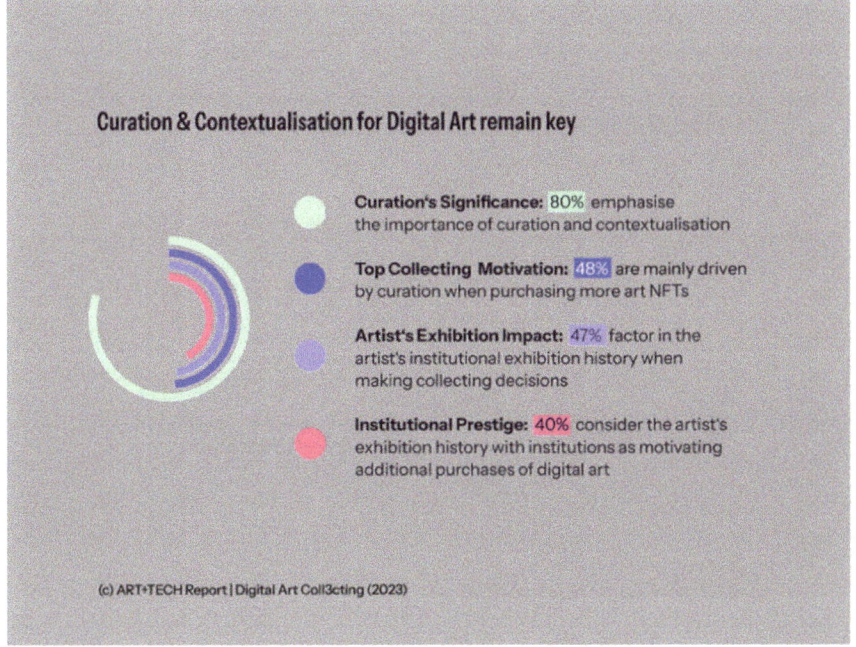

Fig. 6 Curation and contextualization for digital art remain key [16]

7 Intellectual Property and Legal Aspects

The integration of blockchain technology in the art market has brought transformative changes, particularly in the realm of intellectual property (IP) and usage rights with Non-Fungible Tokens (NFTs). A crucial point of contention and confusion arises from the fact that owning an NFT does not necessarily grant the owner the IP rights of the associated artwork. This distinction is often misunderstood, leading to uncertainties among art NFT owners about their legal rights, especially concerning the commercial use of the artwork.

Digital certificates linked to NFTs serve to authenticate artworks and identify their creators, thereby acknowledging Digital Artists as the original authors. However, the purchase of an NFT usually conveys ownership of the digital token itself, rather than the underlying IP of the artwork it represents. Consequently, while collectors own the digital asset, they may not hold the rights to reproduce, distribute, or commercially exploit the artwork. This situation emphasizes the importance of differentiating between owning the NFT and holding the copyright of the artwork it represents. This distinction mirrors the circumstances faced by collectors of traditional, physical art.

In the realm of NFT transactions, it is imperative to establish clear and concise terms and conditions. This clarity not only facilitates smoother transactions, but also serves as a cornerstone for protecting intellectual property rights associated with NFT sales. While smart contracts offer a robust mechanism for managing rights and

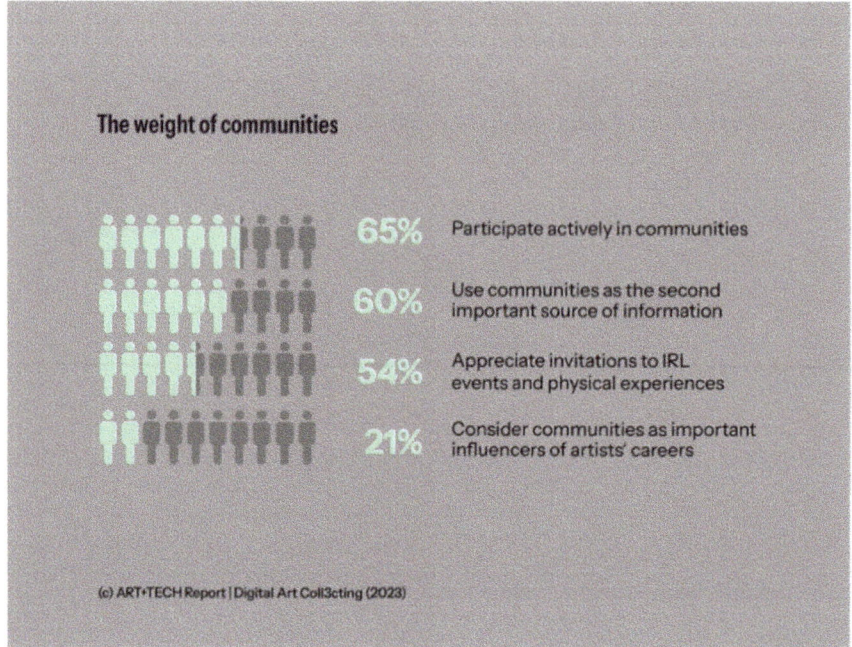

Fig. 7 The weight of communities [16]

automating processes, especially in secondary sales, it is crucial to recognize that they do not inherently possess legal binding power. Therefore, understanding and delineating the legal boundaries and enforceability of smart contracts in the context of NFT transactions becomes a vital aspect of safeguarding intellectual property rights in the Crypto Art landscape.

Fractional Ownership, another complexity introduced by blockchain, democratizes art ownership but also complicates the division of usage rights among multiple owners of fractional shares of an NFT. This situation requires a nuanced understanding of usage rights and copyright in the context of blockchain and Digital Art.

The rise of Artificial Intelligence (AI) in art creation further challenges traditional concepts of authorship, necessitating adaptations in legal frameworks to address the evolving landscape of Digital Art ownership and IP rights. These developments highlight the critical need for harmonization of blockchain technology with existing copyright laws, ensuring clarity and protection for all parties involved in the Digital Art ecosystem. The implementation of effective regulation is now imperative to address these emerging challenges, ensuring that IP rights evolve in tandem with the revolutionary capabilities of AI, while fostering innovation and protecting the rights of creators and inventors in this dynamic digital era.

In conclusion, while blockchain and AI have revolutionized the art world, they also raise pivotal questions about the nature of ownership, usage rights, and copyright in the digital era. Artists, collectors, and legal professionals must navigate these complexities to fully embrace the potential of Digital Art and NFTs.

8 Summary and Outlook

8.1 Summary

In conclusion, this chapter provides a comprehensive exploration of the evolution of Crypto Art and its consequential impact on the traditional art market. Delving into its historical roots, we have scrutinized its transformative effects across various dimensions—artists, collectors, intermediaries, and legal frameworks. The origins, classifications, various blockchain protocols, and the challenges posed to traditional paradigms by Crypto Art have been thoroughly examined. Notably, Non-Fungible Tokens (NFTs) have emerged as a pivotal force, imparting tangible value to Digital Art and bridging the divide between physical and Digital Art domains. The discussion delineates the distinctions between 'Digital Fine Art' and formats like 'NFT-Collectibles', shedding light on their unique attributes and roles within the expansive Crypto Art spectrum.

Further, we have explored how NFTs can challenge the traditional art market by contrasting its mechanisms with the unique dynamics of the NFT market. A significant emphasis has been placed on the transformative impacts of NFTs for artists, collectors, and intermediaries. For artists, NFTs have heralded a renaissance in Digital Art as an art form, enabling greater autonomy and direct engagement with collectors. Collectors benefit from increased accessibility, transparency, and enhanced value in their acquisitions. Intermediaries, particularly traditional galleries, encounter both challenges and opportunities for adaptation in this evolving market.

The chapter also addresses initial legal aspects and intellectual property implications in Crypto Art. This includes discussions on the nuances of owning an NFT versus the intellectual property rights of the underlying artwork and the role of smart contracts in defining usage rights.

This extensive exploration highlights the need for continual adaptation and collaboration to fully leverage the potential of Crypto Art in a rapidly evolving landscape. Looking forward, its journey is just unfolding, heralding a future rich in innovation, transformation, and creative exploration. With technologies like AI continually advancing, the prospect for Crypto Art is not only promising but also thrilling. The integration of these advanced technologies is set to further revolutionize the art world, opening up new possibilities for artistic expression and interaction.

8.2 Outlook

Moving forward, the realm of Crypto Art is evolving into a dynamic and multifaceted space that represents the convergence of artistic expression and cutting-edge technology. This progression highlights the ongoing interaction and fusion of these two distinct yet interrelated spheres.

While the initial hype surrounding Crypto Art may have subsided, the underlying technology of blockchain remains robust and is here to stay. Importantly, we are still in the early stages of discovering its full potential, with meaningful applications in the art world continually emerging. Above all, the future will witness novel approaches in how art, be it physical or digital, is created, bought, sold, distributed, and exhibited. This evolution will reshape the art market's very fabric, introducing new paradigms in art interaction.

Significant changes are on the horizon for the art market in the coming years. As Web3 technologies such as blockchain, cryptocurrency, and NFTs become more deeply integrated and widely adopted, they are paving the way for a new, more inviting ecosystem. Consequently, this environment will be characterized by greater accessibility, enhanced transparency, and a trustless nature, leading to a democratization of art creation and collecting [27].

The rise of new forms of art is imminent, marking an evolution of contemporary art with blockchain emerging as a creative medium of our times. Concurrently, Digital Art will continue to ascend, with innovative forms like Generative Art, Dynamic NFTs, and AI-driven art, promising an exciting future [9]. In terms of art collecting and ownership, we are moving towards more transparent, immutable authenticity, and ownership records. This transition signifies a safer, more secure, direct, and autonomous approach to art collecting, free from traditional gatekeeping. Direct relationships between collectors and artists will flourish, alongside new forms of collectors' participation and co-creation.

Moreover, the rising power of community and collective experiences with like-minded individuals can foster collective ownership through tokenization, introducing novel ways of owning and collecting art. The field of art collecting will see the entry of new collector segments. Younger, digitally native, and crypto-native individuals are diversifying and segmenting the collecting space [6]. As our world becomes increasingly digital and virtual, the demand for digital and virtual objects like art NFTs is set to rise.

In this evolving landscape, it is important to consider some critical perspectives while acknowledging the potential and positive changes Crypto Art brings. The increasing involvement of new individuals in the art space, both as creators and collectors, is noteworthy. However, as we contemplate the democratizing influence of the technology behind Crypto Art, questions arise about its long-term impact. Will it genuinely foster inclusivity and equal opportunity, or might it inadvertently replicate established structures? Networks, thus ultimately social capital, continue to play crucial roles in success within the Crypto Art world. Furthermore, the question of whether the ultimate success in Crypto Art will be predominantly influenced by financial means is also a point of consideration. This observation prompts reflection on whether the new art ecosystem might still mirror traditional power dynamics and influence structures. These considerations are vital as they encourage a balanced understanding of Crypto Art's trajectory and its broader implications within the art world and society. As we finally look forward to the future, the journey of Crypto Art continues to be an exciting and dynamic field to discover, especially with contemporary art constantly evolving by reflecting on our world and times, and new emerging technologies undoubtedly shaping its course.

Acknowledgements We extend our deepest gratitude to Dr. Annette Doms for her insightful reflections and impulses. Her contributions were instrumental in elucidating key aspects at crucial junctures of this chapter, greatly enhancing the clarity and depth of the subject matter discussed.

Our sincere thanks go to Dr. Nike Schmidt for her meticulous revisions and valuable legal perspectives in the subsection 'Intellectual Property and Legal Aspects.' Her expertise has been vital in ensuring the accuracy and comprehensiveness of the legal analysis presented.

We are also grateful to Dr. Wiebke Fröhlich for providing an initial legal overview and clarification of terms, laying a solid foundation for the chapter's exploration of complex legal themes.

Furthermore, we would like to extend our heartfelt thanks to Wolf Lieser, Director and Founder of DAM Projects GmbH and Paul Victor Schmidt, COO of fx(hash), as well as Kenny Schachter for their generous provision of artworks that significantly enriched this chapter. Their willingness to share these pieces has allowed for a more vivid and tangible exploration of the themes discussed

Our appreciation goes to the team behind the Art + Tech Report for allowing the use of their images. These visual representations have been crucial in illustrating and enhancing the narrative and arguments presented in our work.

Additionally, we would like to express our appreciation to Katy Schaper for the linguistic adjustments of this chapter

The collaboration and support of each of these individuals and organizations have been invaluable in the completion of this chapter, and we are profoundly grateful for their contributions.

References

1. C. Paul, Digital art, in *World of art* (Thames & Hudson, New York, 2003) Zugegriffen: 17 Nov 2023 [Online]. Verfügbar unter: http://bvbr.bib-bvb.de:8991/F?func=service&doc_library=BVB01&doc_number=010505560&line_number=0001&func_code=DB_RECORDS&service_type=MEDIA
2. L. Manovich, Data science and digital art history. Int. J. Digit. Art Hist., Bd. **1** (2015). https://doi.org/10.11588/DAH.2015.1.21631
3. L. Manovich, Ten key texts on digital art: 1970–2000. Leonardo **35**(5), 567–575 (2002)
4. The Rise of Digital Art, ArtDiction. Zugegriffen: 17 Nov 2023 [Online]. Verfügbar unter: https://www.artdictionmagazine.com/the-rise-of-digital-art/
5. D. Challis, Shaping the future market for digital art. Pursuit. Zugegriffen: 17 Nov 2023 [Online]. Verfügbar unter: https://pursuit.unimelb.edu.au/articles/shaping-the-future-market-for-digital-art
6. D. Duray, F. Abdessamad, the Editors, The digital innovators reshaping the art world. Observer. Zugegriffen: 12 Dec 2023 [Online]. Verfügbar unter: https://observer.com/list/the-digital-innovators-reshaping-the-art-world/
7. Centre Pompidou, L'actualité des expositions. Centre Pompidou. Zugegriffen: 12 Dec 2023 [Online]. Verfügbar unter: https://www.centrepompidou.fr/fr/programme/expositions

8. C. McAndrew, *The Art Market 2023. A Report by Art Basel & UBS*, Art Basel and UBS (2023) [Online]. Verfügbar unter: https://www.ubs.com/global/en/our-firm/art/collecting/art-market-survey/download-survey-report-2023.html
9. A. Tremayne-Pengelly, How NFT artist Tyler Hobbs survived the crypto crash. Observer. Zugegriffen: 12 Dez 2023. [Online]. Verfügbar unter: https://observer.com/2023/06/nft-tyler-hobbs-auction-crypto-crash/
10. C. McAndrew, *The Art Market 2022. A Report by Art Basel & UBS*, Art Basel and UBS (2022) [Online]. Verfügbar unter: https://d2u3kfwd92fzu7.cloudfront.net/Art%20Market%202022.pdf
11. B. Boucher, *Collectors Are Spending More, Not Less, On NFT Art, According to the Art Market Report, Art Basel*. Zugegriffen: 21 Nov 2023 [Online]. Verfügbar unter: https://www.artbasel.com/stories/art-market-report-2022-collectors-cryptocurrency-nfts
12. W. Chen, Z. Xu, S. Shi, Y. Zhao, J. Zhao, A survey of blockchain applications in different domains. Dez. 17–21 (2018). https://doi.org/10.1145/3301403.3301407
13. F.-Y. Wang, R. Qin, Y. Yuan, B. Hu, Nonfungible tokens: constructing value systems in parallel societies. IEEE Trans. Comput. Soc. Syst. Bd. **8**(5), 1062–1067 (2021). https://doi.org/10.1109/TCSS.2021.3109359
14. M. Connor, *Collecting Contemporary Art Means Collecting Digital Art, Rhizome*. Zugegriffen: 17 Nov 2023 [Online]. Verfügbar unter: https://rhizome.org/editorial/2013/oct/11/collecting-contemporary-art-means-collecting-digit/
15. S. Shilina, *Blockchain, Creativity and Arts Intertwine, Paradigm*. Zugegriffen: 20 Nov 2023 [Online]. Verfügbar unter: https://medium.com/paradigm-research/blockchain-creativity-and-arts-intertwine-d3c42739312f
16. K. Gold, K. Leipold, J. Neuschäffer, A. Schwanz, *ART+TECH Report Digital Art Coll3cting Report 2023*, ART+TECH Report, Berlin, 3 Oct 2023
17. C. McAndrew, *A Survey of Global Collecting 2023—A Report Published by Art Basel & UBS Prepared by Dr. Clare McAndrew, Arts Economics*, Art Basel and UBS (2023). Zugegriffen: 17 Nov 2023 [Online]. Verfügbar unter: https://theartmarket.artbasel.com/
18. K. Reichert, *NFT-Art in der Krise: Wenn Sie sich für Kunst interessieren, müssen Sie sich mit NFTs nicht beschäftigen*, 16. Februar 2023. Zugegriffen: 17 Nov 2023 [Audio]. Verfügbar unter: https://www.swr.de/swr2/kunst-und-ausstellung/nft-art-in-der-krise-wenn-sie-sich-fuer-kunst-interessieren-muessen-sie-sich-mit-nfts-nicht-beschaeftigen-100.html
19. SuperRare | CryptoArt | NFT Art Marketplace | Digital Art, SuperRare | CryptoArt | NFT Art Marketplace | Digital Art. Zugegriffen: 17 Nov 2023 [Online]. Verfügbar unter: https://superrare.com
20. J. Exmundo, *Quantum: The Story Behind the World's First NFT, nft now*. Zugegriffen: 20 Nov 2023 [Online]. Verfügbar unter: https://nftnow.com/art/quantum-the-first-piece-of-nft-art-ever-created/
21. Definition of TECHIE. Zugegriffen: 17 Nov 2023 [Online]. Verfügbar unter: https://www.merriam-webster.com/dictionary/techie
22. Vineet, *From Comic, Alt-Right Symbol, to Meme NFT, to Viral Coin—The Pepe Journey*, NFT Evening. Zugegriffen: 20 Nov 2023 [Online]. Verfügbar unter: https://nftevening.com/from-comic-alt-right-symbol-to-meme-nft-to-viral-coin-the-pepe-journey/
23. Beeple (1981), *Everydays: The First 5000 Days | Christie's*. Zugegriffen: 20 Nov 2023 [Online]. Verfügbar unter: https://onlineonly.christies.com/s/beeple-first-5000-days/beeple-b-1981-1/112924
24. Right Click Save Magazine, Zugegriffen: 8 Dec 2023 [Online]. Verfügbar unter: https://www.rightclicksave.com/
25. Chainleft, *What Does "On-Chain" Really Mean?* 23 June 2023. Zugegriffen: 20 Nov 2023 [Text]. Verfügbar unter: https://www.rightclicksave.com/article/what-does-on-chain-really-mean
26. S. Guskin, *Beeple's 'HUMAN ONE' NFT Sells for $29 Million at Christie's*. Zugegriffen: 20 Nov 2023 [Online]. Verfügbar unter: https://ocula.com/magazine/art-news/beeples-human-one-nft-sells-for-29-million/

27. H. Hallak, *The NFT Renaissance Beyond Digital Collectibles*, Observer. Zugegriffen: 13 Dec 2023 [Online]. Verfügbar unter: https://observer.com/2023/10/nfts-digital-art-renaissance/
28. Particle Collection, *Mint Your Particle Now for Love is in the Air by Banksy*. Zugegriffen: 20 Nov 2023 [Online]. Verfügbar unter: https://www.particlecollection.com/gallery/love-is-in-the-air
29. Crypto.com, *Bitcoin NFTs—How the Ordinals Protocol Works*. Zugegriffen: 20 Nov 2023 [Online]. Verfügbar unter: https://crypto.com/university/bitcoin-nfts-ordinals-protocol
30. D. Jones, *Bitcoin Digital Culture Expands With Ordinals*. Zugegriffen: 20 Nov 2023 [Online]. Verfügbar unter: https://medium.com/@itsdono/bitcoin-digital-culture-expands-with-ordinals-dc5582fb636b
31. R. Polleit Riechert, *Die schöne Rendite an der Wand bleibt*. Februar 2022
32. S.P. Fraiberger, R. Sinatra, M. Resch, C. Riedl, A.-L. Barabási, Quantifying reputation and success in art. Science **362**(6416), 825–829 (2018). https://doi.org/10.1126/science.aau7224
33. C. McAndrew, *A Survey of Global Collecting 2022—A Report Published by Art Basel & UBS Prepared by Dr. Clare McAndrew, Arts Economics, Art Basel and UBS* (2022) [Online]. Verfügbar unter: https://d2u3kfwd92fzu7.cloudfront.net/A_Survey_of_Global_Collecting_in_2022.pdf
34. K. Vasan, M. Janosov, A.-L. Barabási, Quantifying NFT-driven networks in crypto art. Sci. Rep., Bd. **12**(1), 2769 (2022). https://doi.org/10.1038/s41598-022-05146-6
35. Derek, *Storing Value in Digital Objects*. Collab+Currency. Zugegriffen: 8 Dec 2023 [Online]. Verfügbar unter: https://medium.com/collab-currency/storing-value-in-digital-objects-a92f54fa98cc
36. C. Thompson, *The Untold Story of the NFT Boom*. The New York Times, 12 May 2021. Zugegriffen: 21 Dec 2023 [Online]. Verfügbar unter: https://www.nytimes.com/2021/05/12/magazine/nft-art-crypto.html
37. J. Thaddeus-Johns, NFTs Promised to Revolutionize the Art World—But Are Galleries on Board? Artsy. Zugegriffen: 22 Nov 2023 [Online]. Verfügbar unter: https://www.artsy.net/article/artsy-editorial-nfts-promised-revolutionize-art-galleries-board
38. M. Winkelmann, *Beeple: An Interview With the Top Crypto Artist in the World*, 15 May 2022. Zugegriffen: 22 Nov 2023 [Text]. Verfügbar unter: https://nftevening.com/nft-artist-beeple-interview-with-nftevening-at-nft-in-america/
39. G. Bak, Dimitria, H. Barrows, *The Digital Art Collector #2, Georg's Substack*. Zugegriffen: 22 Nov 2023 [Online]. Verfügbar unter: https://thedigitalartcollector.substack.com/p/the-digital-art-collector-2
40. T. Locke, *These millennial creators are making 6 figures selling NFTs: 'It changed the trajectory of my career and my life'*. CNBC. Zugegriffen: 22 Nov 2023 [Online]. Verfügbar unter: https://www.cnbc.com/2021/05/12/meet-the-millennial-creators-making-six-figures-selling-nfts.html
41. A. Langer, *Die Kunstklärerin der Gen Z. Bus. Punk* Nr. 02/2022, S. 27 (2022)
42. C. Nast, *Why the World's Most Talked-About New Art Dealer Is Instagram*. Vogue. Zugegriffen: 17 Nov 2023 [Online]. Verfügbar unter: https://www.vogue.com/article/buying-and-selling-art-on-instagram
43. A. Kakar, *Art Collector Insights 2023* (ARTSY, Report, 2023)
44. J.W. Lee, S.H. Lee, The legitimation of young and emerging artists in digital platforms: the case of Saatchi art. J. Arts Manag. Law Soc. **53**(1), 19–41 (2023). https://doi.org/10.1080/10632921.2022.2080136
45. We Are Museums, *Online Community*. Zugegriffen: 8 Dec 2023 [Online]. Verfügbar unter: https://wearemuseums.com/community

Su-Zeong Fröhlich combined her passions for art, science, and social impact in Web3 after years of working as an internal medicine physician in Berlin. Since entering the crypto space in 2017, she launched VISIBLR, a Web3 art project for BIPoC, which received a grant from the Tezos Foundation. She has showcased her art in various galleries, including Bright Moments Gallery. In addition, she has hosted blockchain workshops for children in Berlin and Los Angeles during Outer Edge LA. She was part of the DLT Talents scholarship program at the Blockchain Center of the Frankfurt School of Finance, won the WomanHack Hackathon 2022, and was a semifinalist at the Outer Edge Los Angeles Hackathon 2023. She is also the co-founder of the Web3 consulting firm Fox and Happy Blocks and has spoken at prominent events such as the Internationale Funkausstellung Berlin (IFA), DMEXCO, and NFT.NYC.

Kerstin Gold is a Berlin-based Strategy Advisor who works at the intersection of art and technology. With an indepth understanding of the art industry, a keen interest in emerging technologies, and a strong passion for (digital) art, she provides strategic guidance to the art ecosystem. She advises artmarket players such as galleries, cultural institutions, and foundations on digital transformation and business model innovation, and she serves as a start-up advisor to culturetech- and arttech ventures and as approved mentor at VC accelerator programmes.

Kerstin is co-founder and author of the ART + TECH Report, which has been analyzing the impact of technology on the art industry since 2020. The independent research initiative explored the dynamics of web3 and collecting digital art in 2023, examined the buying patterns and collecting motivations of art NFT collectors in 2022 and surveyed collectors online art buying behaviour in 2021.

Prior to her career as an independent strategy advisor, Kerstin gained +12 years of strategy and consultancy expertise in the international creative industry in London and Berlin, advising global creative brands such as adidas, Absolut Vodka and PlayStation, leading global interdisciplinary teams and managing an agency's branch office with +80 employees. She holds an MA in business administration, trained as a systemic executive business coach and was accepted into two full-scholarship programs ("DLT Talents"; "NFT Talents") at Blockchain Center-Frankfurt School of Finance.

As a strong advocate for greater diversity in the tech space, Kerstin's expertise extends to speaking engagements bridging the art, business, and technology realms. She lives in Berlin with her family.

Introduction to Decentralization and Ownership in Web3

Gustav Hemmelmayr

1 Introduction

At present (in 2023/2024, when this article is written), large international companies own and operate Web 2.0 platforms that dominate our online experience. These platforms have become powerful entities due to platform effects, which draw an increasing number of people, and due to these platforms' ability to monetize their—our—data. Web3 has emerged with the aim of disrupting this dominance and providing alternative solutions that are similarly convenient and user-friendly, while shifting power into the hands of users who now have the autonomy to shape their own digital journey.

To fully understand the feasibility of this significant shift, it is essential to explore the technological foundations of Web3 and the series of innovations that paved its way.

Web3 is defined as a decentralized online ecosystem based on blockchain [1].

The blockchain technology was created by an anonymous individual or group by the pseudonym Satoshi Nakamoto with their Bitcoin paper [2]. Bitcoin was invented as a permissionless, tamperproof, peer-to-peer electronic cash system that uses a proof-of-work mechanism to operate a distributed ledger network. The objective behind Bitcoin was to establish a payment mechanism that challenges traditional banking systems by substituting the trust typically placed in these institutions with a verifiable system of integrity. This is achieved through a decentralized network of ledgers that cannot be manipulated by any central authority. This setup ensures an inherently trustworthy infrastructure for payments.

This technology lays the groundwork for numerous blockchains and other forms of distributed networks of ledgers—together called distributed ledger technologies or DLT—and is fundamental to the development of the Web3.

G. Hemmelmayr (✉)
BOTLabs GmbH (Schreibt Ausdrücklich Unter Nennung Seines Namens), Berlin, Germany
e-mail: gustav_hemmelmayr@yahoo.com

While the Bitcoin blockchain's first and essential feature was the ability to create and transfer digital values, the advent of the so-called smart contract in the Ethereum blockchain opened blockchain technology up to a higher variation of use cases and business models such as standardized and easily created new tokens.

Subsequent initiatives, such as Polkadot, have evolved from this foundation, with the example of Polkadot forging a comprehensive ecosystem of diverse, interoperable blockchains that are scalable and underpinned by a complex governance system [3].

This constant and ongoing transition towards decentralization eliminates the need for middlemen and institutional gatekeepers in more and more use cases. Thus, it becomes a system that can be directly controlled by its users. It brings about a digital landscape where the power and control over technology, data, and digital assets are placed back into the hands of individual users. Through decentralization, Web3 is set to facilitate a new age of digital ownership, autonomy, and privacy.

2 Technical Roots for Decentralization

Interestingly, the underpinnings of decentralized systems, while fundamental to Web3, are grounded in longstanding concepts in computer science.

In regard to decentralization, Paul Baran introduced the concept of Distributed Communications in 1962, elucidating the distinctions among Centralized, Decentralized, and Distributed Networks [4]. In 1985, Fisher and others [5] demonstrated that achieving consensus in a decentralized system is not always reliable, or at least under the conditions which their research was based on.

Cryptography is one of the most critical elements for decentralized and public networks. There was a paper by Whitfield Dieffie and Martin Hellman as early as 1976, that introduced the concept of key cryptography [6], and in 1978 the RSA public key cryptosystem was invented by Ron Rivest, Adi Shamir and Leonard Adleman. In 1980 the concept of protocols for public key cryptosystems was introduced by Ralph Mercle.

In the following sections, we will delve into a few scientific concepts that laid the ground for blockchain technology and the challenges they presented for the development of the first blockchain.

2.1 Centralized, Decentralized and Distributed Systems

Centralized, decentralized, and distributed networks represent different ways information is exchanged and managed within a system.

Paul Baran in his above-mentioned article "On Distributed Communication Networks" [4] from 1962 points out that although there are a large number of possible networks, they always have one of two components, namely the centralized ones,

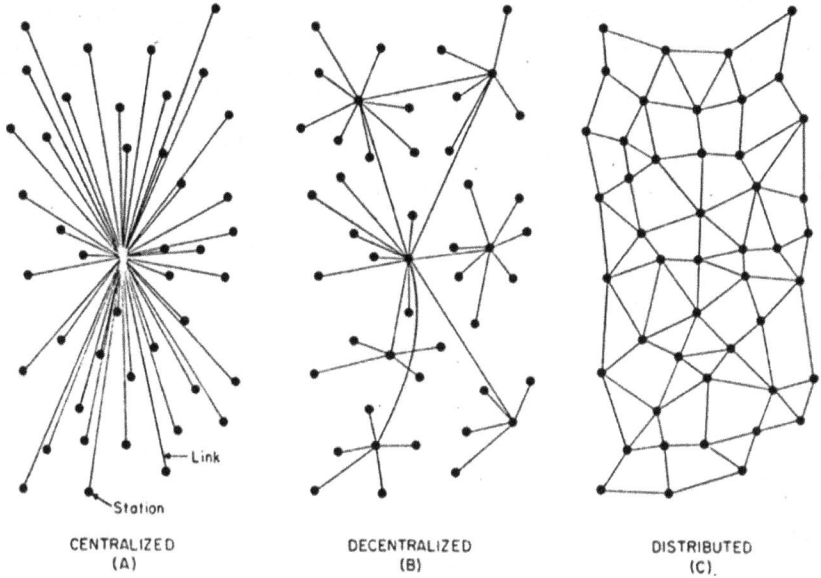

Fig. 1 Centralized, decentralized, distributed by Baran [4]

which have the shape of a star, and the distributed ones, which form a grid or mesh, as shown below in Fig. 1 as types (a) and (c).

In a centralized network, all communication passes through a single central node or server. This central node controls the flow of information and manages the network operations. While this setup facilitates control and coordination, it can be vulnerable. If the central node fails, the whole network can be affected.

Decentralized networks, on the other hand, incorporate multiple central nodes or servers. Information can flow through different routes, reducing the dependency on a single point. This structure enhances the network's resilience as the failure of one node does not halt the entire system. However, the network's efficiency can be affected due to the complexity of its structure.

Distributed networks take decentralization a step further. In this design, all nodes are interconnected, and there is no central authority. Information can take multiple paths to reach its destination, enhancing the network's robustness and survivability. In case of a node failure, the network can automatically reroute the data. This high level of redundancy and flexibility makes distributed networks ideal for situations requiring high survivability and adaptability.

Each network type has its strengths and weaknesses, and the choice depends on the specific use case and requirements.

2.2 Consensus in a Decentralized System (Byzantine Generals Problem)

The Byzantine Generals Problem [7] refers to the challenge in a distributed computer system where components may provide conflicting information due to malfunctions. This issue is abstractly illustrated through a scenario where a group of generals located at different places must agree on a common battle plan, but some may be traitors sending misleading messages. The problem is to develop an algorithm that ensures loyal generals reach a consensus, despite the existence of traitors. The problem can be solved only if more than two-thirds of the generals are loyal, or if unforgeable written messages are used, with practical applications in ensuring reliability in computer networks.

In the context of blockchain technology, the Byzantine Generals Problem refers to the situation where nodes in a distributed network must agree on a certain value (such as whether a transaction is valid), even though some nodes might be faulty or malicious.

Blockchain technology solves this problem with the help of consensus mechanisms like Proof of Work (PoW) in Bitcoin or Proof of Stake (PoS) in Ethereum 2.0. These mechanisms ensure that all honest nodes in the network reach agreement (consensus) about the current state of the distributed ledger, even if some nodes are trying to spread false information or are not working correctly.

The consensus mechanism and its Byzantine Fault Tolerance (BFT) are essential components of blockchain technology, effectively preventing double spending, where a user might attempt to spend the same digital currency twice. Resolving this double-spending issue is not only pivotal for safeguarding the security and integrity of the blockchain, but also serves as a key enabler in unlocking the potential for value creation within the blockchain technology landscape.

2.3 FLP Impossibility

The FLP Impossibility theorem [5], named after its authors Michael J. Fischer, Nancy A. Lynch, and Michael S. Paterson, is a fundamental theorem in distributed computing. It highlights a critical issue in the design of decentralized systems, particularly in achieving consensus in the presence of failures.

The theorem mathematically proofs that in an asynchronous distributed system (a system where there are no limits on how long a computation or message transfer can take), it is impossible to design a deterministic consensus protocol that can guarantee progress in every situation, even if only a single process may fail or crash. This implies that in a network of communicating nodes, if there's a chance that one node might fail, there is no absolute guarantee that all nodes will reach a common agreement.

Simplified, a system that operates with reliable messaging and functions asynchronously—meaning there are no parameters dictating when a computation must be completed, and there is no synchronized time—cannot distinguish between a crashed process and a delayed one. As a result, a crash can incapacitate the entire system because it continues to wait for the delayed process, even if it will never respond.

This finding has profound implications for the design of blockchain and other distributed ledger technologies [8]. Blockchain operates on the principle of consensus, where nodes must agree on the validity of transactions to prevent double-spending and maintain the integrity of the network. The FLP Impossibility theorem indicates that achieving this consensus in a completely fault-tolerant manner in asynchronous systems is impossible.

In response to this challenge, blockchain networks often implement a variety of strategies to reach consensus despite the possibility of node failures (Fig. 2). These include using probabilistic consensus mechanisms like Proof of Work (PoW) or Proof of Stake (PoS), or designing the system to be partially synchronous, where the network operates under the assumption that messages are typically delivered within some known time bound. The FLP Impossibility theorem serves as a reminder of the inherent complexities and challenges in designing decentralized systems. It underscores the need for careful design and robust fault-tolerant mechanisms to maintain the reliability, security, and integrity of blockchain networks in the face of potential failures.

3 Decentralization in Blockchain

Blockchain technology is subject to constant evolution, both in the development of the technology itself as well as in the applications surrounding this technology. This evolution extends to numerous other projects that have diverged from Bitcoin's original blueprint. These projects, each operating under their unique set of principles, aim to address and resolve various challenges inherent in the existing blockchain systems. Given the constant stream of innovations and improvements in this space, it remains a complex task to provide a succinct overview. A significant aspect of this complexity stems from the fact that the advancements in one project often inspire and influence the development trajectory of others.

Despite these complexities and interdependencies, one can still identify three distinct waves of innovation [9] within this evolving landscape. The first wave, represented by Bitcoin, introduced the concept of a decentralized monetary system and digital store of value, offering an alternative to traditional financial institutions. The second wave, led by Ethereum, expanded on Bitcoin's foundations by adding functionalities such as Smart Contracts and Decentralized Applications (DApps), thereby creating a platform for global development. The third wave, featuring platforms like Cardano, Polkadot, and Solana, aims to address previous shortcomings by focusing on issues like scalability, interoperability, and on-chain governance, with the ultimate goal of catering to billions of users and conducting millions of transactions per second.

3.1 Bitcoin as a Decentralized Network

Bitcoin's fundamental design is rooted in the principle of decentralization [2]. This decentralization is evident in multiple aspects of Bitcoin's technology, such as its Peer-to-Peer (P2P) network, the network of nodes, its open-source nature, and its consensus-driven changes, as well as its global usage.

At the heart of Bitcoin's technology is a decentralized network of miners. These miners play a crucial role in maintaining the security of the Bitcoin network. They validate and confirm transactions, ensuring that all transactions are legitimate and preventing fraudulent activities such as double spending. This system of decentralization creates a network where power and control are distributed among numerous participants, rather than centralized in a single authority.

Bitcoin's decentralized network of nodes also plays a vital role in maintaining the integrity of the system. Each node in the network contains a copy of the entire blockchain, which is the distributed ledger that records all Bitcoin transactions. This means that even if a single node fails, the network can continue to function normally, enhancing its robustness and reliability.

On the usage side, Bitcoin's decentralization allows it to be accessible to anyone with an internet connection, anywhere in the world. It facilitates direct transactions between users without the need for a central authority, enabling borderless transactions and a truly global financial system.

On the other hand, the Bitcoin funds are controlled in a decentralized manner. Even though all coins are found on the blockchain, there is no central control over these coins, but the control lies with each user. Acess to funds is safeguarded only through cryptography, with the address serving as the public key and the seed phrase functioning as the private key. Access is exclusively managed through that key pair mechanism, and there is no "password reset" functionality available—"*not your key, not your coins*", as it is commonly said.

On the license side, Bitcoin's code is open source, meaning anyone can review, audit, or contribute to it. Changes to the Bitcoin protocol require consensus among network participants, and decisions about these changes are made collectively by the community of miners. This decentralized governance model ensures that no single entity can unilaterally control or manipulate the system. If anyone is unhappy with the decisions of the miners, anyone can fork [10] their own bitcoin blockchain and continue this fork with all the other miners who agree with them. Even if anyone should want to use the Bitcoin code to spin off their own bitcoin blockchain, they are free to do that.[1]

One of the most significant implications of Bitcoin's decentralization is that it creates a system that is resistant to censorship or interference by governments or other central authorities. Because control is distributed among a multitude of miners and nodes around the world, no single entity can shut down the network. Even if a government were to attempt to ban or restrict Bitcoin within its jurisdiction, the decentralized nature of the network would allow it to continue operating in other parts of the world. This resilience to external interference enhances Bitcoin's value as a decentralized, global currency.

The decentralization of Bitcoin is further enhanced by layered solutions like the Lightning Network [11] and rollups [12], which improve transaction efficiency and reduce costs. The Lightning Network allows for off-chain transactions, bolstering Bitcoin's scalability, while Rollups consolidate multiple transactions into one, optimizing on-chain data storage. Additionally, Bitcoin wallets [13] simplify the user experience, thereby promoting wider adoption. These elements not only augment Bitcoin's usability but also reinforce decentralization by distributing network activity across diverse platforms and making them as easy and secure to access as possible.

In summary, decentralization is a key feature of the Bitcoin blockchain's design and the ecosystem around it, and it is this decentralization that has enabled Bitcoin to become the world's first truly global, censorship-resistant currency [14].

3.2 Smart Contracts and Decentralized Applications in Ethereum

The next landmark progression in the evolution of decentralization comes through Ethereum [15][2] with Smart Contracts and Decentralized Applications (DApps).

---

[1] Such usage of the Bitcoin technology would of course be seen as something different than the original code and it would need to onboard enough miners and certainly have different trading values for the coins created on it.

[2] For the purpose of showing the three different waves of blockchain development, Ethereum is here portrayed as the initial concept, without taking Ethereum 2.0 into consideration.

Ethereum 2.0 [16] emerged during the third wave of blockchain development and represents a significant change in the technology. The upgrade process for Ethereum 2.0 was divided into three phases. Phase 0, launched in December 2020, introduced the Beacon Chain, establishing the proof-of-stake (PoS) consensus mechanism, marking a significant shift from the previous proof-of-work (PoW) system. Phase 1, known as "The Merge," successfully occurred in September 2022. This phase transitioned Ethereum from PoW to PoS, converting Ethereum 1.0 into a shard—a fraction of the newly created larger overall system—within the Ethereum 2.0 ecosystem while preserving transaction history and smart contract functionality. The final stage, Phase 2, scheduled for 2023, involves the introduction of Shard Chains, which will serve as decentralized storage spaces for application data, significantly enhancing Ethereum's transaction handling and data storage capacity. Ethereum holders will experience an automatic transition to Ethereum 2.0, requiring no action on their part.

These two pivotal features substantially broadened the potential applications of this technology and extended the capacity of blockchain technology beyond its initial parameters.

Smart Contracts enable the execution of value transactions, but also incorporate the concept that transactions and counter-transactions can be automatically processed directly within the blockchain without the need for intermediaries. Functioning (in its most basic form) akin to a digital vending machine, a trustworthy Smart Contract executes a specific action only once the requested amount is paid or another set of preconditions fulfilled. The versatile nature of Smart Contracts means they can cater to a diverse range of applications.

One of the first notable applications of Smart Contracts was arguably the issuance of new coins, particularly in standardized forms [17] such as the ERC-20 token for fungible tokens [18] or ERC-721 for non-fungible tokens (NFTs) [19]. These innovations sparked a wave of Initial Coin Offerings (ICOs) in 2017 and catalyzed a substantial market for NFTs in 2021 [20].

A Decentralized Application (DApp) in Ethereum is a software application that operates on the Ethereum blockchain. It is open-source, autonomous, and operates with a consensus mechanism, rendering it independent of any single controlling entity. In comparison to traditional apps, they offer increased security, transparency, and decentralization, using the advantages of blockchain technology. This removes a single point of failure and allows a more democratic control for a broader range of use cases.

Smart contracts and decentralized applications represent transformative technologies with extensive real-world applications [21]. In finance, they're fundamental to Decentralized Finance (DeFi) mechanisms, facilitating trustless transactions, such as lending, borrowing, and trading. In the gaming industry, they are paired with Non-Fungible Tokens (NFTs) to enhance in-game asset ownership and transferability. In real estate, smart contract-enabled tokenization allows fractional ownership and efficient transaction processing. Moreover, they are used in Decentralized Autonomous Organizations (DAOs) to automate corporate governance, reducing administrative costs. In healthcare, they could enable secure, trustless, and transparent data sharing. They could also be used in the Internet of Things (IoT) field for enhanced security and transparency. These examples underscore the versatility and transformative potential of smart contracts and DApps.

In conclusion, Ethereum has marked a significant turning point in the landscape of decentralized technologies. It has expanded the initial blockchain infrastructure established by Bitcoin, and broadened the potential applications of blockchain technology, creating new opportunities for digital ownership, decentralized finance and decentralized autonomous organizations.

3.3 Decentralized, Autonomous and Interoperable with Polkadot

Polkadot, an open-source blockchain network based on the Polkadot Whitepaper from 2017 [3] and launched in 2020 [22], introduced several novel features not seen in previous blockchain networks. Polkadot's objective is to establish a genuine Web3 infrastructure.

Polkadot, as conceptualized in the Whitepaper [22], aims to redefine the architecture of blockchain systems by promoting decentralization, autonomy, and interoperability. The vision is to develop a heterogeneous multi-chain framework that addresses the challenges of scalability and extensibility, which is achieved by separating the consensus architecture into two distinct components: canonicality (agreement upon a single valid history) and validity (execution of transactions following shared rules).

Decentralization is facilitated via a divide-and-conquer approach. Polkadot compartmentalizes the roles of security and transport, creating a system that balances its core through incentivized participation of public nodes. This allows for the integration of multiple, diverse consensus systems, promoting decentralization and enhancing scalability.

Autonomy from the founders is prioritized through a governance structure built into Polkadot. The first envisioned structure with a Council and a Technical Committee, likely to resemble stable political systems with a bicameral aspect, placed the ultimate authority with the holders of the stakeable token,[3] who have "referendum" control. This structure was designed to be robust, flexible, and capable of evolution and adaptation, and even includes a Treasury to fund projects in the Polkadot ecosystem. This governance has been updated in 2023, when OpenGov [23] abolished the Council and Technical Committee and put all decisions of the network under direct control of the members of the PolkadotDAO. Each referendum is subject to different voting times and requirements based on its potential impact. To ensure security, proposal suggestions are categorized according to their operation and origin. Additionally, the bonding of tokens for Parachains – separate blockchains within the Polkadot system that gain their security from the Polkadot Relay chain— functions as a voting system as well and ensures that the chain's purpose aligns with the network's goals.

Interoperability is another cornerstone of the Polkadot vision. The platform's heterogeneous nature allows varied consensus systems to interoperate within a trustless, fully decentralized "federation".

Within the Polkadot architecture, four main roles are identified: collator, fisherman, nominator, and validator. Each role has distinct responsibilities, whether

---

[3] Stakebale token are token that can be used in a blockchain with a proof-of-stake algorithm to secure the network by backing someone trustworthy who is building blocks or finalizing them. Normally, some kind of reward for staking will then also lead to passive income for the staker.

In a governance system, such functionality can also be used for backing a referendum or for voting.

it's proposing new blocks, monitoring and reporting misconduct, contributing to a validator's security bond, or sealing new blocks on the network. This division of labor reinforces decentralization and autonomy while maintaining the integrity of the system.

In summary, Polkadot's approach to blockchain architecture aims to address the existing limitations of these systems by emphasizing decentralization, autonomy from the founders, and interoperability. Through its unique consensus architecture, robust governance structure, and division of roles, Polkadot seeks to provide a practical, scalable, global infrastructure.

3.4 Summary: Three Waves of Decentralization in Web3

In the three waves of development outlined above, we observe the decentralized online ecosystem based on blockchain – Web3 – developing varying levels of decentralization that manifest in distinct application forms. In its initial form, ownership is confined to the holding of tokens, whether Bitcoin or other tokens generated in later stages of other projects.

Through Ethereum's smart contracts, additional forms of tokens were introduced. These are not native blockchain tokens, but can be generated by anyone with Ethereum's technology and are based on standardization for a specific use case or project, such as ERC-20 tokens or NFTs.

With Ethereum's DApps, entire applications were added, many of which fall under the umbrella of decentralized finance (DeFi), aimed at disrupting the finance sector with decentralized alternatives. A part of this decentralization of finance, and a higher development stage, was the invention of the DAO, which tried to enable a whole organization to operate without central leadership by allowing voting mechanisms directly on the blockchain. With this, a decentralized company that makes investments and generates profits like a traditional company can be generated.

In the third stage of development, we see the evolution of a full governance system that reserves the final decision on all project matters to token holders – on all project's matters such as who gets the Parachains, software updates and financing of projects in the ecosystem. Here the DAO-like structures can also be used to collectively take care of a software infrastructure, commonly used by its token holders.

From the connection of Polkadot's Relay Chain and the technological infrastructure of the Parachains, an interoperability of the Parachains with each other, with Polkadot, and additionally with the outside world via Bridges can be achieved, leading to an ecosystem that spans an even larger arc.

4 Four Levels of Decentralized Ownership and Control in Web3

From the current state of development of blockchain and DLT technology, four levels of Decentralized Ownership and Control in Web3 can be derived, as shown in Fig. 3. in the box on the right—the ownership (1) referring to the native token as a value within and/or outside the blockchain, (2) referring to the software, as the intellectual property including the right to use, change, etc. (3) referring to the governance, which could be voting rights in regards to evolution of the software, funding for new projects, etc. and (4) referring to the use cases that are made on top of the initial DLT-project (like a ERC-20 token or a DAO on top of Ethereum or a Parachain in Polkadot). The first three of them are all strongly connected to each other and part of the blockchain infrstructure, while the 4th one is using this infrastructure to create their own use case on top of this infrastructure—potentially using the features of the original project, like smart contracts, for creating their own use case such as a token, a parachain or a DAO.

All these assets—token, software, governance, and use case with its own assets—can be owned directly by its users and therefore in a decentralized way, while they are also fully open for the public to access, use, and modify, embodying the fundamental blockchain principle of permissionless participation in ownership structures.

The Ownership of Native Tokens of a Blockchain

On this level, the DLT project creates tokens with its native blockchain and these tokens form an integral part of the usage through the token economy of this blockchain. Examples include the original Bitcoin (BTC), Ethereum's Ether (ETH) or Polkadot's Dot (DOT). Essentially, these tokens are pieces of software generated within a blockchain infrastructure, in accordance with the token economy of their respective projects. There are typically three ways of distribution: If the tokens had been sold through a traditional purchase agreement or a SAFT (Simple Agreement for Future Tokens) before the blockchain went live, they are typically generated with the genesis block and then distributed to their owners. They can also be generated and distributed as incentives for mining, staking, or other actions crucial to the network. A third option is the random distribution to people in the network, called air drop. The decentralized ownership on tokens is a fundamental characteristic that contributes to the appeal and functionality of blockchain technology and everyone can aquire or earn token to participate.

The Ownership Regarding the Software

As part of a native blockchain, software is developed to construct the blockchain, implement the consensus algorithm, and incorporate all essential features.

Public permissionless blockchains are based on open-source software, that everyone can use without needing to ask, pay or fulfill any criteria. Anyone can download the Bitcoin blockchain and start to become a full node or a miner using the Bitcoin software [24]. The Bitcoin software and the Ethereum software [25] are under the MIT License [26], where permission is granted to deal in the software without restriction, *"including the rights to use, copy, modify, merge, publish, distribute, sublicense, and/or sell copies of the Software, and to permit persons to whom the Software is furnished to do so"*, subject to including the copyright notice and the permission notice. Polkadot is under the GNU General Public License v3.0, a strong copyleft license that guarantees end users the freedom to run, study, share, and modify the software.

To participate in the decentralized system actively for example to contribute as a miner, collator, validator or just to run a full node, anyone can download and use the software. Therefore, the ownership of the software is truly decentralized and anyone with an internet connection and a computer can download and use it.

Governance as Ownership and Control Over a Technical Infrastructure

At this third level of the native blockchain, a token—akin to the token in the first level—assumes a unique role within a system, empowering its holders to shape the trajectory and future of a project. Consequently, the blockchain software encompasses an all-inclusive system that enables users to propose and vote on projects, upgrades, and other pertinent decisions. Essentially, projects incorporating a governance function operate similarly to a cooperative as defined by the Coop [27]; they act as an autonomous association of individuals voluntarily united to collaboratively manage a technical infrastructure intended for collective use.

This governance model allows for a decentralized decision-making process, where each token holder has a voice in the development and management of the project. Token holders can submit proposals for new features, changes to existing protocols, or funding for new initiatives. These proposals are then voted on by the community, ensuring that the direction of the project aligns with the collective will of its participants.

The decentralized nature of such governance frameworks ensures that no single entity can monopolize control over the project. This democratization of power not only fosters a more equitable ecosystem but also enhances the resilience and adaptability of the project. By distributing decision-making authority across a broad base of stakeholders, the project can better withstand external pressures and internal conflicts, leading to a more stable and sustainable development trajectory.

Moreover, the use of blockchain technology in governance ensures transparency and accountability. All proposals, votes, and decisions are recorded on the blockchain, creating an immutable ledger that can be audited by anyone. This transparency builds trust among participants and reduces the potential for corruption or manipulation.

In addition to technical and developmental decisions, governance tokens can also play a role in financial decisions, such as the allocation of funds from a project's treasury. This can include funding for development, marketing, community initiatives, and more. By involving the community in these financial decisions, projects can ensure that resources are allocated in a manner that best supports the project's long-term goals and the interests of its participants.

Overall, governance as ownership and control over a technical infrastructure represents a significant evolution in how projects can be managed and operated. It aligns the interests of all stakeholders, promotes active participation, and leverages the strengths of decentralized networks to create robust, community-driven ecosystems.

Ownership on the Use Case Level

Additionally, we observe tokens or other assets that, unlike native blockchain coins, are the result of a specific use case that a blockchain adheres to. In Fig. 3, you can see that they derive from a DLT project with that project's token, using its software and potentially even its governance to create their own use case, for example in the form of a token, a connected chain or a DAO.

While these usecases themselves might consist of all three—tokens, software and governance systems—they normally have no direct effect to the native project. For example a Parachain token in Polkadot is a usecase of the native Polkadot blockchain, but does not allow for participation in the Polkadot governance. If the Parachain has its own governance, the Parachain token might have voting rights there.

With multichain technology like Polkadot, such assets on a blockchain can also consist of a new blockchain—a Parachain connected to the Polkadot's Relay Chain, that includes features of a stand-alone blockchain like governance and native token plus the possibilities of assets on this blockchain. This level also includes ERC-20 tokens or Non-Fungible Tokens (NFTs) such as ERC-721 and ERC-1155 in Ethereum, that live on the Ethereum blockchain but are typically created and issued via smart contract by other bodies than Ethereum itself was. This level also incorporates tokenized real-world assets, examples of which include tokenized properties. Additionally, this level includes wrapped coins that symbolize value from another blockchain, as well as Decentralized Finance (DeFi) products created through smart contracts. Other use cases form their own blockchains with specific functionalities, such as the KILT blockchain and its identity features.

5 Examples of Decentralized Ownership and Control in Web3

This chapter entails an in-depth exploration of several applications that have been architected atop the Ethereum or Polkadot ecosystems. Particular attention will be paid to their specific employment of decentralized structures to innovate novel use cases or business models within the Web3 framework. Furthermore, the examination will elucidate how these applications engender a sense of ownership for the stakeholders and users interacting with this data, thus creating a new paradigm of personal control and autonomy in the digital realm.

FLP Impossibility vs. Blockchain Consensus

Assumptions of the FLP Paper	Blockchain Consensus Mechanisms
Byzantine problem does not occur	Byzantine problem occurs (nevertheless remains out of consideration here)
reliable messaging system	reliable messaging system
Asynchronous	Partially synchronous system
- No assumptions about time	- Timestamp, - regular blocks, - temporal chaining of the blocks
- no synchronized clock	- Synchronized clock
- crashed processes are not detected (no possibility to distinguish between delay and crash)	- crashed or delayed processes are not considered in the chain
Process crash endangers the system	Process crash at one participant does not affect the security of the system

Fig. 2 Overview FLP assumptions versus blockchain solutions

5.1 Decentralized Finance (DeFi)

Decentralized finance (DeFi) [28] decentralizes traditional financial systems by leveraging blockchain technology and cryptocurrency. By cutting out intermediaries like banks and exchanges, DeFi enables peer-to-peer financial transactions.

DeFi applications, built predominantly on the Ethereum blockchain, are being utilized across various financial activities. These include traditional financial transactions such as payments, trading securities, and lending. They also encompass the operation of decentralized exchanges (DEXs), digital wallets, and stable coins, which are tied to non-cryptocurrencies like the U.S. dollar to counteract volatility.

Currently the most successful DApps on Ethereum are tools for Decentralized Finance, such as Uniswap, being thrice under the top 15 applications, next to other decentralized financial services such as OpenSea, Metamask Swap or SushiSwap [29] (Fig. 4).

Innovative DeFi offerings, like yield harvesting, non-fungible tokens (NFTs), and flash loans demonstrate the potential for DeFi to transform financial transactions. For instance, flash loans, which are executed and repaid within the same transaction, capitalize on decentralized arbitrage opportunities.

DeFi operates on blockchain—where transactions are recorded in encrypted code, ensuring security and transparency. This decentralized nature makes DeFi resistant to fraud[4] and provides users with a level of anonymity.

---

[4] This resistance to fraud appears to be one of the fundamental strengths of DeFi, potentially fuelling its adoption, particularly in light of the challenges encountered with Centralized Exchanges in recent years. Owing to a lack of transparency regarding the crypto assets stored within these exchanges, central actors have been able to utilize these assets for their own speculative purposes.

Introduction to Decentralization and Ownership in Web3

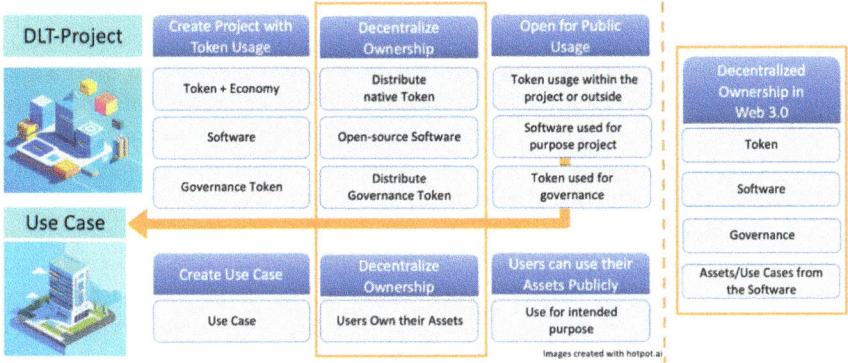

Fig. 3 Four levels of decentralized ownership and control in Web3

However, while DeFi presents exciting prospects, it comes with risks, including hacking threats, absence of consumer protections, and significant collateral requirements for loans. Despite these challenges, DeFi's potential to revolutionize the financial sector is substantial, offering greater transparency, security, and accessibility in financial transactions (Fig. 4).

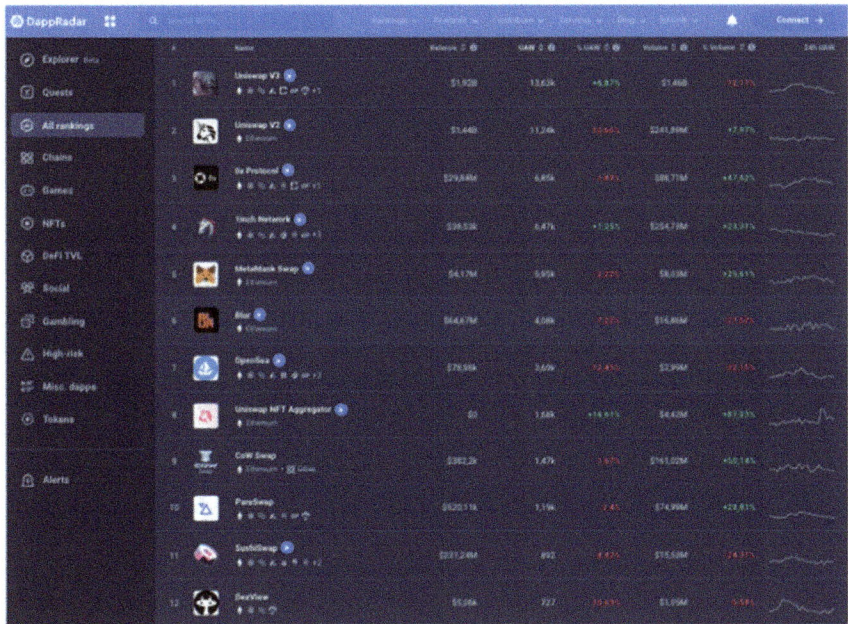

Fig. 4 Screenshot most successful DApps on Ethereum according to DAppRadar, Dec 6th 2023

Decentralized Finance (DeFi) significantly enhances ownership and control for stakeholders by allowing them to directly manage and benefit from their financial assets. Through DeFi platforms, users can engage in borrowing, lending, and trading activities without relying on traditional intermediaries like banks. This autonomy not only reduces dependency on centralized institutions but also empowers users to make decisions that best suit their financial goals. By utilizing smart contracts, DeFi ensures transparent and trustless transactions, giving stakeholders full control over their assets and the terms of their financial engagements.

The democratization of financial services through DeFi extends ownership and control to a broader range of participants, including those who are underserved by traditional finance. By providing open access to financial tools and services, DeFi platforms enable users from all backgrounds to participate in the global financial system. This inclusivity fosters a more equitable financial landscape where control is distributed among a diverse group of stakeholders. Additionally, the transparency and immutability of blockchain technology ensure that all participants have equal access to information, further reinforcing the principles of democratized ownership and control in the financial sector.

5.2 Decentralized Autonomous Organizations (DAOs)

Another noteworthy usage of Ethereum are Decentralized Autonomous Organizations (DAOs), with "the DAO" [30] being the first and most prominent trying to establish an organization that is executed directly through DLT-technology. The DAO, as detailed in [31], was a pioneering application of blockchain technology, acting as a decentralized venture capital fund for crypto and blockchain-focused projects. It operated on the Ethereum blockchain via smart contracts. DAO token holders were granted the ability to participate in the governance of the organization by using their tokens to vote on project proposals. If these voted-upon projects turned profitable, the token holders were set to receive financial returns. This innovative model tried to democratize the venture capital process by distributing decision-making power and potential profits to individual investors. However, as events unfolded, the implementation of this model also revealed significant vulnerabilities [32] and regulatory challenges [33] that have influenced the evolution of DAOs and blockchain technology eversince.

Currently, most DAOs [34], can be classified into the following categories. Protocol DAOs administer decentralized protocols, such as lending applications or exchanges, embodying a decentralized governance model often influenced by token holders. Grant DAOs and Philanthropy DAOs serve as decentralized philanthropic mechanisms, allocating funds or advancing social causes, respectively, in a manner that democratizes decision-making and resource distribution. Social DAOs coalesce individuals with shared interests, implementing a form of decentralized social governance, while Collector DAOs pool resources for collective investment in high-value items, demonstrating a unique model of shared ownership. Venture DAOs exemplify collective decision-making in investment, targeting early-stage Web3 companies. Media DAOs decentralize content creation, shifting control from traditional media houses to community members. Lastly, SubDAOs represent a model where subsets of DAO members oversee specific functions, enabling more efficient decentralized governance. Each type manifests unique aspects of decentralized control and ownership, revolutionizing traditional organizational structures with blockchain technology.

The concept of Decentralized Autonomous Organizations (DAOs) is not confined to the Ethereum ecosystem and is being further developed and expanded across various blockchain platforms. For instance, Polkadot claims to be the largest DAO globally, showcasing its commitment to decentralized governance through its Polkadot OpenGov framework [35]. Polkadot OpenGov exemplifies the broader adoption and evolution of DAOs beyond Ethereum, highlighting how different blockchain projects are leveraging decentralized governance to foster more inclusive and democratic ecosystems.

According to the DAO Whitepaper [36] by the German Blockchain Bundesverband, the legal status of DAOs in their different shapes remains unclear, leading to discussions about their legal form and the concept of self-regulation through code. Although some organizations use "Legal Wrappers" or operate as "Decentralized in Name Only" (DINOs) to navigate existing legal structures, the majority of DAOs still face significant legal and regulatory challenges, particularly in Germany and Europe, where the adoption of DAOs is affected by legal uncertainty and a lack of established legal concepts to limit the liability of involved actors.

Regardless of the specific applications and legal classifications of DAOs, their emergence is fundamentally revolutionary because they enable the creation of new forms of corporations, associations, and cooperatives for various purposes. The direct and immediate participation enabled by blockchain technology elevates our social and commercial interactions to a new level. Users can now connect and collaborate independently of third parties, transcending geographical boundaries to establish enterprises where they collectively and autonomously make all significant decisions. This paradigm shift allows individuals to directly engage in governance and operational processes, fostering a decentralized and democratized environment that redefines traditional organizational models. The ability to form and manage such entities without intermediaries not only enhances efficiency but also promotes a more equitable distribution of ownership and control of resources, fundamentally transforming the landscape of collaborative endeavors.

5.3 Decentralized Data Assets with Ocean

Ocean Protocol [37] primarily operates on the Ethereum blockchain and the main token deployment of Ocean (OCEAN) remains on the Ethereum mainnet. It utilizes

Ethereum's smart contracts and token standards (ERC20 and ERC721) for its functionalities. However, it can also be deployed on other Ethereum Virtual Machine (EVM) networks such as Polygon, Binance Smart Chain (BSC), Moonriver, and Energy Web Chain (EWC).

At its core, Ocean Protocol uses blockchain technology to facilitate the secure publishing, sharing, and selling of data while preserving privacy. It employs data tokens and data non-fungible tokens (NFTs) to represent ownership or access rights to data. This allows individuals to monetize their data by selling tokens representing the data, creating a potential income stream from their digital assets.

Beyond individual data monetization, Ocean Protocol also enables collective utilization of data from multiple sources. Its "Compute-to-Data" feature allows computations to be performed on privately held data without exposing the data itself, thereby preserving privacy while enabling insights to be derived from diverse datasets. This fosters collaborative research and innovation while ensuring data security and privacy.

Moreover, Ocean Protocol includes a community marketplace for data and tools to build custom marketplaces. These marketplaces provide a platform for buying and selling data, enabling data providers to monetize their data and data consumers to access a wide range of datasets.

The OceanDAO, a decentralized autonomous organization, plays a crucial role in the Ocean ecosystem. It funds projects that contribute to the growth and development of the Ocean Protocol, such as software development, outreach, and more. OceanDAO operates with a democratic voting system, allowing the community to decide on the allocation of funds.

In summary, Ocean Protocol provides a robust infrastructure for the ownership and control on monetization of individual data, the collective utilization of data from various sources, and the democratic governance of its ecosystem. It can be used for a multitude of purposes, from research and development to business analytics, machine learning, and more, resulting in a truly decentralized and democratized data economy.

5.4 Decentralized Peer-to-Peer Storage with IPFS

The InterPlanetary File System (IPFS) [38] is a decentralized, peer-to-peer data storage and retrieval system that connects all computing devices with the same system of files. Similar to the web, IPFS can be viewed as a single BitTorrent swarm within a Git repository, offering a high throughput, content-addressed storage model. It does not use an underlying blockchain technology but can be used with any blockchain.

IPFS combines a distributed hash table, incentivized block exchange, and a self-certifying namespace, eliminating single points of failure and the requirement for mutual trust. It models all data as part of a singular Merkle Directed Acyclic Graph (DAG), enabling effective versioning and the tracking of changes.

In terms of ownership and control, IPFS decentralizes data distribution, leaving control not with a centralized authority, but with individual nodes or users. Each user maintains full control over their data, deciding what to share or store. This decentralization fosters data sovereignty and empowers users to effectively manage their data.

IPFS also enables the creation of a "Permanent Web" where links do not expire, ensuring data longevity and preventing accidental loss of crucial information.

To manage data more efficiently, IPFS uses data sharding, a process of breaking large pieces of data into smaller chunks. Each shard can be stored on different nodes within the network, which reduces the burden on any single node and enhances data retrieval speed. Not only does this improve system performance, but it also increases data redundancy and resilience, as multiple copies of the same shard can be stored across the network.

In summary, IPFS offers a decentralized, efficient, and reliable platform for data storage and distribution, providing users with unprecedented control and ownership of their data, while using data sharding to enhance performance, redundancy, and resilience.

5.5 Decentralized Identity with KILT

KILT Protocol [39] is a Parachain in Polkadot and serves as a decentralized infrastructure for self-sovereign identity, an approach to digital identity, that gives individuals control over the information they use. KILT's identity concept transforms offline-approaches of identity management—inherently decentralized with paper credentials that are owned and carried individually—to the digital realm. In contrast to Web 2.0 digital identities, which often involve centralized storage of personal data, KILT allows individuals to self-manage their identities in a decentralized manner, using the blockchain to anchor and verify such data and cryptographic keys to control the data in the user's wallet. This approach not only preserves privacy but also enhances security (Fig. 5).

In KILT infrastructure, users generate their unique decentralized identifiers (DIDs) and attach verifiable credentials to these DIDs. These credentials are stored by the users themselves, effectively eliminating the need for centralized data storage. By decentralizing the storage of personal data, KILT removes the attractive target that centralized databases often present to malicious actors, thereby mitigating the risk of large-scale data breaches.

The shift towards self-sovereign identity represents a significant advancement in the preservation of individual privacy and security. By placing control directly in the hands of individuals, KILT promotes the principle of data minimization, a key tenet of modern data protection regulations. Furthermore, by eliminating the need for central data storage, KILT reduces the potential attack vectors for hackers, thereby enhancing the overall security of personal data.

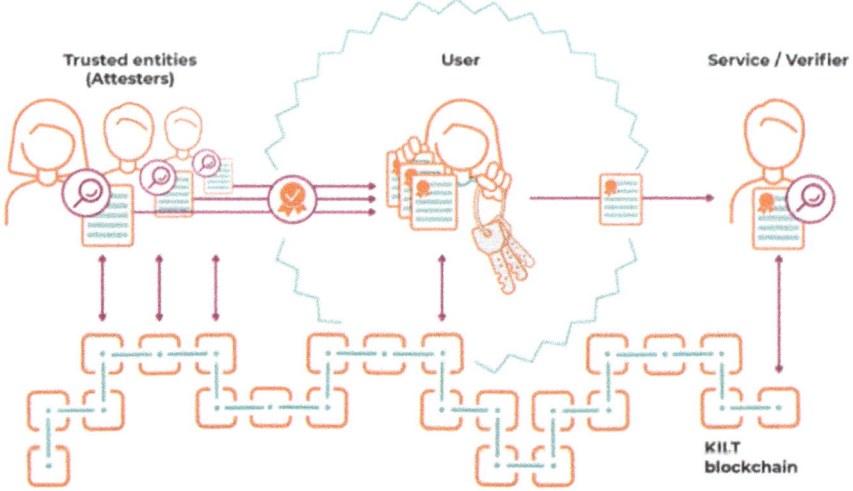

Fig. 5 Decentralized digital identity with KILT

This paradigm shift facilitated by KILT aligns with the broader movement towards Web3 and its prioritization of decentralization, privacy, and individual control, marking a significant departure from the centralized models of Web 2.0 digital identities.

In addition to the advantages that KILT presents for individual users, entities that have successfully established public trust in their respective sectors find substantial potential to strategically leverage this trust to generate and monetize novel business opportunities. This process would involve verifying users' properties when requested by the users themselves, and subsequently issuing credentials to authenticate these properties. The culmination of these operations would be the creation of a comprehensive new ecosystem, predicated on the principles of Web3 and self-sovereign identities. It's important to note that this represents a significant shift in the landscape of digital identity management, opening up new possibilities and applications.

One of these use cases of KILT is its integration with Deloitte for creating blockchain-readable, reusable digital credentials for Know Your Customer (KYC) and Know Your Business (KYB) processes [40, 41]. Traditionally paper-based and single-use, Deloitte has leveraged the KILT blockchain to revolutionize the issuance of reusable KYC and KYB credentials, enhancing efficiency and granting users sovereignty over their personal data. Furthermore, Deloitte, in collaboration with Hapag-Lloyd AG, Vodafone, and Nexxiot, launched the KYX project, which aims to boost global trade compliance and operational efficiency in the logistics industry.

KILT also introduces the Decentralized Identity Provider (DIP), enabling cross-chain decentralized digital identities [42]. DIP allows any blockchain that supports cross-consensus messages (XCM) to delegate users' identity management and verification to KILT or other third-party identity providers. This feature enhances interoperability both within the Polkadot ecosystem and beyond, fostering a seamless user experience across multiple platforms and bolstering the overall efficiency and utility of blockchain-based identity management.

All these features in KILT are designed with the overarching goal of returning ownership and control over users' data directly into their hands while enabling interoperability between different players. The decentralized approach eliminates the need for centralized data storage, thereby reducing vulnerabilities associated with large-scale data breaches. Moreover, KILT's infrastructuresupports cross-chain decentralized digital identities, fostering seamless interoperability within the Polkadot ecosystem and beyond. This capability allows users to manage and verify their identities across multiple platforms effortlessly, promoting a more integrated and user-centric digital environment.

6 Conclusion

Decentralization inherent in blockchain technology and DLT introduces a novel paradigm for ownership and control of values, assets, data, and identities. This shift facilitates the creation of new ecosystems and business models, underpinned by a reimagined form of collaboration. The transformative power of decentralization disrupts established industries, challenging traditional structures and systems. This disruption is particularly evident in the context of Web 2.0, where exploitation of personal data has been a pervasive concern. The advent of blockchain technology and Web3 marks a significant departure from this paradigm, offering a more equitable alternative where users govern their own digital experiences.

This fundamental shift in digital ownership and control represents a move toward a more inclusive, transparent, and user-centric environment. The democratization inherent in this process fosters new and improved forms of collaboration, which could potentially disrupt entire industries. Importantly, this transformation is not merely a theoretical possibility but is already being realized in various applications built on platforms such as Ethereum and Polkadot.

Therefore, the exploration of blockchain and Web3 technologies is not just the study of technological advancement, but of a societal evolution toward greater digital autonomy and privacy. As we continue to explore this promising landscape, we anticipate further innovative developments that will persist in redefining our concepts of ownership and control, further advancing the transition toward a digital realm driven by its users.[5]

---

[5] For the purposes of linguistic revision and translation, the AI-based tools Spellbook (https://www.spellbook.legal/) and DeepL (https://www.deepl.com) were utilized in this article.

References

1. G. Edelman, The Father of Web3 Wants You to Trust Less. Gavin Wood, who coined the term Web3 in 2014, believes decentralized technologies are the only hope of preserving liberal democracy. 29 Nov 2021. In: WIRED. Available under https://www.wired.com/story/web3-gavin-wood-interview/
2. S. Nakamoto, Bitcoin: A Peer-to-Peer Electronic Cash System (2009). Available under https://bitcoin.org/en/bitcoin-paper
3. G. Wood, Polkadot. Vision for a Heterogeneous Multi-Chain Framework. Draft 1. As committed on GitHub between November 14 and December 8, 2016 and available under https://github.com/polkadot-io/polkadot-white-paper/blob/master/PolkaDotPaper.pdf
4. P. Baran, On Distributed Communications 1962. Available under https://www.rand.org/pubs/research_memoranda/RM3420.html
5. M.J. Fisher, N.A. Lynch, M.S. Paterson, Impossibility of distributed consensus with one faulty process. J. Assoc. Comput. Mach. 32(2) (1985)
6. T. Lindseth, Bitcoin Prehistory. 29 Jan 2024. In Kryptolabs. Available under https://www.kryptolabs.no/post/bitcoin-prehistory
7. L. Lamport, R. Shostak, M. Pease, The byzantine generals problem. ACM Trans. Programming Languages Syst. 4(3), (1982). Received April 1980; revised November 1981; accepted November 1981
8. Melodies Sim: Practical Understanding of FLP Impossibility for Distributed Consensus. How are distributed consensus algorithms such as Raft implemented in the real world despite the FLP Theorem? Published on Medium. In: Level Up Coding. May 9th, 2021. Available under https://levelup.gitconnected.com/practical-understanding-of-flp-impossibility-for-distributed-consensus-8886e73cdfe5
9. Clarity in Crypto: 3 Waves of Crypto. Where we've been, where we are, and where we're going. 2022. Published under: https://www.clarityincrypto.com/3-waves-of-crypto
10. Wikipedia: List of bitcoin forks. https://en.wikipedia.org/wiki/List_of_bitcoin_forks
11. J. Poon, T. Dryja, The Bitcoin Lightning Network: Scalable Off-Chain Instant Payments. DRAFT Version 0.5.9.2. 14 Jan 2016. Published under http://lightning.network/docs/
12. Bitcoin Rollups: The basics. Rollups provide different execution environments for Bitcoin. Available under: https://www.bitcoinrollups.io/the-basics
13. Bitcoin.org: Choose your Bitcoin wallet. Select a wallet to store your bitcoin so you can start transacting on the network. Available under: https://bitcoin.org/en/choose-your-wallet
14. Bitcoin.com: What is censorship resistance? Available under: https://www.bitcoin.com/get-started/what-is-censorship-resistance/
15. B. Vitalik, Ethereum: A Next-Generation Smart Constract and Decentralized Application Platform. 2014. Pulished under: https://ethereum.org/669c9e2e2027310b6b3cdce6e1c52962/Ethereum_Whitepaper_-_Buterin_2014.pdf
16. M. Lennard, Ethereum 2.0 Roadmap—Die technische Implementierung. 13 Nov 2023. In: Blockchainwelt.de. Published under: https://blockchainwelt.de/ethereum-2-0/roadmap/
17. S. Corwin, Token Standards. Last edited 11 June 2024. In: Ethereum Website. Available under https://ethereum.org/en/developers/docs/standards/tokens/
18. V. Fabian, B. Vitalik, Token Standard ERC-20. November 19, 2015. Published under: https://github.com/ethereum/ercs/blob/master/ERCS/erc-20.md
19. E. William, D. Dieter, E. Jacon S. Nastassia, ERC-721: Non-Fungible Token Standard. 24 Jan 2018. Published under: https://eips.ethereum.org/EIPS/eip-721
20. NFT tech: Comparing the ICO Boom of 2017/18 and the NFT Boom of 2021. 11 Jan 2022. Published under: https://www.nfttech.com/newsroom/comparing-the-ico-boom-of-2017-18-and-the-nft-boom-of-2021
21. Crypto Council for Innovation (CCI): Real-World Use Cases for Smart Contracts and dApps. In: Explainers. Published under: https://cryptoforinnovation.org/real-world-use-cases-for-smart-contracts-and-dapps/

22. CoinDCX Blog: What is Polkadot (DOT) | Whitepaper Summary. 9 Aug 2022. Available under https://coindcx.com/blog/cryptocurrency/polkadot-whitepaper-summary/?__cf_chl_tk=wlcBmElsaGgokKuhNE30Vt4Swlj.rM.xhxa1NqBG9xM-1701861041-0-gaNycGzNDdA
23. Polkadot: Welcome to OpenGov. https://polkadot.network/features/opengov/
24. Bitcoin https://github.com/bitcoin/bitcoin
25. Ehereum License in github: https://github.com/ethereum/eth-account/tree/master
26. The MIT License https://opensource.org/license/mit/
27. International Co-operative Alliance: Cooperative identity, values and principles. https://www.ica.coop/en/cooperatives/cooperative-identity
28. E. Napoletano, What Is DeFi? Understanding Decentralized Finance. In: Forbes Advisor. Reviewed by Michael Adams. 28 Apr 2023. Published under: https://www.forbes.com/advisor/investing/cryptocurrency/defi-decentralized-finance/
29. DappRadar. Rankings. Protocol: Ethereum. https://dappradar.com/rankings/protocol/ethereum
30. Staff, The Economist: The DAO of accrue. A new, automated investment fund has attracted stacks of digital money. 21 May 2016 https://www.economist.com/finance-and-economics/2016/05/19/the-dao-of-accrue
31. F. Samuel, The Story of the DAO—Its History and Consequences. 24 Dec 2017. Published in Medium. Available under: https://medium.com/swlh/the-story-of-the-dao-its-history-and-consequences-71e6a8a551ee
32. Crypto 101: Ethereum DAO Hack. In Bitstamp Learn. 17 July 2023. Published under: https://www.bitstamp.net/learn/crypto-101/ethereum-dao-hack/
33. U.S. Securities and Exchange Commission: SEC Issues Investigative Report Concluding DAO Tokens, a Digital Asset, Were Securities. U.S. Securities Laws May Apply to Offers, Sales, and Trading of Interests in Virtual Organizations. 25 July 2017. Available under: https://www.sec.gov/news/press-release/2017-131
34. WingRiders: Types of Decentralized Autonomous Organization (DAO). In: Medium.com. 22 Sept 2022. Published under: https://medium.com/@wingriderscom/types-of-decentralized-autonomous-organization-dao-c364e8288a81
35. Polkadot: Polkadot is the largest DAO globally. 15th April 2024. On: X(formerly Twitter), @Polkadot. Available under: https://x.com/Polkadot/status/1779930843273707818
36. Blockchain Bundesverband: Working Group-DAO Whitepaper. 21 Mar 2023. Published under: https://bundesblock.de/dao-whitepaper/
37. Ocean Protocol Foundation with BigchainDB GmbH: Ocean Protocol: Tolls for the Web3 Data Economy, Technical Whitepaper, Version 2022-Sept-01.
38. B. Juan, IPFS—Content Addressed, Versioned, P2P File System. (DRAFT 3). 14 July 2014. Published under: https://arxiv.org/abs/1407.3561
39. BOTLabs GmbH: KILT Whitpaper. Version 15th January 2020. Published under: https://kilt-protocol.org/files/KILT-White-Paper.pdf
40. Deloitte Switzerland: Press Releases: Deloitte Integrates KILT Identity Blockchain, Creating New Markets with Reusable Digital Credentials. Published May 4th 2023 under https://www2.deloitte.com/ch/en/pages/press-releases/articles/deloitte-integrates-kilt-identity-blockchain-creating-new-markets-with-reusable-digital-credentials.html April 18
41. KILT Protocol: Deloitte KYC Credentials Are Live. 9 Nov 2023. Published in Medium under https://medium.com/kilt-protocol/deloitte-kyc-credentials-are-live-b7df4e86a7f0
42. KILT Protocol: Unchaining Identity: Decentralized Identity Provider (DIP) Enables Cross-Chain Solutions. 18 Apr 2024. Published in Medium under https://medium.com/kilt-protocol/unchaining-identity-decentralized-identity-provider-dip-enables-cross-chain-solutions-ba0baa1e68d0

Gustav Hemmelmayr is an experienced Austrian lawyer, having graduated from the University of Vienna in 2000. He also holds additional Master's degrees in European Integration (University for Continuing Education, Krems, 2001), Political Communication (Freie Universität, Berlin, 2014), and Blockchain & DLT (University of Applied Science, Mittweida, 2023).

In his current (2024) position as General Counsel for BOTLabs GmbH and KILT Protocol in Berlin, he oversees legal, compliance and regulatory issues related to open-source technology, cryptocurrency and decentralization. His primary areas of expertise include decentralized identities, regulatory compliance and governance for decentralized autonomous organizations (DAOs).

Following his master thesis on Polkadot Governance and several years of professional experience in Web3, Gustav Hemmelmayr began publishing articles in legal journals such as REthinkin: Law and Recht Digital (RDi), where he aligns decentralized technical concepts with regulatory frameworks. He is also a regular speaker at Web3 events, such as Blockchance (2023), the Nordic Blockchain Conference (2023), and Conf3rence in Dortmund (2024).

Gustav Hemmelmayr sees strong potential in decentralized systems to create new forms of digital identities, governance, and global interactions. As a knowledgeable voice in regulatory frameworks and decentralized technological architecture, Gustav Hemmelmayr bridges the gap between legal and technical aspects in his publications.

Enforcement of Rights in the Metaverse

Simon J. Heetkamp and Ida Holschbach

1 Conflict in and with the Metaverses

Despite substantial investment, metaverse technology is not yet widely deployed, nor considered technologically ready. [1] Nonetheless, regulators have begun to pay attention, not only looking at how to apply existing legal frameworks, but also laying the groundwork for new legislation and preparing for the metaverse's implications [3]. The metaverse is projected to be an extended universe similar to video games, creating vast walkable immersive virtual worlds. It's also meant to be a place for other activities, enabling one to work in whatever digital reality one would like, e.g. enabling office workers to move their desk to a digital beach, go shopping, or pursue leisure activities like sports or cinema visits.

This brings with it an enormous potential for disputes between users, and users and the platform. As a result, two categories of dispute and thereby dispute resolution can be envisioned: The first is the regulation *of* the metaverse, encompassing competition, data protection and cyber security law. The second, and the one on which this chapter will focus, is regulation and its enforcement *inside* the metaverse.

There have already been multiple instances where (particularly) people using female avatars reported being sexually harassed in virtual reality. In Decentraland, a browser-based platform, virtual plots of land have been sold for enormous[2] sums (Chirinos) [17], which might lead to neighborhood disputes. In computer games, in-game items are often stolen or defrauded, a problem that is sure to be translated into the metaverse.

---

[1] The European Commission is already laying the groundwork for legislation on the metaverse [3], and preparing for its implications.

[2] Given the volatility of cryptocurrencies, these numbers are to be taken with a grain of salt.

S. J. Heetkamp (✉) · I. Holschbach
Institut Für Versicherungswesen (ivwKöln), Technische Hochschule Köln, Cologne, NRW, Deutschland
e-mail: simon.heetkamp@th-koeln.de

A comparison to video games also helps when looking at metaverse governance: The developers of virtual worlds have almost god-like power and control not only the existence and supply of certain items, but they can also freely take from or give them to users or delete their profiles altogether. Blockchain technology, decentralization and by extension cryptocurrencies and non-fungible tokens can help: Writing contracts and ownership into a publicly stored ledger might free contractual relations inside the metaverse from the all-powerful developers, and make them more akin to the real world and its legal concepts [18].

Metaverses thus bring unique challenges to the enforcement of rights and the delivery of justice, but existing legal frameworks could offer solutions. These shall be discussed in the following sections.

2 Key Challenges of Rights Enforcement in Virtual Worlds

Natural and legal persons possess given rights. These rights can be distinguished by their origin: Rights might originate directly in the law, or stem from contracts (contractual rights) or be inherent in all human beings (human rights). Rights enforcement therefore is the process of realizing a situation described by a particular right. This is traditionally a matter for the courts: If a debtor does not pay what he was contractually obligated to deliver, the creditor can appeal and, if necessary, obtain enforcement.

The metaverse however has several characteristics that present unique challenges to the current system: anonymity, questions of jurisdiction, applicable law, court proceedings, and the possibility of alternative dispute resolution (ADR). In the following section, we present those key challenges and discuss potential solutions.

2.1 Avatars and Anonymity

Anonymity or pseudonymity

The law, and in particular its enforcement, is always determined by its subject: the person or legal entity that enters into legal relationships, i.e. concludes contracts or violates the rights of others.

This proves a challenge in the metaverse. Among internet users, anonymity and pseudonymity are common. Many people, especially gamers who are currently at the forefront of the metaverse (Verwaltungswissenschaft et al. 2023) only know and refer to each other by their chosen alias (or username). In metaverses, users interact using a digital embodiment: an avatar. Username and avatar might correspond to the real name and looks of a person, but they can be modified at will. Using this digital identity, users exchange goods and services without revealing their real identity [4]. While not using their real name might be freeing for a lot of users, the law requires

Enforcement of Rights in the Metaverse

the providers of a service to identify their customers in certain cases. This "Know-your-customer-Principle" applies to protect minors or to prevent money laundering, for example [4].

But even in areas where there is no obligation to identify the other party, e.g. in many contractual relations, the enforcement of rights and obligations might require it: A statement of claim, and therefore a law suit, must be addressed to the defendant, whose identity must therefore be known.

There are multiple different approaches to deal with the problem, all of which have advantages and disadvantages.

Collaboration with platforms

One solution to the problem of anonymity is for the State to collaborate with the platforms, either voluntarily or by law,[3] as it is already the case with social media and online marketplaces. The creditor or victim of a crime in the metaverse would therefore have the right to ask the operating platform to hand over the personal data of the debtor, or alleged perpetrator. A simple idea, the execution is complicated by the metaverse's structure: the approach puts the burden on the creditor to get this information from the platform, for which a court order might be necessary, but it also depends on the platform having identifiable information on that user. Signing up for the metaverse precursor Meta Horizons requires only an email address and few personal information (name and birth date), none of which is independently verified, meaning that it could be entirely fictional.[4]

An additional obstacle is that some metaverse platforms, like Decentraland,[5] claim decentralization. In this context it means that the necessary resources, both hardware and software, are in multiple locations, and owned and run by different individuals and organizations, which makes it more difficult to identify the responsible person [14].

Additionally, the username and avatar are disconnected from the person wielding them. There is no guarantee that whoever signed up to a service with their real information is actually the person using the account and corresponding avatar. Identifying the opponent would therefore prove difficult or impossible in many cases.

---

[3] An example of this can be the German Network Enforcement Act (Netzwerkdurchsetzungsgesetz), which introduced compliance rules with fines for social network providers regarding the handling of user complaints about hate crime and other criminal content online, and gave victims a right to information about the infringer's personal information on the basis of a court order.

[4] Users might choose not to disclose their personal information for security or privacy concerns. Consequently, the contractual obligation to sign up under a person's real name has been declared legally void in the case of Facebook in Germany (BGH Urt. v. 27.1.2022–III ZR 3/21, MMR 2023, 75 Rn. 31 ff.).

[5] Decentraland claims that "Content within Decentraland is hosted and served to users via a network of community-owned content servers", https://docs.decentraland.org/player/general/faq/ (14.12.2023).

2.2 Digital Identities

Another potential solution to anonymity are the so called digital identities. Those can be divided in three groups: digital identities issued by the State, identity as a service, and self-sovereign identities.

State-issued digital identities

Identity (ID) and its physical representation as ID card and passport is traditionally issued by the State. Thus, it seems only natural that states might also offer digital identities that can be used in the metaverse, with such efforts already underway. The European eIDAS regulation calls it "electronic identification means" and defines it as "a material and/or immaterial unit containing person identification data and which is used for authentication for an online service".[6] The electronic identification means is used to identify a natural or legal person or the origin and integrity of data. Following the eIDAS-Regulation, several states already offer rudimentary digital identities: German citizens can use the Smart-eID function on certain services to identify themselves by tapping their ID-card against compatible smartphones (2023).

Identity as a service

Identity as a service describes a number of identification methods offered by private actors. Single-sign-on-services are one of the most common already wide-spread forms: Users can register an account with Google, and then use this account to log into a number of other online-services. For users, this is convenient: They don't have to create separate accounts with separate passwords. The service provider trusts Google to know the identity of the user, and Google profits by building data profiles about their customers [13]. Another prominent example are phone numbers used in two-factor-authentication; especially when registering a new number requires identification with an ID: The mobile company provides verified identification as a service [4].

Self-sovereign identities

The difference between self-sovereign identity (SSI) and other digital identities is that with SSI, the user is controlling what personal data they reveal, and when they do so. The holder of an SSI adds attributes, meaning personal information like their name, age, address or profession. These attributes can be verified by a third party, or self-attested [4].

Security and privacy

Digital identities could be the key to transferring personhood into the metaverse. However, they do not come without security and privacy concerns.

---

[6] Art. 3 para. 2 Regulation (EU) No 910/2014 of the European Parliament and of the Council of 23 July 2014 on electronic identification and trust services for electronic transactions in the internal market and repealing Directive 1999/93/EC (eIDAS).

Digital identities are often discussed in connection with blockchain technology [4]. It can make digital identities tamper-resistant and easily verifiable, but there are also numerous other technological solutions.

One important element of digital identities are Zero-Knowledge-Proofs. Personal information can be revealed selectively: To enter a virtual casino, holders must only reveal their age, thereby allowing a verification of the legal age, but no other personal information. This provides advantages from a privacy and security standpoint, but these claims have been disputed.[7]

2.3 *(International) Jurisdiction*

Before a court can hear a case, it has to establish whether it has jurisdiction. Traditionally, it is territorial, meaning that German courts are responsible for disputes arising on German soil. However, the question of jurisdiction posed by the metaverse is more complicated: Which court has jurisdiction if a Brazilian and a French national have a dispute in a metaverse run on American servers?

As jurisdiction is territorial, the forum usually depends on the nationality or domicile, or others factors, like the location where a tort occurs or a contract ought to be fulfilled [11]. In the metaverse, this is complicated by users' wish for anonymity. Additionally, it is hard to locate the metaverse: Is jurisdiction determined by the location of the user(s), the location of the servers or is a new "metaverse" forum required?

In many countries, especially the US, companies will include arbitration clauses in their terms and conditions, thereby requiring the parties to resolve their dispute through an arbitration process outside the court system. Arbitration can make disputes costly for users and there are allegations of bias, for example in the selection of the procedural rules or jurors [18]. Even so, arbitration clauses might be subject to formal requirements that are hard to fulfill with electronic contracts[8] or completely void [25], thereby paving the way for state courts.

---

[7] Memorably, privacy activist Lilith Wittman criticized the German SSI-program, describing the process as "you don't show your ID, as you might do in a store, but hand it to the online store and they make a copy of it, which they can prove is genuine" (translated from German), creating a security hazard [27].

[8] Section 1031 subsection 5 of the German Procedural Code (ZPO) requires that arbitration contracts to which a consumer is a party (which the majority of users will be) must be contained in a separate document and signed either by hand or electronically.

2.4 Remote Court

Video proceedings

Traditionally, court hearings have been conducted in court rooms, with the parties and/or their representatives as well as judges, witnesses and jury present. In the metaverse, this is more difficult, especially if the parties involved are separated by large distances.

While remote court proceedings aren't new, court proceedings via video transmission only became wide-spread during the Covid-19 pandemic.[9] Although remote court proceedings have so far been conducted using videoconferencing applications like Zoom or Skype, metaverse proceedings using avatars are not in the distant future. In fact, in February 2023 a Columbian judge conducted the first court hearing in the metaverse [2] and there are reports of a Chinese metaverse court [30].

Certain requirements must be met in order to successfully conduct remote court hearings: First of all, the legal framework must be in place, meaning that the national procedural law must accommodate virtual proceedings. Secondly, courts and parties must have access to the technology required for remote proceedings, including fast broadband connection, computers and know-how. And thirdly, courts must be willing to conduct hearings remotely, as the proceedings are often left to the discretion of the court.

Reservations against remote court hearings include the need for expensive technical equipment and proficiency of use (which in turn has implications for access to justice) a fear of technical outages and the possibility to follow facial expressions and body language [22]. In the case of the Covid-19 pandemic and the response to it, these reservations against video proceedings were overruled by the need for the justice system to function. Technological improvements and wider use might do the same for metaverse proceedings: In September 2023, Mark Zuckerberg presented Meta's new photorealistic VR-avatars that display realistic facial expressions, a big improvement on the cartoonish avatars from before (Carter).

2.5 Enforcement

Rights enforcement, and especially foreclosure in the metaverse is fraught with special problems. In the real world, enforcement measures are usually carried out by state institutions such as bailiffs who enforce against the debtor's assets. In the metaverse, this will prove more difficult.

---

[9] In Germany, while video trials have been possible since 2002, but is in the discretion of the court, which led to their use only as measures to fight the pandemic hit (ZRP 2023, 66, beck-online). The Society for Computers and Law (UK) tracks the international rise of remote courts since 2020 (https://remotecourts.org/commentary.htm).

If the identity of a person behind an avatar is known, enforcement can proceed according to existing rules. Difficulties only ensue when it happens to be a cross-border lawsuit: Cross-border enforcement is dependent on the recognition and enforcement by the other state. States can refuse to recognize and enforce decisions from other states for example for reasons of violations of the public order [26].

Seizing assets proves difficult when the identity of the person behind the avatar is unknown. It seems likely that future legislation would allow enforcement within the virtual world, for example seizing the avatar itself, or its virtual assets. This of course only is effective if those things have value, meaning that a secondary market exists.

Enforcement could also make use of smart contracts based on blockchain technology. A blockchain is a digital representation of a transaction, which is validated by the community and stored on replicated ledgers. Smart contracts in turn build on blockchains: A smart contract is an algorithmic program stored on the blockchain that is executed when certain conditions are met, thereby automating the outcome [23]. For example, the condition could be that the creditor informs the system that a payment is due, which could then automatically trigger a transaction from the debtor to the creditor's benefit.

Similarly, metaverses could require users to provide a security or deposit, which would then be used to compensate the creditors. Securities are already common in procedural law, making this a realistic option for the near future [26].

2.6 Evidence

The metaverse and, in particular VR-technology as one of its components could have a major impact on the taking of evidence. Virtual technologies can make it possible to walk through crime scenes in three dimensions and make complex disputes easier to untangle.

Virtual reality as evidence

Photographic and video evidence has been used in courts for decades. Virtual reality is an immersive representation of reality, and lifts the former technologies to a new level.

The potential of virtual reality in court cases has already been demonstrated: In Germany, a judge surveyed the site of policemen killings using VR-glasses, his view projected on a screen for the public. In an investigation against a speeding driver, police recreated the events and created a 360 degree video to be viewed with VR-glasses [12]. Virtual reality has been used internationally as evidence material and has the potential to revolutionize evidence proceedings.

Virtual recreations are not only of relevance in criminal proceedings, as these examples might suggest, but can also be used in matters of civil or administrative law: Using VR, complicated product designs can be explored three-dimensionally, and building plans can be visualized from every perspective.

Visualizations like this can not only replace lengthy documents, but lead to a better and more intuitive understanding of the presented evidence [7].

Virtual reality can also replace on-site visits, making proceedings more cost-effective: Instead of having to schedule an on-site inspection, the visit can be conducted virtually and with all parties present, thus making proceedings shorter and more flexible.

Training in VR-environments

Virtual reality also has educational value: Police and law enforcement officers are already using virtual environments. Trainees can work through VR-scenarios, which allows them to practice difficult situations at a lower cost, and with guiding feedback from supervisors [8, 31].

Virtual reality can also be used as a learning environment for judges-in-training. We are developing a VR-witness application, in which law students and judges in training can hear a witness about the facts of a case and solve practice cases. The witness is an avatar with voice response that outputs answers generated by a large language model (ChatGPT) trained with the facts of the case. Training in the metaverse could not only be cost-effective, but also better prepare trainees, which could improve law and rights enforcement.

Evidence in metaverse proceedings

If a dispute arises in the metaverse, the evidence given by the parties will most likely include electronically conducted contracts, direct messages and screenshots. This has implications for evidence hearings.

In many jurisdictions, digital documents only have the same legal standing as a handwritten signature when they have an electronic signature that adheres to the requirements of the specific national regulation.[10] Other documents and screenshots only have the value of prima facie evidence; meaning that the court itself decides whether to admit the evidence and how it evaluates it. While electronic signatures can be used to make screenshots more credible,[11] it remains an open question how to prove things that happen in a virtual world.

Technology and new legislation may be the answer. Some arbitration rules already contain new rules on the taking of evidence in remote hearings, and procedural law might follow suit [26]. Additionally, smart contracts and blockchain technology may become admissible as evidence. Transactions on a blockchain are stored publicly, and hence tamper-resistant, which could give them probative value. Once procedural rules for the taking of evidence admit blockchain products as evidence, proving the existence and content of contracts and their violations in the metaverse may become much easier.

---

[10] eIDAS in the EU, NIST-DSS in the US or ZertES in Switzerland.

[11] One example of this is NetzBeweis.com, a browser plugin that makes it possible to save website or chat screenshots with a timestamp and digital signature.

3 Justice Delivery in the Metaverse

3.1 The Future of Courts

The countless possibilities of the metaverse, continuous digitalization and web3-technology might even change justice systems at their core.

Traditionally, the delivery of justice was dominated by two actors: state courts and arbitration tribunals. In many states, courts are firmly integrated into the State as a third power, the judiciary. Judges are appointed by the State, and procedure is based on national law. Decisions and judgments have the weight of the State on their side, and can be enforced using the State's monopoly on the use of force.

Arbitration however is a process in which parties agree to have a neutral tribunal or panel (the arbitrators) reach a binding decision based on rules they previously agreed on. The New York Convention[12] requires states to recognize and enforce arbitral decisions.

These systems of justice delivery can be understood as competitors in an open market [9]: State courts, perhaps especially in commercial and civil law, provide a service on a market open to competition, and more convenient or cost-effective deliverers of justice may emerge. Two such competitors could be decentralized justice and a privatization of justice.

Decentralized justice

The term "decentralized justice" describes various systems that offer justice as a combination of crowd sourcing, game theory and blockchain. They promise economic efficiency and easy enforcement, all while serving sectors currently under-served by the justice system.

In simple terms, decentralized justice describes a process where other users of the service are chosen (at random) as jurors, and then presented with the dispute. The jurors make their decisions independently and without conferring with each other. The final judgment is then decided purely by what the majority of jurors decided on, and only those voting with the majority are rewarded financially, which is designed to keep jurors honest and impartial [1].

Undeniably, this process is very different from traditional dispute resolution, and brings its share of problems [21]. Nevertheless, there are areas of conflict underserved by existing dispute resolution mechanisms: Online commerce for example frequently leads to conflicts with a low value, thereby making expensive court proceedings an inefficient and rarely used option [1]. Decentralized justice may be a cheap and efficient solution for such comparatively simple disputes with binary solutions.

---

[12] United Nations Convention on the Recognition and Enforcement of Foreign Arbitral Awards, done at New York, June 10, 1958, available at https://uncitral.un.org/en/texts/arbitration/conventions/foreign_arbitral_awards.

Privatisation of justice and hybrid justice

The privatization of justice is the trend towards more private actors acting as courts and making and enforcing decisions. It is best exemplified by Meta's Oversight Board.

In 2018, as a response to pressure from users and civil society, as well as criticism from governments, Meta introduced the Meta Oversight Board. Composed of several public intellectuals,[13] it provides independent review of content decisions made by Meta, functioning as a "supreme court" [16] of content moderation [10]. Users of Meta's services can submit their appeal to the board, out of which the board selects the cases that will be heard. A panel then hears the case, and the board comes to a decision, which is implemented by Meta, and has precedential value (Oversight [24]). Prominent examples include the Oversight Board's decision that the removal of Donald Trump's posts on the 6th of January 2021 had been justified (Meta Oversight Board). While the oversight board has been criticized for its limited jurisdiction and lack of independence from Meta [28], it has been praised for its transparency and quality of private justice [10].

Overall, it is not hard to imagine that the idea of an Oversight Board can be translated into the metaverse. Driven by criticism from the public, a need for transparency and threats of regulation by governments, companies will want to create their own oversight boards. In Meta's metaverse for example, Meta could establish its own court for disputes arising between its users and user and platform, modeled after the Oversight Board.

This could signify a trend towards privatization of justice, at least in the digital sphere: It is no longer just the State that issues laws, but also the platforms that issue guidelines. They could also create tribunals that make decisions based on these.

In effect, what might emerge is not a full privatization of justice, but a new hybrid justice: Platforms are allowed to self-govern, but state governments audit and regulate. The European Digital Service Act (DSA), which applies to metaverse operators as hosting services [15], can serve as an example: Art. 14 point 4 stipulates that platforms shall act with "due regard to the rights and legitimate interests of all parties involved, including the fundamental rights of the recipients of the service", thereby placing them in the framework of European and national law. Similarly, in Art. 17 point 3 f) it requires large platforms to provide out-of-court dispute settlement bodies, but guarantees the right of the user concerned to redress against the decision before a state court.

3.2 Conclusion

The metaverse, as well as its components like VR and the blockchain are still rapidly developing fields with changing dynamics. Technological advancement

---

[13] The board comprises for example a former prime minister of Denmark, a Nobel Peace Price Laureate as well as representatives from NGOs and law professors.

forces constant adaption of the legal framework, which often only occurs as a reaction to developments rather than anticipating them. But whilst it might seem as if the law is constantly mending occurring problems, there is potential for innovative forces to shape regulation and lay the legal groundwork for the metaverse.

References

1. Y. Aouidef, F. Ast, B, Deffains, Decentralized justice: a comparative analysis of blockchain online dispute resolution projects. Frontiers in Blockchain **4** (2021)
2. C. Bello, Colombia makes history by hosting court hearing in the metaverse. Euronews (2023)
3. T. Breton, People, technologies and infrastructure—Europe's plan to thrive in the metaverse (STATEMENT/22/5525). In: Blog of Commissioner Thierry Breton (2022). https://ec.europa.eu/commission/presscorner/api/files/document/print/en/statement_22_5525/STATEMENT_22_5525_EN.pdf. Accessed 28 Nov 2023
4. C. Busch, § 16 Digitale Identitäten im metaverse. in ed. by H. Steege, K.J. Chibanguza, Metaverse: Rechtshandbuch, 1st edn. Nomos Verlagsgesellschaft (2023)
5. T. Carter, Mark Zuckerberg just previewed Meta's new VR avatars—and they don't suck. In: Business Insider (2023). https://www.businessinsider.com/mark-zuckerberg-new-metaverse-avatars-are-now-humanlike-2023-9. Accessed 27 Dec 2023
6. C. Chirinos, Someone just paid $450,000 to be Snoop Dogg's neighbor in the metaverse. Here's how you can live by a celebrity too | Fortune. In: Fortune. https://fortune.com/2022/02/02/how-to-buy-metaverse-real-estate-snoop-dogg-celebrity-neighbor/. Accessed 21 Dec 2023
7. C. Donalek, S.G. Djorgovski, A. Cioc, A. Wang, J. Zhang, E. Lawler, S. Yeh, A. Mahabal, M. Graham, A. Drake, S. Davidoff, J.S. Norris, G. Longo, Immersive and collaborative data visualization using virtual reality platforms, in *2014 IEEE International Conference on Big Data (Big Data)*. (IEEE, Washington, DC, USA, 2014), pp. 609–614
8. L. Giessing, M.O. Frenkel, Virtuelle Realität als vielversprechende Ergänzung im polizeilichen Einsatztraining – Chancen, Grenzen und Implementationsmöglichkeiten. Springer Books: pp. 677–692 (2022)
9. R. Guise-Rübe, S. Kuhnke-Fröhlich, Zivilprozess: Die Videoverhandlung als Normalität. Legal Tribune Online (2021)
10. R. Gulati, Meta's oversight board and transnational hybrid adjudication—what consequences for international law? (2022). https://doi.org/10.17169/refubium-35394
11. T.C. Hartley, Basic principles of jurisdiction in private international law: the European Union, the United States and England. Int. Comparat. Law Quart. **71**(1), 211–226 (2022). https://doi.org/10.1017/S0020589321000427
12. S. Heetkamp, G. Irskens, Digitalisierung der Beweisaufnahme—neue Technologien in der Praxis. RDi **8**, 382–389 (2023)
13. T.B.M. Indenhuck, M. Britz, The rise of the online-you: Auf dem Weg zu digitalen Identitäten. RDi **6**, 289–295 (2023)
14. N. John, J. Müller, J. Rennert J, John/Müller/Rennert: Plattformhaftung und dezentrale Netzwerke: Die Haftung auf Mastodon. GRUR **10**, 691–698 (2023)
15. M.C. Kettemann, C. Böck, M. Müller M, Regulatory approaches to immersive worlds: an introduction to metaverse regulation. Project Immersive Democracy (2023)
16. K. Klonick, *Inside the Making of Facebook's Supreme Court*. The New Yorker (2021)
17. W. Lavin, NFT collector spends $450,000 to live as Snoop Dogg's virtual neighbour. NME (2012)
18. A. Martínez Sánchez, Law and dispute resolution inside video games. ZERL (2021)
19. M. Martini, J. Botta J, Der Staat und das Metaversum. MMR 2023(MMR-Beilage) 887–904 (2023)

20. Meta Oversight Board Former President Trump's suspension—Case decision 2021-001-FB-FB (2021)
21. J. Metzger, Decentralized justice in the era of blockchain. Int. ODR **5**(1–2), 69–81 (2018)
22. B.R. Muir, E.J. Newman, M. Rossner, The role of video background cues in the virtual court: a psychological perspective. Psychol. Crime Law 1–19 (2023). https://doi.org/10.1080/1068316X.2023.2224493
23. S. Navale, B. Chengappa, An Arbitrator's toolkit: blockchain, cryptocurrency, and smart contract dispute resolution. DRJ **77**(2023), 1–27 (2023)
24. Oversight Board, Oversight Board Bylaws and Code of Conduct (2023)
25. B. Quarch, § 31 Rechtsdienstleistung/Anwaltliches Berufsrecht. in ed. by H. Steege, K.J. Chibanguza metaverse: Rechtshandbuch, 1st edn. (2023)
26. E. Wagner, M. Holm-Haduella, M. Ruttloff, Metaverse und Recht (2023)
27. L. Wittmann, Mit dem Personalausweis zum Onlineshopping: Wie selbstbestimmt sind "selbstbestimmte Identitäten"? In: Medium (2022). https://lilithwittmann.medium.com/mit-dem-personalausweis-zum-onlineshopping-wie-selbstbestimmt-sind-selbstbestimmte-identit%C3%A4ten-f096a5bdd55a. Accessed 27 Dec 2023
28. D. Wong, L. Floridi, Meta's oversight board: a review and critical assessment. Minds Mach. **33**, 261–284 (2023). https://doi.org/10.1007/s11023-022-09613-x
29. eGovernment Monitor, *Use and Acceptance of Digital Administrative Services from the Citizens' Perspective: A Comparison of the German Federal States* (Austria, and Switzerland, Germany, 2023), p. D21
30. A Chinese local court recently opened a hearing in metaverse, saying it helps drive the digitization of the judicial system. PingWest (2022a)
31. Axon's virtual reality simulator is the future of police training. In: Police1 (2022b). https://www.police1.com/police-products/virtual-reality-training-products/articles/axons-virtual-reality-simulator-is-the-future-of-police-training-NNI4OD3FyE3WVCQY/. Accessed 20 Dec 2023

Simon J. Heetkamp, LLM. is Professor of Commercial Law, Mobility and Insurance Law at the TH Cologne. He previously worked as a judge at the Regional Court of Cologne and the Local Court of Cologne in the North Rhine-Westphalian judiciary. At the beginning of 2022, Simon Heetkamp initiated the "digitale richterschaft", which is an exchange platform for digitalisation topics in the judiciary (www.digitale-richterschaft.de). Before becoming a judge, Simon Heetkamp worked for several years in the litigation department of a large German, internationally active commercial law firm.

Simon Heetkamp has been working on topics at the interface of law and digitalisation for many years. As part of his comparative law PhD thesis, he focussed on the regulation and structuring of online dispute resolution procedures. As a lawyer, he designed an IT tool to manage mass action proceedings and to support the handling of these proceedings. His Master's thesis focussed on the use of VR technology in civil proceedings. Simon Heetkamp's current research focusses on legal issues in connection with artificial intelligence, virtual reality and the metaverse.

Ida Holschbach is a lawyer in training and a researcher. She holds dual Bachelor's degrees in French and German law from Paris I Panthéon-Sorbonne University and the University of Cologne. Ms Holschbach passed his first German bar exam in 2024. As a research assistant to Prof. Dr. Heetkamp at TH Deutz, she investigated the application of virtual reality and large language models (LLMs) in law. Currently, she is researching the significance of artificial intelligence for open educational resources (OER) in law at the VEStOR project at Leibniz University Hannover.

Procurement of Industrial Machinery in the Metaverse

Potentials of Successful Strategy Development

Natalia Broza-Abut, Tobias Jornitz, and Axel T. Schulte

1 Introduction

1.1 Transformation of Procurement

In a constantly changing digital landscape, companies are continuously challenged to rethink and realign their business models. The evolution of the internet has taken us from static websites to social media to a new promising era in the form of the Metaverse. In this immersive, connected environment, physical and digital realities merge, creating entirely new opportunities and challenges for business strategies.

One area where this transformation is particularly noticeable is the procurement process, e.g. of industrial machinery. Traditionally characterized by physical meetings and showrooms, industrial companies are now faced with the challenge of entering a world where business transactions can take place in virtual space.

The following article provides an insight into the transformation of the procurement process in the Metaverse and its opportunities and challenges. It examines how the purchasing cycle can be automated by digital progress and which criteria are decisive for a purchasing decision and thus also its automation. Furthermore, a roadmap will be presented to help companies prepare for these changes and fully exploit the potential that the Metaverse offers.

While technology and data should be at the center of any business strategy, this article aims to provide a clear understanding and guidance for those companies that are ready to take the next step into the future of digital procurement.

N. Broza-Abut (✉) · T. Jornitz · A. T. Schulte
Fraunhofer IML, Dortmund, Germany
e-mail: natalia.broza-abut@iml.fraunhofer.de

T. Jornitz
e-mail: tobias.jornitz@iml.fraunhofer.de

1.2 Case Study

To illustrate the transformative power of the Metaverse in the industrial machinery procurement process, let's look at the fictional example of Mr. Müller, head of procurement for a mechanical engineering company in the agricultural sector. He uses the innovative possibilities of the Metaverse to make the procurement process more efficient, time-saving and cost-effective.

Wearing VR glasses, Mr. Müller enters an interactive showroom. It felt as real as if he was physically walking through an exhibition hall. He could look at each machine from all angles, test its functions and even see how it would work in a simulated production environment.

While he was looking at the machines, colleagues from other countries joined him in the virtual room. Together, they were able to discuss and compare specifications, prices and delivery times in real time. The Metaverse provided them with a platform for seamless communication without having to leave their offices.

During his visit, Mr. Müller was able to customize the machine to his specific requirements. He changed components, selected special software add-ons and immediately saw how these changes affected performance and price.

In another area of the showroom, the manufacturer offered interactive training sessions. Mr. Müller was able to see how his employees would be trained before he made the purchase, which gave him additional confidence in his purchase decision.

He was impressed when he was invited into a virtual conference room to negotiate with the manufacturer's sales team. All documents, contracts and specifications were shared, edited and finalized in real time.

Through the entire process in the Metaverse, Mr. Müller was not only able to save travel costs and time, but also work more efficiently. Instant collaboration, real-time training, and transparent negotiations enabled him to make an optimal purchasing decision.

The technology had not only simplified the shopping process, but also created an enhanced shopping experience through innovative features and seamless interactions.

2 Procurement Process in the Metaverse

2.1 Key Technologies and Data

In a world that is increasingly digitalized, the Metaverse reveals new horizons for the procurement process. But why is the Metaverse crucial in this context at all? Why can't the procurement process be automated and optimized by individual technologies, independent of the Metaverse? To answer these questions, we need to systematically analyze and understand the unique characteristics and opportunities that the Metaverse offers.

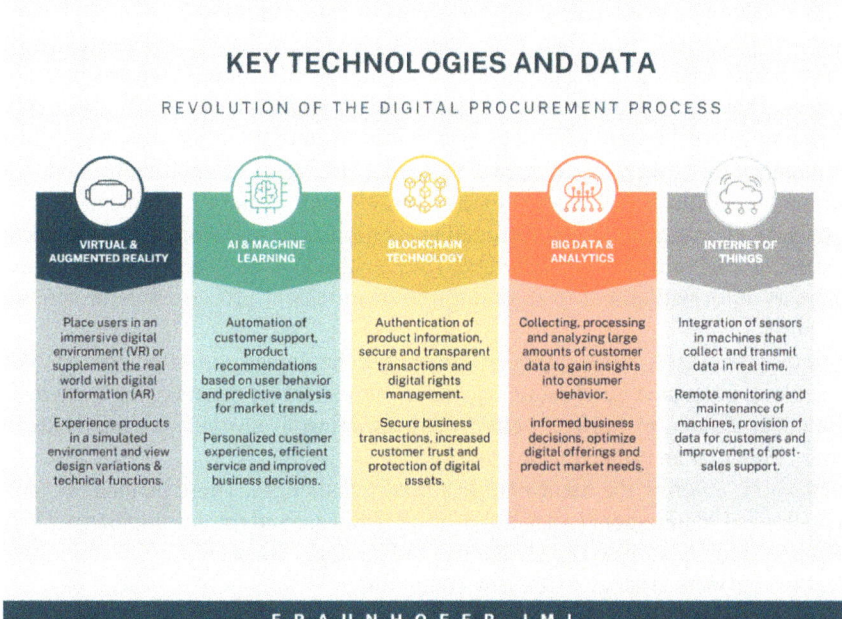

Fig. 1 Key technologies and data

The Metaverse combines a variety of technologies, including Virtual and Augmented Reality, artificial intelligence, blockchain, and cloud computing, to name just a few. These technologies are not just individual tools, in the Metaverse they are combined synergistically to create an integrated, immersive and interactive space for business processes. The following section discusses the most relevant among them, referred to as key technologies (Fig. 1), and highlights their potentials for the procurement process.

Virtual and Augmented Reality enable immersive experiences that will revolutionize the initiation and negotiation in procurement. They open spaces for virtual product demonstrations and allow customers to inspect and test products in a virtual environment, leading to greater transparency and informed purchasing decisions.

Artificial Intelligence and Machine Learning contribute to automating the procurement process by assisting in data collection and analysis, supplier selection, and price negotiations. By predicting market trends and offering personalized purchasing recommendations, they help make the procurement process more efficient and intelligent.

Blockchain technology will ensure transparency and traceability in procurement. It is used to ensure that all transactions and negotiations are secure, transparent, and immutable, which is especially relevant in the context of the Metaverse to guarantee security and trust.

By collecting and analyzing large amounts of data, purchasers can make better-informed decisions. Big Data and Analytics enable a better understanding of market trends, make accurate predictions, and strategically optimize the procurement process. They also form the basis for successful AI applications.

The so-called Internet of Things contributes to improving the networking and communication between machines and systems in the supply chain. It enables smooth and automated interaction between the different elements of the procurement process.

Cloud Computing and Edge Computing are particularly relevant for making data processing and storage more efficient. They enable fast and secure access to the necessary information and tools during the procurement process, thus promoting flexibility.

In the Metaverse, the merging of different technologies and principles enables a seamless, integrated purchasing experience. Procurement benefits here from real-time collaboration, global connectivity, and an immersive experience that leads to better evaluation and selection of products.

However, entering the Metaverse also brings challenges. These include not only technological requirements, such as powerful hardware and the development of VR/AR applications, but also organizational and strategic considerations to optimally adapt the business strategy to the new conditions.

The decision to enter the Metaverse should be made strategically, weighing the costs, benefits, and long-term potentials for the procurement processes carefully. The added value of the Metaverse will become clear compared to traditional or other digital procurement methods.

2.2 Automation of the Procurement Process

In the future landscape of the Metaverse, the procurement process can be largely automated to increase efficiency and minimize human error. The use of artificial intelligence, machine learning and other advanced technologies will make it possible to create automated systems capable of making independent purchasing decisions based on a set of predefined criteria and algorithms. In the following, these criteria, shown in Fig. 2, are explained and data sources for information acquisition are also examined.

The technical specifications of the product are central to the automated procurement process. Detailed information such as performance, efficiency, reliability and technological compatibility are evaluated here. Automated systems must access databases updated with the latest product specifications and perform comparative analyses to determine the best options based on the specific requirements of the purchaser. The integration of real-time data feeds from manufacturers and other sources will be crucial to ensure that the analyzed information is always current and reliable.

Additionally, an efficient automated purchasing system should be able to conduct a comprehensive assessment of the price-performance ratio. By accessing current

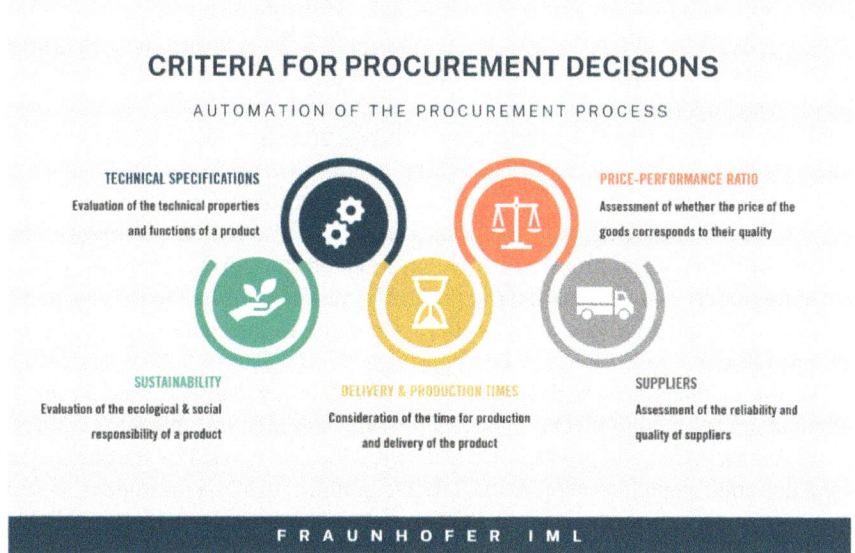

Fig. 2 Criteria for procurement decisions

price information and conducting a thorough analysis of associated performance and specifications, the system can identify products that offer value and quality. For a dynamic assessment, algorithms must be developed to ensure that purchasing decisions optimally consider financial considerations.

Delivery and production times are often critical factors in the procurement process. Automated systems should be configured to analyze and evaluate delivery time estimates, production capacities, and associated logistical considerations. By aggregating real-time data from various sources, these systems will help create realistic and efficient schedules that contribute to avoiding delays and associated costly consequences.

Additionally, sustainability is becoming an increasingly important factor in modern industrial production. Therefore, access to environmental data, energy consumption figures, and other relevant information is needed to assess the sustainability profiles of various products. By considering criteria such as energy efficiency, material consumption, and emissions data, automated systems can help make purchasing decisions that align with the ecological goals and responsibilities of the company.

Choosing the right suppliers is also a central aspect of the purchasing decision. A sophisticated system would not only analyze the direct attributes of the products but also perform a holistic assessment of the suppliers. Factors such as reliability, ethical standards, and sustainability play a crucial role. Suppliers who focus on sustainable practices and compliance with ethical standards should be preferred to build long-term business relationships that go beyond mere transactional exchange.

Such an approach not only promotes responsible corporate practices but also helps create a network of suppliers based on trust and quality. This, in turn, can strengthen the resilience and adaptability of the procurement process in a constantly changing global economic landscape.

Of course, the successful automation of the procurement process is essential for collecting high-quality and secure data. The quality of data forms the foundation for reliable, accurate, and insightful analyses that contribute to information gathering and decision-making. Prioritizing data security is also relevant to preserve the integrity of the data and protect sensitive information flow. To achieve this, robust data management systems and security protocols must be implemented to ensure the confidentiality, integrity, and availability of the data. Various data sources will be examined next.

The use of manufacturer data is relevant to obtain detailed and specific information about selected products. This data provides essential insights into technical specifications, prices, and production capacities, which are needed for the automated evaluation and comparison of products. The integration of this data into the automated procurement process enables an efficient and accurate assessment of the available options, based on actual specifications and costs.

Independent evaluations and reviews offer an objective view of the performance and reliability of the product and suppliers. They complement the manufacturer data with practical insights and experiences that help gain a more comprehensive understanding of the performance and supplier behavior in real operation. The automation systems should be able to collect and analyze these evaluations to ensure a versatile and objective assessment.

Specialized IoT tools and dashboards should be developed and integrated to deliver continuous, real-time updated data about the performance, efficiency, and sustainability of the product. These tools can automate data collection, aggregation, and analysis to provide an immediate understanding of the performances and support informed purchasing decisions.

Market research reports are also an essential resource to obtain deep analyses and assessments of market trends, prices, and future developments. These reports can help the automated system understand the context and strategic direction for the purchase, by bringing broader market knowledge and strategic analyses into the decision-making process.

By utilizing and integrating these tools and data sources into a unified system, automated purchasing systems could be developed that are capable of making comprehensive and informed purchasing decisions in the Metaverse, thus revolutionizing the procurement process.

2.3 Gap-Analysis

In the current business world, the procurement process is increasingly digitalized and automated. Companies use Enterprise Resource Planning (ERP) systems, e-procurement platforms, and artificial intelligence to process orders, evaluate suppliers, and achieve cost savings. These tools have made the process more efficient, transparent, and cost-effective. However, we are on the threshold of a new era in which the Metaverse has the potential to elevate the automation of purchasing to the next level.

Currently, automated purchasing systems are capable of routine tasks such as reordering inventory, sending purchase contracts, and verifying invoices. AI systems can evaluate suppliers by analyzing data on delivery times, quality, and price-performance ratios. However, these processes often operate in isolation and require human intervention to manage exceptions and make complex decisions.

Despite advances in automation, many systems do not work seamlessly together, and issues with integration and interoperability often arise. Data silos frequently exist, hindering information flow. The Metaverse offers a platform where data and systems can converge in an integrated environment, creating a holistic and responsive purchasing ecosystem.

Most current negotiations and business dealings require in-person meetings or are limited to two-dimensional communication methods. In the Metaverse, buyers and sellers could come together in real-time in a virtual space to inspect products, discuss contract terms, and make complex business agreements. This will greatly advance global collaboration.

While some companies are beginning to use blockchain for transparent and secure transactions, the technology is not yet widely adopted in the procurement sector. In the Metaverse, blockchain and smart contracts could be used as standard to automatically execute and manage contracts, further reducing manual effort.

Currently, purchasers must actively seek training opportunities to keep up with the latest technologies. In the Metaverse, training and development programs could be directly integrated into the work environment, allowing learning and application to go hand in hand.

The automation of procurement has already come a long way, but the Metaverse not only offers the possibility of a fully integrated, interactive, and autonomous procurement process, but also provides a completely new and personalized shopping experience. Implementing the Metaverse in procurement will lead to a fundamental change that not only increases efficiency but also redefines the way we initiate, negotiate, and conclude business. For companies, this means a journey on which they must adapt and develop to fully utilize the range of possibilities the Metaverse offers.

2.4 Forecast for the Procurement Process in the Metaverse

The forecast for the growth of the procurement process in the Metaverse is extremely positive and will be significantly driven by disruptive technologies such as blockchain and Web3. It is expected that blockchain technologies will be massively adopted in areas such as currency, gaming, and social interactions, with a focus on integration into users' daily activities, often without them being aware of it. In the realm of purchasing, micropayments, including those in the Metaverse, are expected to increasingly rely on blockchain.

Also notable in this development is the growth of the market for virtual goods. Experts anticipate that the Metaverse market could reach a volume of over 400 billion US dollars by 2030. This reflects the immense potential inherent in the digital economy of the Metaverse and underscores the increasing importance of virtual goods and services [1, 2].

The tokenization of various assets will also revolutionize value transfer and liquidity in the procurement process. Additionally, it is expected that up to 5 trillion US dollars could flow into new digital money formats, such as CBDCs (Central Bank Digital Currencies) and stablecoins [1].

It is also significant to mention that the Metaverse is expected to reach the "Early Majority" phase of the innovation adoption curve by 2030, as shown in Fig. 3, based on [Ghose and Bantanidis [1], p. 10]. This development marks an important turning point, as it suggests that the Metaverse is moving beyond early adopters and beginning to be embraced by a broader mass. This transition to the "Early Majority" is crucial for the widespread acceptance and normalization of the technology.

The technology will evolve in a way that enables a simpler, more cost-effective, and more efficient purchasing experience. Blockchain and Web3 play a key role in the evolution and expansion of the procurement process in the Metaverse. The integration of these technologies into everyday life, the enormous growth of the market for virtual goods, and reaching the 'Early Majority' phase in the adoption cycle will fundamentally shape the Metaverse by 2030 and enhance its influence on the procurement process.

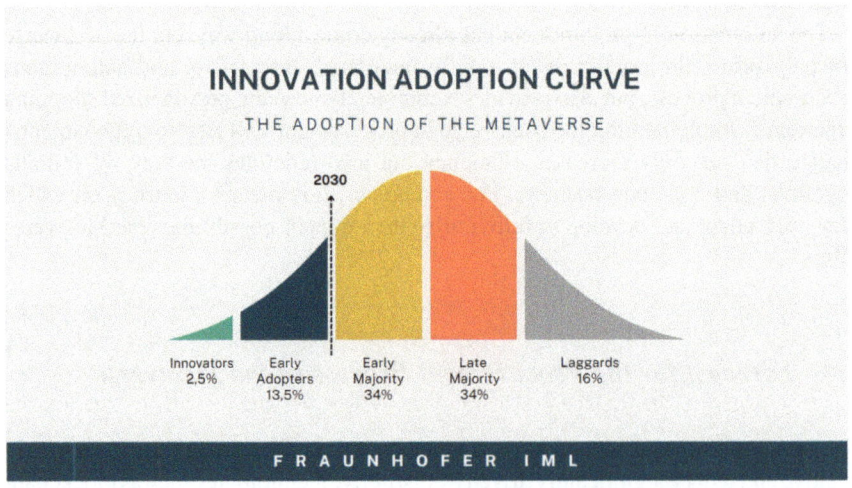

Fig. 3 Innovation adotion curve

3 Roadmap—Strategy of the Digital Procurement Process in the Metaverse

3.1 Strategy Development

Adapting to the digital procurement process in the Metaverse requires more than just technological innovations. It involves rethinking business models, redefining customer relationships, and creating a culture of innovation. This chapter serves as a roadmap that can assist companies in successfully developing strategies at each stage of the procurement process, as shown in Fig. 4, by highlighting various potentials.

The development of an effective strategy for the Metaverse is often closely linked to existing digitalization strategies. Fundamental elements such as building a robust IT infrastructure and promoting core IT competencies among employees are essential for success in the Metaverse. These aspects form the foundation on which innovative digital business models can be developed.

Moreover, establishing digital products and business models plays a crucial role. Companies that already offer digital services and products successfully find a natural extension space in the Metaverse. The experience accumulated in the digital context enables these companies to make the transition of their offerings into the Metaverse smoother. This provides a decisive advantage, as familiarity with digital business

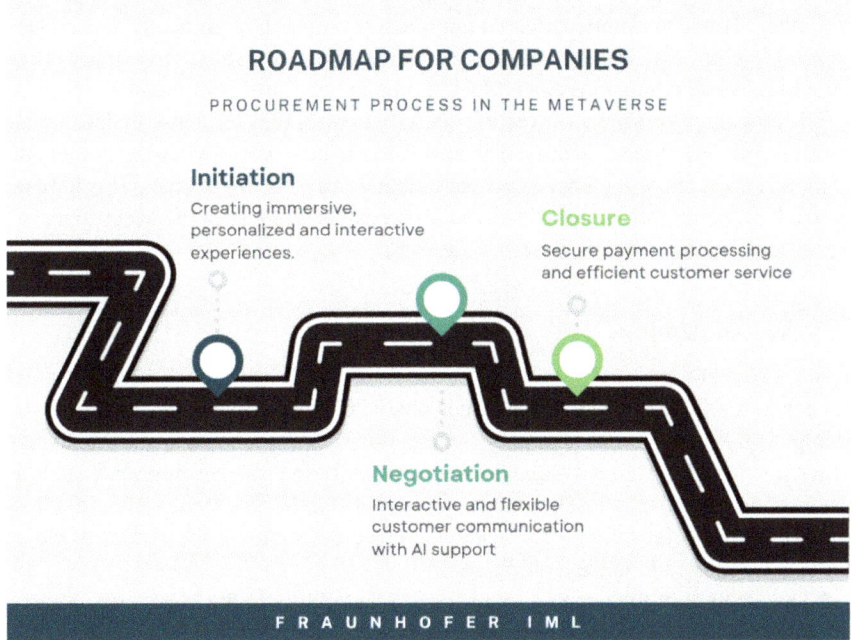

Fig. 4 Roadmap for companies

models facilitates the adoption and adaptation to the specific requirements and opportunities of the Metaverse. Companies that are already established in the digital world can strengthen their market position in the Metaverse.

3.2 Initiation

During the initiation phase of the procurement process in the Metaverse, companies experience a significant change in how they can interact with customers. Therefore, it is crucial that they not only establish a presence in the Metaverse but also create an atmosphere that specifically attracts and engages customers. The goal is to create an immersive experience that blurs the boundaries between physical and digital reality. This can be achieved through the development of interactive presentation platforms or virtual showrooms, where products and services are showcased in innovative ways.

A key aspect in the initiation phase is the personalization of the shopping experience. In the Metaverse, this personalization can go far beyond what is possible in the physical world. By analyzing data with AI, companies can offer tailored recommendations based on the individual preferences of the customer. This not only generates a deeper understanding of the customer but also creates a stronger emotional connection to the brand.

Another critical factor is building a community in the Metaverse. Brands can create platforms where customers can interact not only with the company but also with each other. This community-oriented approach promotes brand loyalty and creates a sense of belonging. Through events, exclusive offers, and interactive experiences in the Metaverse, companies can build a strong and engaged community.

For companies, this means investing not only in technology but also in developing a strong, authentic brand identity. The aim is to create a world where customers do not merely shop but also gather experiences that go beyond the product. This holistic approach in the initiation phase is key to a successful strategy in the Metaverse.

3.3 Negotiation

The negotiation phase in the Metaverse is characterized by interactive and dynamic pricing. Unlike traditional shopping, where prices are often fixed, the Metaverse allows for flexible pricing systems. These can be based on real-time data such as customer demand, behavioral patterns, or market trends. Companies should adjust prices, create special offers, or provide exclusive discounts in the virtual world to motivate customers to make a purchase.

A key element of the negotiation phase is adaptability. In the Metaverse, companies must be able to quickly respond to the needs and desires of customers. This can be achieved through customizable product configurators or personalized offers.

Flexibility in pricing and offerings is crucial to effectively manage the negotiation process and satisfy customers.

The use of artificial intelligence and analytical tools is indispensable in this phase. They enable companies to understand customer behavior and preferences and to create tailored offers based on this information. By analyzing data, companies can identify trends and adjust their negotiation strategies accordingly. This ensures that offers are not only more attractive but also more relevant to individual customers.

Moreover, maintaining transparency and building trust are important in the negotiation phase. This involves clear information about prices, availability, and delivery conditions. In the Metaverse, where physical products are not tangible, customer trust becomes even more significant. Companies must ensure that their virtual sales processes are reliable and understandable. The use of blockchain technology can play a key role here by ensuring transparency and trust in the transaction process. This helps to strengthen customer trust in the company and build long-term customer loyalty.

The negotiation phase in the Metaverse challenges companies to effectively use digital communication channels, respond flexibly to customer needs, and provide a high level of personalization and transparency. By using AI and analytical tools, they can create tailored offers that are both attractive and relevant. This leads to stronger customer engagement and successful sales negotiations in the Metaverse.

3.4 Closure

The closing phase in the Metaverse's procurement process requires smooth and secure transaction handling. The integration of advanced payment systems is crucial. Companies must offer a variety of payment methods, ranging from traditional currencies to cryptocurrencies, to accommodate the diverse preferences of customers. The payment process should be easy, fast, and above all, secure, to ensure customer trust and facilitate the purchasing decision.

After successful payment, optimizing delivery and service processes is a vital part of the conclusion. In the Metaverse, companies need to provide innovative solutions for delivering physical products that go beyond conventional online shopping. This could include using augmented reality for interactive product guides. For digital products or services, access should be immediate and uncomplicated.

The closing phase is not the end of the customer relationship, but a crucial step in ensuring customer satisfaction. Companies must provide effective support systems to assist customers with any questions or problems. This includes a straightforward and transparent process for returns or exchanges. Providing high-quality customer support in the Metaverse, possibly through virtual assistants or interactive help centers, is critical for long-term customer loyalty.

Finally, it's about offering customers a sustainable experience that extends beyond the purchase itself. Companies should develop strategies to stay connected with customers after the sale, whether through personalized follow-up messages, exclusive offers, or invitations to virtual events.

The closing phase in the Metaverse's procurement process is characterized by the challenge of ensuring efficient, secure, and customer-friendly transaction handling. Simultaneously, it is crucial to secure customer satisfaction through excellent service and support and sustainable relationship management. Companies that successfully implement these aspects can establish themselves in the Metaverse as trustworthy and customer-oriented players.

Adapting to the digital procurement process is not a one-time project but an ongoing process of learning and adjusting. The Metaverse opens many opportunities and offers companies a platform to create innovative and enriching shopping experiences. By understanding and leveraging the unique features of the Metaverse, they can build a deep and lasting relationship with their customers.

4 Summary

4.1 Outlook

Insights into the potential of the Metaverse have shown that digital transformation is not just an option but a necessity for modern businesses. Mr. Müller's case study illustrates how the Metaverse can revolutionize the procurement process. The integration of virtual reality, artificial intelligence, blockchain, and other technologies creates an immersive and efficient platform for business transactions. This development represents not only technological advancement but also a cultural shift in how businesses and customers interact.

Moreover, the automation of the procurement process in the Metaverse has the potential to significantly surpass traditional methods. It allows for faster, more accurate, and efficient responses to customer needs while saving costs and time. Through the synergistic integration of various technologies, new standards for security, transparency, and customer engagement are being set. Forecasts indicate that the Metaverse is not just a passing phenomenon but a fundamental change in the way we conduct business.

The ongoing integration of blockchain and Web3 technologies will further enhance transparency and security while opening new opportunities for microtransactions and digital commerce. The increasing acceptance of the Metaverse will also promote the development of new business models and strategies tailored to this unique environment.

Therefore, it is particularly relevant for companies to continuously adapt and evolve. Investing in technology, customer experience, and brand identity is crucial to succeed in this new ecosystem. The roadmap for the digital procurement process in the Metaverse provides a strategic foundation that helps companies navigate the challenges and potentials.

Entering the Metaverse requires courage, creativity, and a willingness to question traditional business practices and think anew. The companies that embrace this challenge will not only survive but will also be able to significantly shape the future of digital shopping.

References

1. R. Ghose, S. Bantanidis, *Money, Tokens, and Games Blockchain's Next Billion Users and Trillions in Value* (Citi GPS: Global Perspectives and Solutions, 2023). pp. 8–12
2. Metaverse: Marktdaten and analyse. Statista (2023)
3. R. Illing, Metaverse Shopping: Wie Einzelhändler die virtuelle Welt für sich nutzen (2023). www.e-commerce-magazin.de

Natalia Broza-Abut was teamlead Procurement and Finance in Supply Chain Management at the Fraunhofer Institute for Materialflow and Logistics (IML). With over seven years of research experience in the field of Blockchain and Distributed Ledger Technology (DLT), she has made significant contributions to the advancement of these technology in industrial applications. Currently, she is employed at Westenergie in the area of IT Integration Management, combining her research expertise with practical applications in the energy sector.

Tobias Jornitz is a research associate at Fraunhofer IML in the "Procurement and Finance in Supply Chain Management" department. As part of his work, he deals with the effects of new technologies on current procurement processes, in particular AI and blockchain. Before joining Fraunhofer IML, he worked in strategic procurement in the automotive industry and successfully studied international business administration and logistics. He also teaches organizational theory at Bochum University of Applied Sciences.

Axel T. Schulte is a Department Head in the area of Enterprise Logistics at the Fraunhofer Institute. This applied research performed by Axel and his team of 20 + Full-time researchers represents the increasing demand for the stronger incorporation of digital technologies and solutions in managing todays' global Supply Chains. The application of Blockchain, Artificial Intelligence (AI) and Internet of Things (IoT) technologies are at the center of the current research. In addition, the department is responsible for the transfer of terrestrial logistics solutions to the space exploration sector.

Axel joined Fraunhofer in 2013 after working for nearly 15 years in the Procurement and Supply Chain area. His professional activities during this time include various international assignments in Switzerland, Russia, and South America.

In addition, until 2022, Axel held an Associate Professorship at St. Petersburg State University in St. Petersburg, Russia and was heading the Research Center for International Logistics CIL.

Prior to his professional career Axel Schulte gained an engineering degree and a doctorate degree in Business and Economics, both at the Technical University of Munich.

SuppliedTrust: A Blockchain-Based Governance Framework to Establish More Trust in Consumer Products

Timucin Korkmaz

1 Use Case and Stakeholder

1.1 Use Case: A Trustworthy Production Process

The key question in this project was whether new technological developments can help to create more trust in a sensitive but unregulated end product such as baby wipes? To answer this question, the manufacturing process within production was examined in detail. This limitation of the use case was also the first step in defining the target vision in more detail. However, the resulting concept and its implementation should explicitly include a possible extension to other participants in the supply chain.

The next level of detail clarified what exactly should be trusted. Pharmaceutical products have strict regulations regarding their manufacturing, marketing and distribution to ensure safety and efficacy for consumers. Automobiles and other vehicles are heavily regulated to ensure they meet strict safety standards. Baby wipes do not have these strict and measurable criteria, yet the success of these products depends on their quality and trust in the manufacturer. Parents are looking for products that work well, are reliable and meet their baby's needs. While brand image and trust in the company help to establish long-term success on the market. Therefore, these qualitative statements about the respective product were examined and it was determined which quality assurance measures during production supported these statements.

The main objective was to make these measurement results available to the end consumer in an understandable and comprehensible way without disclosing sensitive internal company information. Another question was whether it was possible to

T. Korkmaz (✉)
Fraunhofer IML, Bereich Unternehmenslogistik, Dortmund, NRW, Germany
e-mail: Timucin.Korkmaz@iml.fraunhofer.de

use this collected and reliable information from the blockchain for other internal purposes, for example to make forecasts for production. In particular, with the perspective that supplier and distributor information could possibly be added in the future.

1.2 Stakeholder: Required Information Flow Along the Supply Chain

The second half of the target vision was to identify which stakeholders have a legitimate interest in the information flow. It was necessary to analyze who can contribute information and who is allowed to request information. This formed the basis from which the requirements for a blockchain infrastructure were derived.

In research or pilot applications, they often use a Blockchain-as-a-Service (BaaS) [1] in which a provider offers a preconfigured blockchain on its own infrastructure for use. Even if this approach offers many small and medium-sized companies the opportunity to use blockchain technology without great effort, the question arises where the decentralization from which the blockchain derives many of its distinctive features remains. In order not to leave this question unanswered, a dedicated blockchain infrastructure was set up here. So, the node operators formed a group of stakeholders. They provided a infrastructure and thus formed a distributed data storage system, established by different stakeholders. As a permissionless blockchain framework, such as Bitcoin or Ethereum, where all infrastructure tasks are also open aside the consensus mechanism, is not required here, a permissioned blockchain framework was chosen.

In addition to the node operators, the use case also required another category of stakeholders, the participants. These were further divided into the identifiable and the anonymous. In order to be able to trace the information flow, the participants who contribute information to the blockchain must been identifiable. This was the case if, for example, the ERP system wanted to write a new data record to the blockchain, or a sensor wanted to add additional information to an existing data record. Communication endpoints that request information from the blockchain must also been identified in order to ensure which entity is requesting information so that it was possible to determine the scope of the information that could be provided. This was also the way how an anonymous information request worked. An authorized entity could request the blockchain and forward the information itself via a microservice, for example. So if an anonymous end user requests information about their product via the manufacturer's website, the website accepts this request and sends a request to the blockchain itself. The website is granted only limited access to the blockchain and receives an abstracted form of the information, which it can then present to the end customer. These stakeholder categories were sufficient and generic enough to include other participants along the entire supply chain at a later stage.

1.3 Target Vision: The Whole Story in One Picture

The aspects outlined so far helped to define the target vision of the project and to develop and implement the technical concepts along these lines. At this point the focus will be set on the conceptual an implementation of the blockchain related part of the project.

The designated goal was to enable consumers to scan a QR code with a Progressive Web App (PWA) [2] when buying a pack of baby wipes, so that they receive information about the quality of the item in their hand. The information states which quality tests the product has undergone. An interpretation by pictogram shows the customer how the result turned out. A "thumbs up" indicates a good result, a "thumbs in the middle position" indicates an adequate result and a "thumbs down" indicates a poor result. In this way, values such as pH value measurements, which should remain within a certain spectrum, are presented in an abstract form that is easy for the consumer to understand. The information presented here comes from a blockchain and is therefore protected against manipulation. The consumer receives a reliable statement about their product, which does not come from a system that is solely within the control of one participant and could possibly be manipulated with a simple data change.

2 Technical Concept for the Infrastructure and Communication Endpoints

2.1 Blockchain Infrastructure and Data Model

As already mentioned, a permissioned blockchain was used here. The requirements analysis showed that the blockchain framework Tendermint/Cosmos was the most suitable choice at the given time. It is not as complex as the Hyper Ledger Fabric, for example, but offers an efficient consensus mechanism as it implements the Byzantine Fault Tolerant (BFT) approach. At the same time, it involves the implementation of the Light Client concept. In this concept, a digital identifier, known as a mnemonic, is issued from the blockchain, allowing an external component such as a sensor to identify itself to the blockchain. For communication, an external component must also be familiar with the communication protocol of the blockchain. For this purpose, a software library is provided, which, together with the mnemonic, enables the communication endpoint to interact with the blockchain. Due to its versatile functions, a Tendermint/Cosmos blockchain was established across national borders within the project consortium.

The data storage itself was structured like a key-value store. For production, a batch number was generated, and all additional information generated was attached under this number. Therefore, the producer's ERP system had to be connected to the blockchain using a light client. In this way the producer was able to generate a new dataset.

As in other SMEs, the IT infrastructure had grown over time and was therefore highly heterogeneous. Values from test results were sometimes even recorded manually. Therefore, an additional communication endpoint was needed, through which employees could attach a test result to a batch.

For the implementation of the use case, the previously mentioned anonymous communication endpoint for consumers was also provided. From the perspective of the blockchain, the actual communication partner is the producer's website. This website has both a mnemonic and the library to communicate with the blockchain. It takes one step further and passes on the received information to the customer.

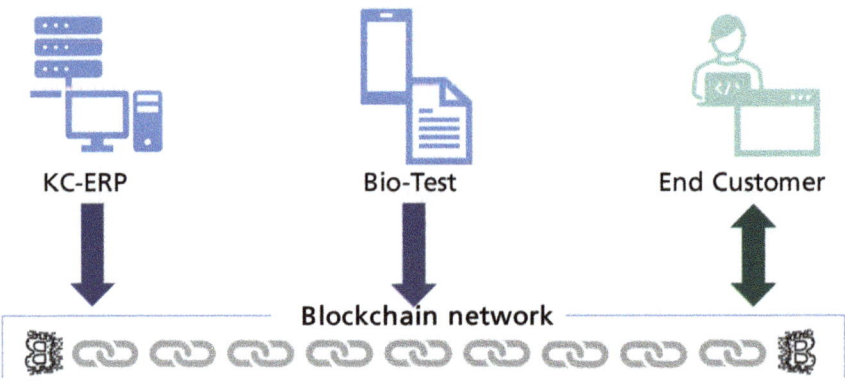

2.2 Communication Endpoints

The principle of Light Clients with access via mnemonics is implemented in a simple and storage-efficient manner, making it possible to integrate cyber-physical systems or simple sensors with a blockchain. In this project, the blockchain's own components were addressed through a microservice. This enables the implementation of APIs according to needs and processing incoming information to be forwarded to the blockchain. Additionally, the entire Light Client was consolidated into a Docker container, further simplifying its applicability.

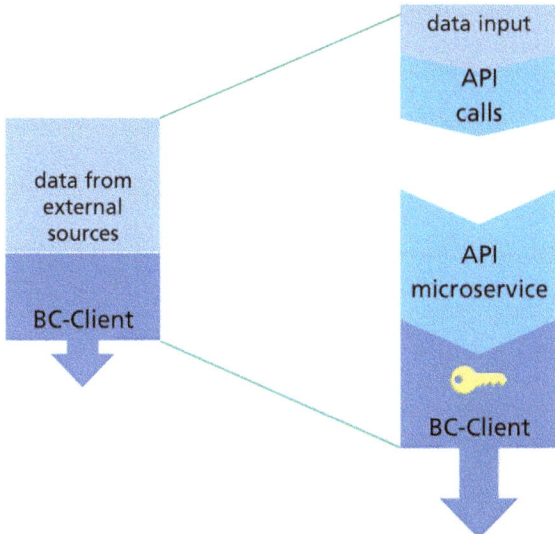

3 Conclusion and Outlook

The project has demonstrated that even a broad statement like "building trust" can yield good results through careful delineation of the use case and the step-by-step detailing of the goal. Even though blockchain technology itself is very complex, when applied correctly, it can create significant value in the supply chain, particularly for sensitive yet unregulated products. Moreover, it has been demonstrated that a variety of systems can be seamlessly connected to a blockchain in a native manner. This ensures not only the protection of stored information from manipulations but also establishes a reliable proof of their origin. The developed concepts and components are sufficiently generic to allow for the inclusion of suppliers and distributors in a subsequent phase. This way, the journey of each product from manufacturing to the point of sale could be traced. This enhanced transparency has the potential to further increase consumer trust and can also be utilized to comply with legal requirements, such as the Supply Chain Due Diligence Act, for example.

References

1. K. Alan, B. Ulrik, B. Roman, Blockchain out of the box—where is the blockchain in blockchain-as-a-service? in *Proceedings of the 54th Hawaii International Conference on System Sciences* (2021)
2. Web.dev, https://web.dev/explore/progressive-web-apps?hl=de. Accessed 23 Dec 2024

GPSR Compliance
The European Union's (EU) General Product Safety Regulation (GPSR) is a set of rules that requires consumer products to be safe and our obligations to ensure this.

If you have any concerns about our products, you can contact us on

ProductSafety@springernature.com

In case Publisher is established outside the EU, the EU authorized representative is:

Springer Nature Customer Service Center GmbH
Europaplatz 3
69115 Heidelberg, Germany

www.ingramcontent.com/pod-product-compliance
Ingram Content Group UK Ltd.
Pitfield, Milton Keynes, MK11 3LW, UK
UKHW022203230426
470311UK00001BA/3